Praise for *The Scientists*

"In many ways *The Scientists* also serves as a handy reference work. Each scientist's story, succinct and entertaining, can be perused and appreciated individually. Historians may quibble over a particular detail or analysis, but no matter. Gribbin's work offers general audiences an engaging and informative view of modern science's prodigious accomplishments since the Renaissance." —*The Washington Post*

"Beginning with Copernicus and the shift from mysticism to reason, Gribbin tracks 500 years of Western-science history through the life stories of the people who charted the course. The text is enlivened by anecdotes that define the characters and their achievements. . . . All fields of science are included. Gribbin carefully illustrates how each . . . accomplishment, rather than being an isolated advance, has been part of a burgeoning scientific revolution that continues today." —*Science News*

"*The Scientists* is best read for its insights into scientific personalities and how science builds on itself. The author, an astrophysicist but better known as a science historian, takes pains to record contributions of relatively unknown scientists whose discoveries led to the insights of science's stars." —*The Columbus Dispatch*

"The prolific Gribbin has written the *human* history of the physical sciences. From the Renaissance to today, the cast of major scientific characters includes the exalted and the obscure, and their personal stories make this comprehensive history a page-turner."
—*Library Journal,* Best Science-Technology Books 2003

"Populated by [colorful] characters and replete with scientific clarity, Gribbin's work is the epitome of what a general-interest history of science should be." —*Booklist* (starred, boxed review)

"A thoroughly readable survey of scientific history, spiced by a brilliant and memorable cast of characters." —*Kirkus Reviews*

"As expansive (and as massive) as a textbook, this remarkably readable popular history explores the development of modern science through the individual stories of philosophers and scientists both renowned and overlooked. . . . The real joy in the book can be found in the way Gribbin . . . revels not just in the development of science but also in the human details of his subjects' lives." —*Publishers Weekly*

"Science buffs will love this book for the nuggets of hard-to-find information. . . . But the book also serves tyros as an excellent introduction to science." —*The News & Observer* (Raleigh, NC)

"A terrific read . . . Tell[s] the story of science as a sequence of witty, information-packed tales . . . complete with humanizing asides, glimpses of the scientist's personal life and amusing anecdotes."
—*The Sunday Times* (London), Books of the Year

"Excels at making complex science intelligible to the general reader . . . If you're looking for a book that captures the personal drama and achievement of science, then look no further." —*The Guardian*

Photo: Ellen Warner

JOHN GRIBBIN trained as an astrophysicist at Cambridge University and is currently Visiting Fellow in Astronomy at the University of Sussex. His many books include *In Search of Schrödinger's Cat, Schrödinger's Kittens and the Search for Reality, Q is for Quantum,* and *Deep Simplicity*.

1. *A mythical meeting of minds – Aristotle, Hevelius and Kepler arguing about the orbits of comets. From Hevelius's* Cometographia, *1668.*

THE SCIENTISTS

A History of Science Told Through the Lives of Its Greatest Inventors

JOHN GRIBBIN

RANDOM HOUSE
TRADE PAPERBACKS
NEW YORK

2004 Random House Trade Paperback Edition

 Published in the United States by Random House Trade Paperbacks, an imprint of The Random House Publishing Group, a division of Random House, Inc., New York, and simultaneously in Canada by Random House of Canada Limited, Toronto.

This book was originally published in hardcover in the United Kingdom by Allen Lane, an imprint of Penguin Books, in 2002 and in the United States by Random House, an imprint of The Random House Publishing Group, a division of Random House, Inc., in 2003.

LIBRARY OF CONGRESS CATALOGING-IN-PUBLICATION DATA

Gribbin, John R.
The scientists: a history of science told through the lives of its greatest inventors / John Gribbin.
p. cm.
Includes bibliographical references and index.
ISBN 0-8129-6788-7
1. Scientists—Biography. 2. Science—History. I. Title.
Q141.G79 2003
509.2'2—dc21 2003046607
[B]

Random House website address: www.atrandom.com

Printed in the United States of America

6 8 9 7 5

Contents

Book Two: THE FOUNDING FATHERS

Book Three: THE ENLIGHTENMENT

List of Illustrations

Acknowledgements

I am grateful to the following institutions for providing access to their libraries and other material: Académie Française and Jardin des Plantes, Paris; Bodleian Library, Oxford; British Museum and Natural History Museum, London; Cavendish Laboratory, Cambridge; Geological Society, London; Down House, Kent; Linnaean Society, London; Royal Astronomical Society; Royal Geographical Society; Royal Institution; Trinity College, Dublin; University of Cambridge Library. As always, the University of Sussex provided me with a base and support, including Internet access. It would be invidious to single out any of the many individuals who discussed aspects of the project with me, but they know who they are, and all have my thanks.

Both singular and plural forms of the personal pronoun appear in the text. 'I', of course, is used where my own opinion on a scientific matter is being presented; 'we' is used to include my writing partner, Mary Gribbin, where appropriate. Her help in ensuring that the words which follow are comprehensible to non-scientists is as essential to this as to all my books.

Introduction

The most important thing that science has taught us about our place in the Universe is that we are not special. The process began with the work of Nicolaus Copernicus in the sixteenth century, which suggested that the Earth is not at the centre of the Universe, and gained momentum after Galileo, early in the seventeenth century, used a telescope to obtain the crucial evidence that the Earth is indeed a planet orbiting the Sun. In successive waves of astronomical discovery in the centuries that followed, astronomers found that just as the Earth is an ordinary planet, the Sun is an ordinary star (one of several hundred billion stars in our Milky Way galaxy) and the Milky Way itself is just an ordinary galaxy (one of several hundred billion in the visible Universe). They even suggested, at the end of the twentieth century, that the Universe itself may not be unique.

While all this was going on, biologists tried and failed to find any evidence for a special 'life force' that distinguishes living matter from non-living matter, concluding that life is just a rather complicated form of chemistry. By a happy coincidence for the historian, one of the landmark events at the start of the biological investigation of the human body was the publication of *De Humani Corporis Fabrica* (*On the Structure of the Human Body*) by Andreas Vesalius in 1543, the same year that Copernicus eventually published *De Revolutionibus Orbium Coelestium* (*On the Revolutions of Celestial Bodies*). This coincidence makes 1543 a convenient marker for the start of the scientific revolution that would transform first Europe and then the world.

Of course, any choice of a starting date for the history of science is arbitrary, and my own account is restricted geographically, as well as

Parte.

le faltaua poco para llegar ala cola / d̃ manera
q̃ touiese latitud septẽtrional / los q̃ estuuiesen
enlos climas septẽtrionales veriã la luna eclip
sar a todo el sol: y los d̃la eq̃nocial veriã eclipsa
da la p̃te septẽtrional d̃l sol: y los meridionales
veriã el sol sin eclipsi. Assi q̃ aunq̃ el eclipsi del
sol sea total o particular no puede ser vniuer-
sal en todo la tierra. Notase que para la quãti
dad destos eclipsis el diametro assi del sol co-
mo dela luna diuidẽ los astrologos en doze p̃-
tes yguales: y a estas
p̃tes llamã dedos / o pũ
tos. Y segũ los puntos
del diametro de la luna
que cubre la sombra de
la tierra / o las partes d̃l
diametro del sol que cu-
bre la luna: tantos de-
dos / o pũtos se dira eclip
sar. Si seis / medio. Si
tres quarto. si quatro /
tercio. si nueue / tres q̃r-
tos. si ocho / dos tercios.
¶ Ha se tambien d̃ no-
tar que aun q̃ el sol sea
mayor que la luna alas
vezes pesce la luna ma-
yor quel sol: y esto sera
q̃ndo el sol estuuiere en
el auge d̃l escẽtrico: y la
luna en el opposito del
auge del epiciclo. y quã
do assi parece lo puede

Doze p̃tes del diametro llamã dedos o puntos.

Diametro visual del sol y d̃la luna.

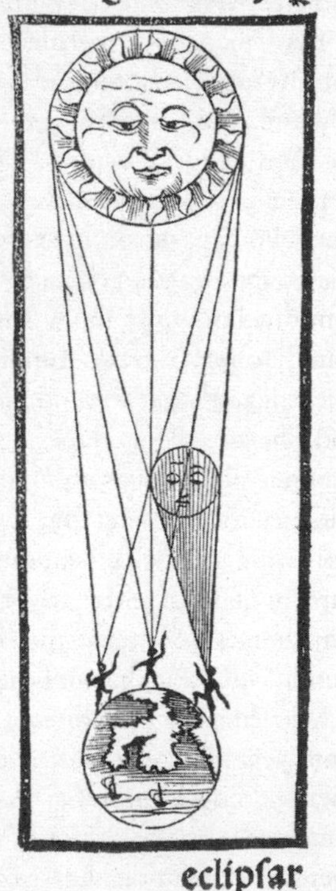

eclipsar

2. *A plate from Martin Cortes de Albacar's*
Breve compendio de la esfera y de la arte de navigar, *1551.*

in the time span it covers. My aim is to outline the development of *Western* science, from the Renaissance to (roughly) the end of the twentieth century. This means leaving to one side the achievements of the Ancient Greeks, the Chinese, and the Islamic scientists and philosophers who did so much to keep the search for knowledge about our world alive during the period that Europeans refer to as the Dark and Middle Ages. But it also means telling a coherent story, with a clear beginning in both space and time, of the development of the world view that lies at the heart of our understanding of the Universe, and our place in it today. For human life turned out to be no different from any other kind of life on Earth. As the work of Charles Darwin and Alfred Wallace established in the nineteenth century, all you need to make human beings out of amoebas is the process of evolution by natural selection, and plenty of time.

All the examples I have mentioned here highlight another feature of the story-telling process. It is natural to describe key events in terms of the work of individuals who made a mark in science – Copernicus, Vesalius, Darwin, Wallace and the rest. But this does not mean that science has progressed as a result of the work of a string of irreplaceable geniuses possessed of a special insight into how the world works. Geniuses maybe (though not always); but irreplaceable certainly not. Scientific progress builds step by step, and as the example of Darwin and Wallace shows, when the time is ripe, two or more individuals may make the next step independently of one another. It is the luck of the draw, or historical accident, whose name gets remembered as the discoverer of a new phenomenon. What is much more important than human genius is the development of technology, and it is no surprise that the start of the scientific revolution 'coincides' with the development of the telescope and the microscope.

I can think of only one partial exception to this situation, and even there I would qualify the exception more than most historians of science do. Isaac Newton was clearly something of a special case, both because of the breadth of his scientific achievements and in particular because of the clear way in which he laid down the ground rules on which science ought to operate. Even Newton, though, relied on his immediate predecessors, in particular Galileo Galilei and René Descartes, and in that sense his contributions followed naturally from what went before. If Newton had never lived, scientific progress might have been held back by a few decades. But *only* by a few decades. Edmond Halley or Robert Hooke might well have come up with the famous inverse square law of gravity; Gottfried Leibniz actually did invent calculus independently of Newton (and made a better job of it); and Christiaan Huygens's superior wave theory of light was held back by Newton's espousal of the rival particle theory.

None of this will stop me from telling much of my version of the history of science in terms of the people involved, including Newton. My choice of individuals to highlight in this way is not intended to be comprehensive; nor are my discussions of their individual lives and work intended to be complete. I have chosen stories that represent the development of science in its historical context. Some of those stories, and the characters involved, may be familiar; others (I hope) less so. But the importance of the people and their lives is that they reflect the society in which they lived, and by discussing, for example, the way the work of one specific scientist followed from that of another, I mean to indicate the way in which one *generation* of scientists influenced the next. This might seem to beg the question of how the ball got rolling in the first place – the 'first cause'. But in this case it is easy to find the first cause – Western science got started because the Renaissance happened. And once it got started, by giving a boost to technology it ensured that it would keep on rolling, with new scientific ideas leading to improved technology, and improved technology providing the scientists with the means to test new ideas to greater and greater accuracy. Technology came first, because it is possible to make machines by trial and error without fully understanding the principles on which they operate. But once science and technology got together, progress really took off.

I will leave the debate about why the Renaissance happened when and where it did to the historians. If you want a definite date to mark the beginning of the revival of Western Europe, a convenient one is 1453, the year the Turks captured Constantinople (on 29 May). By then, many Greek-speaking scholars, seeing which way the wind was blowing, had already fled westwards (initially to Italy), taking their archives of documents with them. There, the study of those documents was taken up by the Italian humanist movement, who were interested in using the teaching found in classical literature to re-establish civilization along the lines that had existed before the Dark Ages. This does rather neatly tie the rise of modern Europe to the death of the last vestige of the old Roman Empire. But an equally important factor, as many people have argued, was the depopulation of Europe by the Black Death in the fourteenth century, which led the survivors to question the whole basis of society, made labour expensive and encouraged the invention of technological devices to replace manpower. Even this is not the whole story. Johann Gutenberg's development of moveable type in the mid-fifteenth century had an obvious impact on what was to become science, and discoveries brought back to Europe by another technological development, sailing ships capable of crossing the oceans, transformed society.

Dating the end of the Renaissance is no easier than dating the beginning – you could say that it is still going on. A convenient round number is 1700; but from the present perspective an even better choice of date might be 1687, the year Isaac Newton published his great work *Philosophiae Naturalis Principia Mathematica* (*The Mathematical Principles of Natural Philosophy*) and, in the words of Alexander Pope, 'all was light'.

The point I want to make is that the scientific revolution did not happen in isolation, and certainly did not start out as the mainspring of change, although in many ways science (through its influence on technology and on our world view) became the driving force of Western civilization. I want to show how science developed, but I don't have space to do justice to the full historical background, any more than most history books have space to do justice to the story of science. I don't even have space to do justice to all of the science here, so if you want the in-depth story of such key concepts as quantum theory, evolution by natural selection or plate tectonics, you will have to look in other books (including my own). My choice of events to highlight is necessarily incomplete, and therefore to some extent subjective, but my aim is to give a feel for the full sweep of science, which has taken us from the realization that the Earth is not at the centre of the Universe and that human beings are 'only' animals, to the theory of the Big Bang and a complete map of the human genome in just over 450 years.

In his *New Guide to Science* (a very different kind of book from anything I could ever hope to write), Isaac Asimov said that the reason for trying to explain the story of science to non-scientists is that:

> No one can really feel at home in the modern world and judge the nature of its problems – and the possible solutions to those problems – unless one has some intelligent notion of what science is up to. Furthermore, initiation into the magnificent world of science brings great esthetic satisfaction, inspiration to youth, fulfillment of the desire to know, and a deeper appreciation of the wonderful potentialities and achievements of the human mind.[1]

I couldn't put it better myself. Science is one of the greatest achievements (arguably *the* greatest achievement) of the human mind, and the fact that progress has actually been made, in the most part, by ordinarily clever people building step by step from the work of their predecessors makes the story more remarkable, not less. Almost any of the readers of this book, had they been in the right place at the right time, could have made the great discoveries described here. And since the progress of science has by no means come to a halt, some of you may yet be involved in the next step in the story.

John Gribbin

1. Details of books mentioned in the text can be found in the bibliography.

Book One

OUT OF THE DARK AGES

1 Renaissance Men

Emerging from the dark – The elegance of Copernicus – The Earth moves! – The orbits of the planets – Leonard Digges and the telescope – Thomas Digges and the infinite Universe – Bruno: a martyr for science? – Copernican model banned by Catholic Church – Vesalius: surgeon, dissector and grave-robber – Fallopio and Fabricius – William Harvey and the circulation of the blood

Emerging from the dark

The Renaissance was the time when Western Europeans lost their awe of the Ancients and realized that they had as much to contribute to civilization and society as the Greeks and Romans had contributed. To modern eyes, the puzzle is not that this should have occurred, but that it should have taken so long for people to lose their inferiority complex. The detailed reasons for the hiatus are outside the scope of this book. But anyone who has visited the sites of classical civilization around the Mediterranean can get a glimpse of why the people of the Dark Ages (in round terms, roughly from AD 400 to 900) and even those of the Middle Ages (roughly from AD 900 to 1400) felt that way. Structures such as the Pantheon and the Colosseum in Rome still inspire awe today, and at a time when all knowledge of how to build such structures had been lost, it must have seemed that they were the work almost of a different species – or of gods. With so much physical evidence of the seemingly god-like prowess of the Ancients around, and with newly discovered texts demonstrating the intellectual prowess of the Ancients emerging from Byzantium, it would have been natural to accept that they were intellectually far superior to the ordinary people who had followed them, and to accept the teaching of ancient philosophers such as Aristotle and Euclid as a kind of Holy Writ, which could not be questioned. This was, indeed, the way things were at the start of the Renaissance. Since the Romans contributed very little to the discussion of what might now be called a scientific view of the world, this meant that by the time of the Renaissance the received wisdom about the nature of the Universe had been essentially unchanged since the great days of Ancient

Greece, some 1500 years before Copernicus came on the scene. Yet, once those ideas were challenged, progress was breathtakingly rapid – after fifteen centuries of stagnation, there have been fewer than another five centuries from the time of Copernicus to the present day. It is something of a cliché, but nonetheless true, that a typical Italian from the tenth century would have felt pretty much at home in the fifteenth century, but a typical Italian from the fifteenth century would find the twenty-first century more unfamiliar than he or she would have found the Italy of the Caesars.

The elegance of Copernicus

Copernicus himself was an intermediate figure in the scientific revolution, and in one important way he resembled the Ancient Greek philosophers rather than the modern scientist. He did not carry out experiments, or even make his own observations of the heavens (at least, not to any significant degree), and did not expect anyone else to try to test his ideas. His great idea was purely that – an idea, or what is today sometimes called a 'thought experiment', which presented a new and simpler way of explaining the same pattern of behaviour of heavenly bodies that was explained by the more complicated system devised (or publicized) by Ptolemy. If a modern scientist has a bright idea about the way the Universe works, his or her first objective is to find a way to test the idea by experiment or observation, to find out how good it is as a description of the world. But this key step in developing the scientific method had not been taken in the fifteenth century, and it never occurred to Copernicus to test his idea – his mental model of how the Universe works – by making new observations himself, or by encouraging others to make new observations. To Copernicus, his model was better than that of Ptolemy because it was, in modern parlance, more elegant. Elegance is often a reliable guide to the usefulness of a model, but not an infallible one. In this case, though, it eventually turned out that Copernicus's intuition was right.

Certainly, the Ptolemaic system lacked elegance. Ptolemy (sometimes known as Ptolemy of Alexandria) lived in the second century AD, and was brought up in an Egypt that had long since come under the cultural influence of Greece (as the very name of the city he lived in records). Very little is known about his life, but among the works he left for posterity there was a great summary of astronomy, based

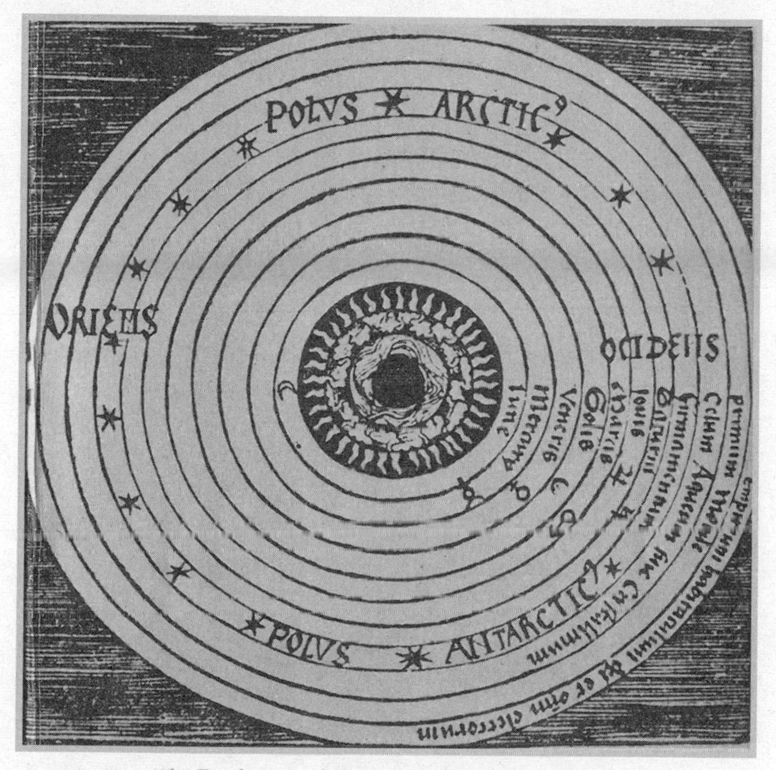

3. *The Earth-centred Ptolemaic model of the Universe. From Reisch's* Margarita Philosophica, *1503.*

on 500 years of Greek astronomical and cosmological thinking. The book is usually known by its Arabic title, the *Almagest*, which means 'the Greatest', and gives you some idea of how it was regarded in the centuries that followed; its original Greek title simply describes it as 'The Mathematical Compilation'. The astronomical system it described was far from being Ptolemy's own idea, although he seems to have tweaked and developed the ideas of the Ancient Greeks. Unlike Copernicus, however, Ptolemy does seem to have carried out his own major observations of the movements of the planets, as well as drawing on those of his predecessors (he also compiled important star maps).

The basis of the Ptolemaic system was the notion that heavenly objects must move in perfect circles, simply because circles are perfect (this is an example of how elegance does not necessarily lead to truth!). At that time, there were five known planets to worry about (Mercury, Venus, Mars, Saturn and Jupiter), plus the Sun and Moon, and the stars. In order to make the observed movements of these objects fit the requirement of always involving perfect circles, Ptolemy had to make two major adjustments to the basic notion that the Earth lay at the centre of the Universe and everything else revolved around it. The first (which had been thought of long before) was that the motion of a particular planet could be described by saying that it revolved in a perfect little circle around a point which itself revolved in a perfect big circle around the Earth. The little circle (a 'wheel within a wheel', in a sense) is called an epicycle. The second adjustment, which Ptolemy seems to have refined himself, was that the large crystal spheres, as they became known ('crystal' just means 'invisible' in this context), which carry the heavenly bodies round in circles, didn't actually revolve around the Earth, but around a set of points slightly offset from the Earth, called 'equant points' (around different equant points, to explain details of the motion of each individual object). The Earth was still regarded as the central object in the Universe, but everything else revolved around the equant points, not around the Earth itself. The big circle centred on the equant point is called a deferent.

The model worked, in the sense that you could use it to describe the way the Sun, Moon and planets seem to move against the background of fixed stars (fixed in the sense that they all kept the same pattern while moving together around the Earth), which were themselves thought to be attached to a crystal sphere just outside the set of nested crystal spheres that carried the other objects around the relevant equant points. But there was no attempt to explain the physical processes that kept everything moving in this way, nor to explain the nature of the crystal spheres. Furthermore, the system was often criticized as being unduly complicated, while the need for equant points made many thinkers uneasy – it raised doubts about whether the Earth ought really to be regarded as the centre of the Universe. There was even speculation (going right back to Aristarchus, in the third century BC, and revived occasionally in the centuries after Ptolemy) that the Sun might be at the centre of the Universe, with the Earth moving around it. But such ideas failed to find favour, largely because they flew in the face of 'common sense'. Obviously, the solid Earth could not be moving! This is one of the prime examples of the need to avoid acting on the basis of common sense if you want to know how the world works.

There were two specific triggers that encouraged Copernicus to come up with something better than the Ptolemaic model. First, that each planet, plus the Sun and Moon, had to be treated separately in the model, with its own offset from the Earth and its own epicycles. There was no coherent overall description of things to explain what was going on. Second, there was a specific problem which people had long been aware of but which was usually brushed under the carpet. The offset of the Moon's orbit from the Earth, required to account for changes in the speed with which the Moon seems to move across the sky, was so big that the Moon ought to be significantly closer to the Earth at some times of the month than at others – so its apparent size ought to change noticeably (and by a calculable amount), which it clearly did not. In a sense, the Ptolemaic system *does* make a prediction that can be tested by observation. It fails that test, so it is not a good description of the Universe. Copernicus didn't think quite like that, but the problem of the Moon certainly made him uneasy about the Ptolemaic model.

Nicolaus Copernicus came on the scene at the end of the fifteenth century. He was born in Torun, a Polish town on the Vistula, on 19 February 1473, and was originally known as Mikolaj Kopernik, but later Latinized his name (a common practice at the time, particularly among the Renaissance humanists). His father, a wealthy merchant, died in 1483 or 1484, and Nicolaus was brought up in the household of his mother's brother, Lucas Waczenrode, who became the bishop of Ermeland. In 1491 (just a year before Christopher Columbus set off on his first voyage to the Americas), Nicolaus began his studies at the University of Krakow, where he seems to have first become seriously interested in astronomy. In 1496, he moved on to Italy, where he studied law and medicine, as well as the usual classics and mathematics, in Bologna and Padua, before receiving a doctorate in canon law from the University of Ferrara in 1503. Very much a man of his time, Copernicus was strongly influenced by the humanist movement in Italy and studied the classics associated with that movement. Indeed, in 1519, he published a collection of poetic letters by the writer Theophylus Simokatta (a seventh-century Byzantine), which he had translated from the original Greek into Latin.

By the time he completed his doctorate, Copernicus had already been appointed as canon at Frombork Cathedral in Poland by his uncle Lucas – a literal case of nepotism that gave him a post amounting to a sinecure, which he held for the rest of his life. But it was not until 1506 that he returned permanently to Poland (giving you some idea of just how undemanding the post was), where he worked as his uncle's physician and secretary until Lucas died in 1512. After his uncle's death, Copernicus gave more attention to his duties as a canon, practised medicine and held various minor civil offices, all of which gave him plenty of time to maintain his interest in astronomy. But his revolutionary ideas about the place of the Earth in the Universe had already been formulated by the end of the first decade of the sixteenth century.

The Earth moves!

These ideas did not appear out of thin air, and even in his major contribution to scientific thought (sometimes regarded as *the* major contribution to scientific thought) Copernicus was still a man of his time. The continuity of science (and the arbitrariness of starting dates for histories) is clearly highlighted by the fact that Copernicus was strongly influenced by a book published in 1496, the exact time when the 23-year-old student was becoming interested in astronomy. The book had been written by the German Johannes Mueller (born in Königsberg in 1436, and also known as Regiomontanus, a Latinized version of the name of his birthplace), and it developed the ideas of his older colleague and teacher Georg Peuerbach (born in 1423), who had (of course) been influenced by other people, and so on back into the mists of time. Peuerbach had set out to produce a modern (that is, fifteenth-century) abridgement of Ptolemy's *Almagest*. The most up-to-date version available was a Latin translation made in the twelfth century by Gerard of Cremona, which was translated from an Arabic text which had itself been translated from the Greek long before. Peuerbach's dream was to update this work by going back to the earliest available Greek texts (some of which were now in Italy following the fall of Constantinople). Unfortunately, he died in 1461, before he could carry out the task, although he had begun a preliminary book summarizing the edition of the *Almagest* that was available. On his deathbed, Peuerbach made Regiomontanus promise to complete the work, which he did, although the new translation of Ptolemy was not carried out. But Regiomontanus did something that was in many ways even better, producing his book the *Epitome*, which not only summarized the contents of the *Almagest*, but added details of later observations of the heavens, revised some of the calculations Ptolemy had carried out and included some critical commentary in the text (in itself a sign of the confidence of Renaissance man in his own place as an equal of the Ancients). This commentary included a passage drawing attention to a key point that we have already mentioned, the fact that the apparent size of the Moon on the sky does not change in the way that the Ptolemaic system requires. Although Regiomontanus died in 1476, the *Epitome* was not published for another twenty years, when it set the young Copernicus thinking. Had it appeared before Regiomontanus died, there is every likelihood that someone else, rather than Copernicus (who was only three in 1476), would have picked up the baton.

Copernicus himself did not exactly rush into print with his ideas. We know that his model of the Universe was essentially complete by 1510, because not long after that he circulated a summary of those ideas to a few close friends, in a manuscript called the *Commentariolus* (*Little Commentary*). There is no evidence that Copernicus was greatly concerned about the risk of persecution by the Church if he published his ideas more formally – indeed, the *Commentariolus* was described in a lecture at the Vatican attended by Pope Clement VII and several cardinals, given by the papal secretary Johan Widmanstadt. One of the cardinals, Nicholas von Schönberg, wrote to Copernicus urging him to publish, and the letter was included at the beginning of his masterwork, *De Revolutionibus Orbium Coelestium* (*On the Revolution of the Celestial Spheres*), when Copernicus did eventually publish his ideas, in 1543.

So why did he delay? There were two factors. First, Copernicus was rather busy. Reference to his post as canon as a sinecure may be accurate, but it doesn't mean that he was willing to sit back and enjoy the income, dabble in astronomy and let the world outside go by. As a doctor, Copernicus worked both for the religious community around

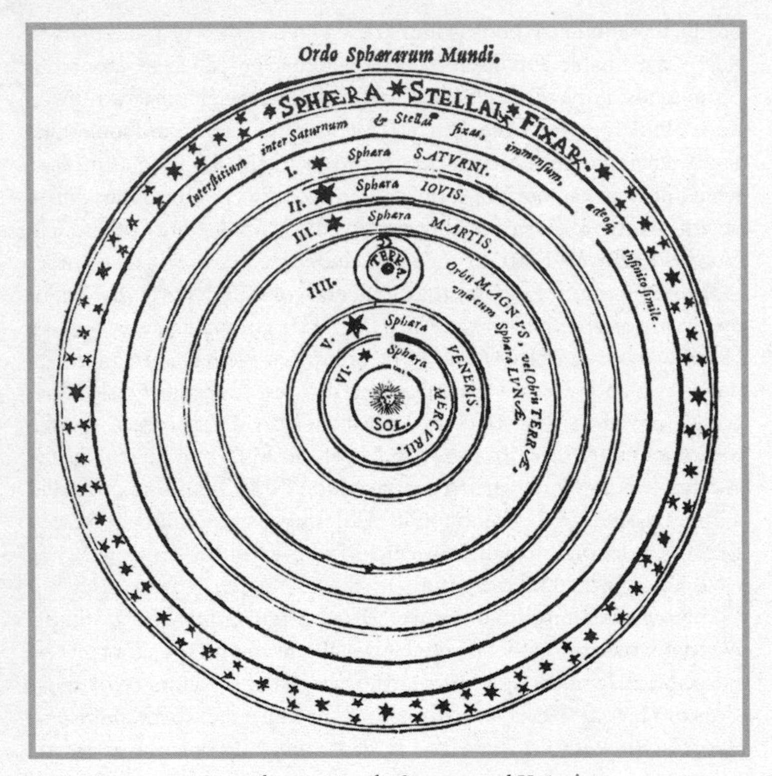

4. *An early version of a Sun-centred Universe, from Rheticus's* Narratio Prima, *1596*

Frombork Cathedral and (unpaid, of course) for the poor. As a mathematician, he worked on a plan for the reform of the currency (not the last time a famous scientist would take on such a role), and his training in law was put to good use by the diocese. He was also pressed into unexpected service when the Teutonic Knights (a religious-military order, akin to the Crusaders, who controlled the eastern Baltic states and Prussia) invaded the region in 1520. Copernicus was given command of a castle at Allenstein, and held the town against the invaders for several months. He was, indeed, a busy man.

But there was a second reason for his reluctance to publish. Copernicus knew that his model of the Universe raised new questions, even if it did answer old puzzles – and he knew that it didn't answer all of the old puzzles. As we have said, Copernicus didn't do much observing (although he did oversee the construction of a roofless tower to use as an observatory). He was a thinker and philosopher more in the style of the Ancient Greeks than a modern scientist. The thing that most worried him about the Ptolemaic system, typified by the puzzle of the Moon, was the business of equants. He couldn't accept the idea, not least because it needed different equants for different planets. Where, in that case, was the true centre of the Universe? He wanted a model in which everything moved around a single centre at an unvarying rate, and he wanted this for aesthetic reasons as much as anything else. His model was intended as a way of achieving that and, on its own terms, it failed. Putting the Sun at the centre of the Universe was a big step. But you still had to have the Moon orbiting around the Earth, and you still needed epicycles to explain why the planets seem to slow down and speed up in their orbits.

Epicycles were a way of having deviations from perfectly circular motion while pretending that there were no deviations from perfectly circular motion. But the biggest problem with the Copernican world view was the stars. If the Earth were orbiting around the Sun and the stars were fixed to a crystal sphere just outside the sphere carrying the most distant planet, then the motion of the Earth should cause an apparent motion in the stars themselves – a phenomenon known as parallax. If you sit in a car travelling down a road, you seem to see the world outside moving past you. If you sit on a moving Earth, why don't you see the stars move? The only explanation seemed to be that the stars must be very much further away than the planets, at least hundreds of times further away, so that the parallax effect is too small to see. But why should God leave a huge empty space, at least hundreds of times bigger than the gaps between the planets, between the outermost planet and the stars?

There were other troubling problems with a moving Earth. If the Earth moves, why isn't there a constant gale of wind blowing past, like the wind in your hair that rushes past if you are in an open-topped car travelling on a motorway? Why doesn't the motion cause the oceans to slop about, producing great tidal waves? Indeed, why doesn't the motion shake the Earth to bits? Remember that in the sixteenth century, motion meant riding on a galloping horse or in a carriage being pulled over rutted roads. The notion of smooth motion (even as smooth as a car on a motorway) must have been very difficult to grasp without any direct experience of it – as late as the nineteenth century there were serious concerns that travelling at the speed of a railway train, maybe as high as 15 miles per hour, might be damaging to human health. Copernicus was no physicist and didn't even attempt to answer these questions, but he knew they cast doubts (from the perspective of the sixteenth century) on his ideas.

There was another problem, which lay completely outside the scope of sixteenth-century knowledge. If the Sun lay at the centre of the Universe, why didn't everything fall into it? All Copernicus could say was that 'Earthy' things tended to fall to Earth, solar things tended to fall to the Sun, things with an affinity for Mars would fall to Mars, and so on. What he really meant was, 'we don't know'. But one of the most important lessons learned in the centuries since Copernicus is that a scientific model doesn't have to explain everything to be a good model.

After the arrival of Georg Joachim von Lauchen (also known as Rheticus) in Frombork in the spring of 1539, Copernicus, despite his doubts and his busy life, was persuaded to put his thoughts into a form that could be published. Rheticus, who was the professor of mathematics at the University of Wittenberg, knew of Copernicus's work, and had come to Frombork specifically to learn more about it; he realized its importance, and was determined to get the master to publish it. They got on well together, and in 1540, Rheticus published a pamphlet *Narratio Prima de Libtus Revolutionum Copernici* (*The First Account of the Revolutionary Book by Copernicus*, usually referred to simply as the *First Account*) summarizing the key feature of the Copernican model, the motion of the Earth around the Sun. At last, Copernicus agreed to publish his great book, although (or perhaps because) he was now an old man. Rheticus undertook to oversee the printing of the book in Nuremberg, where he was based, but (as has often been told) things didn't quite work out as intended. Rheticus had to leave to take up a new post in Leipzig before the book was completely ready to go to press and deputed the task to Andreas Osiander, a Lutheran minister, who took it upon himself to add an unsigned preface explaining that the model described in the book was not intended as a description of the way the Universe really is, but as a mathematical device to simplify calculations involving the movements of the planets. As a Lutheran, Osiander had every reason to fear that the book might not be well received, because even before it was published Martin Luther himself (an almost exact contemporary of Copernicus – he lived from 1483 to 1546) had objected to the Copernican model, thundering that the Bible tells us that it was the Sun, not the Earth, that Joshua commanded to stand still.

Copernicus had no chance to complain about the preface because he died in 1543, the year his great work was published. There is a touching, but probably apocryphal, tale that he received a copy on his deathbed, but whether he did or not, the book was left without a champion except for the indefatigable Rheticus (who died in 1576).

The irony is that Osiander's view is quite in keeping with the modern scientific world view. All of our ideas about the way the Universe works are now accepted as simply models which are put forward to explain observations and the results of experiments as best we can. There is a sense in which it is acceptable to describe the Earth as the centre of the Universe, and to make all measurements relative to the Earth. This works rather well, for example, when planning the flight of a rocket to the Moon. But such a model becomes increasingly complicated as we try to describe the behaviour of things further and further away from Earth, across the Solar System. When calculating the flight of a spaceprobe to, say, Saturn, NASA scientists in effect treat the Sun as being at the centre of the Universe, even though they know that the Sun itself is in orbit around the centre of our galaxy, the Milky Way. By and large, scientists use the simplest model they can which is consistent with all the facts relevant to a particular set of circumstances, and they don't all use the same model all the time. To say that the idea of the Sun being at the centre of the Universe is just a model which aids calculations involving planetary orbits, is to say what any planetary scientist today would agree with. The difference is that Osiander was not expecting his readers (or rather, Copernicus's readers) to accept the equally valid view that to say that the Earth is at the centre of the Universe is just a model which is useful when calculating the apparent motion of the Moon.

It is impossible to tell whether Osiander's preface soothed any ruffled feathers at the Vatican, but the evidence suggests that there were no ruffled feathers there to soothe. The publication of *De Revolutionibus* was accepted essentially without a murmur by the Catholic Church, and the book was largely ignored by Rome for the rest of the sixteenth century. Indeed, it was largely ignored by most people at first – the original edition of 400 copies didn't even sell out. Osiander's preface certainly didn't soothe the Lutherans, and the book was roundly condemned by the European protestant movement. But there was one place where *De Revolutionibus* was well received and its full implications appreciated, at least by the cognoscenti – England, where Henry VIII married his last wife, Catherine Parr, the year the book was published.

The orbits of the planets

What was particularly impressive about the full Copernican model of the Universe was that by putting the Earth in orbit around the Sun it automatically put the planets into a logical sequence. Since ancient times, it had been a puzzle that Mercury and Venus could only be seen from Earth around dawn and dusk, while the other three known planets could be seen at any time of the night. Ptolemy's explanation (rather, the established explanation that he summarized in the *Almagest*) was that Mercury and Venus 'kept company' with the Sun as the Sun travelled once around the Earth each year. But in the Copernican system, it was the Earth that travelled once around the Sun each year, and the explanation for the two kinds of planetary motion was simply that the orbits of Mercury and Venus lay inside the orbit of the Earth (closer to the Sun than we are), while the orbits of Mars, Jupiter and Saturn lay outside the orbit of the Earth (further from the Sun than we are). By making allowance for the Earth's motion, Copernicus could work out the length of time it took each planet to orbit once around the Sun, and these periods formed a neat sequence from Mercury, with the shortest 'year', through Venus, Earth, Mars and Jupiter to Saturn, with the longest 'year'.

But this wasn't all. The observed pattern of behaviour of the planets is also linked, in the Copernican model, to their distances from the Sun relative to the distance of the Earth from the Sun. Even without knowing any of the distances in absolute terms, he could place the planets in order of increasing distance from the Sun. The order was the same – Mercury, Venus, Earth, Mars, Jupiter and Saturn. This clearly indicated a deep truth about the nature of the Universe. There was much more to Copernican astronomy, for those with eyes to see, than the simple claim that the Earth orbits around the Sun.

Leonard Digges and the telescope

One of the few people whose eyes clearly saw the implications of the Copernican model soon after the publication of *De Revolutionibus* was the English astronomer Thomas Digges. Digges was not only a scientist, but one of the first popularizers of science – not quite *the* first, since he followed, to some extent, in the footsteps of his father, Leonard. Leonard Digges was born around 1520, but very little is known about his early life. He was educated at the University of Oxford and became well known as a mathematician and surveyor. He was also the author of several books, which were written in English – very unusual at the time. The first of his books, *A General Prognostication*, was published in 1553, ten years after *De Revolutionibus*, and partly thanks to its accessibility in the vernacular it became a best seller, even though in one crucial respect it was already out of date. Leonard Digges provided in his book a perpetual calendar, collections of weather lore and a wealth of astronomical material, including a description of the Ptolemaic model of the Universe – in some ways, the book was not unlike the kind of farmers' almanacs that were popular in later centuries.

In connection with his surveying work, Leonard Digges invented the theodolite around 1551. About the same time, his interest in seeing accurately over long distances led him to invent the reflecting telescope (and almost certainly the refracting telescope as well), although no publicity was given to these inventions at the time. One reason for the lack of development of these ideas was that the elder Digges's career was brought to an abrupt end in 1554, when he took part in the unsuccessful rebellion led by the Protestant Sir Thomas Wyatt against England's new (Catholic) Queen Mary, who had come to the throne in 1553 on the death of her father, Henry VIII. Originally condemned to death for his part in the rebellion, Leonard Digges had his sentence commuted, but he forfeited all his estates and spent what was left of his life (he died in 1559) struggling unsuccessfully to regain them.

When Leonard Digges died, his son Thomas was about 13 years old (we don't know his exact date of birth), and was looked after by his guardian, John Dee. Dee was a typical Renaissance 'natural philosopher'; a good mathematician, student of alchemy, philosopher and (not quite typical!) astrologer to Queen Elizabeth I (who came to the throne in 1558). He may, like Christopher Marlowe, have been a secret agent for the Crown. He was also, reputedly, an early enthusiast for the Copernican model, although he published nothing on the subject himself. Growing up in Dee's household, Thomas Digges had access to a library containing more than a thousand manuscripts, which he devoured before publishing his own first mathematical work in 1571, the same year that he saw to publication a posthumous book by his father (*Pantometria*), which gave the first public discussion of Leonard Digges's invention of the telescope. In the preface to the book, Thomas Digges describes how:

> My father, by his continuall painfull practises, assisted with demonstrations mathematicall, was able, and sundry times hath, by proportionall glasses duely situate in convenient angles, not onely discovered things farre off, read letters, numbred peeces of money with the very coyne and superscription thereof cast by some of his freends on purpose upon downes in open fields but also seven miles off declared what hath been doone at that instant in private places.

Thomas also studied the heavens himself, and made observations of a supernova seen in 1572, some of which were used by Tycho Brahe in his analysis of that event.

Thomas Digges and the infinite Universe

Thomas Digges's most important publication, though, appeared in 1576. It was a new and greatly revised edition of his father's first book, now titled *Prognostication Everlasting*, and it included a detailed discussion of the Copernican model of the Universe – the first such description in English. But Digges went further than Copernicus. He stated in the book that the Universe is infinite, and included a diagram which showed the Sun at the centre, with the planets in orbit around it, and indicated a multitude of stars extending to infinity in all directions. This was an astonishing leap into the unknown. Digges gave no reason for this assertion, but it seems highly likely that he had been looking at the Milky Way with a telescope, and that the multitude of stars he saw there convinced him that the stars are other suns spread in profusion throughout an infinite Universe.

But Digges did not devote his life to science any more than Copernicus did, and he didn't follow up these ideas. With his background as the son of a prominent Protestant who had suffered at the hands of Queen Mary, and his links with the Dee household (under the protection of Queen Elizabeth), Thomas Digges became a Member of Parliament (serving on two separate occasions) and adviser to the government. He also served as Muster-Master General to the English forces in The Netherlands between 1586 and 1593, where they were helping the Protestant Dutch to free themselves from the rule of Catholic Spain. He died in 1595. By that time, Galileo Galilei was already an established professor of mathematics in Padua and the Catholic Church was turning against the Copernican model of the Universe because it had been taken up by the heretic Giordano Bruno, who was embroiled in a long trial which would end with him being burned at the stake in 1600.

Bruno: a martyr for science?

It's worth mentioning Bruno here, before we go back to pick up the threads of the work of Tycho, Johannes Kepler and Galileo, which followed on from the work of Copernicus, because it is often thought that Bruno was burned because of his support for the Copernican model. The truth is that he really *was* a heretic and was burned for his religious beliefs; it was just unfortunate that the Copernican model got tangled up in the whole business.

The principal reason that Bruno, who was born in 1548, came into conflict with the Church was because he was a follower of a movement known as Hermetism. This cult based its beliefs on their equivalent of holy scripture, documents which were thought in the fifteenth and sixteenth centuries to have originated in Egypt at the time of Moses, and were linked with the teaching of the Egyptian god Thoth (the god of learning). Hermes was the Greek equivalent of Thoth (hence Hermetism), and to followers of the cult he was Hermes Trismegistus, or Hermes the Thrice Great. The Sun, of course, was also a god to the Egyptians, and there have been suggestions that Copernicus himself may have been influenced by Hermetism in putting the Sun at the centre of the Universe, although there is no strong evidence for this.

This is no place to go into the details of Hermetism (especially since the documents on which it was based later turned out not to originate from Ancient Egypt), but to fifteenth-century believers the documents were interpreted as, among other things, predicting the birth of Christ. In the 1460s, copies of the material on which Hermetism was based were brought to Italy from Macedonia and stirred great interest for well over a century, until it was established (in 1614) that they had been written long after the start of the Christian era and so their 'prophecies' were produced very much with the benefit of hindsight.

The Catholic Church of the late sixteenth century was able to tolerate ancient texts that predicted the birth of Jesus, and such thoroughly respectable Catholics as Philip II of Spain (who reigned from 1556 to 1598, married England's Queen Mary and was a staunch opponent of Protestantism) subscribed to these beliefs (as, incidentally, did John Dee, Thomas Digges's guardian). But Bruno took the extreme view that the old Egyptian religion was the true faith and that the Catholic Church should find a way of returning to those old ways. This, needless to say, did not go down too well in Rome, and after a chequered career wandering around Europe (including a spell in England from 1583 to 1585) and stirring up trouble (he joined the Dominicans in 1565 but was expelled from the order in 1576, and while in England he made so many enemies he had to take refuge in the French Embassy), he made the mistake of visiting Venice in 1591, where he was arrested and handed over to the Inquisition. After a long imprisonment and trial, it seems that Bruno was finally condemned on the specific charges of Arianism (the belief that Christ had been created by God and was not God incarnate) and carrying out occult magical practices. We cannot be absolutely sure, because the records of the trial have been lost; but rather than being a martyr for science, as he is occasionally represented, Bruno was actually a martyr for magic.

Copernican model banned by Catholic Church

Although his fate may seem harsh by modern standards, like many martyrs, Bruno to some extent brought it on himself, since he was given every opportunity to recant (one reason why he was held for so long before being condemned). There is no evidence that his support for Copernicanism featured in the trial at all, but it is clear that Bruno was a keen supporter of the idea of a Sun-centred Universe (because it fitted with the Egyptian view of the world), and that he also enthusiastically espoused Thomas

Digges's idea that the Universe is filled with an infinite array of stars, each one like the Sun, and argued that there must be life elsewhere in the Universe. Because Bruno's ideas made such a splash at the time, and because he was condemned by the Church, all these ideas got tarred with the same brush. Moving with its customary slowness, it still took the Church until 1616 to place *De Revolutionibus* on the Index of banned books (and until 1835 to take it off the Index again!). But after 1600 Copernicanism was distinctly frowned upon by the Church, and the fact that Bruno was a Copernican and had been burned as a heretic was hardly encouraging for anyone, like Galileo, who lived in Italy in the early 1600s and was interested in how the world worked. If it hadn't been for Bruno, Copernicanism might never have received such adverse attention from the authorities, Galileo might not have been persecuted and scientific progress in Italy might have proceeded more smoothly.

But Galileo's story will have to wait, while we catch up with the other great development in science in the Renaissance, the study of the human body.

Vesalius: surgeon, dissector and grave-robber

Just as the work of Copernicus built on the rediscovery by Western Europeans of the work of Ptolemy, so the work of Andreas Vesalius of Brussels built on the rediscovery of the work of Galen (Claudius Galenus). Of course, neither of these great works from ancient times was ever really lost, and they were known to the Byzantine and Arabic civilizations even during the Dark Ages in Western Europe; but it was the resurgence of interest in all such writings (typified by the humanist movement in Italy and linked with the fall of Constantinople and the spread of original documents and translations westwards into Italy and beyond associated with the Renaissance) that helped to stir the beginnings of the scientific revolution. Not that this seemed like a revolution to those taking part in its early stages – Copernicus himself, and Vesalius, saw themselves as picking up the threads of ancient knowledge and building from it, rather than overturning the teaching of the Ancients and starting anew. The whole process was much more evolutionary than revolutionary, especially during the sixteenth century. The real revolution, as I have mentioned, lay in the change of mentality which saw Renaissance scholars regarding themselves as the equals of the Ancients, competent to move forward from the teachings of the likes of Ptolemy and Galen – the realization that the likes of Ptolemy and Galen were themselves only human. It was only with the work of Galileo and in particular Newton that, as we shall see, the whole process of investigation of the world really changed in a revolutionary sense from the ways of the ancient philosophers to the ways of modern science.

Galen was a Greek physician born around AD 130 in Pergamum (now Bergama), in the part of Asia Minor that is now Turkey. He lived until the end of the second century AD, or possibly just into the beginning of the third century. As the son of a wealthy architect and farmer living in one of the richest cities in the Greek-speaking part of the Roman Empire, Galen had every advantage in life and received the finest education, which was steered towards medicine after his father had a dream when the boy was 16, foretelling his success in the field. He studied medicine at various centres of learning, including Corinth and Alexandria, was chief physician to the gladiators at Pergamum for five years from AD 157, then moved to Rome, where he eventually became both the personal physician and friend of the emperor Marcus Aurelius. He also served Commodus, who was the son of Marcus Aurelius and became emperor when his father died in AD 180. These were turbulent times for Rome, with more or less constant warfare on the borders of the Empire (Hadrian's Wall was built a few years before Galen was born), but it was still long before the Empire went into serious decline (the Empire was not divided into Eastern and Western parts until AD 286, and Constantinople wasn't founded until AD 330). Galen, secure at the heart of the Empire, whatever the troubles on its borders, was a prolific writer and, like Ptolemy, summed up the teachings of earlier men who he admired, notably Hippocrates (indeed, the modern idea of Hippocrates as the father of medicine is almost entirely a result of Galen's writings). He was also an obnoxious self-publicist and plagiarist – one of the kindest things he says about his fellow physicians in Rome is to refer to them as 'snotty-nosed individuals'.[1] But his unpleasant personality shouldn't be allowed to obscure his real achievements, and Galen's greatest claim to fame lay in his skill at dissection and the books he wrote about the structure of the human

1. Quoted by Vivian Nutton, in Conrad *et al.* See Bibliography.

body. Unfortunately (and bizarrely, given the attitude to slaves and gladiatorial games), human dissection was frowned upon at the time, and most of Galen's work was carried out on dogs, pigs and monkeys (although there is evidence that he did dissect a few human subjects). So his conclusions about the human body were mostly based on studies of other animals and were incorrect in many ways. Since nobody seems to have done any serious research in anatomy for the next twelve or thirteen centuries, Galen's work was regarded as the last word in human anatomy until well into the sixteenth century.

The revival of Galen was part of the humanist obsession with all things Greek. In religion, not only the Protestant movement of the sixteenth century but also some Catholics believed that the teaching of God had been corrupted by centuries of interpretation and amendment to Biblical writing since the time of Jesus, and there was a fundamentalist move to return to the Bible itself as the ultimate authority. Part of this involved studying the earliest Greek versions of the Bible rather than translations into Latin. Although the suggestion that nothing worthwhile had happened since ancient times was a little extreme, there was certainly some truth in the idea that a medical text that had been corrupted by passing through several translations (some of those translations had been made from Arabic texts translated from the Greek) and copied by many scribes might be less accurate than one might wish, and it was a landmark event in medicine when Galen's works were published in the original Greek in 1525. Ironically, since hardly any medical men could read Greek, what they actually studied were new Latin translations of the 1525 edition. But, thanks to these translations and the printing press, Galen's work was disseminated more widely than ever before over the next ten years or so. Just at this time, the young Andreas Vesalius was completing his medical education and beginning to make a name for himself.

Vesalius was born in Brussels on 31 December 1514, a member of a family with a tradition of medicine – his father was the royal pharmacist to Charles V, the so-called Holy Roman Emperor (actually a German prince). Following in the family tradition, Vesalius went first to the University of Louvain then, in 1533, enrolled to study medicine in Paris. Paris was at the centre of the revival of Galenism, and as well as being taught the works of the master, Vesalius also learned his skill

5. *Andreas Vesalius. From Vesalius's* De Humani Corporis Fabrica, *1543.*

at dissection during his time there. His time in Paris came to an abrupt end in 1536, because of war between France and the Holy Roman Empire (which, as historians are fond of pointing out, was neither holy, Roman, nor an empire; but the name has passed into history), and he returned to Louvain, where he graduated in medicine in 1537. His enthusiasm for dissection and interest in the human body are attested by a well-documented occasion in the autumn of 1536, when he stole a body (or what was left of it) from a gibbet outside Louvain and took it home for study.

By the standards of the day, the medical faculty at Louvain was conservative and backward (certainly compared with Paris), but with the war still going on, Vesalius could not return to France. Instead, soon after he graduated, Vesalius went to Italy, where he enrolled as a graduate student at the University of Padua at the end of 1537. This seems to have been merely a formality though, since after being given an initial examination which he passed with flying colours, Vesalius was almost immediately awarded the degree of doctor of medicine and appointed to the faculty at Padua. Vesalius was a popular and successful teacher in the still-new Galenic 'tradition'. But, unlike Galen, he was also an able and enthusiastic dissector of human beings, and in striking contrast to his grave-robbing activities in Louvain, these researches were aided by the authorities in Padua, notably the judge Marcantonio Contarini, who not only supplied him with the bodies of executed criminals, but sometimes delayed the time of execution to fit in with Vesalius's schedule and need for fresh bodies. It was this work that soon convinced Vesalius that Galen had had little or no experience of human dissection and encouraged him to prepare his own book on human anatomy.

The whole approach of Vesalius to his subject was, if not exactly revolutionary, a profound step forward from what had gone before. In the Middle Ages, actual dissections, when undertaken at all, would be carried out for demonstration purposes by surgeons, who were regarded as inferior medical practitioners, while the learned professor would lecture on the subject from a safe distance, literally without getting his hands dirty. Vesalius performed his dissection demonstrations himself, while also explaining to his students the significance of what was being uncovered, and thereby raised the status of surgery first at Padua and gradually elsewhere as the practice spread. He also employed superb artists to prepare large diagrams used in his teaching. Six of these drawings were published in 1538 as the *Tabulae Anatomica Sex* (*Six Anatomical Pictures*) after one of the demonstration diagrams had been stolen and plagiarized. Three of the six drawings were by Vesalius himself; the other three were by John Stephen of Kalkar,

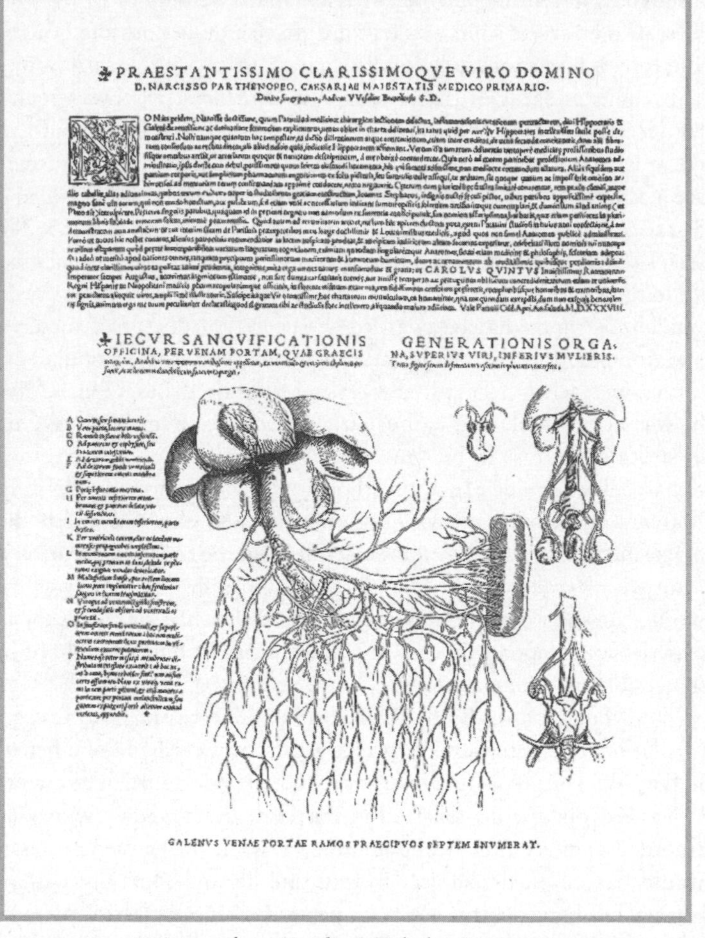

PRAESTANTISSIMO CLARISSIMOQVE VIRO DOMINO
D. NARCISSO PARTHENOPEO, CAESARIAE MAIESTATIS MEDICO PRIMARIO.

IECVR SANGVIFICATIONIS OFFICINA, PER VENAM PORTAM, QVAE GRAECIS

GENERATIONIS ORGANA, SVPERIVS VIRI, INFERIVS MVLIERIS.

GALENVS VENAE PORTAE RAMOS PRAECIPVOS SEPTEM ENVMERAT.

6. *A page from Vesalius's* Tabulae Sex, *1538.*

a highly respected pupil of Titian, which gives you some idea of their quality. It is not known for sure, but Stephen was probably also the main illustrator used for the masterwork *De Humani Corporis Fabrica* (usually known as the *Fabrica*), published in 1543.

Apart from the accuracy of its description of the human body, the importance of the *Fabrica* was that it emphasized the need for the professor to do the dirty work himself, instead of delegating the nitty gritty of the subject to an underling. In the same vein, it stressed the importance of accepting the evidence of your own eyes, rather than believing implicitly the words handed down from past generations – the Ancients were not infallible. It took a long time for the study of human anatomy to become fully respectable – there remained a lingering disquiet about the whole business of cutting people up. But the process of establishing that the proper study of man is man, in the wider sense, began with the work of Vesalius and the publication of the *Fabrica*. The *Fabrica* was a book for the established experts in medicine, but Vesalius also wanted to reach a wider audience. He produced alongside it a summary for students, the *Epitome*, which was also published in 1543. But having made this mark on medicine and laid down a marker for the scientific approach in general, Vesalius suddenly abandoned his academic career although still not 30 years old.

He had already been away from Padua for a considerable time in 1542 and 1543 (mostly in Basle) preparing his two books for publication, and although this seems to have been an officially sanctioned leave of absence, he never returned to his post. It is not entirely clear whether he had simply had enough of the criticisms his work had drawn from unreconstructed Galenists, or whether he wanted to practise medicine rather than teach it (or a combination of these factors), but armed with copies of his two books Vesalius approached Charles V and was appointed as court physician – a prestigious post which had one principal disadvantage in that there was no provision for the physician to resign during the lifetime of the Emperor. But Vesalius can hardly have regretted his decision, since when Charles V allowed him to leave his service in 1556 (shortly before Charles abdicated) and granted him a pension, Vesalius promptly took up a similar post with Philip II of Spain, the son of Charles V (the same Philip who later sent the Armada to attack England). This turned out not to be such a good idea. Spanish physicians lacked the competence that Vesalius was used to, and initial hostility to him as a foreigner became exacerbated by the growth of the independence movement in the Netherlands, then ruled by Spain. In 1564, Vesalius obtained permission from Philip to go on a pilgrimage to Jerusalem, but this seems to have been an excuse to stop off in Italy and open negotiations with the University of Padua, with a view to taking up his old post there once again. But on his way back from the Holy Land, the ship Vesalius was travelling in encountered severe storms and was delayed sufficiently for supplies to run low, while the passengers also suffered severe seasickness. Vesalius became ill (we don't know exactly what with) and died on the Greek island of Zante, where the ship ran aground, in October 1564, in his fiftieth year. But although Vesalius himself contributed little directly to the achievements of science after 1543, he had a profound influence through his successors in Padua, which led directly to one of the greatest insights of the seventeenth century, William Harvey's discovery of the circulation of the blood.

In a way, Harvey's story belongs in the next chapter. But the line from Vesalius to Harvey is so clear that it makes more sense to follow it to its logical conclusion now, before returning to the development of astronomy in the sixteenth century. Just as this is not a book about technology, I do not intend to dwell on the strictly medical implications of the investigation of the human body. But Harvey's special contribution was not what he discovered (though that was impressive enough) but the way in which he proved that the discovery was real.

Fallopio and Fabricius

The direct line from Vesalius to Harvey involves just two other people. The first was Gabriele Fallopio (also known as Gabriel Fallopius), who was a student of Vesalius in Padua, became professor of anatomy in Pisa in 1548 and came back to Padua as professor of anatomy – Vesalius's old post – in 1551. Although he died in 1562 at the early age of 39, he made his mark on human biology in two ways. First, he carried out his own research on the systems of the human body, very much in the spirit of Vesalius, which, among other things, led to him discovering the 'Fallopian tubes' which still bear his name. Fallopio described these links between the uterus and the ovaries as flaring out at the end like a 'brass trumpet' – a *tuba*. This accurate description somehow got mistranslated as 'tube', but modern medicine seems to be stuck with the inaccurate version of the term.[1] But perhaps

1. Incidentally, another piece of tubing in the body, the Eustachian tube linking the middle ear to the pharynx, was also described around this time, by Bartolomeo Eustachio. This is no coincidence, but rather a sign of the way a new generation of dissectors were eagerly setting about their work.

Fallopio's greatest contribution to anatomy was his role as the teacher of Girolamo Fabrizio, who became known as Hieronymous Fabricius ab Aquapendente, and succeeded Fallopio to the chair in Padua when Fallopio died.

Fabricius was born on 20 May 1537, in the town of Aquapendente, and graduated from Padua in 1559. He worked as a surgeon and taught anatomy privately until he was appointed to the chair in Padua in 1565 – the post had been left vacant for three years following Fallopio's death, so Fabricius was Fallopio's direct successor, in spite of the gap. It was during this gap that Vesalius opened negotiations to take up the post, and if it hadn't been for his ill-fated trip to Jerusalem he would probably have got the job ahead of Fabricius. A lot of Fabricius's work concerned embryology and the development of the foetus, which he studied in hens' eggs, but with the benefit of hindsight we can see that his most important contribution to science was the first accurate and detailed description of the valves in the veins. The valves were already known, but Fabricius investigated them thoroughly and described them in detail, first in public demonstrations in 1579 and later in an accurately illustrated book published in 1603. But his skill as an anatomist in describing the valves was not matched by any notable insight into their purpose – he thought they were there to slow down the flow of blood from the liver to allow it to be absorbed by the tissues of the body. Fabricius retired in 1613, because of ill health, and died in 1619. By then, however, William Harvey, who studied under Fabricius in Padua from some time in the late 1590s to 1602, was well on the way to explaining how the blood circulation system really worked.

William Harvey and the circulation of the blood

Before Harvey, the received wisdom (going right back to Galen and earlier times) was that blood was manufactured in the liver and carried by the veins throughout the body to provide nourishment to the tissues, getting used up in the process, so that new blood was constantly being manufactured. The role of the arterial system was seen as carrying 'vital spirit' from the lungs and spreading it through the body (actually, not so far from the truth given that oxygen would not be discovered until 1774). In 1553, the Spanish theologian and physician Michael Servetus (born in 1511, and christened Miguel Serveto) referred in his book *Christianismi Restitutio* to the 'lesser' circulation of the blood (as it was later known) in which blood travels from the right-hand side of the heart to the left-hand side of the heart via the lungs, and not through tiny holes in the dividing wall of the heart, as Galen had taught. Servetus reached his conclusion largely on theological grounds, not through dissection, and presented them almost as an aside in a theological treatise. Unfortunately for Servetus, the theological views he expressed here (and in earlier writings) were anti-Trinitarian. Like Giordano Bruno, he did not believe that Jesus Christ was God incarnate and he suffered the same fate as Bruno for his beliefs, but at different hands. John Calvin was at the height of his reforming activity at the time, and Servetus had written to him (in Geneva) about his ideas. When Calvin stopped replying to his letters, Servetus, based in Vienna, continued to send a stream of increasingly vituperative correspondence. This was a big mistake. When the book was published Calvin contacted the authorities in Vienna and had the heretic imprisoned. Servetus escaped and headed for Italy, but made the further mistake of taking the direct route through Geneva (you would have thought he would have had more sense), where he was recognized, recaptured and burned at the stake by the Calvinists on 27 October 1553. His books were also burned, and only three copies of *Christianismi Restitutio* survive. Servetus had no influence on the science of his times, and Harvey knew nothing of his work, but the story of how he met his end is an irresistible insight into the world of the sixteenth century.

Ever since Galen, it had been thought that the veins and the arteries carried different substances – two kinds of blood. The modern understanding is that the human heart (like the hearts of other mammals and birds) is really two hearts in one, with the right-hand half pumping deoxygenated blood to the lungs, where the blood picks up oxygen and returns to the left-hand half of the heart, which pumps the oxygenated blood on around the body. One of Harvey's key discoveries was that the valves in the veins, described so accurately by his teacher Fabricius, are one-way systems, which allow blood to flow only towards the heart, and that this blood must originate as arterial blood, which is pumped away from the heart and travels through tiny capillaries linking the arterial and venous systems to enter the veins. But all that lay far in the future when Harvey started out on his medical career.

Harvey was born in Folkestone, Kent, on 1 April 1578. The eldest of seven sons of a yeoman farmer, William was educated at King's School, Canterbury, and Caius College in Cambridge, where he obtained his BA in 1597 and probably began to study medicine. But he soon moved on to Padua, where he was taught by Fabricius and graduated as a doctor of medicine in 1602. As a student in Padua, Harvey must have known about Galileo, who was teaching there at the time, but as far as we know the two never met. Having returned to England in 1602, Harvey married Elizabeth Browne, the daughter of Lancelot Browne, physician to Elizabeth I, in 1604. Moving in royal circles, Harvey had a distinguished medical career; he was appointed physician at St Bartholomew's Hospital in London in 1609, having already been elected as a Fellow of the College of Physicians in 1607, and in 1618 (two years after William Shakespeare died) became one of the physicians to James I (who succeeded Elizabeth in 1603). In 1630, Harvey received an even more prestigious appointment as personal physician to James's son Charles I, who came to the throne in 1625. His reward for this service was an appointment as Warden of Merton College, Oxford, in 1645 (when he was 67). But with the Civil War raging in England, Oxford came under the sphere of influence of the Parliamentary forces in 1646 and Harvey retired from this post (although technically retaining his post as Royal Physician until Charles was beheaded in 1649), leading a quiet life until his death on 3 June 1657. Although elected President of the College of Physicians in 1654, he had to decline the honour on the grounds of age and ill health.

So the great work for which Harvey is now remembered was actually carried out in his spare time, which is one reason why it took him until 1628 to publish his results, in his landmark book *De Motu Cordis et Sanguinis in Animalibus* (*On the Motion of the Heart and Blood in Animals*). The other reason is that even fifty years after the publication of the *Fabrica*, there was still strong opposition in some quarters to attempts to revise Galen's teaching. Harvey knew that he had to present an open-and-shut case in order to establish the reality of the circulation of the blood, and it is the way he presented that case which makes him a key figure in the history of science, pointing the way forward for scientists in all disciplines, not just medicine.

Even the way Harvey became interested in the problem shows how things had changed since the days when philosophers would dream up abstract hypotheses about the workings of the natural world based on principles of perfection rather than on observation and experience. Harvey actually measured the capacity of the heart, which he described as being like an inflated glove, and worked out how much blood it was pumping into the arteries each minute. His estimates were a little inaccurate, but good enough to make the point. In modern units, he worked out that, on average, the human heart pumped out 60 cubic centimetres of blood with every beat, adding up to a flow of almost 260 litres an hour – an amount of blood that would weigh three times as much as an average man. Clearly, the body could not be manufacturing that much blood and there must really be a lot less continuously circling through the veins and arteries of the body. Harvey then built up his case, using a combination of experiments and observation. Even though he could not see the tiny connections between the veins and the arteries, he proved they must exist by tightening a cord (or ligature) around an arm. Arteries lie deeper below the surface of the arm than veins, so by loosening the ligature slightly he allowed blood to flow down the arm through the arteries while the cord was still too tight to allow blood to flow back up the arm through the veins, and so the veins below the ligature became swollen. He pointed out that the rapidity with which poisons can spread throughout the entire body fitted in with the idea that the blood is continually circulating. And he drew attention to the fact that the arteries near the heart are thicker than those further away from the heart, just as would be required to withstand the greater pressure produced near the heart by the powerful ejection of blood through its pumping action.

But don't run away with the idea that Harvey invented the scientific method. He was, in truth, more of a Renaissance man than a modern scientist, and still thought in terms of vital forces, an abstract conception of perfection and spirits that kept the body alive. In his own words (from the 1653 English translation of his book):

In all likelihood it comes to pass in the body, that all the parts are nourished, cherished, and quickned with blood, which is warm, perfect, vaporous, full of spirit, and, that I may so say, alimentative: in the parts the blood is refrigerated, coagulated, and made as it were barren, from thence it returns to the heart, as to the fountain or dwelling-house of the body, to recover its perfection, and there again by natural heat, powerful and vehement, it is melted, and is dispens'd again through the body from thence, being fraught with spirits, as with balsam, and that all things do depend upon the motional pulsation of the heart: So the heart is the beginning of life, the Sun of the Microcosm, as proportionably the Sun deserves to be call'd the heart of the world, by whose virtue, and pulsation, the blood is mov'd perfected, made vegetable, and is defended from corruption, and mattering: and this familiar household-god doth his duty to the whole body, by nourishing, cherishing, and vegetating, being the foundation of life, and author of all.

This is very far from the common misconception that Harvey was the person who first described the heart as *only* a pump that keeps the blood circulating around (it was actually René Descartes who took that step, suggesting in his *Discourse on Method*, published in 1637, that the heart is a purely mechanical pump). Nor is it the whole truth simply to say, as many books do, that Harvey saw the heart as the source of the blood's heat. His views were more mystical than that. But Harvey's work was still a profound step forward, and throughout his surviving writings (many of his papers were lost, unfortunately, when his London rooms were ransacked by Parliamentary troops in 1642) there is a repeated emphasis on the importance of knowledge derived from personal observation and experience. He specifically pointed out that we should not deny that phenomena exist just because we do not know what causes them, so it is appropriate to look kindly on his own incorrect 'explanations' for the circulation of the blood and to focus on his real achievement in discovering that the blood does circulate. Although Harvey's idea was by no means universally accepted at first, within a few years of his death, thanks to the development of the microscope in the 1650s, the one gap in his argument was plugged by the discovery of the tiny connections between arteries and veins – a powerful example of the connection between progress in science and progress in technology.

But if Harvey was, as far as scientific history is concerned, one of the last of the Renaissance men, that doesn't mean we can draw a neat line on the calendar after his work and say that proper science began then, in spite of the neat coincidence of the timing of his death and the rise of microscopy. As the example of the overlap of his publications with those of Descartes highlights, history doesn't come in neat sections, and the person who best fits the description of the first scientist was already at work before Harvey had even completed his studies in Padua. It's time to go back to the sixteenth century and pick up the threads of the developments in astronomy and the mechanical sciences which followed on from the work of Copernicus.

2 The Last Mystics

The movement of the planets – Tycho Brahe – Measuring star positions – Tycho's supernova – Tycho observes comet – His model of the Universe – Johannes Kepler: Tycho's assistant and inheritor – Kepler's geometrical model of the Universe – New thoughts on the motion of planets: Kepler's first and second laws – Kepler's third law – Publication of the Rudolphine star tables – Kepler's death

The movement of the planets

The person who most deserves the title of 'first scientist' was Galileo Galilei, who not only applied what is essentially the modern scientific method to his work, but fully understood what he was doing and laid down the ground rules clearly for others to follow. In addition, the work he did following those ground rules was of immense importance. In the late sixteenth century, there were others who met some of these criteria – but the ones who devoted their lives to what we now call science were often still stuck with a medieval mindset about the relevance of all or part of their work, while the ones who most clearly saw the, for want of a better word, philosophical significance of the new way of looking at the world were usually only part-time scientists and had little influence on the way others approached the investigation of the world. It was Galileo who first wrapped everything up in one package. But Galileo, like all scientists, built on what had gone before, and in this case the direct link is from Copernicus, the man who (himself drawing on the work of his predecessors such as Peuerbach and Regiomontanus) began the transformation of astronomy in the Renaissance, to Galileo, via Tycho Brahe and Johannes Kepler (and on, as we shall see, from Kepler and Galileo to Isaac Newton). Tycho, as he is usually known, also provides a particularly neat example of the way in which profoundly significant scientific work could still, at that time, be mixed up with distinctly old-fashioned and mystical interpretations of the significance of that work. Strictly speaking, Brahe and Kepler weren't quite the last mystics – but they certainly were, in astronomy at least, transitional figures between the mysticism of the Ancients and the science of Galileo and his successors.

Tycho Brahe

Tycho Brahe was born in Knudstrup, on the southern tip of the Scandinavian peninsula, on 14 December 1546. This is now in Sweden but was then part of Denmark. The baby was christened Tyge (he was even a transitional figure in the way that he later Latinized his first name, but not his surname). He came from an aristocratic family – his father, Otto, served the King as a Privy Counsellor, was the lieutenant of several counties in turn and ended his career as governor of Helsingborg Castle (opposite Elsinore, later made famous by William Shakespeare in *Hamlet*, first performed in 1600). As Otto's second child but eldest son, Tycho was born with the proverbial silver spoon in his mouth, but his life almost immediately took a twist which might have come right out of a play. Otto had a brother, Joergen, an admiral in the Danish navy, who was married but childless. The brothers had agreed that if and when Otto had a son, he would hand the infant over to Joergen to raise as his own. When Tycho was born, Joergen reminded Otto of his promise, but received a frosty response. This may not have been unrelated to the fact that Tycho had a twin brother who was stillborn, and his parents may well have feared that Otto's wife, Beate, might not be able to have more children. Biding his time, Joergen waited until Tycho's first younger brother was born (only a little over a year later) and then kidnapped little Tycho and took him off to his home at Tostrup.

With another healthy baby boy to raise (Otto and Beate eventually produced five healthy sons and five healthy daughters) this was accepted by the family as a *fait accompli*, and Tycho was indeed raised by his paternal uncle. He received a thorough grounding in Latin as a child before being sent to the University of Copenhagen in April 1559, when he was not yet 13 years old – not unusually young in those days for the son of an aristocrat to begin the education aimed at fitting him for high office in the state or Church.

Joergen's plans for Tycho to follow a career of service to the King in the political field began to fall apart almost at once, because on 21 August 1560 there was an eclipse of the Sun. Although total in Portugal, it was only a partial eclipse in Copenhagen. But what caught the imagination of 13-year-old Tycho Brahe was not the less-than-spectacular appearance of the eclipse, but the fact that the event had been predicted long before, from the tables of observations of the way the Moon seems to move among the stars – tables going back to ancient times but modified by later observations, particularly by Arabian astronomers. It seemed to him 'as something divine that men could know the motions of the stars so accurately that they could long before foretell their places and relative positions'.[1]

Tycho spent most of the rest of his time in Copenhagen (just over eighteen months) studying astronomy and mathematics, apparently indulged by his uncle as a phase he would grow out of. Among other things, he bought a copy of a Latin edition of the works of Ptolemy and made many notes in it (including one on the title page recording that he purchased the book, on the last day of November 1560, for two thaler).

In February 1562, Tycho left Denmark to complete his education abroad, part of the usual process intended to turn him into an adult fit for his position in society. He went to the University of Leipzig, where he arrived on 24 March, accompanied by a respectable young man called Anders Vedel, who was only four years older than Tycho but was appointed by Joergen as his tutor, to act as a companion and (it was clearly understood) to keep the younger man out of mischief. Vedel was partly successful. Tycho was supposed to study law in Leipzig, and he did this work with reasonable diligence. But his great academic love was still astronomy. He spent all his spare money on astronomical instruments and books, and stayed up late making his own observations of the heavens (conveniently, when Vedel was asleep). Even though Vedel held the purse strings, and Tycho had to account to him for all his expenditure, there was little the elder man could do to curb this enthusiasm, and Tycho's skill as an observer and knowledge of astronomy increased more rapidly than his knowledge of law.

Measuring star positions

When Tycho became more knowledgeable about astronomy, though, he realized that the accuracy with which men seemed to 'know the positions of the stars' was much less impressive

1. Quoted by Dreyer, from Gassendi's biography of Tycho, first published in 1654, which drew on Tycho's personal papers.

than he had thought at first. In August 1563, for example, a conjunction of Saturn and Jupiter took place – a rare astronomical event in which the two planets are so close together on the sky that they seem to merge. This had great significance for astrologers,[1] had been widely predicted and was eagerly anticipated. But while the actual event occurred on 24 August, one set of tables was a whole month late in its prediction and even the best was several days in error. At the very start of his career in astronomy, Tycho took on board the point which his immediate predecessors and contemporaries seemed unwilling to accept (either out of laziness or too great a respect for the Ancients) – that a proper understanding of the movement of the planets and their nature would be impossible without a long series of painstaking observations of their motions relative to the fixed stars, carried out to a better accuracy than any such study had been carried out before. At the age of 16, Tycho's mission in life was already clear to him. The only way to produce correct tables of the motions of the planets was by a prolonged series of observations, not (as Copernicus had) by taking the odd observation now and then and adding them more or less willy-nilly to the observations of the Ancients.

Remember that the instruments used to make observations in those days, before the development of the astronomical telescope, required great skill in their construction and even greater skill in their use (with modern telescopes and their computers, it is the other way around). One of the simplest techniques used by Tycho in 1563 was to hold a pair of compasses close to his eye, with the point of one leg of the pair on a star and the other point on a planet of interest – say, Jupiter. By using the compasses set with this separation to step off distances marked on paper, he could estimate the angular separation of the two objects on the sky at that time.[2] But he needed much better accuracy than this could provide. Although the details of the instruments he used are not crucial to my story, it is worth mentioning one, called a cross-staff or radius, which Tycho had made for him early in 1564. This

1. Some astronomers now think that a series of similar conjunctions at the time of the birth of Jesus may have been the phenomenon known as the 'star of Bethlehem'.

2. Telling the time accurately was, of course, another big problem in the 1560s, long before the development of accurate clocks – one of the many examples of the interdependence of science and technology.

was a standard kind of instrument used in navigation and astronomy in those days, consisting basically of two rods forming a cross, sliding at right angles to one another, graduated and subdivided into intervals so that by lining up stars or planets with the ends of the cross pieces it was possible to read off their angular separation from the scale. It turned out that Tycho's cross-staff had not been marked up correctly, and he had no money to get it recalibrated (Vedel was still trying to do his duty by Joergen Brahe and keep Tycho from spending all of his time and money on astronomy). So Tycho worked out a table of corrections for the instrument from which he could read off the correct measurement corresponding to the incorrect reading obtained by the cross-staff for any observation he made. This was an example that would be followed by astronomers trying to cope with imperfect instruments right down the centuries, including the famous 'repair' made to the Hubble Space Telescope by using an extra set of mirrors to correct for flaws in the main mirror of the telescope.

As an aristocrat with a (seemingly) secure future, there was no need for Tycho to complete the formality of taking a degree, and he left Leipzig in May 1565 (still accompanied by Vedel) because war had broken out between Sweden and Denmark, and his uncle felt he should return home. Their reunion was brief. Tycho was back in Copenhagen by the end of the month, where he found that Joergen had also just returned from fighting a sea battle in the Baltic. But a couple of weeks later, while the King, Frederick II, and a party that included the admiral were crossing the bridge from Copenhagen castle into the town, the King fell into the water. Joergen was among those who immediately went to his rescue, and although the King suffered no long-term ill effects, as a result of his immersion, Joergen Brahe contracted a chill, complications developed and he died on 21 June. Although the rest of the family (with the exception of one of his maternal uncles) frowned upon Tycho's interest in the stars and would have preferred him to follow a career fitting his station in society, he had an inheritance from his uncle and there was nothing they could do (short of another kidnap) to tie him down. Early in 1566, soon after his nineteenth birthday, Tycho set off on his travels, visiting the University of Wittenburg and then settling for a time and studying at the University of Rostock, where he did eventually graduate.

These studies included astrology, chemistry (strictly speaking alchemy) and medicine, and for a time Tycho made few observations of the stars. The breadth of his interests is not surprising, since so little was known about any of these subjects that there wasn't much point in trying to be a specialist, while the astrological influence meant that there was thought to be, for example, a strong connection between what went on in the heavens and the workings of the human body.

Tycho was, like his peers, a believer in astrology and became adept at casting horoscopes. Not long after his arrival in Rostock, there was an eclipse of the Moon, on 28 October 1566. On the basis of a horoscope he had cast, Tycho declared that this event foretold the death of the Ottoman Sultan, Sulaiman, known as the Magnificent. In truth, this wasn't a very profound prediction, because Sulaiman was 80 years old. It was also a popular one in Christian Europe, since he had earned his sobriquet the Magnificent partly by conquering Belgrade, Budapest, Rhodes, Tsabriz, Baghdad, Aden and Algiers, and had been responsible in 1565 for a massive attack on Malta, successfully defended by the Knights of St John. The Ottoman Empire was at its peak under Sulaiman and a serious threat to the eastern parts of Christian Europe. When news reached Rostock that the Sultan had indeed died, Tycho's prestige soared – although the shine was taken off his achievement when it turned out that the death had occurred a few weeks before the eclipse.

Later the same year, one of the most famous incidents in Tycho's life occurred. At a dance held on 10 December, Tycho quarrelled with another Danish aristocrat, Manderup Parsbjerg. The two ran into one another again at a Christmas party on 27 December, and the row (we don't know for sure what it was about, but one version of the story is that Parsbjerg mocked Tycho's prediction of the death of a Sultan who was already dead) reached such a pitch that it could only be settled by a duel. They met again at 7 pm on 29 December in pitch darkness (such an odd time to choose that it may have been an accidental encounter) and lashed out at each other with swords. The fight was inconclusive, but Tycho received a blow which cut away part of his nose, and he concealed this disfigurement for the rest of his life using a specially made piece manufactured out of gold and silver. Contrary to most popular accounts, it wasn't the tip of the nose that Tycho lost, but a chunk out of the upper part; he also used to carry a box of ointment about with him and could often be seen rubbing it on to the afflicted region to ease the irritation.

Apart from its curiosity value, the story is important because it correctly portrays Tycho, now just past his twentieth birthday, as a bit of a firebrand, arrogantly aware of his own abilities and not always willing to follow the path of caution. These traits would surface in later life to bring him a lot more grief than a damaged nose.

During his time in Rostock, Tycho made several visits to his homeland. Although he was unable to convince his family that he was doing the right thing by following his interests in things like astronomy, in other quarters, his increasing stature as a man of learning did not go unnoticed. On 14 May 1568, Tycho received a formal promise from the King, still Frederick II, that he could have the next canonry to become vacant at the Cathedral of Roskilde, in Seeland. Although the Reformation of the church had taken place more than thirty years before, back in 1536, and Denmark was staunchly Protestant, the income that had formerly gone to the canons of the cathedral was now spent on providing support for men of learning. They were still called canons and they still lived in a community associated with the cathedral, but they had no religious duties and the posts were entirely in the gift of the King. Frederick's offer certainly reflected Tycho's potential as a 'man of learning', but it is also worth remembering, if the promise seems rather generous to one so young, that Tycho's uncle had died, all too literally, in the service of the King.

Having completed his studies in Rostock, and with his future secured by the promise of a canonry, in the middle of 1568 Tycho set off on his travels again. He visited Wittenburg once more, then Basle, before settling for a spell in Augsburg, early in 1569, and beginning a series of observations there. To assist in this work, he had a huge version of the instrument called a quadrant made for him. It had a radius of about 6 metres, big enough so that the circular rim could be calibrated in minutes of arc for accurate observations, and it stood on a hill in the garden of one of his friends for five years, before being destroyed by a storm in December 1574. But Tycho had left Augsburg in 1570, returning to Denmark when news came that his father was seriously ill. In spite of this, Tycho was not to be distracted from his

7. *Tycho's great quadrant, 1569.*

life's work, and was making observations from Helsingborg Castle by the end of December that year.

Otto Brahe died on 9 May 1571, just 58 years old, and left his main property at Knudstrup jointly to his two eldest sons, Tycho and Steen. Tycho went to live with his mother's brother, also called Steen, the only person in the family who had ever encouraged his interest in astronomy, and who, according to Tycho himself, had been the first person to introduce papermaking and glass manufacture on a large scale to Denmark. Until late in 1572, perhaps under the influence of the elder Steen, Tycho devoted himself mainly to chemical experiments, although he never abandoned his interest in astronomy. But in the evening of 11 November 1572 his life was changed again by one of the most dramatic events that the Universe can provide.

Tycho's supernova

That evening Tycho was returning to the house from his laboratory, and taking in the panorama of the stars along the way, when he realized that there was something odd about the constellation Cassiopeia – the W-shaped constellation that is one of the most distinctive features of the northern sky. There was an extra star in the constellation. Not only that, but it was particularly bright. To appreciate the full impact of this on Tycho and his contemporaries, you have to remember that at that time stars were regarded as fixed, eternal and unchanging lights attached to a crystal sphere. It was part of the concept of the perfection of the heavens that the constellations had been the same for all eternity. If this really were a new star, it would shatter that perfection – and once you accepted that the heavens were imperfect, who could say what might follow?

One observation, though, did not prove that what Tycho had seen was a new star. It might be a lesser object, such as a comet. At that time, comets were thought to be atmospheric phenomena occurring only a little way above the surface of the Earth, not even as far away as the Moon (although for all anyone knew, the atmosphere itself extended at least as far as the Moon). The way to tell would be to measure the position of the object relative to the adjacent stars of Cassiopeia, and see if it changed its position, like a comet or meteor, or was always in the same place, like a star. Fortunately, Tycho had just completed the construction of another very large sextant, and whenever the clouds cleared in the nights that followed, he concentrated his attention on the new star. It stayed visible for eighteen months, and in all that time it never moved relative to the other stars. It was, indeed, a new star, so bright at first (as bright as Venus) that it could be seen in daylight, although it gradually faded from December 1572 onwards. Of course, many other people saw the star as well, and many fanciful accounts of its significance were circulated in 1573. Tycho had written his own account of the phenomenon. Although he was at first reluctant to publish it (possibly because he was concerned at how others might react to the shattering of heavenly perfection, but also because the star was still visible so his account was necessarily incomplete and not least because it might be regarded as unseemly for a nobleman to be seen to be involved in such studies), he was persuaded by friends in Copenhagen that he ought to do so to set the record straight. The result was a little book *De Nova Stella* (*The New Star*), which appeared in 1573 and gave us a new astronomical word, nova.[1] In the book, Tycho showed that the object was not a comet or meteor and must belong to the 'sphere' of the fixed stars, discussed the astrological significance of the nova (in vague and general terms) and made a comparison with an object reported to have been seen in the heavens by Hipparchus around 125 BC.

It was quite easy to read astrological significance into anything visible in the heavens at that time, since much of Europe was in turmoil. Following the initial success of the Reformation movement, the Catholic Church was fighting back, notably through the activities of the Jesuits in Austria and the southern German states. In France, the Protestant Huguenots were suffering severe setbacks in the middle part of what became known as the French Wars of Religion, and there were bloody battles in The Netherlands between the independence fighters and the Spanish. Tycho could hardly write a book about a new star appearing in the midst of all that turmoil without at least nodding in the direction of astrology. But the key facts were clear from *De Nova Stella* – that the object was fixed among the fixed stars and met

1. We now know that there are two kinds of 'new star', one bright and relatively common, one very much brighter still and much rarer. The super-bright novae are called, logically enough, supernovae. The new star of 1572 was actually one of the super-bright objects, and is now regarded as a supernova. But what mattered most in Tycho's day was not its brightness but its newness.

every criterion to be regarded as a genuinely new star. Many other astronomers studied the object (including Thomas Digges, whose position closely matched Tycho's own), but Tycho's measurements were clearly the most accurate and reliable.

There is one irony in all this. Tycho in particular made an intensive study of the star to see if there was any trace of the parallax shift that could be expected if the Earth really did move around the Sun. Because Tycho was such a superb observer and had built such accurate instruments, this was the most sensitive search for parallax yet undertaken. He could find no evidence of parallax, and this was an important factor in convincing him that the Earth is fixed, with the stars rotating about it on their crystal sphere.

Tycho's life was not immediately transformed by his work on the new star (now sometimes referred to as Tycho's star or Tycho's supernova), but in 1573 it did change significantly for personal reasons. He formed a permanent liaison and settled down with a girl called Christine (or Kirstine). Very little is known about Christine except that she was a commoner – some accounts say she was the daughter of a farmer, others the daughter of a clergyman and others that she was a servant at Knudstrup. Probably because of the difference in status, the couple never went through a formal marriage ceremony. In sixteenth-century Denmark, however, such a wedding was regarded as something of an optional extra, and the law said that if a woman openly lived with a man, kept his keys and ate at his table, after three years she was his wife. Just in case there might be any doubt, some time after Tycho's death several of his relatives signed a legal declaration that his children were legitimate and that their mother had been his wife. Whatever the formal status, the marriage was a successful and seemingly happy one, producing four daughters and two sons who survived childhood, and two more babies who died in infancy.

In 1574, Tycho spent part of his time in observing, but most of the year in Copenhagen, where, at the behest of the King, he gave a series of lectures at the university. But although, as this request shows, his reputation was on the increase, he was not happy with the conditions in Denmark and felt that he could get more support for his work if he went abroad. After extensive travels during 1575 he seems to have decided to settle in Basle, and returned to Denmark at the end of the year to put his affairs in order for the move. By now, though, there was an awareness at court that Tycho's presence in Denmark added to the prestige of the whole country, and the King, already sympathetic, was urged to do something to keep the now-famous astronomer at home. Tycho turned down the offer of a royal castle for his base – perhaps wisely, given the administrative duties and responsibilities that would have been involved, but not the kind of offer that most people would refuse. Undaunted, King Frederick hit on the idea of giving Tycho a small island, Hveen, located in the sound between Copenhagen and Elsinore. The proposal included an offer to pay for the construction of a suitable house on the island out of the royal purse, plus an income. This really was an offer Tycho couldn't refuse, and on 22 February 1576 he made his first visit to the island where he would make most of his observations, fittingly carrying out an observation of a conjunction of Mars and the Moon on the island that evening.[1] The formal document assigning the island to Tycho was signed by the King on 23 May. At the age of 29, Tycho's future seemed secure.

As long as Frederick remained on the throne, Tycho was able to enjoy an unprecedented amount of freedom to run his observatory just as he liked. The island was small – roughly oblong in shape and just three miles from shore to shore along its longest diagonal – and the highest point on it, chosen as the site for Tycho's new residence and observatory, was only 160 feet above sea level. But at first money was no object, as in addition to his other income, Tycho was granted more lands on the mainland. He neglected his duties as Lord of the Manor in connection with these lands abominably, and this would eventually lead to problems, but at first he seemed to have all the benefits with none of the responsibilities. Even the long-promised canonry finally fell into his lap in 1579. The observatory was christened Uraniborg, after Urania, the Muse of astronomy, and over the years developed into a major scientific institution with observing galleries, library and studies. The instruments were the best that money could provide, and

1. A conjunction is when one astronomical object moves in front of (or behind) another; a solar eclipse is a conjunction in which the Moon passes in front of the Sun.

as the work of observing developed and more assistants came to the island to work with Tycho, a second observatory was built near by. Tycho established a printing press in Uraniborg to ensure the publication of his books and astronomical data (and his rather good poetry), and when he had difficulty obtaining paper he built a papermaking works as well. But don't run away with the idea that Uraniborg was entirely the forerunner of a modern observatory and technological complex. Even here, Tycho's mysticism was reflected in the layout of the buildings, itself intended to reflect the structure of the heavens.

Tycho observes comet

Most of Tycho's work on the island over the next twenty years can be glossed over, because it consisted of the dull but essential task of measuring the positions of the planets relative to the fixed stars, night after night, and analysing the results. To put the task in perspective, it takes four years of observing to track the Sun's movements 'through' the constellations accurately, twelve years for each of Mars and Jupiter, and thirty years to pin down the orbit of Saturn. Even though Tycho had started observing at the age of 16, his earlier measurements were incomplete, and less accurate than those he could now make; even twenty years on, Hveen was barely enough for the job in hand. This work would not come to fruition until Johannes Kepler drew upon Tycho's tables to explain the orbits of the planets, years after Tycho had died. But in 1577, alongside his routine work Tycho observed a bright comet, and his careful analysis of how the comet moved showed once and for all that it could not be a local phenomenon, closer to the Earth than the Moon is, but must travel among the planets themselves, actually crossing their orbits. Like the observations of the supernova of 1572, this was a shattering blow to the old ideas about the heavens, this time destroying the notion of crystal spheres, since the comet moved right through the places where these spheres were supposed to be.

Tycho first saw the comet on 13 November 1577, although it had already been noticed in Paris and London earlier that month. Other European observers also calculated that the comet must be moving among the planets, but it was universally acknowledged that Tycho's observations were more accurate than those of anyone else, and it was his work that clinched the matter in the minds of most of his contemporaries. Several other, fainter, comets were studied in the same way over the next few years, confirming his conclusions.

His model of the Universe

The comet studies and his earlier observations of the supernova encouraged Tycho to write a major book, *Astronomiae Instauratae Progymnasmata* (*Introduction to the New Astronomy*), which appeared in two volumes in 1587 and 1588.[1] It was in this book that he laid out his own model of the Universe, which looks to modern eyes like something of a backward step, because it is a kind of halfway house between the Ptolemaic system and the Copernican system. But there were elements of the Tychonic model that broke new ground, and it deserves more credit than it is usually given.

Tycho's idea was that the Earth is fixed at the centre of the Universe and that the Sun, the Moon and the fixed stars orbit around the Earth. The Sun itself was seen as being at the centre of the orbits of the five planets, with Mercury and Venus moving in orbits smaller than the orbit of the Sun around the Earth, and with Mars, Jupiter and Saturn moving in orbits which are centred on the Sun, but which include both the Sun and the Earth within those orbits. The system did away with epicycles and deferents, and it explained why the motion of the Sun was mixed up with the motion of the planets. In addition, by displacing the centre of the planetary orbits from the Earth, Tycho filled up most of the space out to the assumed position of the fixed stars – which, in Tycho's model, was just 14,000 Earth radii away from us (there was no problem with parallax, of course, because in this model the Earth did not move). But the really significant, modern-looking idea in all this is that Tycho did not regard the orbits as being associated with anything physical like crystal spheres, but saw them merely as geometrical relationships which describe the movement of the planets. Although he did not state it this way, he was the first astronomer to imagine the planets hanging unsupported in empty space.

But in other ways Tycho was less modern. He could not accept what

1. 'Appeared' in so far as printing of at least a large part of the book took place on Hveen. But only a few copies were circulated to Tycho's contacts at that time. Full publication did not take place until 1603, under the editorship of Johannes Kepler.

he called the 'physical absurdity' of letting the Earth move, and he was convinced that if the Earth were rotating on its axis then a stone dropped from a tall tower would fall far to one side of the tower as the Earth moved underneath it while it was falling. It is also relevant to note that at this time the most virulent opposition to the Copernican system still came from the Protestant churches of northern Europe, while it was largely ignored by the Catholic Church (Bruno had yet to stir up their opposition to these ideas). Religious tolerance was not a feature of Denmark in the late sixteenth century, and anyone whose position depended utterly on the patronage of the King would have been mad to promote Copernican ideas, even if he did believe in them (which, it is clear, Tycho did not).

While the routine observations (so important to science, but utterly boring to describe) continued, Tycho's position at Hveen came under threat, just at the time his book was being printed, with the death of Frederick II in 1588. Frederick's successor, his son Christian, was only 11 when the King died, and the Danish nobles elected four of their number to serve as Protectors until he reached the age of 20. At first, there was little change in the government's attitude to Tycho – indeed, more money was provided later that year to cover debts that he had incurred building up the observatory. During his last years at Hveen, Tycho was clearly regarded as a great national institution, and he received many distinguished visitors, including James VI of Scotland (later, on the death of Elizabeth, to become James I of England), who had come to Scandinavia to marry Anne, one of King Christian's sisters. The two hit it off, and James granted Tycho a thirty-year copyright for all his writings published in Scotland. Other visitors were not so congenial, and Tycho clearly did not always relish his role as a kind of performing poodle. He managed to offend several members of the nobility with his offhand manner towards visitors he did not like, and by his flouting of protocol by allowing his low-born common-law wife a place of honour at table. Although we don't know all the reasons, it is clear that Tycho was becoming dissatisfied with the arrangements for his work at Hveen as early as 1591, when he wrote in a letter to a friend that there were certain unpleasant obstacles to his work which he hoped to resolve, and commented that 'any soil is a country to the brave, and the heavens are everywhere overhead'.[1] Tycho also quarrelled with some of his tenants on the mainland and got into trouble with the ruling Council for neglecting the maintenance of a chapel that formed part of his estates. But none of these distractions seems to have affected his observations, which included a major catalogue of the positions of the fixed stars, which he said reached a thousand in 1595, although just 777 of the best positions were eventually published in the first volume of Kepler's edition of Tycho's *Progymnasmata*.

A year later, King Christian IV was crowned and soon began to make his presence felt. Christian saw a need to make economies in just about every area of state activity, and among many other things immediately withdrew the mainland estates granted to Tycho by Frederick II from his stewardship. Most of Tycho's friends at court had died by now (Tycho himself was nearing 50), and the King was probably right in thinking that with Uraniborg long since built and running smoothly, it ought to be possible to keep things ticking over there on a greatly reduced budget. But Tycho was used to being given considerable indulgence and saw any reduction in his income as an insult, as well as a threat to his work. If he couldn't maintain Uraniborg at the level he wanted, with many assistants, printers, papermakers and all the rest, he wouldn't maintain it at all.

Things came to a head in March 1597, when the King cut off Tycho's annual pension. Although he was still a wealthy man in his own right, Tycho felt this to be the last straw, and made immediate plans to move on. He left the island in April 1597 and spent a few months in Copenhagen before setting off on his travels, initially to Rostock, accompanied by an entourage of about twenty people (students, assistants and so on), his most important portable instruments and his printing press.

There, Tycho seems to have had second thoughts, and wrote what he probably regarded as a conciliatory letter to King Christian, in which he said (among many other things) that if he had a chance to continue his work in Denmark he 'would not refuse to do so'. But this only made the situation worse. Christian was offended at Tycho's high

1. Quoted by Dreyer. Other quotes in this chapter from the same source, unless otherwise specified.

tone and the way he treated the King as an equal, and not least by this haughty phrase, which implied that Tycho might refuse a royal request. In his reply, he said that 'it is very displeasing to us to learn that you seek help from other princes, as if we or the kingdom were so poor that we could not afford it unless you went out with woman and children to beg from others. But whereas it is now done, we have to leave it so, and not trouble ourselves whether you leave the country or stay in it'. I must admit to having rather more sympathy for Christian than he is usually given, and a less arrogant individual than Tycho might well have been able to reach an accommodation with the King without leaving Hveen. But then a less arrogant individual than Tycho would still have had his nose intact and might never have become such a great astronomer in the first place.

His boats home well and truly burned, Tycho moved on to Wandsbeck, near Hamburg, where he resumed his observing programme (the heavens were, indeed, everywhere overhead) while he sought a new permanent base for his work. This led to an invitation from the Holy Roman Emperor, Rudolph II, a man much more interested in science and art than he was in politics. This was good for Tycho but bad for most of middle Europe, with Rudolph's reign leading, partly through his poor qualities as a politician (some historians think he was actually mad), to the Thirty Years War. Tycho arrived in Prague, the capital of the Empire, in June 1599 (having left his family in Dresden). After an audience with the Emperor, he was appointed Imperial Mathematician, granted a good income and offered a choice of three castles in which to set up his observatory. Tycho chose Benatky, 35 kilometres to the northeast of Prague, and left the city itself with some relief – a contemporary account describes the walls of the city as:

> Less than strong, and except the stench of the streetes drive backe the Turks . . . there is small hope in the fortifications thereof. The streets are filthy, there be divers large market places, the building of some houses is of free stone, but the most part are of timber and clay, and are built with little beauty or art, the walles being all of whole trees as they come out of the wood, the which the barke are laid so rudely as they may on both sides be seen.

A far cry from the peace and comfort of Uraniborg. It is no surprise that towards the end of 1599 Tycho spent several weeks at a secluded Imperial residence in the countryside to avoid an outbreak of plague. But with the threat gone and his family arrived from Dresden, Tycho began to settle in at the castle, sending his eldest son back to Denmark to fetch four large observing instruments from Hveen. It took a long time to get the instruments to Benatky, and the castle had to be adapted to make a suitable observatory. It is hardly surprising that Tycho, now in his fifties, didn't make any significant observations here in the short time that remained before his death. But even before arriving in Prague, he had entered into a correspondence that would ensure that his life's work would be put to the best possible use by the ablest member of the next generation of astronomers, Johannes Kepler.

Johannes Kepler: Tycho's assistant and inheritor

Kepler had none of the advantages of birth which had given Tycho a head start in life. Although he came from a family that had once ranked among the nobility and had its own coat of arms, Johannes's grandfather, Sebald Kepler, was a furrier, who moved from his home town of Nuremberg to Weil der Stadt, not far from Stuttgart in the southern part of Germany, some time around 1520. Sebald was a successful craftsman who rose high in the community, serving for a time as mayor (burgomeister). This was no mean achievement, since he was a Lutheran in a town dominated by Catholics; Sebald was clearly a hard worker and a pillar of the community. The same could hardly be said of his eldest son, Heinrich Kepler, who was a wastrel and drinker whose only steady employment was as a mercenary soldier in the service of whichever prince needed hired hands. He married young, a woman called Katherine, and the couple shared a house with several of Heinrich's younger brothers. The marriage was not a success. Apart from Heinrich's faults, Katherine herself was argumentative and difficult to live with, and she also had great faith in the healing powers of folk remedies involving herbs and such like – scarcely an uncommon belief at the time, but one which was to contribute to her eventual imprisonment as a suspected witch, and cause much grief to Johannes.

Johannes had a distinctly disturbed and rather lonely childhood (his only brother, Christoph, was much younger than him). He was born on 27 December 1571, but when he was only 2 his father went off to fight in the Netherlands, and Katherine followed, leaving the infant in the care of his grandfather. Heinrich and Katherine returned in 1576 and moved the family to Leonberg, in the Duchy of Württemberg. But in 1577 Heinrich was off to war again. On his return, he tried his hand at various businesses, including, in 1580, the favourite of the drunkard, running a tavern, this time in the town of Ellmendingen. Not surprisingly, he lost all his money. Eventually, Heinrich set off to try his luck as a mercenary again and disappeared from his family altogether. His fate is not known for sure, although he may have taken part in a naval campaign in Italy; whatever, his family never saw him again.

In all this turmoil, Johannes was tossed about from household to household and school to school (but at least his family was still far enough up the social ladder that he did go to school, with the aid of scholarships from a fund established by the Dukes of Württemberg). As if this weren't bad enough, while staying with his grandfather he caught smallpox, which left him with bad eyesight for the rest of his life, so that he could never have become an observer of the heavens like Tycho. But his brain was unaffected, and although he was often set back by having to change schools when his family moved, by the age of 7 he was allowed to enter one of the new Latin schools in Leonberg. These schools had been introduced after the Reformation, primarily to prepare men for service in the Church or the state administration; only Latin was spoken in the schools, in order to inculcate the pupils with the language of all educated men at the time. With all the interruptions, it took Johannes five years to complete what should have been three years' worth of courses – but as a graduate of a Latin school, he was entitled to take an examination to be admitted to a seminary and train for the priesthood, the obvious and traditional route out of poverty and a life of toil for an intelligent young man. Although Kepler's interest in astronomy had already been stirred as a child when he saw (on two separate occasions) a bright comet (the same one studied by Tycho in 1577) and an eclipse of the Moon, his future in the Church seemed clearly mapped out when he passed the examination in 1584 and was admitted to a school in Adelberg at the age of 12. Once again, the language of the school was Latin, in which Kepler became fluent.

Although the discipline at the school was harsh and Kepler was a sickly youth who was often ill, he showed such promise academically that he was soon moved to a more advanced school at Maulbronn and prepared by his tutors for entry to the University of Tübingen to complete his theological studies. He passed the entrance examination for the university in 1588, then had to complete a final year at Maulbronn before he could take up his place at the university, at the age of 17. Although training to become a priest, the courses Kepler was required to attend in his first two years at Tübingen included mathematics, physics and astronomy, in all of which he was an outstanding pupil. He graduated from this part of the course in 1591, second out of a class of fourteen, and moved on to his theological studies described by his tutors as an exceptional student.

Along the way, he had also learned something that was not in the official curriculum. The university's professor of mathematics was Michael Maestlin, who dutifully taught his students in public the Ptolemaic system approved by the Reformed Church. In private, though, Maestlin also explained the Copernican system to a select group of promising pupils, including Kepler. This made a deep impression on the young man, who immediately saw the power and simplicity of the Sun-centred model of the Universe. But it wasn't just in his willingness to accept the Copernican model that Kepler deviated from the strict Lutheran teaching of his time. He had grave doubts about the religious significance of some of the Church rituals, and although he believed firmly in the existence of God, he never found a formally established Church whose teachings and rituals made sense to him, and he persisted in worshipping in his own way – a distinctly dangerous attitude in those troubled times.

Just how Kepler would have reconciled his own beliefs with a role as a Lutheran clergyman we will never know, because in 1594, the year he should have completed his theological studies, his life was changed by a death in the distant town of Graz, in Austria. In spite of its distance, there was a seminary in Graz that had always had close academic connections with the University of Tübingen, and when the mathematics professor there died, it was natural for the authorities to ask Tübingen to suggest a replacement. The Tübingen authorities recommended Kepler, who was rather startled to be offered the post just when he was thinking about starting life as a clergyman. Although initially reluctant, he allowed himself to be persuaded that he really was the best man for the job, and left on the understanding that if he wanted, he could come back to the university in a couple of years, finish his training and become a Lutheran minister.

The 22-year-old professor of mathematics arrived in Graz on 11 April 1594. Although still within the Holy Roman Empire, he had crossed a significant invisible border, from the northern states where the Reformed Churches held sway to the southern region where the Catholic influence was dominant. But this invisible border was constantly changing, since under the treaty known as the Peace of Augsburg, settled in 1555, each prince (or duke, or whatever) was free to decide the appropriate religion in his domain. There were dozens of princes ruling individual statelets within the 'empire', and the state religion sometimes changed literally overnight when one prince died, or was overthrown, and was replaced by one of a different religious persuasion. Some princes were tolerant and allowed freedom of worship; others insisted that all their subjects convert to the new flavour of the month or forfeit their property at once. Graz was the capital of a statelet called Styria, ruled by Archduke Charles, who was determined to crack down on the Protestant movement, although at the time Kepler arrived exceptions such as the Lutheran seminary in Graz were still being tolerated.

Kepler was a poor man with no financial resources from his family – his university studies had been paid for by a scholarship and he had to borrow money for the journey to Graz. His situation wasn't improved when the authorities at the seminary decided to put him on a three-quarter salary until he proved his worth. But there was one way in which he could both make some money and endear himself to the top people in the Graz community – by casting horoscopes. Throughout his life, Kepler used astrology as a means to supplement his always inadequate income. But he was well aware that the entire business was utter tosh, and while he became skilful at the art of talking in vague generalities and telling people what they wanted to hear, in private letters he referred to his clients as 'fatheads' and described the astrology business as 'silly and empty'. A good example of Kepler's skill in this despised art came when he was commissioned to produce a calendar for 1595 predicting important events for the year ahead. His successful predictions included rebellious activity by the peasants in Styria, incursions into Austria by the Turks in the east and a cold winter. His skill in dressing these common-sense predictions up in astrological mumbo jumbo not only established his reputation in Graz, but got his salary increased to the full level appropriate for his post.

But although Kepler may have been less superstitious than many of his peers, he was still too mystically inclined to be called the first scientist. This is clearly highlighted by his first important contribution to the cosmological debate, which spread his reputation far beyond the confines of Styria.

Kepler's geometrical model of the Universe

Kepler was never able to be an effective observer of the heavens because of his bad eyesight, and in Graz he had no access to observational data. So he was left to follow in the mental footsteps of the Ancients, using pure reason and imagination to try to come up with an explanation for the nature of the cosmos. The question that particularly intrigued him at that time was why there should be six, and only six, planets in the Universe, accepting that Copernicus was right and the Earth itself is a planet. After puzzling over this for some time, Kepler hit on the idea that the number of planets might be related to the number of regular solid figures that can be constructed using Euclidean geometry. We are all familiar with the cube, which has six identical square faces. The other four regular solids are: the tetrahedron, made up of four identical triangular sides; the dodecahedron, made of twelve identical pentagons; the icosahedron (a more complicated twenty-sided figure made of identical triangular faces); and the octahedron (made from eight triangles).

The bright idea that Kepler came up with was to nest these (imaginary) figures one inside the other, so that in each case the corners of the inner figure just touched the surface of a sphere surrounding the solid, and that sphere in turn just touched the inner sides of the surfaces of the next solid figure out in the nest. With five Euclidean solids to play with, and one sphere inside the innermost solid as well as one outside the outermost solid, this defined six spheres – one for each of the orbits of the planets. By putting the octahedron in the middle, surrounding the Sun and just enclosing a sphere with the orbit of Mercury, followed

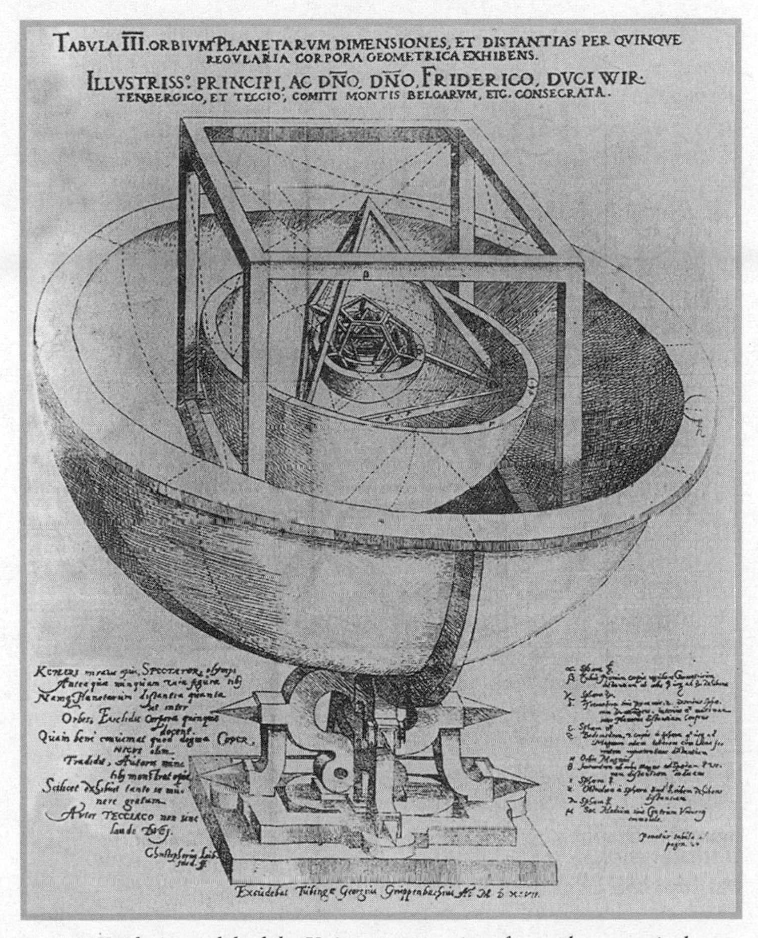

8. *Kepler's model of the Universe as a series of nested geometrical shapes. From Kepler's* Mysterium Cosmographicum, *1596.*

by an icosahedron, a dodecahedron, a tetrahedron and a cube, he got a spacing between the corresponding spheres that more or less corresponded to the spacing of the orbits of the planets around the Sun.

The agreement was never more than approximate, and it was based on a mystical belief that the heavens must be governed by geometry, not on anything that we would now call science. The model fell apart as soon as Kepler himself showed that the orbits of the planets are elliptical, like an elongated circle, not circular; and, in any case, we now know that there are more than six planets, so the geometry cannot be made to work even on its own terms. But when Kepler came up with the idea late in 1595 it seemed to him like a Divine revelation – which is ironic, since by espousing the Copernican model with the Sun at the centre of the Universe, Kepler's idea flew in the face of Lutheran teaching, and he was still a Lutheran, of sorts, himself.

Kepler spent the winter of 1595/6 working out his idea in detail, and corresponded with his old teacher Michael Maestlin about it. Early in 1596, he was granted leave of absence from his teaching duties to visit his ailing grandparents, and took the opportunity to call in on Maestlin in Tübingen. Maestlin encouraged Kepler to develop his ideas in a book and oversaw the printing of the book, which appeared in 1597, not long after Kepler returned (rather late, but trailing clouds of glory from his now widely discussed model) to his duties in Graz. The book is usually known as *Mysterium Cosmographicum* (*The Mystery of the Universe*), and it contained an idea which, with hindsight, is even more significant than the model of nested solids it described. Kepler picked up on the observation by Copernicus that the planets move slower in their orbits the further they are from the Sun, and suggested that they were kept moving in those orbits by a force (he called it 'vigour') reaching out from the Sun and pushing them along. He argued that the vigour would be less vigorous (so to speak) further from the Sun, and would only be able to push more distant planets more slowly. This idea, which was partly stimulated by the work of William Gilbert on magnetism (more of this in the next chapter) was an important step forward because it suggested a physical cause for the motion of the planets, where previously the best idea anyone had come up with was that the planets were pushed around by angels. Kepler specifically said that 'my aim . . . is to show that the machine of the universe is not similar to a divine animated being, but similar to a clock'.[1]

Kepler sent copies of his book to the most eminent thinkers of his

1. Quoted by Shapin.

day, including Galileo (who didn't bother to reply, but mentioned the new model in his lectures) and, most significantly, Tycho, at that time based in Germany. Tycho replied to Kepler with a detailed critique of the work, and was impressed by the mathematical skill of the author of the book, even though the idea of a Sun-centred Universe was still anathema to him. Indeed, Tycho was sufficiently impressed that he suggested that Kepler might care to join the team of assistants surrounding the older man. The offer soon proved extremely opportune.

In April 1597, Kepler married Barbara Müller, a young widow and daughter of a wealthy merchant. Although his need for financial security may have been a factor in the marriage, everything went well at first, with Kepler now on full salary and enjoying a happy home life. But two children died in infancy (although three others later survived), Barbara's family, feeling that she had married beneath her status, withheld money she was entitled to, and life with Kepler on a teacher's salary (even a full one) proved to be much tougher than life as the daughter of a successful merchant. Another problem blew up out of Kepler's eagerness to consolidate his new reputation by associating with other mathematicians and discussing his ideas with them. He wrote a letter to the then Imperial Mathematician, Reimarus Ursus, seeking his opinion on his own work, and sycophantically praising Ursus as the greatest mathematician of all time. Ursus didn't bother to reply, but took Kepler's praise out of context and published it as a kind of endorsement of some of his own work – which, as it happens, was critical of Tycho. It took a lot of tactful correspondence before Kepler was able to soothe Tycho's offended feelings and restore friendly relations with the great astronomer. Increasingly, Kepler longed for an opportunity to get his hands on Tycho's by now legendary wealth of observational data and test his ideas about planetary orbits using these accurate figures for the movement of the planets.

While all this was going on, the political situation in Styria deteriorated. In December 1596, Archduke Ferdinand, a devout Catholic, became the ruler of Styria. At first, he moved slowly to reform (or counter-reform) the state more to his liking, but after a few months the Protestant community, upset by changes in taxation which favoured the Catholics at their expense, and by other 'reforms', submitted an official list of complaints about their treatment under the new regime. This was a big mistake – probably the very response that Ferdinand had been trying to provoke, so that he could represent the Protestants as unruly troublemakers. After a visit to Italy in the spring of 1598, when he had an audience with the Pope and visited holy shrines, Ferdinand came back determined to wipe out the Protestant influence in Styria. In September, an edict was published telling all Protestant teachers and theologians to leave the state within two weeks or convert to Catholicism. There was no choice but to obey, and Kepler was among the many ejected Lutherans who took refuge in neighbouring states – although most left wives and families behind in the hope that they would be allowed to return. Out of the entire contingent of refugees from Graz, however, Kepler alone, for reasons that are not entirely clear but may have owed much to his increasing stature as a mathematician, was allowed back within a month. After all, in addition to his teaching post he was the district mathematician, a post which required its holder to live in Graz (the Archduke could, though, have simply sacked him and appointed another district mathematician). But the severity of the conditions Kepler now had to live under is highlighted by the fact that when his baby daughter died and he evaded the ceremony of last rites, he was not allowed to bury the infant until he had paid a fine for this omission.

In 1599, when the situation in Graz was becoming intolerable for Kepler, Tycho was establishing himself some 320 kilometres away near Prague, where people were free to worship in their own manner. In January 1600, an offer that was to transform Kepler's life came along. A Styrian nobleman called Baron Hoffman, who was impressed by Kepler's work and liked the mathematician, was also a Counsellor of the Emperor, Rudolph II, and had met Tycho. He had to go to Prague on court business, and offered to take Kepler with him and introduce him to Tycho. As a result, the first meeting of the two men who were between them to lay the foundations of scientific astronomy took place at Benatky Castle on 4 February 1600. Tycho was now 53, Kepler 28. Tycho had the greatest body of accurate astronomical data yet assembled, but was tired and in need of help to analyse the material. Kepler had nothing but his mathematical ability and a burning zeal to unlock the mysteries of the Universe. It might seem a marriage made in heaven, but there were still hurdles to be overcome before Kepler could achieve the breakthrough that made him a key figure in the history of science.

Although Kepler had intended paying a fairly brief visit to Tycho at this time (he had left his wife and stepdaughter behind in Graz and had not resigned his posts there), it became an extended sojourn. The impoverished Kepler desperately needed an official post with an income so that he could work with Tycho, and he equally desperately needed to get his hands on the data, which Tycho doled out only in driblets, cautious about giving a relative stranger a free hand with his life's work. Tycho's extensive entourage and the construction work going on at the castle to turn it into an observatory made it difficult for Kepler to settle down to work anyway, and he inadvertently offended one of Tycho's key assistants, who had been struggling with the problem of calculating the orbit of Mars, by offering to take over the task (an offer interpreted as an arrogant gesture by Kepler, setting himself up as a superior mathematician). Realizing that Tycho would never part with a copy of his data that he could take away to work on at home, and that the only way to get to grips with the puzzle was to stay for a year or more, Kepler (who was also well aware that his mathematical skills were second to none) drew up a list of his requirements if he were to be able to stay at the castle. Kepler gave the list to a friend, asking him to mediate with Tycho – but Tycho got hold of the list itself and took exception to what he saw as Kepler's high-handed demands, even though he had, in fact, been negotiating with Rudolph to obtain an official post for Kepler. Eventually, things were smoothed out to the point where Tycho offered to pay Kepler's moving expenses from Graz and assured him that the Emperor would come through with a paid position soon.

In June 1600, Kepler returned to Graz to try to sort out his affairs there – only to be confronted with an ultimatum from the city officials, tired of his long absences, who wanted him to go to Italy and study to be a physician, so that he would be more useful to the community. Before Kepler had time to make any decision, a deterioration in the religious situation made the decision for him. In the summer of 1600, all citizens of Graz who were not already Catholics were required to change their faith at once. Kepler was among sixty-one prominent citizens who refused to do so, and on 2 August he was dismissed from both his posts and, like the other sixty, given six weeks and three days to leave the state, forfeiting virtually all of what little property he had. Kepler wrote to the only two good contacts he had, Maestlin and Tycho, asking for help. Tycho's reply came almost by return, assuring him that negotiations with the Emperor were going well and urging that Kepler should head for Prague at once, with his family and what goods he was allowed to take.

The family arrived in the stinking, unhealthy city of Prague in mid-October, and were housed by Baron Hoffman through a winter which saw both Johannes and Barbara severely ill with fever, while their limited supply of money diminished rapidly. Still with no appointment from the Emperor, in February 1601 the Keplers moved in with Tycho's household at a new residence provided by Rudolph for the astronomer. Their relationship remained uneasy – Kepler unhappy at being dependent on Tycho, Tycho unhappy with what he saw as Kepler's ingratitude. But eventually Kepler was formally introduced to the Emperor, who appointed him as Tycho's official (and paid!) assistant in compiling a new set of tables of planetary positions which was to be called, in the Emperor's honour, the Rudolphine Tables.

At last Kepler's position had been regularized, although Tycho continued to dole out his wealth of data in penny packets, as and when he thought Kepler needed it, rather than giving him free access. It was hardly a close and friendly relationship. But then, on 13 October, Tycho was taken ill. After ten days when he was frequently delirious and close to death, and heard to cry out on more than one occasion that he hoped he should not seem to have lived in vain, on the morning of 24 October his mind cleared. With his younger son and his pupils, as well as a visiting Swedish nobleman in the service of the King of Poland, gathered around what was obviously going to be his deathbed, Tycho handed over the task of completing the Rudolphine Tables, and with it the responsibility for the vast treasury of planetary data, to Kepler – although he urged him to use the data to demonstrate the truth of the Tychonic model of the world, not the Copernican model.

Tycho's mind was certainly clear at that point, as he realized that for all their disagreements, Kepler was the most able mathematician in his entourage, the person most likely to make best use of Tycho's data and to ensure that, indeed, he had not lived in vain. He died soon after making this bequest of his life's work to the stunned younger man, who only weeks before had been a penniless refugee. Kepler must have been even more stunned a couple of weeks later to be appointed as Tycho's successor as Imperial Mathematician to the Court of Rudolph II, with responsibility for all of Tycho's instruments and unpublished work. It was a far cry from his early life in Germany. Although his life would still not be easy, and he would often have trouble getting his full salary out of the Emperor, at least Kepler would be able, at long last, to get to grips with the puzzle of planetary motion.

Kepler's work during his years in Prague was hampered by many factors. There were the continuing financial difficulties; there was interference from Tycho's heirs who were both eager to see the Rudolphine Tables and Tycho's other posthumous publications in print (not least in the hope of getting money from the books) and concerned that Kepler might distort (in their view) Tycho's data to lend credence to Copernican ideas; and there were his duties as Imperial Mathematician (meaning Imperial Astrologer), requiring him to spend much of his time in what he knew to be the fatuous task of advising Rudolph on the significance of heavenly portents for the prospects of war with the Turks, bad harvests, the progress of the religious troubles and so on. In addition, the calculations themselves were laborious and had to be checked and rechecked for arithmetical slips – surviving pages of Kepler's interminable calculations show sheet after sheet packed with arithmetical calculations of planetary orbits, a labour almost unimaginable in these days of pocket calculators and portable computers.

New thoughts on the motion of planets: Kepler's first and second laws

Not surprisingly, it took years to solve the riddle of the orbit of Mars, with Kepler moving in stages away from the idea of a perfectly circular orbit centred on the Sun. First, he tried an offset (but still circular) orbit, so that Mars was closer to the Sun in one half of the circle than the other – this matched up to some degree with the discovery that Mars moved faster in one half of its orbit (the half nearer the Sun). Along the way, Kepler made the now seemingly obvious, but then highly significant, step of carrying out some of his calculations from the perspective of an observer on Mars, looking at the Earth's orbit – a huge conceptual leap which presages the idea that all motion is relative. It was actually while still working with his 'eccentric' circular orbit, in 1602, that Kepler came up with what is now known as his second law – that an imaginary line joining the Sun to a planet moving in its orbit around the Sun sweeps out equal areas in equal times. This is a precise way of expressing just how much faster the planet moves when it is closer to the Sun, since a shorter radius line has to sweep across a bigger angle to cover the same area that a longer radius line sweeps out when moving across a smaller angle. It was only after this discovery that Kepler realized (after trying out other possibilities) that the shape of the orbit is actually elliptical, and in 1605, having been distracted from the task by other work, he came up with what is now known as his first law, that each planet moves in its own elliptical orbit around the Sun, with the Sun at one of the two foci (the same focus for each of the planets) of the ellipse. With those two laws, Kepler had done away with the need for epicycles, equants and all the complicating baggage of earlier models of the Universe, including his own mystical idea of nested solids (although he never accepted this).

Although news of Kepler's discoveries spread, the full discussion of his ideas didn't appear in print until his book *Astronomia Nova* was published in 1609 – publication was delayed by printing problems and lack of finance. But even the publication of the book did not bring the kind of instant acclaim from his peers that you might expect. People didn't like the idea of elliptical orbits (many people still had not accepted that the Earth was not at the centre of the Universe), and only a skilled mathematician could appreciate that Kepler's model was not just another piece of mystical thinking (like his nested solids or Tycho's model) but was securely founded on observational fact. It's no surprise, really, that Kepler only achieved the stature he deserves in the eyes of historians after just such a mathematician, Isaac Newton, used Kepler's laws in combination with his own theory of gravity to explain *how* the planets moved in elliptical orbits. Indeed, in his own time, Kepler was more famous as an astrologer than as an astronomer, although the distinction between the two was rather blurred. This is highlighted by one of the distractions from his planetary work, which occurred in 1604, when another 'new' star, as bright as the planet Jupiter, appeared in the sky during the summer and remained visible to the naked eye until well into 1606. To most people, this was an event of dramatic astrological importance, and Kepler had to interpret its significance as part of his duties as Imperial Mathematician. Although judiciously noncommittal about the implications of the event in his report to the Emperor, Kepler stuck his neck out sufficiently to suggest that in spite of its brightness the star must be at the same distance as the other stars, and was not a phenomenon in the region of the Universe occupied by the planets. Like Tycho before him, he saw the supernova as undermining the Aristotelian notion of literally fixed and eternal stars.

Not all of Kepler's 'distractions' from his planetary work lacked scientific significance. Also in 1604, he published a book about optics, analysing the way the eye works by refracting light rays that enter the pupil to focus them on the retina, so that all the rays coming from a single point on an illuminated object are focused at a single point on the retina. He then used this idea to explain that some people have bad eyesight (clearly a subject close to his heart) because imperfections in the eye cause the rays to be focused at a point either in front of or behind the retina – and he went on to describe how eyeglasses worked to correct these defects, something nobody had previously understood, even though glasses had been used for more than 300 years on an empirical basis. After Galileo applied the telescope to astronomy and news of his discoveries spread, Kepler developed his ideas about optics to explain how the telescope works. His scientific interests could be thoroughly down-to-earth, not just concerned with the heavenly spheres.

The years following the supernova saw a deterioration in the political and religious situation in central Europe, as the rival religious groups formed the political alliances which were to become involved in the Thirty Years War. Apart from the impact it had on the rest of Kepler's life, this struggle is significant in the history of science since the turmoil in central Europe, combined with the suppression of Galileo's ideas by the Catholic Church, were at least contributory factors in stunting the growth of scientific ideas in the region and ensuring that the full flowering of the seed that Kepler planted took place in England, where (in spite of the Civil War) there was a more settled academic environment in which people like Newton could work.

In 1608, several Protestant states joined together as the Protestant

Union, while their rivals formed the Catholic League the following year. Rudolph was by now a semi-recluse, obsessed by his art treasures, and distinctly odd, if not completely mad. Even in peacetime, he was in no condition to rule the Holy Roman Empire effectively (in so far as any Emperor actually did rule this disparate collection of states), he had run out of money and power gradually passed into the hands of his brother, Matthias, who became Emperor when Rudolph died in 1612. Kepler had long seen which way the wind was blowing and sought an appointment at his old University, Tübingen, but was rejected out of hand because of his unorthodox religious beliefs. At the same time, there were troubles at home. In 1611, Barbara became ill with epilepsy and one of their three children died of smallpox. Anxious to get out of Prague before everything fell apart politically, Kepler travelled to Linz, where he applied for a job as district mathematician and was accepted in June. But on hurrying back to Prague to make arrangements for the move, he found his wife seriously ill once again. She died of typhus a few days after his return. Depressed and uncertain about his future, Kepler lingered in Prague until Rudolph died, when, to his surprise, Matthias confirmed him in the post of Imperial Mathematician, offered him an annual salary (not that Kepler would ever see much of it), but gave him leave to go to Linz and take up the post there as well. Leaving his remaining children with friends for the time being, Kepler, still only 40 years old, set off on his travels again.

Even in Linz, though, his troubles continued. That part of Styria was firmly in the grip of the extreme orthodox Lutheran Church; the chief priest was a Tübingen man who knew of Kepler's non-mainstream views and refused to allow him to receive Holy Communion, a source of deep distress to Kepler, who was profoundly religious, albeit in his own way. Repeated appeals to the church authorities failed to resolve the situation, but took up time that Kepler might have spent on his planetary work. He also had his duties as district mathematician, and he soon remarried, a young woman of 24 who bore him six children, three of whom died in infancy. Kepler was also involved in religious work of another kind, using an eclipse of the Moon recorded in Herod's time to show that Jesus had actually been born in 5 BC, and he was involved in calendar reform (it was only in 1582 that Pope Gregory XIII had introduced the modern calendar, and many states in Protestant Europe were reluctant to make the change). The greatest distraction, though, came in the years after 1615, when Kepler's mother was accused of witchcraft. To put the seriousness of this in perspective, that year six so-called witches were burned in the town where she was now living, Leonberg. This was not a situation that Kepler could ignore,[1] and over the next few years he made repeated visits to Leonberg and was involved in lengthy petitions to the authorities on her behalf while the threat of a trial hung over her. It was only in August 1620 that the old woman was finally arrested and imprisoned. Although she was tried later that year, the judges found that there was insufficient evidence to convict her, but enough to cast doubt. She was held in prison until October 1621, when it was decided that she had suffered enough and was released. She died six months later.

Kepler's third law

In view of these particular troubles, and his troubled personal life in general, it is ironic that one of Kepler's last great works carried the title *Harmonice Mundi* (*Harmony of the World*), although, of course, it referred to the world of the planets, not to the troubled Earth. It was in that book (mostly a mystical volume of no scientific significance) that he described how, on 8 March 1618, the idea that has become known as Kepler's third law came to him, and how it was completed later that year. The law relates the time it takes a planet to go around the Sun once (its period, or year) to its distance from the Sun in a very precise way, quantifying the general pattern that Copernicus had discovered. It says that the squares of the periods of any two planets are proportional to the cubes of their distances from the Sun. For example (using modern measurements), Mars is 1.52 times as far from the Sun as the Earth is, and 1.52^3 is 3.51. But the length of the 'year' on Mars is 1.88 times the length of the year on Earth, and 1.88^2 is 3.53 (the numbers don't quite match because I rounded them off to two decimal places).

Publication of the Rudolphine star tables

Harmonice Mundi was published in 1619, by which time the Thirty Years War was in full swing. Because of difficulties caused by the war, and his mother's trial for witchcraft, the other great work published by Kepler around this time,

1. Apart from his natural feelings for his mother, if she were convicted of witchcraft he would be unlikely to be able to hold on to his Imperial appointment.

his *Epitome of Copernican Astronomy*, appeared in three volumes, produced in 1618, 1620 and 1621. As well as boldly making the case for the Sun-centred Universe of Copernicus, this more accessible book brought Kepler's ideas to a wider readership, and in some ways signalled the end of his great contributions to astronomy. But there was still one outstanding commitment to be resolved. Thanks in no small measure to the invention of logarithms by John Napier (1550–1617) in England, which had recently been published and greatly eased the burden of arithmetical calculation for Kepler, the Rudolphine Tables were at last published in 1627 (delayed by war, riots and even a siege of Linz), ending Kepler's obligations to the Holy Roman Empire. The tables made it possible to calculate the positions of the planets to an accuracy thirty times better than the tables Copernicus had compiled, and remained the standard used for generations. Their value was highlighted in 1631, when the French astronomer Pierre Gassendi observed a transit of Mercury (when Mercury passes in front of the Sun), which had been predicted by Kepler using the new tables. This was the first transit of Mercury ever observed.

Printing wasn't the only part of Kepler's life disrupted by the war. In 1619, Ferdinand II had become Emperor after Matthias died, and this was the same fervently Catholic Ferdinand who had caused Kepler so much grief in Styria earlier in his career. Having already been persecuted in Linz by his own Lutheran Church for not being sufficiently orthodox, after 1625 the changing political situation under Ferdinand brought Catholic dominance across all of Austria, and now he was persecuted for being too Lutheran. There was no longer any prospect of him retaining his court position unless he converted to Catholicism, and this he still would not contemplate (although it seems that Ferdinand was well disposed towards Kepler personally and would have been happy to have him back in Prague if he would even pay lip service to such a conversion). In 1628, Kepler managed to secure a position with the Duke of Wallenstein, a man who tolerated all forms of religious worship (provided they were Christian) and who never made a move without consulting his astrologers. He knew Kepler from his time in Prague, when he had cast a horoscope which, in the eyes of the Duke, was remarkably accurate in its prophecies. Wallenstein seemed an ideal benefactor and protector, a powerful man whose positions included commander of the army for Ferdinand.

Kepler's death

The Kepler family arrived in the Silesian town of Sagan to start their new lives in July 1628. The best thing about the new job was that Kepler was paid regularly. The most curious is that he had time to complete one of the first science-fiction stories, *Dream of the Moon*. The most unfortunate is that soon after he arrived, Duke Wallenstein decided to go along with the Counter Reformation in order to curry favour with the Emperor, and although Kepler, as an employee of the Duke, was exempt from the new laws that were promulgated, once again he saw his Protestant neighbours ruined and living in fear. In spite of his efforts to please the Emperor, in the summer of 1630 Wallenstein fell from favour and was dismissed from his key post as commander of the army. Once again, Kepler's future looked uncertain, and he needed to draw on every resource he had in anticipation of another move. For some time he had been trying to get his hands on some money that was owed to him in Linz, and he was summoned to a meeting with the authorities there to settle the matter. In October, he set out from Sagan to keep this appointment (for 11 November), travelling slowly via Leipzig and Nuremberg as far as Regensburg, where he arrived on 2 November. There, he was struck down by a fever and took to his bed. On 15 November 1630, a few weeks short of his fifty-ninth birthday, Kepler died. He was a man of his time, poised between the mysticism of the past (which even coloured his own thinking about the Universe) and the logical science of the future, but whose stature as a voice for reason stands even higher in the context of a world where princes and emperor's still depended on the prognostications of astrologers, and where his own mother was tried for witchcraft. At the same time as Kepler was carrying out his great work, an even more powerful voice of scientific reason was being heard further south in Italy, where although there was just as much superstition and religious persecution as in central Europe, at least there was some measure of stability and the persecution always came from the same Church.

3 The First Scientists

William Gilbert and magnetism – Galileo on the pendulum, gravity and acceleration – His invention of the 'compass' – His supernova studies – Lippershey's reinvention of the telescope – Galileo's developments thereon – Copernican ideas of Galileo judged heretical – Galileo publishes Dialogue on the Two Chief World Systems *– Threatened with torture, he recants – Galileo publishes* Two New Sciences *– His death*

William Gilbert and magnetism

There was no single moment in history when science replaced mysticism as a means of explaining the workings of the world; but the lives of two men neatly circumscribe the transition, which occurred (for them at least) more or less as the sixteenth century became the seventeenth. Of course, there were mystically inclined scientists later, including (as we have seen) as eminent a scientific figure as Johannes Kepler, and (as we shall see shortly) the alchemists. But after the first decade of the seventeenth century, the scientific method of comparing hypotheses with experiments and observation to weed out the wheat from the chaff had been clearly expressed in the work of William Gilbert, in England, and Galileo Galilei, in Italy, and was there as an example to be followed by those who had eyes to see.

Although Galileo is one of the towering figures in science, known by name to every educated person today, and Gilbert is less well known than he deserves, Gilbert had the earlier birth date and, chronologically speaking at least, deserves the title of the first scientist. William Gilbert is the name that he has become known by in the history books, although the preferred contemporary spelling used by his own family was Gilberd. He was born on 24 May 1544[1] in Colchester, in the county of Essex, where he was a member of a prominent local family – his father, Jerome Gilberd, was the recorder of the borough, an

1. Some sources say 1540, but this seems to be a mistake.

important officer of local government. William had a comfortable place in a settled society and suffered none of the hardships experienced by Kepler; he studied at the local grammar school, then went up to Cambridge in 1558. Very little is known about his early life, but some accounts say that he also studied at Oxford, although there is no formal record of this. He completed his BA in 1560 and became a Fellow of his college (St John's), proceeding to the degrees of MA in 1564 and MD in 1569. He then travelled on the Continent for several years, before settling in London, where he became a Fellow of the Royal College of Physicians in 1573.

Gilbert was an extremely successful and eminent physician, who held just about every office in the Royal College in turn, culminating in his election as President in 1599. The following year, he was appointed as the personal physician of the Queen, Elizabeth I, and later knighted by her. When she died in May 1603 he was appointed as physician by her successor, James I (who, as James VI of Scotland had journeyed to Denmark to find a bride, and met Tycho during his sojourn there). But Gilbert outlived Elizabeth by only a few months, and died on 10 December 1603. In spite of his fame as a medical man, Gilbert made his mark on science in physics, through his thorough investigation of the nature of magnetism.

These investigations were, strictly speaking, the work of a gentleman amateur, a man wealthy enough to have spent, according to contemporary accounts, some £5000 of his own money on his scientific work in the thirty years after he settled in London.[1] At first, he was interested in chemistry, but soon (after convincing himself that the alchemical belief in the transmutation of metals was a fantasy) switched to the study of electricity and magnetism, a feature of the world which had remained essentially neglected since the investigations (or rather, speculations) of the Greek philosophers some 2000 years earlier. This work culminated in 1600, after some eighteen years of study, with the publication of a great book, *De Magnete Magneticisque Corporibus, et de Magno Magnete Tellure* (*Concerning Magnetism, Magnetic Bodies, and the Great Magnet Earth*), usually known simply as *De Magnete*.

1. This is equivalent to several million pounds in modern terms, and it is hard to see what he could have spent it on, so the number may be an exaggeration!

It was the first significant work in the physical sciences to be produced in England.

Gilbert's work was comprehensive and thorough. He disproved, by experiment, many old mystical beliefs about magnetism – such as the notion that lodestone, a naturally occurring magnetic ore, could cure headaches, and the idea that a magnet could be deactivated by rubbing it with garlic – and invented the technique of magnetizing pieces of metal using lodestone. He discovered the laws of magnetic attraction and repulsion that are so familiar to us today from school, showed that the Earth itself acts like a giant bar magnet, and gave the names 'north pole' and 'south pole' to the two extremities of a bar magnet. His investigations were so thorough and complete that, after Gilbert, nothing new was added to scientific knowledge of magnetism for two centuries, until the discovery of electromagnetism in the 1820s and the subsequent work of Michael Faraday. Gilbert's interests also extended, necessarily in a more speculative way, into the heavens. He was a supporter of the Copernican world model, partly because he thought that the planets might be held in their orbits by magnetism (an idea which influenced Kepler). But Gilbert's originality also shone through in his discussion of Copernican astronomy, where he pointed out how easy it is to explain the precession of the equinoxes (a phenomenon by which the point on the sky where the Sun crosses the celestial equator in spring and autumn seems to drift slowly westwards as the centuries pass) in terms of the wobble of a spinning Earth (like the wobble of a spinning top), and how difficult it is to explain the phenomenon in terms of crystal spheres centred on the Earth (requiring truly horrendous complications which we shall not go into here). He also suggested that the stars are at different distances from the Earth (not all attached to a single crystal sphere) and might be Sun-like bodies orbited by their own habitable planets. His investigation of static electricity, produced by rubbing objects made of things like amber or glass with silk, was less complete than his study of magnetism, but Gilbert realized that there was a distinction between electricity and magnetism (indeed, he coined the term 'electric' in this context), although it was not until the 1730s that the French physicist Charles Du Fay (1698–1739) discovered that there are two kinds of electric charge, dubbed 'positive' and 'negative', which behave in some ways like magnetic poles, with like charges repelling one another and opposite charges attracting one another.

The most important feature of *De Magnete*, though, was not *what* Gilbert had discovered, but *how* he had discovered it, and the clear way in which he set out this scientific method as an example for others to follow. *De Magnete* had a direct influence on Galileo, who was inspired by the book to carry out his own investigations of magnetism, and who described Gilbert as the founder of the experimental method of science. Right at the beginning of the preface to his book, Gilbert sets out his scientific stall: 'In the discovery of secret things, and in the investigation of hidden causes, stronger reasons are obtained from sure experiments and demonstrated arguments than from probable conjectures and the opinions of philosophical speculators'.[1] Practising what he preaches, Gilbert goes on to describe his experiments in loving detail, so that any reasonably careful worker can reproduce them for themselves, while spelling out the power of this approach:

> Even as geometry rises from certain slight and readily understood foundations to the highest and most difficult demonstrations, whereby the ingenious mind ascends above the aether: so does our magnetic doctrine and science in due order first show forth certain facts of less rare occurrence; from these proceed facts of a more extraordinary kind; at length, in a sort of series, are revealed things most secret and privy in the earth, and the causes are recognized of things that, in the ignorance of those of old or through the heedlessness of the moderns, were unnoticed or disregarded.

And among many asides castigating those philosophers 'discoursing . . . on the basis of a few vague and indecisive experiments' Gilbert exclaims 'how very easy it is to make mistakes and errors in the absence of trustworthy experiments', and urges his readers, 'whosoever would make the same experiments', to 'handle the bodies carefully, skilfully and deftly, not heedlessly and bunglingly; when an experiment fails, let him not in his ignorance condemn our discoveries, for there is naught in these Books that has not been investigated and again and again done and repeated under our eyes.'

1. This and other quotations from the translation by P. Fleury Mottelay.

This must have been music to the soul of Galileo when he read Gilbert's words. For, quite apart from the importance of the discoveries he made, Galileo's key contribution to the birth of science lay precisely in emphasizing the need for accurate, repeated experiments to test hypotheses, and not to rely on the old 'philosophical' approach of trying to understand the workings of the world by pure logic and reason – the approach that had led people to believe that a heavier stone will fall faster than a lighter stone, without anyone bothering to test the hypothesis by actually dropping pairs of stones to see what happened. From their habit of strolling around the campus of a great university or through the streets of the city while discussing such issues, the old school of 'scientific' philosophy was known as the peripatetic school.[1] Like Gilbert, Galileo practised what he preached, and it was the peripatetic approach that was blown apart by his work in Italy late in the sixteenth century and early in the seventeenth century.

Galileo on the pendulum, gravity and acceleration

Galileo Galilei was born in Pisa on 15 February 1564 – the same year that William Shakespeare was born and in the same month that Michelangelo died. The double name came about because a fifteenth-century ancestor of Galileo, called Galileo Bonaiuti, became such an important figure in society as an eminent physician and magistrate that the family changed their name to Galilei in his honour. 'Our' Galileo was given the ancestor's Christian name as well, and is almost always referred to solely by that Christian name, with the mildly ironic twist that the once-famous Galileo Bonaiuti is now remembered solely as the ancestor of Galileo Galilei. At the time Galileo was born, the family had good connections and a respected place in society, but finding the money to maintain that position would always be a problem. Galileo's father, Vincenzio, who had been born in Florence in 1520, was an accomplished professional musician who was keenly interested in mathematics and musical theory. He married a young woman called Giulia in 1562, and Galileo was the eldest of seven children, three of whom seem to have died in infancy. His surviving siblings were Virginia, born in 1573, Michelangelo (1575) and Livia (1587), making Galileo by some way

1. The original peripatetics were the followers (literally!) of Aristotle, but the name was also used by the Italian philosophers of the late sixteenth century.

the oldest of the survivors and head of the family after his father's death, which was to cause him considerable worry.

But all that lay far ahead in 1572, when Vincenzio decided to return to Florence, taking Giulia with him but leaving Galileo with relatives in Pisa for two years while he re-established himself in his home town. This was at a time when the whole region of Tuscany, and Florence and Pisa in particular, was flourishing in the Renaissance. The region was ruled by the Duke of Florence, Cosimo de' Medici, who had also been crowned Duke of Tuscany by the Pope in 1570 for his part in successful military campaigns against the Moors. In Florence, the capital of Tuscany, Vincenzio became a court musician, and his family mixed with dukes and princes at the artistic and intellectual heart of reborn Europe.

Until he was 11, Galileo was educated at home, largely by his father but with the help of an occasional tutor. He became an excellent musician in his own right, but never followed in his father's professional footsteps and played (chiefly the lute) only for pleasure throughout his life. Vincenzio was something of a freethinker, by the standards of the day, and had no great love of the forms and rituals of the Church. But when the time came to send Galileo away for more formal schooling, in 1575, the obvious place to send him, purely on educational grounds, was to a monastery (Vincenzio chose one at Vallombrosa, some 30 kilometres to the east of Florence). Like many an impressionable youth before (and after) him, Galileo fell in love with the monastic way of life, and at the age of 15 he joined the order as a novice. His father was horrified, and when the boy developed an eye infection he whisked him away from the monastery to see a doctor in Florence. The eye recovered, but Galileo never returned to the monastery and no more was said about becoming a monk. Although his education in Florence continued for another couple of years under the supervision of monks from the same order as those in Vallombrosa, he lived at home under the careful eye of his father. In the records at Vallombrosa Abbey, Galileo Galilei became officially listed as a defrocked priest.

Although Vincenzio had managed to get by as a musician, he was aware of the insecurity of his calling and planned for his eldest son to be set up in a respectable and financially rewarding career. What could be better than to have him train as a physician, like his illustrious namesake? In 1581, at the age of 17, Galileo was enrolled as a medical student at the University of Pisa, where he lived with the same relatives of his mother who had looked after him in the early 1570s. As a student, Galileo was argumentative and unafraid to question the received (largely Aristotelian) wisdom of the time. He became known to the other students as 'the wrangler', for his love of argument, and when he looked back on those years late in his life he recounted how he had immediately thought of a way to refute the Aristotelian idea, enshrined in peripatetic teaching, that objects with different weights fall at different speeds. Hailstones come in many different sizes and they all reach the ground together. If Aristotle was right, heavy hailstones would have to be manufactured higher in a cloud than lighter hailstones – *exactly* the right distance higher up so that, falling at a greater speed, they reached the ground alongside the lighter hailstones formed at lower altitudes. It seemed rather unlikely to Galileo, and he delighted in pointing out to his fellow students and the teachers at the university that a much simpler explanation is that all the hailstones are made in the same place inside a cloud, so they all fall together at the same speed, whatever their weight.

This kind of argument was essentially a distraction from Galileo's medical studies, although in any case he did not pursue these with any enthusiasm. Early in 1583, however, any prospect of a career in medicine vanished. During the winter months at that time, the court of the Grand Duke of Tuscany always took up residence in Pisa, from Christmas to Easter. With his court connections through his father, Galileo got to know the court mathematician, Ostilio Ricci, socially, and early in 1583 he called on his new friend just as Ricci was giving a lecture on mathematics to some students. Rather than go away and come back later, Galileo sat in on the lecture and became fascinated by the subject – his first encounter with mathematics proper, rather than mere arithmetic. He joined Ricci's students on an informal basis and began to study Euclid instead of his medical textbooks. Ricci realized that Galileo had an aptitude for the subject and supported him when he asked Vincenzio if he could switch his studies from medicine to mathematics. Vincenzio refused, on the seemingly reasonable grounds that there were plenty of jobs for doctors but very few for mathematicians. Galileo just carried on studying mathematics anyway, largely ignoring his medical courses, with the result that he left Pisa in 1585 without a degree of any kind and went back to Florence to try to scrape a living as a private tutor in mathematics and natural philosophy.

One other noteworthy event occurred while Galileo was a medical student in Pisa, although the story has been considerably distorted and embellished down the centuries. It is almost certain that it was at this time that Galileo became mesmerized by the slow, steady swing of a chandelier in the cathedral during a rather dull sermon and, for want of anything better to do, timed the swing of the pendulum as the arc the chandelier moved over gradually decreased, using his pulse. This led him to discover that the pendulum always took the same time to complete one swing, whether it was swinging over a short arc or a long arc. Legend has Galileo rushing off home, carrying out some experiments with pendulums of different lengths and inventing the grandfather clock on the spot (like other Galileo legends this story owes much to the writings of Vincenzo Viviani, a young man who became Galileo's scribe and devoted disciple much later, after the old man became blind, and who often got carried away with his descriptions of great moments in the life of his master). In fact, the idea stayed in Galileo's mind until 1602, when the careful experiments were carried out and the proof established that the period of swing of a pendulum depends only on its length, not on the weight of the pendulum nor the length of the arc it is swinging through. But the seed really was planted in that cathedral in Pisa in 1584 or 1585.

Although Galileo began to make a reputation as a natural philosopher, carried out experiments and began making notes that would later be developed in his important writings about science, he barely made a living during the next four years in Florence. Without independent means, the only way he could get the security to carry out his scientific work was to find an influential patron, and Galileo's saviour came in the form of the Marquis Guidobaldo del Monte, an aristocrat who had written an important book on mechanics and was keenly interested in science. It was partly thanks to Del Monte's influence that in 1589, just four years after he had left the University of Pisa without a degree, Galileo returned to the same university as professor of mathematics, with a contract for three years.

In spite of the grand title, this was a very modest first step on the academic ladder. As Vincenzio Galilei no doubt pointed out to his son, the professor of medicine at Pisa then received a salary of 2000 crowns a year, while the professor of mathematics had to make do with 60. Galileo had to supplement his income by taking students who lived with him and had the benefit of his teaching and influence more or less full time, not just in the hours set aside for tutorials. This was a normal procedure at the time, but only the sons of the rich and powerful could afford to have the benefit of this kind of tuition, with the result that as these young men finished their courses and returned home, Galileo's fame spread in just those circles where it could do him most good.

The teaching that these private pupils got in Galileo's home was often very different from the official courses that Galileo was required to teach by the university. Although styled professor of mathematics, his brief included what we would now call physics and was then called natural philosophy. The official syllabus was still largely based on Aristotle, and Galileo dutifully, but without enthusiasm, taught the party line in his lectures. In private, he discussed new and unconventional ideas about the world, and even wrote the first draft of a book spelling out some of those ideas, but decided not to publish it – surely a wise decision for a young man who had yet to make his mark.[1]

Another of the Galileo legends introduced by Viviani refers to Galileo's time as professor of mathematics in Pisa, but is, once again, almost certainly not true. This is the famous story of how Galileo dropped different weights from the Leaning Tower to show that they would arrive at the ground below together. There is no evidence that he ever did any such thing, although in 1586 a Flemish engineer, Simon Stevin (1548–1620; also known as Stevinus), really did carry out such experiments using lead weights dropped from a tower about 10 metres high. The results of these experiments had been published and may have been known to Galileo. The connection between Galileo and weights being dropped from the Leaning Tower, which Viviani has confused with Galileo's time as professor of mathematics in Pisa, actually dates from 1612, when one of the professors of the old

1. The title of the draft book was *On Motion*, but the draft bears little resemblance to the book Galileo published under the same title some years later.

Aristotelian school tried to refute Galileo's claim that different weights fall at the same speed by carrying out the famous experiment. The weights hit the ground at very nearly the same time but not exactly at the same moment, which the peripatetics seized on as evidence that Galileo was wrong. He was withering in his response:

> Aristotle says that a hundred-pound ball falling from a height of one hundred cubits hits the ground before a one-pound ball has fallen one cubit. I say they arrive at the same time. You find, on making the test, that the larger ball beats the smaller one by two inches. Now, behind those two inches you want to hide Aristotle's ninety-nine cubits and, speaking only of my tiny error, remain silent about his enormous mistake.

The true version of the story tells us two things. First, it highlights the power of the experimental method – even though the peripatetics wanted the weights to fall at different speeds and prove Aristotle was right, the experiment they carried out proved that Aristotle was wrong. Honest experiments always tell the truth. Second, the quotation above gives a true flavour of Galileo's style and personality. It is impossible to believe that if he really had carried out the famous experiment himself there would be no mention of this triumph anywhere in his writings. For sure, he never did it.

Galileo never really fitted in at the University of Pisa and soon started looking for another post. He refused to wear the academic gown that was a badge of his office, mocking his fellow professors for being more interested in the trappings of their position than in really investigating how the world worked, and he made a striking figure (at that time with a full head of red hair and a substantial red beard) fraternizing with students in the seedier bars in town. Quite apart from his anti-establishment views (which made it increasingly unlikely that his appointment would be renewed in 1592), the need for a better income became pressing in 1591, when Vincenzio Galilei died. Far from leaving any substantial inheritance for his children, not long before he died Vincenzio had promised a generous dowry for his daughter Virginia and Galileo and Michelangelo Galilei, his younger brother, became legally responsible for the debt. In practice, this meant that Galileo, as head of the family, had to take on the debt, since not only did Michelangelo fail to pay his share, he became an itinerant and impecunious musician who kept coming back to Galileo for 'loans' which were never repaid. All this was particularly taxing for Galileo since he loved spending money himself, enjoying fine wines and good food, and entertaining his friends generously whenever he was in funds.

The post Galileo set out to obtain was the chair of mathematics at the University of Padua. Apart from the fact that this was a more prestigious and better-paid job, Padua was part of the Venetian Republic, a state that was rich and powerful enough to stand up to Rome, and where new ideas were positively encouraged, instead of being frowned upon. Galileo campaigned for the post by visiting the court at Venice itself, where he was helped by the Tuscan Ambassador. When he wanted to, Galileo could be charming and socially adroit, and he made a big impact in Venice, where he became particularly friendly with Gianvincenzio Pinelli, a wealthy intellectual who owned a large library of books and manuscripts, and General Francesco del Monte, the younger brother of Guidobaldo. He got the job, initially for four years at a salary of 180 crowns per year, with a stipulation that the head of the Venetian Republic, the Doge, could renew the appointment for a further two years if he wished. With permission from the Grand Duke of Tuscany, Galileo took up his new post in October 1592, when he was 28 years old. (The Grand Duke was now Ferdinando; Cosimo had died in 1574 and was succeeded by Ferdinando's elder brother Francesco, but he had died in 1587 without a male heir, although his daughter Marie did become Queen of France.) The initial four-year appointment eventually extended to a stay of eighteen years in Padua, which Galileo later recalled as being the happiest years of his life.

Galileo made his mark in Padua in very practical ways, first with a treatise on military fortifications (a matter of considerable importance to the Venetian Republic) and then with a book on mechanics, based on the lectures he was giving at the university. Among other things, Galileo spelt out clearly how systems of pulleys work, so that although at first sight it may seem miraculous – a case of something for nothing – that a 1 kg weight (say) can be used to lift a 10 kg weight, in order to do this the 1 kg weight has to move ten times as far as the 10 kg weight, as if it made ten separate journeys to lift ten individual 1 kg weights. Galileo's social and intellectual life also flourished in Padua, revolving around his new friends, such as Pinelli. Among this new circle there were two men in particular who would play a large part in the story of Galileo's later life – Friar Paolo Sarpi and Cardinal Roberto Bellarmine. Although Sarpi became a close friend of Galileo and Bellarmine was on friendly terms with him (if not much more than an acquaintance), they represented very different religious positions. Sarpi was so unorthodox a Catholic that some of his opponents would later suspect him of being a closet Protestant, while Bellarmine was a leading establishment figure, a theologian and intellectual who would play a major part in the prosecution of Giordano Bruno for heresy.[1]

But although Galileo was now highly regarded professionally and moved in influential circles, he was constantly worried about money. He tried to solve his financial problems by inventing something that would make him rich. One of his early ideas was one of the first thermometers, which worked 'upside down' to modern eyes. A glass tube, open at one end and with a bulbous swelling at the other, was first heated (to expel air) then placed open end downwards upright in a bowl of water. As the air in the tube cooled and contracted, it sucked water up. Once the thermometer had been set up, if it got warmer the remaining air in the bulb would expand, pushing the level of liquid downwards, while if it got cooler the air would contract still more, sucking the water even further up the tube. The invention was not a success, because the height of liquid in the tube also depended on the changing pressure of the air outside. But it does show how ingenious Galileo was, and his skill at practical work.

His invention of the 'compass'

Another idea, developed in the mid-1590s, was a modest success, but didn't make Galileo rich. This was a device known as a 'compass' – a graduated metal instrument that could be used as a calculator. It was initially a device to help gunners calculating the elevations required to fire their guns over different distances, but developed over the next few years into an all-purpose calculating tool, the late sixteenth-century equivalent of a pocket calculator, dealing

1. In 1605, Bellarmine would almost certainly have been elected Pope, but declined to offer himself as a candidate, preferring to be the power behind the throne.

with such practical matters as calculating exchange rates for money and working out compound interest. By the end of the 1590s, the instrument was selling so well that for a short time Galileo had to employ a skilled workman to make them for him. He showed his business acumen by selling the compasses relatively cheaply and charging a healthy tuition fee to anyone who wanted to know how to use it. But this couldn't last – there was no way to stop other people copying the instrument and no way to prevent those who knew how to use it passing on their knowledge.

But although the boost to Galileo's income provided by the invention was short-lived, it came just at the right time. In the second half of the 1590s, his personal commitments had increased when he established a stable relationship with Marina Gamba, a local Paduan woman from a lower social class. The two never married (indeed, they never actually lived together in the same house), but the relationship was openly acknowledged and Marina provided Galileo with three children – two daughters (born in 1600 and 1601) and a son (1606). The son, named Vincenzio after his grandfather, was later formally acknowledged as Galileo's heir and took his name. The daughters were destined to become nuns, a fate which may well have been sealed by Galileo's continuing problems finding the money to pay dowries for his sisters and a determination not to get into the same situation with his daughters. His sister Livia married in 1601, the same year that Galileo's second daughter was born, and Galileo and Michelangelo (who was by now living in Germany) promised her, like Virginia, a large dowry. And, once again, Michelangelo never paid his share.

In 1603, Galilei suffered an illness that was to affect him for the rest of his life. While on a visit to a villa in the hills near Padua with friends, he enjoyed (as he often did) a walk in the hills, followed by a large meal, and went to sleep with his two companions in a room that was fed with cool air from nearby caves through a system of ducts. This early form of air-conditioning was closed off when the three went to sleep, but was later opened by a servant, allowing cool, damp air from the caves to enter the room. All three of the occupants became severely ill, and one of them died. This seems to have been much more than a chill, and it is likely that some form of poisonous gases from the caves had entered the room. Whatever the exact cause, Galileo suffered from repeated bouts of an arthritic ailment for the rest of his life, which sometimes confined him to bed for several weeks at a time. He always believed that the chronic illness resulted from his brush with death in 1603.

By the time he was 40, in 1604, Galileo had established a reputation as a natural philosopher and mathematician, providing practical benefits for the Venetian state, and was leading a full and happy life in Padua. It was there that he carried out his famous experiments with pendulums and with balls rolling down inclined planes, which he used to study acceleration and establish (without actually dropping things vertically) that objects with different weights do accelerate at the same rate under the influence of gravity. It is a key feature of his work that Galileo always carried out experiments to test hypotheses, modifying or abandoning those hypotheses if the outcomes of the experiments did not match their predictions. Galileo also investigated hydrostatics; following the work of Gilbert he studied magnetic phenomena; and he corresponded with other natural philosophers, including Kepler (it was in a letter to Kepler written in May 1597 that Galileo first stated clearly his enthusiasm for the Copernican world model).

Alongside all this, Galileo had a full private life. He studied literature and poetry, attended the theatre regularly and continued to play the lute to a high standard. His lectures were popular (although he began to find them a chore which distracted him from his experimental work and his social life) and his growing reputation as an anti-Aristotelian only enhanced his prestige in the free-thinking Venetian Republic. There had never been any doubts that his position at the university would be renewed every time his contract was up, and his salary had been increased enough so that he could live comfortably, even if he was not able to put anything by for a rainy day, let alone contemplate retirement from his post.

His supernova studies

In 1604, Galileo's stature increased still further when the supernova studied by Kepler appeared in the sky in October. Using the careful surveying techniques that he had developed through his work for the military, Galileo turned himself into an astronomer (for the first time) and established that the new star showed no motion across the sky compared with the other stars. He gave a series of well-received public lectures arguing that it must be as far away from the Earth as the other stars, refuting the Aristotelian notion of an unchanging celestial sphere, and summed up his conclusions in a little poem:

> No lower than the other stars it lies
> And does not move in other ways around
> Than all fixed stars – nor change in sign or size.
> All this is proved on purest reason's ground;
> It has no parallax for us on Earth
> By reason of the sky's enormous girth.[1]

But while Galileo's public reputation increased, his private life began to pose problems. In 1605, both his brothers-in-law were suing him (in Florence) for non-payment of the instalments due on the dowries of his sisters. Galileo's friend Gianfrancesco Sagredo, a Venetian nobleman nine years younger than Galileo, paid the court fees and did his best to delay the legal process, but by the summer of 1605 Galileo had to visit Florence to argue his case. Conveniently, just at this time the Grand Duchess of Tuscany, Christina, invited Galileo to instruct her teenage son Cosimo in the use of Galileo's military compass and tutor him in mathematics generally. This visible sign of Galileo's privileged position at court (perhaps coupled with some direct pressure from Christina on the judiciary) resulted in the claims against Galileo being quietly dropped, at least for the time being. But the visit also revived a desire in Galileo to return to Tuscany for the latter part of his life, preferably with a court appointment that would free him from any lecturing duties.[2] This was a real prospect, since the court mathematician in Florence (Ostilio Ricci, who had first introduced Galileo to mathematics) had died in 1603 and the post was still vacant. He began to campaign for such a return, and published the instruction manual for his compass in a limited edition book, dedicated to Prince Cosimo de' Medici, in 1606. Although Galileo was reappointed to his post in

1. Translation taken from Reston.

2. Galileo may also have been trying to cover his back by keeping all options open, since his discussion of the new star had roused some opposition in Padua, and he also asked the Medicis to support the upcoming renewal of his post there.

Padua (with another increase in salary), he kept the lines of communication with Tuscany very much open.

While Galileo was contemplating major changes in his personal life, and also gathering together the material from his years of experimental work for a planned book, the political situation in Italy changed dramatically. In 1605, Paul V had been elected Pope, and he made a determined effort to extend the authority of the Church and tighten the papal grip on Catholic states. The snag, as far as the Pope was concerned, was that he lacked any powerful armies of his own, and extending his influence meant either relying on the temporal powers of others or on exercising his spiritual authority (with the aid of the Inquisition). Venice was a particular thorn in his flesh, not least because Paolo Sarpi, by now a theological adviser to the Doge, openly argued that the road to Heaven lay through spiritual works alone and denied the so-called 'divine right' of kings and popes to exercise political power in the name of God. On the other side of the debate, the chief intellectual support for the notion of this divine right came from Cardinal Roberto Bellarmine, now very much the power behind the throne in Rome, not least since Paul V knew that he owed his position to Bellarmine's decision not to allow his own name to be put forward for consideration. There were other aspects of the dispute that we shall not go into here, since they have less direct bearing on Galileo's life. The result was that in 1606, the Pope excommunicated the Doge of Venice and all his officials, including Sarpi. Although there was some soul-searching among the priests in Venice, the upshot was that by and large the Republic ignored the excommunication and went about its business (including religious business) as usual. All Jesuits were expelled from the Venetian Republic in retaliation. Spiritual influence, even the threat of hell fire, had clearly failed to extend the authority of the Pope in this instance, and for a time the only alternative – war – seemed a real possibility, with Catholic Spain lining up to support the Pope, and France (at this time largely Protestant) offering aid to Venice.

The crisis passed after a few months, however, and as the tension eased, Sarpi was invited to Rome to argue his theological views with Bellarmine, where, he was told, 'he shall be caressed and well received'. Telling his friends that he knew only too well that the arguments likely to be used by the Vatican would involve the rope and the flame, Sarpi refused, saying that he was too busy with state affairs in Venice. To back him up, the Venetian Senate officially forbade him to leave the Republic. The Vatican, unable to burn Sarpi, burned Sarpi's books instead; the Venetian Senate promptly doubled his salary. Venice had won the political battle with Rome and Sarpi's influence in the Republic was stronger than ever. But on the night of 7 October 1607, Sarpi was savagely attacked in the street by five men who stabbed him fifteen times and left him with a stiletto embedded in his head, which had entered through his right temple and exited through his right cheek. Astonishingly, Sarpi survived (as did the would-be assassins, who escaped to Rome).

The attempt on Sarpi's life made a deep impression on Galileo, who realized that even if the Venetian Republic could stand up to Rome, individuals who failed to toe the Catholic line would be at risk anywhere in Italy. On top of this, the winter of 1607/8 was unusually severe, with heavy falls of snow in Padua, and during March and April 1608 Galileo was severely afflicted with his arthritic complaint. In spite of all these difficulties, he continued to prepare his epic book on mechanics, inertia and motion. It was around this time that Galileo realized, and proved, that when a bullet is fired from a gun, or an object is thrown in the air, the trajectory that it follows is a parabola, a curve related to an ellipse but with one open end. Even at the beginning of the seventeenth century, many people still thought that if a ball were fired horizontally from a cannon, it would fly a certain distance in a straight line, then fall vertically to the ground; more observant people had noticed (or surmised) that the actual flight of the ball followed a curved path, but until Galileo's work nobody knew the shape of the curve, or even whether it was always the same sort of curve, regardless of the speed and weight of the cannonball. He also showed that if the ball hits a target which is at the same height as the gun above sea level, then it strikes the target with the same speed that it leaves the gun (neglecting air resistance).

Galileo's worries about money and his ill health allowed him to be distracted from this work in the summer of 1608, when he was summoned by Christina to Florence to oversee the construction of a great wooden stage over the river Arno, to be used for the wedding of her son, who became Grand Duke Cosimo II in 1609 when Ferdinando died. No matter how important his current project was, Galileo couldn't turn down a summons from Christina,[1] and this was a welcome sign that he was still in favour in Florence, where the post of Court Mathematician remained vacant. But back in Padua, as he turned 45 early in 1609, Galileo was still beset by those financial worries, nervous that he might be a target of the Vatican as a known Copernican and friend of Sarpi, and still longing for the one big idea that he could turn to practical advantage to secure his financial position for what remained of his life. It is at this point that most stories of Galileo's contribution to science begin.

Lippershey's reinvention of the telescope

Galileo first heard rumours of the invention of the telescope (strictly speaking, a reinvention, but news of the Digges's telescopes had never spread in the sixteenth century) in July 1609, on a visit to Venice. News had been rather slow to travel to Italy on this occasion, since Hans Lippershey, a spectacle maker based in Holland, had come up with the discovery by chance the previous autumn and in the spring of 1609 telescopes with a magnifying power of three times were being sold as toys in Paris. When Galileo heard rumours of this amazing instrument he asked his old friend Sarpi for advice, and was surprised to learn that Sarpi had heard the stories some months earlier and discussed them in correspondence with Jacques Badovere, a French nobleman based in Paris, who had once been a pupil of Galileo. But Sarpi had not passed the news on to Galileo – their correspondence had lapsed, partly due to Sarpi's time-consuming duties as adviser to the Senate and partly as a result of the tiredness Sarpi felt after his recovery from the assassination attempt. Although Sarpi may have been slow to realize its importance, Galileo immediately realized that an instrument that could make distant objects visible would be of enormous military and trade importance to Venice, where fortunes often depended on being first to identify which ships were approaching the port. He must have

1. Christina's importance can be judged from the fact that she retained the title Grand Duchess even after Cosimo II married (his wife became merely an Archduchess), and when Cosimo died in 1621 she was joint Regent with his widow during the minority of his son, Ferdinando II.

imagined that his own boat had at last come in, as he considered how best to turn the news to his advantage.

Galileo's developments thereon

But he was almost too late. At the beginning of August, while Galileo was still in Venice, he heard that a Dutchman had arrived in Padua with one of the new instruments. Galileo rushed back to Padua, only to find that he had missed the stranger, who was now in Venice intending to sell the instrument to the Doge. Distraught at the possibility that he might lose the race, Galileo frantically set about building one of his own, knowing nothing more than that the instrument involved two lenses in a tube. One of the most impressive features of Galileo's entire career is that within 24 hours he had built a telescope better than anything else known at the time. Although the Dutch version used two concave lenses, giving an upside-down image, Galileo used one convex lens and one concave lens, giving an upright image. On 4 August, he sent a coded message to Sarpi in Venice telling him of the success; Sarpi, as adviser to the Senate, delayed any decision on what to do with the Dutch visitor, giving Galileo time to build a telescope with a magnifying power of ten times, set in a tooled leather case. He was back in Venice before the end of August, where his demonstration of the telescope to the Senate was a sensation. Being an astute politician, Galileo then presented the telescope to the Doge as a gift. The Doge and the Senate, delighted, offered to give Galileo tenure in his post at the University of Pisa for life, with his salary doubled to 1000 crowns a year.

Galileo accepted, even though the increase in salary would only take effect from the following year, and even though it would commit him to burdensome teaching duties. But he then took himself off to Florence to demonstrate another telescope to Cosimo II. By December 1609, he had made a telescope with a magnifying power of twenty times (and would make at least nine more of comparable power by March 1610; he sent one of these to the Elector of Cologne, for Kepler, the only astronomer honoured in this way, to use to verify Galileo's discoveries). Using his best instrument, Galileo discovered the four brightest (and largest) moons of Jupiter early in 1610. The moons were named by him the 'Medician stars', in honour of Cosimo, but are known to astronomers today as the Galilean satellites of Jupiter. With

9. *Copernicus, Kepler and Galileo with his telescope and the new model of the Universe, from an early English exposition of these ideas, 1640.*

the same instrument, Galileo found that the Milky Way is made up of myriads of individual stars and that the surface of the Moon is not a perfectly smooth sphere (as the Aristotelians believed) but is scarred by craters and has mountain ranges several kilometres high (he estimated the heights of the mountains from the lengths of their shadows on the surface of the Moon). All of these discoveries were presented in a little book, *The Starry Messenger* (*Siderius Nuncius*), in March 1610. The book was dedicated to – who else? – the Grand Duke Cosimo II de' Medici.

The author of *The Starry Messenger* became famous throughout the educated world (the book was translated into Chinese within five years of its publication), and would clearly bring honour to any state he served, especially the state in which he was born. In May 1610, Galileo was offered, and accepted, the post of Chief Mathematician at the University of Pisa, Philosopher and Mathematician to the Grand Duke of Tuscany for life, with a salary of 1000 crowns per year. There would be no teaching duties whatsoever. And as a sweetener, he was released from any obligation to keep up payments on Michelangelo's part of the two outstanding dowries, since he had already more than paid his own share.

Galileo felt no obligation to the Venetian Republic, arguing that since he had not yet started to receive the promised increase in salary, the new deal had not come into effect, and he returned to Florence to take up his new duties in October, just as news reached him that Kepler had indeed observed the four satellites of Jupiter. The move brought about big changes in Galileo's personal life. Marina Gamba decided to remain in Padua, where she had lived all her life, and the couple split, it seems amicably. Galileo's two daughters went to live with his mother in Florence, but his son stayed with Marina for the time being, until he would be old enough to rejoin his father. But these personal upheavals were minor compared with the hornets' nest soon to be stirred by Galileo's new scientific discoveries.

The astronomical observations were direct evidence for the accuracy of the Copernican model. One counter-argument that had previously been used by the peripatetics, for example, was that, since the Moon orbits around the Earth, it would not be possible for the Earth to orbit around the Sun at the same time, because the Earth and the Moon would get separated from one another. By discovering four satellites in orbit around Jupiter, which was itself clearly in orbit around something (whether that something was the Earth or the Sun didn't affect the argument), Galileo showed that it was possible for the Earth's Moon to stay in orbit around the Earth even if the Earth were moving. Shortly before Galileo left Padua, he also noticed something odd about the appearance of Saturn, and although the explanation had to await the work of Christiaan Huygens, the oddity clearly showed that Saturn was not a perfect sphere. Soon after he arrived in Florence, Galileo discovered the phases of Venus, changes in its appearance similar to the phases of the Moon, and these changes could only be explained if Venus orbits the Sun. But there is even more to this story, since Galileo had received a letter from a former pupil, Benedetto Castelli, pointing out that if the Copernican model were correct Venus *must* show phases! Although Galileo had already begun observing Venus when he received the letter, and soon replied to Castelli that his prediction was correct, this is a genuine example of a scientific hypothesis being used to make a prediction that was tested by observation and found to support the hypothesis – the most powerful kind of application of the true scientific method.

None of this persuaded the most die-hard Aristotelians, who simply refused to accept that what was seen through the telescope was real, imagining it to be some artefact produced by the lenses themselves. Galileo himself tested this possibility by observing hundreds of objects through the telescope and close up to see if the instrument did anything except magnify, and he concluded that what he saw through the instrument was real. But although the reluctance of the Aristotelians to believe the evidence can seem laughable today, they did have a point, which has considerable bearing on modern science, where astronomers probe the far reaches of the Universe, and particle physicists delve into the inner structure of atoms and smaller entities, and we are entirely reliant on what our instruments tell us and the way in which we interpret what they are saying. As far as Galileo was concerned, though, it is clear that what he saw was real, in the everyday sense of the word. One of the things Galileo also observed around this time, using a telescope, was dark features on the face of the Sun – sunspots. These had already been seen by other astronomers, but Galileo was not aware of that. Visible blemishes on the surface of the Sun seemed to be yet another nail in the coffin of Aristotelian heavenly perfection.

Although all of this evidence was certainly anti-Aristotelian, and could be used to support the Copernican model, Galileo had been very careful not to endorse the Copernican model publicly, only too aware of the fate of Bruno. He preferred to present his evidence and let the observations speak for themselves, convinced that sooner rather than later even the Church of Rome would have to accept the implications. As a first step in this process, in March 1611 Galileo set off on a visit to Rome, as the official scientific ambassador of the Tuscan state. The visit, which lasted until July, was, on the face of it, a triumph. Galileo was not only received by the Pope (still Paul V), but allowed to address His Holiness while standing, instead of on his knees. Cardinal Bellarmine himself looked through Galileo's telescope and appointed what we would now call a scientific subcommittee of learned priests to examine Galileo's claims for the instrument. The Jesuit members of this subcommittee concluded that:

(i) the Milky Way really is made up of a vast number of stars
(ii) Saturn has a strange oval shape, with lumps on either side
(iii) the Moon's surface is irregular
(iv) Venus exhibits phases
(v) Jupiter has four satellites.

It was official. But no mention was made of the implications of these observations.

While in Rome, Galileo also became a member of what is regarded as the first scientific society in the world, a group known as the Lyncean Academy, which had been founded by four young aristocrats in 1603. It was at a banquet held in honour of Galileo by the 'lynxes' that the name 'telescope' was first suggested for his magnifying device. Galileo also exhibited sunspots during his time in Rome, using the now standard device of projecting the image of the Sun through a telescope on to a white screen. But he doesn't seem to have regarded the discovery of these blemishes on the Sun as very significant at that time. He returned to Florence in triumph in June, having brought glory to the name of Tuscany by his reception in Rome, and having, as he thought, received some sort of official approval for his work.

Any brief account of the rest of Galileo's life is inevitably dominated by his subsequent clash with the authorities in Rome. But this is far from being the whole story of his life, and it is worth elaborating on one piece of work, carried out in the summer of 1611, which highlights the breadth of Galileo's interests and the clear way in which he applied the scientific method. In a discussion among the professors at the University of Pisa about condensation, one of Galileo's colleagues argued that ice should be regarded as a condensed form of water, since it is solid and water is liquid. Galileo, on the other hand, argued that since ice floats on water it must be lighter (less dense) than water and is therefore a rarified form of water.[1] Not so, said the other professor. Ice floats because it has a broad, flat base, which cannot push downwards through the water. Galileo refuted that argument by pointing out that if ice is held down under water and then released, its broad, flat shape doesn't stop it pushing upwards through the water. There then followed a debate about whether solid objects made of the same material (and therefore with the same density) could be made to sink or rise in water only by giving them different shapes. The upshot was that Galileo challenged his main opponent in the debate (which by now had roused widespread interest in Pisa) to show by experiment that objects with the same composition but different shapes, initially totally immersed in water, would either rise or stay immersed depending on their shape. On the day the public experiment was to be carried out Galileo's rival failed to show up.

The point is not that Galileo's reasoning was correct (although it was). What mattered was his willingness to test that reasoning by clearly thought-out experiments, in public, and to stand by the results of the experiments – something that was still a novelty even in 1611. This is what makes him, in the eyes of many people, *the* first scientist; it was also what would ultimately bring him into conflict with the Church, in spite of his apparently warm reception in Rome earlier that year.

Copernican ideas of Galileo judged heretical

Although still extremely cautious about what he put down in print, he began to talk more openly about Copernican ideas following his success in Rome. But whatever

1. The reason why ice is lighter than water is a fascinating story in its own right, which we shall come on to later.

his public utterances on the subject, Galileo's inner feelings about Copernicanism at this time are clearly recorded in a letter he sent to the Grand Duchess Christina (it was actually written in 1614): 'I hold that the sun is located at the centre of revolving heavenly orbs and does not change place. And that the Earth rotates itself and moves around the sun.' Nothing could be plainer. But what of Christina's worries that this flew in the face of Biblical teaching? 'In disputes about natural phenomena,' wrote Galileo, 'one must not begin with the authority of scriptural passages, but with sensory experience and necessary demonstrations.'

His public caution cracked just once, in 1613, when he wrote a little book about sunspots (the book was actually published by the Lyncean Academy). There were two unfortunate aspects to this. First, in a rather over-generous preface, the Lynceans gave Galileo credit for discovering sunspots. This led to a bitter row with the Jesuit astronomer Christopher Scheiner, who (probably rightly) claimed to have seen them before Galileo (in fact, an Englishman, Thomas Harriott, and a Dutchman, Johann Fabricius, beat them both to this discovery). Second, in an appendix to the sunspot book, Galileo made his only clear and unambiguous published statement of support for Copernican ideas, using the example of the moons of Jupiter to support his case. This and his unpublished comments in support of Copernicanism began to draw criticism of Galileo. Confident of his case and sure that he had friends in Rome, after a bout of ill health in 1615, Galileo, now approaching his fifty-second birthday, obtained permission to visit Rome at the end of the year in order to clear the air. This was against the specific advice of the Tuscan Ambassador in Rome, who said that there had been bad feeling against Galileo in some quarters there ever since the seemingly successful (in his opponents' eyes, too successful) visit in 1611 and suggested that another visit would make things worse. In spite of these warnings, Galileo became the official guest of the Ambassador at his residence in Rome on 11 December 1615.

Galileo's presence in Rome brought things to a head in a way that he had not anticipated. On the advice of Bellarmine (now 73 but still the power behind the throne of Saint Peter), Paul V set up a papal commission to decide whether Copernican ideas were heretical, and their official conclusion was that the idea that the Sun lies at the centre of the Universe was 'foolish and absurd . . . and formally heretical'. They went on to say that the idea that the Earth moves through space was 'at the very least erroneous in faith'.

Exactly what happened next, as far as Galileo was concerned, has been a matter of dispute among historians, because there is some ambiguity in the surviving records. But Stillman Drake, of the University of Toronto, has come up with what seems the most probable account of the events of late February 1616, in the light of what happened later. On 24 February, Paul V instructed Bellarmine, as the personal representative of the Pope, to tell Galileo that he must not 'hold or defend' either of the two ideas that the commission had passed judgement on. In other words, it was wrong for Galileo to *believe* the Copernican theory and he must not argue in its favour, even from the perspective of, as it were, Devil's advocate. But the Pope's instructions went further. If, and only if, Galileo objected to this instruction, he was to be warned formally by the Inquisition (the notorious judicial arm of the Papacy responsible for combating heresy), in the presence of a notary and witnesses, that he must not 'hold, defend, or *teach*' (our italics) Copernican ideas. The crucial difference is that without that official warning Galileo would still be allowed to teach Copernican ideas to his students, and even write about them, provided that he was careful to explain that these were heretical notions and that he, Galileo, did not subscribe to them.

On 26 February, Bellarmine received Galileo to convey the Pope's decision. Unfortunately, the representatives of the Inquisition, witnesses and all, were present in the same room, ready to step in if Galileo showed any trace of reluctance to go along with what Bellarmine had to say. Bellarmine met Galileo at the door and murmured to him that whatever happened next, he must go along with it and raise no objections. Galileo, who knew only too well who the other people in the room were, listened carefully to the Pope's warning and clearly did not object. At this point the Inquisition stepped in, determined to get their man, and issued the crucial second warning which made the reference to teaching. Bellarmine, furious (or at least giving a good impression of anger to cover his actions), rushed Galileo from the room before any documents could be signed. But this did not prevent the Inquisition depositing an unsigned, un-notarized and unwitnessed set of 'minutes' of the meeting in the official record. Rumours began to spread that Galileo had in some way been punished by the Inquisition and was guilty of some form of (at the very least) misdemeanour, that he had been forced to abjure his former beliefs and do penance before the Inquisition.

It is clear that Bellarmine explained the true situation to Paul V, since on 11 March Galileo had a long and friendly audience with the Pope, who specifically said that Galileo need have no concern about his position as long as Paul V was alive. Still concerned, Galileo again consulted Bellarmine, who wrote out a sworn affidavit that Galileo had not abjured nor done any penance or been punished for his views, but had simply been informed of the new general edict that covered all members of the Catholic faith. Confident that he was safe, for the time being at least, Galileo returned to Tuscany.

Although Galileo's later life was plagued by illness (apart from his arthritic problem, he suffered a severe hernia which often left him incapacitated), and work on his epic book proceeded slowly, he continued to carry out scientific work in his fifties and sixties, including an attempt to use the regular and predictable motions of the moons of Jupiter as a kind of cosmic clock by which navigators could find the true time while at sea and thereby determine their longitude (a fine idea in principle, but the accurate observations required were impractical from the heaving deck of a ship at sea), and significant work on magnetism. But this took place against a background of change in Galileo's personal life. Partly in recognition of his advancing years, in 1617 Galileo moved to a fine villa – almost a palace – known as Bellosguardo, on a hill to the west of Florence. The move was linked with the entry of his daughters, 16-year-old Virginia and 15-year-old Livia, to the nearby convent at Arcetri, where they became members of the order of the Poor Clares. This was not a result of any particularly deep religious convictions on their part; Galileo saw the move as the only way for his illegitimate daughters to have a secure future, since no respectable man would marry them without a large dowry, and he had no intention of getting involved in the dowry business again. On joining the order, Virginia took the new name of Maria Celeste and Livia became known as Arcangela. Galileo remained close to the girls both geographically and emotionally, and often visited the convent;

surviving letters between Galileo and Maria Celeste give a close insight into his later life.

On the scientific side, Galileo had scarcely got settled in Bellosguardo when he became embroiled in a new controversy. Three comets were seen in 1618, and when a group of Jesuits (including Scheiner) published a rather fanciful account of their significance, Galileo replied in withering terms, sarcastically suggesting that they seemed to think that 'philosophy is a book of fiction by some author, like *The Iliad*', going on to say that the book of the Universe:

> Cannot be understood unless one first learns to comprehend the language and to understand the alphabet in which it is composed. It is written in the language of mathematics, and its characters are triangles, circles and other geometric figures, without which it is humanly impossible to understand a single word of it; without these, one wanders about in a dark labyrinth.

He had a point, and this is indeed a distinguishing feature of real science. Unfortunately, on this occasion Galileo's explanation of comets was also wrong, and there is no point in relating the details of the argument here. But by claiming that the Jesuits dealt in fairy tales while he dealt in facts, Galileo was storing up more trouble for himself in Rome.

In the early 1620s, as the Thirty Years War temporarily shifted in favour of the Catholic side, the political situation in Italy changed in ways that would affect Galileo dramatically. In 1621, three of the people closely involved in his conflicts with Rome died – his protector in Tuscany, Cosimo II (at the early age of 30); the Pope himself, Paul V; and one of Galileo's most important contacts in Rome, Cardinal Bellarmine (a few weeks short of his seventy-ninth birthday). The death of Cosimo II left the affairs of Tuscany in the hands of his wife and mother, acting as Regents for the 11-year-old Ferdinando II. Although Galileo was still in favour at court, the succession of a minor to the throne seriously weakened the influence of Tuscany in Italian politics and reduced the ability of the Tuscan state to protect anyone out of favour with Rome. The death of Bellarmine left Galileo without a friendly witness to the crucial events of 1616, although he still had Bellarmine's written statement. But the death of Paul V at first seemed like good news for science. He was succeeded by Gregory XV, an elderly stopgap who himself died in 1623, when things seemed at last to be going better for Galileo.

Just before Gregory XV died, Galileo received formal permission from Rome to publish a new book, *The Assayer*, which had grown out of his work on comets but ended up covering far more ground and stating the scientific case clearly – the famous quote above about the story of the Universe being 'written in the language of mathematics' is taken from this book. Galileo had also been cultivating new friends in high places – one of these, Francesco Barberini, a member of one of the most powerful families in Rome,[1] received his doctorate from the University of Pisa in 1623. In June that year, Galileo received a letter from Cardinal Maffeo Barberini, the uncle of Francesco (and a man who had previously heaped fulsome praise on Galileo, in print, for his scientific achievements), thanking him for the help he had given to the cardinal's nephew. The terms of the letter were more than friendly. The Barberinis, said the cardinal, 'are ready to serve you always'. Two weeks after the letter was written, Gregory XV died. His elected successor was Cardinal Maffeo Barberini, who took the name Urban VIII and also, among other things, soon appointed his nephew Francesco as a cardinal. Moving with almost as much speed and rather more political adroitness, the Lynceans just had time before *The Assayer* was printed to dedicate the book to Pope Urban VIII and adorn its title page with the Barberini arms, three bees. The Pope was delighted and had the book read aloud to him at table, roaring with laughter at the digs at the Jesuits.

In the spring of 1624, Galileo travelled to Rome to visit both the Barberinis. He was granted six audiences with the Pope, was awarded a gold medal and other honours (including a lifetime pension for his son Vincenzio), and the Pope wrote a letter to Ferdinando II praising Galileo to the skies. But the greatest prize was the permission of the Pope to write a book about the two models of the Universe (or systems of the world, as they were called then), the Ptolemaic model and the Copernican model. The only stipulation was that he had to describe the

1. Powerful, but in the end not widely loved. Later generations of Romans would quip that what the Barbarians had failed to destroy, the Barberinis stole.

two models impartially, without arguing in favour of the Copernican system, and restricting himself to the astronomical and mathematical arguments on both sides. He was allowed to *teach* Copernicanism, but not to *defend* it.

Although Galileo had long dreamed of writing such a book (and had secretly begun to draft sections of it), it was almost as long in the writing as it had been in the dreaming. Apart from his continuing ill health and growing frailty, not the least of the reasons why he became distracted was that around this time Galileo was one of the first people to develop an effective compound microscope, involving two lenses each ground with a doubly convex shape (properly 'lens shaped' in the modern use of the term, instead of being flat on one side and bulging out on the other). It was the difficulty of grinding these lenses that had delayed the invention of the microscope, and nothing better demonstrates Galileo's skill at this craft than his pioneering work in microscopy (for the same reason, although he often bemoaned the difficulty of obtaining good enough glass for his lenses, Galileo's telescopes remained among the best in the world throughout his lifetime). The first detailed illustrations of insects from drawings taken by Galileo using a microscope were published in Rome in 1625, although the full impact of the new device took some time to make itself felt, and Galileo's role in its invention is often lost in the glare of all his other achievements.

Galileo publishes Dialogue on the Two Chief World Systems

Galileo's book, *Dialogue on the Two Chief World Systems* (usually referred to as the *Dialogue*), was finished in November 1629. As the title implies, it took the form of an imagined debate between two people, Salviati (arguing the Copernican case) and Simplicio (arguing the Ptolemaic case). The device of such a dialogue was an old one, going back to the Ancient Greeks, and in principle it offered a neat way to teach unconventional (or in this case, heretical) ideas without exactly endorsing them. But Galileo didn't quite follow that tradition. There had been a real Filippo Salviati, a close friend of Galileo, who had died in 1614, and by choosing this name for his Copernican, Galileo was steering dangerously close to identifying himself with that world view. There had also been a real Simplicio (actually Simplicius), an Ancient Greek who had written a commentary on Aristotle's work, so it could be argued that this was a suitable name for the supporter of Ptolemy (and Aristotle) in the *Dialogue*. It could also be argued that the name implied that only a simpleton would believe the Ptolemaic system was correct. A third 'voice' in the book was provided by Sagredo, named after another old friend of Galileo, Giovanfrancesco Sagredo, who had died in 1620. He was supposed to be an impartial commentator, listening in on the debate between Salviati and Simplicio and raising points for discussion – but the character increasingly tended to support Salviati against Simplicio.

In spite of all this, at first everything seemed set fair for the book. To gain the official seal of approval for publication, it had to be passed by a censor in Rome, and the man chosen for the task, Niccoló Riccardi, a Dominican Father, happened to be the censor who had passed *The Assayer* without requesting any changes. Galileo delivered the manuscript to Riccardi in Rome in May 1630, but in June he had to return home because an outbreak of plague spreading south into Italy threatened to reach Florence and disrupt communications. The book was given a conditional imprimatur; Riccardo wanted a new preface and conclusion added to the book, spelling out that the Copernican position was presented only hypothetically, but he was happy with the bulk of the manuscript and, under the circumstances, gave Galileo permission to return home. Riccardi and his colleagues would make the changes and have them sent on to Galileo to insert in the book. When the insertions arrived in Florence, the covering letter from Riccardi included the sentence, 'the author may alter or embellish the wording, so long as the substance is preserved'. Galileo took this at face value, which turned out to be a big mistake.

Apart from the plague, there were other difficulties affecting the publication of the book. It should have been produced in Rome by the Lynceans. But the death of Prince Frederico Cesi, the Chief Lynx, in August 1630, threw all the affairs of the society into confusion (not least since he had been funding its activities) and permission was given by the Church for the printing to be carried out in Florence. Largely because of the difficulties caused by the plague, which did indeed spread and disrupt all normal activities, printing of the *Dialogue* did not begin until June 1631, and finished copies only went on sale in Florence in March 1632. A few copies were sent immediately to Rome

– the first person there to receive a copy was Cardinal Francesco Barberini, the nephew of the Pope, who wrote to Galileo saying how much he enjoyed it. But others were less pleased.

Once again, in the *Dialogue*, Galileo raised the debate about sunspots, and once again he could not resist a few digs at Scheiner, which infuriated the old Jesuit and his colleagues. Then there was the matter of the additional material supplied by the censor. Galileo had had the preface set in a different typeface from the rest of the book, clearly indicating that it did not represent his own views. And the closing words in which the Copernican system is dismissed as a mere hypothesis (essentially, the words of the Pope, conveyed via Father Riccardi) were put in the mouth of Simplicio. In all fairness, there is no other character in the book who could say those words, since Sagredo ends up on the side of Salviati. But it was suggested to His Holiness that Galileo had done this deliberately, to imply that Urban VIII himself was a simpleton, and this infuriated the Pope, who later said of Galileo, 'He did not fear to make sport of me'.[1] The upshot was that a papal commission was set up to look into the affair. Searching through the files for anything they could find on Galileo, the Jesuits came up with what seemed to be damning evidence – the unsigned minutes from the meeting in 1616, which said that Galileo had been instructed not to 'hold, defend, *or teach*' the Copernican world view. This was the clinching evidence that caused Urban VIII to summon Galileo to Rome to stand trial for heresy – for publishing a book which had been passed by the official censor and received the imprimatur! He also tried to have the distribution of the book stopped, but with printing having been carried out in Florence, it was already too late.

Threatened with torture, he recants

Galileo pleaded old age and illness (he really was ill once again) to delay the trip to Rome, knowing as well as his old friend Paolo Sarpi (who had died in 1623) what that kind of invitation from Rome implied. He also tried to get the political help of the Tuscan state in keeping the Inquisition at bay, but although Ferdinando II had formally taken up his duties as Grand Duke in 1629, at the age of 19, his youth and inexperience meant that Tuscany could not provide the same level of support for Galileo that Venice had once given to Sarpi.

1. Quoted by Reston.

In truth, when Galileo did eventually arrive in Rome on 13 February 1633, he was treated well, compared with most of the guests of the Inquisition. Although he had been detained for the best part of three tiring weeks in quarantine on the Tuscan border (a sign of how severely the plague had disrupted communications), once he arrived in Rome he was allowed, at first, to stay at the Tuscan Embassy. Even when the trial began, in April, he was kept in a comfortable suite of rooms (at least, they would have been comfortable apart from the pain of his arthritis, which had him crying out in his sleep night after night), not thrown into a dank dungeon. The trial itself has been described in detail many times, and we need not elaborate on it here. It is an indication of how little the prosecutors had to go on, though, that among Galileo's supposed 'crimes' were the facts that he had written in Italian, not Latin, so that ordinary people could understand his words, and that he had written in praise of the work of William Gilbert, a 'perverse heretic, a quarrelsome and sophistical defender of Copernicus'. But the key issue was whether Galileo had disobeyed a papal injunction not to teach the Copernican system in any way, and on that issue the Jesuit's unsigned minute of the meeting in 1616 was trumped when Galileo produced the signed document written in Cardinal Bellarmine's own hand spelling out that Galileo was not to 'hold or defend' these views but was in no way constrained any more than any other member of the Catholic faith. Nobody, though, escapes the Inquisition, and once a full show trial had begun, the only conceivable verdict was to find Galileo guilty of something and punish him as a warning to others. From the Inquisition's point of view, the problem was that to bring a false charge of heresy was as serious a crime as heresy. If Galileo was not guilty, his accusers were – and his accusers were the highest authorities of the Catholic Church. Galileo had to be made to confess to something.

It took some heavy persuasion from Cardinal Barberini, who acted in Galileo's best interests throughout, to make the old man appreciate that he really had to confess even if he wasn't guilty, or the torturers would be set to work. Eventually, Galileo understood his true position and made the famous statement in which he claimed not to believe in the Copernican system and confessed that his mistake had been to go too far in presenting the case for Copernicanism in his book, out of misplaced pride in his skill at presenting these ideas (for teaching purposes only) in a plausible way. I 'abjure, curse and detest my errors,' he said. He was 69 years old, in pain from his chronic arthritis and terrified of torture. There is no evidence at all that he muttered the famous words '*eppur, si muove*' ('yet, it does move'); if he had, and if he had been overheard, he would certainly have gone to the rack or the stake (possibly both). The Jesuits had their public victory and all that remained was to pass sentence – life imprisonment. In fact, only seven of the ten cardinals sitting as the council of the Inquisition signed the sentence, with Barberini among the three who refused.

Thanks to Barberini, although the sentence was served, its terms were gradually softened, first to house arrest at the Tuscan Embassy in Rome, then to the custody of the Archbishop of Siena (a Galileo sympathizer) and finally to confinement at Galileo's own house near Arcetri, from early in 1634. Not long after Galileo returned home for the last time (he would not be allowed to leave Arcetri even to visit doctors in Florence, although he was allowed to visit the convent), his daughter Maria Celeste died, on 2 April 1634 (her sister, Arcangela, survived Galileo and died on 14 June 1659).

Galileo publishes Two New Sciences

Isolated in Bellosguardo,[1] Galileo completed his greatest book, *Discourses and Mathematical Demonstrations Concerning Two New Sciences* (usually referred to as *Two New Sciences*), which summed up his life's work on mechanics, inertia and pendulums (the science of moving things), and the strength of bodies (the science of non-moving things), as well as spelling out the scientific method. By analysing mathematically subjects which had previously been the prerogative of philosophers, *Two New Sciences* was the first modern scientific textbook, spelling out that the Universe is governed by laws which can be understood by the human mind and is driven by forces whose effects can be calculated using mathematics. Smuggled out of Italy and published in Leiden in 1638 by Louis Elzevir, this book was an enormous influence on the development of science in Europe in the following decades, even more than the widely translated *Dialogue*. Enormously influential everywhere except Italy, that is; as

1. Not quite isolated; among Galileo's visitors in his last years were Thomas Hobbes and John Milton.

a direct result of the condemnation by the Church of Rome of Galileo's works, from the 1630s onwards, Italy, which had seen the first flowering of the Renaissance, became a backwater in the investigation of the way the world works.

His death

By the time *Two New Sciences* was published, Galileo had gone blind. Even after this, he thought up an idea for an escapement for a pendulum clock, which he described to his son Vincenzio, who actually built such a clock after Galileo died. Similar clocks spread across Europe later in the seventeenth century following the independent work of Christiaan Huygens. From late in 1638 onwards, Galileo had Vincenzo Viviani as his assistant – he acted as Galileo's scribe, and would later write the first biography of Galileo, spreading many of the legends about his master that colour the popular view of the great man today. Galileo died peacefully in his sleep on the night of 8/9 January 1642, a few weeks short of his seventy-eighth birthday. Just two years before, in 1640, the Frenchman Pierre Gassendi (1592–1655) had carried out the definitive experiment to test the nature of inertia, when he had borrowed a galley from the French navy (the fastest means of transport available at the time) and had it rowed flat out across the calm Mediterranean Sea while a series of balls were dropped from the top of the mast to the deck. Every ball fell at the foot of the mast; none was left behind by the motion of the galley.

Gassendi had been strongly influenced by Galileo's writings, and this example highlights the revolution that Galileo, more than anyone, brought about in the investigation of the world, by establishing the whole business of testing hypotheses by getting your hands dirty in experiments, instead of strolling about discussing ideas purely in philosophical terms. From this perspective, it's worth noting one occasion on which Galileo got it wrong – precisely because on this occasion he had to extrapolate from known experiments by 'philosophizing', since there was no feasible way at that time to test his ultimate hypothesis by experiment. By rolling balls down inclined planes and allowing them to roll up another plane, Galileo realized that in the absence of friction a ball would always roll up to the same height it had started from, no matter how steep or shallow the inclines were. In itself, this is a key realization, not least because Galileo was the first scientist to grasp and fully understand the idea that our experiments are always an imperfect representation of some idealized world of pure science – friction is always present in the real world, but that doesn't stop scientists working out how things *would* behave in the absence of friction, and then, as their models become more sophisticated, adding in an allowance for friction later. In the centuries after Galileo, this became a standard feature of the scientific approach – breaking complex systems down into simple components obeying idealized rules and, where necessary, accepting that there would be errors in the predictions of those simple models, caused by complications outside the scope of the models. It was just such complications (like wind resistance), as Galileo also realized, that accounted for the slight difference in arrival times of the two balls at the ground when the Leaning Tower experiment really was carried out.

But Galileo saw a deeper truth in the inclined plane experiments. He went on to consider what would happen if the second inclined plane were made shallower and shallower. The shallower the incline, the further the ball would have to roll to get back to its original height. And if the second plane were horizontal, and friction could be ignored, the ball would roll for ever towards the horizon.

Galileo had realized that moving objects have a natural tendency to keep on moving, unless they are affected by friction or some other outside force. This would be a key component in the full flowering of mechanics achieved following the work of Isaac Newton. But there was one imperfection in Galileo's work. He knew that the Earth was round. So horizontal motion (motion towards the horizon) actually means following a curved path around the curved surface of the Earth. Galileo thought that this meant that inertial motion, without any external force at work, must basically involve moving in a circle, and this seemed to him to explain why the planets stayed in orbit around the Sun. It was René Descartes, a key figure in the decades between Galileo and Newton, who first appreciated that any moving object tends, as a result of inertia, to keep moving in a straight line unless it is acted upon by a force. Galileo had laid the foundations of science and pointed the way for others; but there was plenty for others to do in building on those foundations. It is time to look more closely at the work of Descartes and other scientists who built on the foundations laid by Galileo.

Book Two:

THE FOUNDING FATHERS

4

Science Finds its Feet

René Descartes and Cartesian co-ordinates – His greatest works – Pierre Gassendi: atoms and molecules – Descartes's rejection of the concept of a vacuum – Christiaan Huygens: his work on optics and the wave theory of light – Robert Boyle: his study of gas pressure – Boyle's scientific approach to alchemy – Marcello Malpighi and the circulation of the blood – Giovanni Borelli and Edward Tyson: the increasing perception of animal (and man) as machine

Science is written in the language of mathematics, as Galileo realized. But that language was far from being fully developed in Galileo's time, and the symbolic language that we automatically think of as maths today – the language of equations such as $E = mc^2$ and the way in which geometrical curves can be described by equations – had to be invented before physicists could make full use of mathematics to describe the world we live in. The symbols + and – were only introduced to mathematics in 1540, in a book by the mathematician Robert Recorde, called *The Grounde of Artes*. Recorde was born in Tenby, in Pembrokeshire, some time around 1510, and studied at both the University of Oxford and the University of Cambridge, obtaining qualifications in mathematics and in medicine. Something of a polymath, he was a Fellow of All Souls College in Oxford, physician to Edward VI and to Queen Mary, and sometime general surveyor of mines and money for the Crown. In another book, *Whetstone of Witte*, published in 1557, he introduced the equals sign (=), as he put it, 'bicause noe 2 thynges can be moare equalle' than two parallel lines of the same length. All his accomplishments did not save him from an unhappy end – Recorde died in a debtors' prison in 1558 (the year Elizabeth I became Queen). But his mathematical works were used as standard texts for more than a hundred years, even after the death of Galileo.[1] As John

1. To complete the set, it's worth mentioning that the multiplication symbol (×) was introduced in 1631, and the division symbol (÷) only in 1659. William Oughtred, who introduced the × symbol, also invented the slide rule, some ten years earlier.

Aubrey put it more than a century later, Recorde 'was the first that wrote a good arithmetical treatise in English' and 'the first that ever wrote of astronomie in the English tongue'.

As I have already mentioned, the invention (or discovery) of logarithms early in the seventeenth century enormously simplified and speeded up the laborious processes of arithmetical calculation carried out by astronomers and other scientists – this is the system that involves manipulating 'powers of ten' rather than ordinary numbers, so that (to take a very simple example) 100 × 1000 becomes $10^2 \times 10^3$, which (since 2 + 3 = 5) becomes 10^5, or 100,000. All ordinary numbers can be represented in this sort of way (for example, 2345 can be written as $10^{3.37}$, so that the logarithm of 2345 is 3.37), which means all multiplication and division can be reduced to addition and subtraction (once someone has gone to the trouble of preparing tables of logarithms). In the days before pocket calculators (which means right up until the 1970s), logarithms and the associated tool which uses logarithms, the slide rule, were the only things that made complex calculations feasible for most people.

I don't intend to go into much detail about the history of mathematics in this book, except where it impinges directly on the story of the developing understanding of the way the world works, and of our place in it. But there is one other breakthrough that was published while Galileo was still serving the sentence imposed on him by the Inquisition that is not only too important in its own right to omit, but also introduces us properly to one of the other key figures of the time, René Descartes, a man widely known today as a philosopher, but who had interests across the scientific board.

René Descartes and Cartesian co-ordinates

Descartes was born at La Haye, in Brittany, on 31 March 1596. He came from a prominent and moderately wealthy local family – his father, Joachim, was a lawyer and a Counsellor in the Brittany parliament and although René's mother died not long after he was born, she left him an inheritance which was sufficient to ensure that, although by no means rich, he would never starve, and could choose whatever career (or lack of career) he liked without worrying unduly about money. There was a real prospect, though, that he would never live long enough to have a career of any kind – René was a sickly child, who might not have lived to adulthood and often suffered ill health in later life. When he was about 10 years old (possibly a little younger), Descartes was sent by his father (who hoped that the boy might follow in his footsteps as a lawyer, or perhaps become a doctor) to the newly established Jesuit College at La Flèche, in Anjou. This was one of several new schools that Henry IV, the first Bourbon King of France (also known as Henry of Navarre), allowed the Jesuits to open around this time.

Henry's own career (if that is the right word) typifies the kind of turmoil Europe was experiencing at this time. Before becoming King he had been the leader of the Protestant (Huguenot) movement in the French wars of religion, a series of conflicts which lasted from 1562 to 1598. After a major defeat in 1572, known as the massacre of St Bartholomew's Day, he converted to Catholicism to save his life; but he was imprisoned by the King (Charles IX and then his successor, Henry III), who doubted the sincerity of this move. In 1576 he escaped and disowned the conversion, leading an army in several bloody battles during the civil wars. Henry, who was originally rather distantly in line of succession to the French throne, became the Heir Presumptive when Henry III's brother, the Duke of Anjou, died in 1584 (both Henry III and the Duke were childless). This led the Catholic League in France to recognize the daughter of Philip II of Spain, which had been largely fighting the war on the Catholic side, as the heir to the throne; but the move backfired when the two Henrys joined forces, hoping to crush the League and prevent the Spanish effectively taking over France. Henry III was stabbed by an assassin on 1 August 1589, while the two Henrys were besieging Paris, but lived long enough to confirm Henry of Navarre as his heir. With the fighting dragging on, Henry IV was not crowned until 1594, a year after he had once again declared himself to be a Catholic, and even then the conflict with Spain continued. The wars finally ended in 1598, the year that Henry IV both made peace with Spain and signed the Edict of Nantes, which gave Protestants the right to worship as they pleased – no mean combination of achievements. Henry himself also died at the hands of an assassin, in 1610, when René Descartes was 14 years old. His best epitaph is something he said himself: 'Those who follow their consciences are of my religion, and I am of the religion of those who are brave and good.'

Two years after the death of Henry IV (or possibly in 1613; the records are not clear), Descartes left the Jesuit college and lived for a short time in Paris, before studying at the University of Poitiers, where he graduated with qualifications in law in 1616 (he may have studied medicine as well, but never qualified as a doctor). At the age of 20, Descartes now took stock of his life and decided that he was not interested in a career in the professions. His childhood illness had helped to make him both self-reliant and something of a dreamer, who loved his creature comforts. Even the Jesuits had allowed him considerable indulgence, such as permission to get up late in the morning, which became not so much a habit as a way of life for Descartes. His years of education convinced him primarily of his own ignorance and the ignorance of his teachers, and he resolved to ignore the textbooks and work out his own philosophy and science by studying himself and the world about him.

To this end, he took what seems at first sight a rather bizarre decision, upping sticks and moving to Holland, where he signed up for military service with the Prince of Orange. But the comfort-loving Descartes wasn't a fighting soldier; he found his military niche as an engineer, using his mathematical skills rather than his less-developed physical abilities. It was while he was at the Military School at Breda that Descartes met the mathematician Isaac Beeckman, of Dordrecht, who introduced him to higher aspects of mathematics and became a long-term friend. Not much is known about Descartes's military life over the next few years, during which he served in various European armies, including that of the Duke of Bavaria, although we do know that he was present at the coronation of Emperor Ferdinand II in Frankfurt in 1619. Towards the end of that year, however, the most important event of Descartes's life occurred, and we know exactly when and where it happened because he tells us in his book *The Method* – in full, *Discours de la Méthode pour bien conduire la raison et chercher la Vérité dans les sciences* (*A Discourse on the Method of Rightly Conducting Reason and seeking Truth in the Sciences*) – published in 1637. It was 10 November 1619, and the army of the Duke of Bavaria (gathered to fight the Protestants) was in winter quarters on the banks of the river Danube. Descartes spent the whole day snugly tucked up in bed, dreaming (or daydreaming) about the nature of the world, the meaning of life and so on. The room he was in is sometimes referred to as an 'oven', which is a literal translation of the expression used by Descartes, but this does not necessarily mean that he had literally crawled into some hot chamber usually used for, say, baking bread, since the expression could have been metaphorical. Either way, it was on this day that Descartes first saw the road to his own philosophy (which lies largely outside the scope of the present book) and also had one of the greatest mathematical insights of all time.

Idly watching a fly buzzing around in the corner of the room, Descartes suddenly realized that the position of the fly at any moment in time could be represented by three numbers, giving its distance from each of the three walls that met in the corner. Although he instantly saw this in three-dimensional terms, the nature of his insight is now known to every schoolchild who has ever drawn a graph. Any point on the graph is represented by two numbers, corresponding to the distances along the *x* axis and up the *y* axis. In three dimensions, you just have a *z* axis as well. The numbers used in the system of representing points in space (or on a piece of paper) in this way are now known as Cartesian co-ordinates, after Descartes. If you have ever given someone instructions on how to find a location in a city by telling them something along the lines of 'go three blocks east and two blocks north', you have been using Cartesian co-ordinates – and if you go on to specify a certain floor in the building, you are doing so in three dimensions. Descartes's discovery meant that any geometric shape could be represented simply by a set of numbers – in the simple case of a triangle drawn on graph paper, there are just three pairs of numbers, each specifying one of the corners of the triangle. And any curved line drawn on paper (or, for example, the orbit of a planet around the Sun) could also be represented, in principle, by a series of numbers related to one another by a mathematical equation. When this discovery was fully worked out and eventually published, it transformed mathematics by making geometry susceptible to analysis using algebra, with repercussions that echo right down to the development of the theory of relativity and quantum theory in the twentieth century. Along the way, it was Descartes who introduced the convention of using letters at the beginning of the alphabet (*a*, *b*, *c* . . .) to represent known (or specified) quantities, and letters at the end of the alphabet (especially *x*, *y*, *z*) to represent unknown quantities. And it was he who introduced the now familiar exponential notation, with x^2 meaning $x \times x$, x^3 meaning $x \times x \times x$ and so on. If he had done nothing else, laying all these foundations of analytical mathematics would have made Descartes a key figure in seventeenth-century science. But that wasn't all he did.

After his insights in the 'oven', Descartes renounced the military life in 1620, at the end of his service with the Duke of Bavaria, and travelled through Germany and Holland to France, where he arrived in 1622 and sold off the estate in Poitiers that he had inherited from his mother, investing the proceeds to allow him to continue his independent studies. With his security established, he spent several years wandering around Europe and doing a bit of thinking, including a lengthy spell in Italy (where, curiously, he seems to have made no attempt to meet Galileo), before deciding at the age of 32 that the time had come to settle down and put his thoughts into a coherent form for posterity. He visited Holland again in the autumn of 1628, spent the winter of 1628/9 in Paris, then returned to settle in Holland, where he stayed for the next twenty years. The choice of base was a good one. The Thirty Years War still kept central Europe in turmoil, rumblings from the Wars of Religion still surfaced from time to time in France, but Holland was now securely independent; although it was officially a Protestant country Catholics formed a large part of the population, and there was religious tolerance.

Descartes had a wide circle of friends and correspondents in Holland, including: Isaac Beeckman and other academics; Constantijn Huygens, a Dutch poet and statesman (father of Christiaan Huygens), who was secretary to the Prince of Orange; and the family of the Elector Palatine of the Rhine, Frederick V. The latter connection provides a link, of sorts, between Descartes and Tycho, since Princess Elizabeth, the wife of Frederick V, was the daughter of James I of England.[1] Like Galileo, Descartes never married, but, as John Aubrey puts it, 'as he was a man,

1. This is not much of a coincidence, since most of the royal houses of Europe at the time were interconnected by a web of political marriages. But this marriage turned out to be particularly important: Frederick's daughter Sophia married the Elector of Hanover and was the mother of George I of England.

he had the desires and appetites of a man; he therefore kept a good conditioned hansome woman that he liked'. Her name was Hélène Jans, and they had a daughter, Francine (born in 1635), who he doted on, but who died in 1640.

His greatest works

While reinforcing his already good reputation as a thinker and man of learning in conversations and correspondence with these friends, Descartes spent four years, from 1629 to 1633, preparing a huge treatise in which he intended to put forward all of his ideas about physics. The work was titled *Le Monde, ou Traité de la Lumière*, and it was on the verge of being published when news of Galileo's trial and conviction for heresy reached Holland. Although the full story of the trial did not become clear until later, what seemed to be clear at the time was that Galileo had been convicted for Copernican beliefs, and Descartes's manuscript very much supported Copernican ideas. He stopped the publication immediately, and the book never was published, although large parts of it were used as the basis for some of Descartes's later works. Even granted that Descartes was a Catholic, this does seem to have been a rather hasty over-reaction, since there was nothing the Jesuits in Rome could do to harm him in far-away Holland, and it didn't take much urging from his friends and acquaintances, many of whom had seen parts of the work or had it described in letters, to persuade Descartes to publish something sooner rather than later. The first result was *The Method*, which appeared in 1637, accompanied by three essays, on meteorology, optics and geometry. Although not all of the ideas he presented were correct, the important thing about the essay on meteorology is that it attempted to explain all the working of the weather in terms of rational science, not by recourse to the occult or the whim of the gods. The essay on optics described the workings of the eye and suggested ways to improve the telescope. And the essay on geometry considered the revolutionary insights stemming from that day in bed on the banks of the Danube.

Descartes's second great work, *Meditationnes de Prima Philosophia*, appeared in 1641, and elaborated the philosophy built around Descartes's most famous (if not always correctly interpreted) line, 'I think, therefore I am.' And in 1644 he produced his third major contribution to learning, *Principia Philosophiae*, which was essentially a book about physics, in which Descartes investigated the nature of the material world and made the correct interpretation of inertia, that moving objects tend to keep moving in a straight line, not (as Galileo had thought) in a circle. With the publication of that book out of the way, Descartes made what seems to have been his first trip to France since 1629, and he went there again in 1647 – a significant visit, since on that occasion he met the physicist and mathematician Blaise Pascal (1623–1662) and suggested to the younger man that it would be interesting to take a barometer up a mountain and see how the pressure varied with altitude.[1] When the experiments were carried out (by Pascal's brother-in-law in 1648) they showed that the atmospheric pressure drops with increasing altitude, suggesting that there is only a thin layer of air around the Earth and the atmosphere does not extend for ever. Another visit to France, in 1648, was curtailed by the threat of civil war, but it now looks clear that in the late 1640s Descartes, who was 52 in 1648, was, for whatever reason, becoming restless and no longer seemed committed to spending the rest of his life in Holland. So in 1649, when Queen Christina of Sweden invited him to join the circle of intellectuals that she was gathering in Stockholm, Descartes leaped at the chance. He arrived in Stockholm in October that year, but was horrified to discover that in return for the favours he would receive and the freedom to spend most of his time working at what he pleased, he was expected to visit the Queen each day at 5 am, to give her personal tuition before she got on with the affairs of state. The combination of the northern winter and early rising proved too much for Descartes's comfort-loving body. He caught a chill, which developed into pneumonia, which finished him off on 11 February 1650, just short of his fifty-fourth birthday.

Descartes's influence was profound, most importantly because of the way in which (although he believed in God and the soul) he swept away from his thinking any vestige of mystic forces and insisted that both the world we live in and the material creatures that inhabit the world (including ourselves) can be understood in

1. The barometer had only recently been invented, and Descartes was one of the first people to suggest that what it measured was the weight of air pressing down on the surface of the Earth.

terms of basic physical entities obeying laws which we can determine by experiment and observation. This is not to say that Descartes got everything right, by any means, and one of his biggest ideas was so wrong, yet at the same time so influential, that it held back scientific progress in parts of Europe (notably in France) for decades, until well into the eighteenth century. It's worth looking at this false step before discussing Descartes's influence in areas where he was more right than wrong.

Pierre Gassendi: atoms and molecules

The big thing that Descartes got wrong was his rejection of the idea of a vacuum or 'void'. This also led him to dismiss the idea of atoms, which was being revived around this time by Pierre Gassendi, since the atomic model of the world sees everything as made up of little objects (the atoms) moving about in the void and interacting with one another. Although the idea of atoms went back to the work of Democritus in the fifth century BC, and was revived by Epicurus, who lived from about 342 BC to about 271 BC, it was never more than a minority view in Ancient Greece, and Aristotle, the most influential Greek philosopher in terms of his impact on Western ideas before the scientific revolution, specifically rejected atomism because of its association with the idea of void. Gassendi, who was born at Champtercier, in Provence, on 22 January 1592, became a Doctor of Theology in Avignon in 1616, took holy orders the following year and was teaching at the University of Aix when, in 1624, he published a book criticizing the Aristotelian world view. In 1633, he became Provost of Digne Cathedral, and in 1645 professor of mathematics at the Collège Royale in Paris. But ill health forced him to give up teaching in 1648, and from then until 1650 he lived in Toulon, before returning to Paris, where he died on 24 October 1655.

Although he carried out many astronomical observations and the famous test of inertia using a galley, Gassendi's most important contribution to science was the revival of atomism, which he presented most clearly in a book published in 1649. Gassendi thought that the properties of atoms (for example, their taste) depended on their shape (pointed or round, elongated or squat, and so on), and he had the idea that they might join together, to form what he called molecules, by a kind of hook-and-eye mechanism. He also firmly defended the idea that atoms moved about in a void and that there was literally nothing in the gaps between atoms. But as if to prove the old adage that nobody is perfect, among other things Gassendi opposed Harvey's ideas about the circulation of the blood.

The reason that Gassendi and a good many of his contemporaries were willing to accept the idea of a vacuum in the 1640s was that there was experimental evidence that 'the void' existed. Evangelista Torricelli (1608–1647) was an Italian scientist who got to know Galileo in the last few months of his life, and who became professor of mathematics in Florence in 1642. Galileo introduced Torricelli to the problem that water in a well could not be forced by a pump up a vertical pipe more than about 30 feet (some 9 metres) long. Torricelli reasoned that it was the weight of the atmosphere pressing down on the surface of the water in a well (or anywhere else) that produced a pressure capable of supporting the water in the pipe, and that this could only happen if the pressure produced by the weight of water in the pipe was less than the pressure exerted by the atmosphere. He tested the idea in 1643 using a tube of mercury, sealed at the top, inverted over a shallow dish of the liquid metal, with the open end of the tube below the surface of the liquid. Since mercury is roughly fourteen times heavier than the same volume of water, he predicted that the column of mercury in the tube would settle down at a length of just over 2 feet, which it did – leaving a gap between the top of the mercury column and the sealed end of the tube above the mercury. When Torricelli spotted slight variations in the height of the mercury column from day to day, he reasoned that they were caused by changes in atmospheric pressure – he had invented the barometer, and also created a vacuum.

Descartes's rejection of the concept of a vacuum

Descartes knew all about this work – as we have mentioned, he suggested the idea of taking a barometer up a mountain to see how the pressure changed with altitude. But he didn't accept the idea that the gap above the mercury (or water) was a vacuum. He had the notion that everyday substances like air, or water, or mercury were mingled with a much finer substance, a fluid which filled all the gaps and prevented the existence of a vacuum. The mercury in a barometer, for example, could be likened to a column of something like a flexible kind of steel wool, the sort used for scouring pans, with some invisible liquid, like a very fine olive oil, filling all the otherwise empty spaces between the strands of wire and the gap above the top of the column.[1]

Although the experiments carried out for Pascal (who was too ill to do them himself) by his brother-in-law seem to us to imply that the atmosphere thins out as you go higher, and there must be a limit to its extent with a vacuum above it, Descartes said that his universal fluid extended beyond the atmosphere and across the Universe, so that there was no void anywhere. He developed what looks to modern eyes a very curious model in which the planets are carried around by swirling vortices in this fluid, like chips of wood being carried around by eddies in a stream. From this perspective, he was able to argue that the Earth was not really moving, because it was still compared with the fluid it was embedded in – it was just that this surrounding knot of fluid was in an eddy moving around the Sun. It looks almost like a convoluted scheme devised for the express purpose of providing a loophole to embrace Copernicanism while keeping the Jesuits happy – but all the evidence is that Descartes was impelled towards this model not by any fear of the Inquisition, but by his own abhorrence of the vacuum. The whole story would scarcely rate a footnote in the history of science, except for one thing. Descartes's influence was so great in the decades after his death that in France and some other parts of Europe acceptance of Newton's ideas about gravity and planetary motions was considerably delayed because they disagreed with Descartes. There was an element of chauvinism in this – the French supported their own champion and rejected the ideas of the perfidious Englishman, while Newton was, of course, very much a prophet honoured in his own country.

Although Descartes's idea of a Universe devoid of void, so to speak, led him up a blind alley when trying to explain the movement of the planets, when it came to his work on light the idea was more fruitful, even though it ultimately proved incorrect. According to atomists such as Gassendi, light was caused by a stream of tiny particles emerging

1. In the 1650s, the German Otto von Guericke invented an air pump (often called the vacuum pump), which was able to reduce the pressure of air inside a sealed vessel dramatically, snuffing out candle flames and silencing the ringing of a bell as the air was removed from the chamber.

from bright objects, like the Sun, and impinging on the eyes of the viewer. According to Descartes, vision was a phenomenon caused by pressure in the universal fluid, so that the Sun, for example, pushed on the fluid, and this push (like poking something with a stick) was instantaneously translated into a pressure on the eyes of anyone looking at the Sun.[1] Although the original version of this idea envisaged a steady pressure on the eye, it was only a short step from this to the idea that there might be pulses of pressure spreading out from a bright object – not quite like ripples spreading out on the surface of a pond, but more like the pressure waves that would reverberate through the body of water in the pond if you slapped its surface hard. The person who developed this idea most fully in the second half of the seventeenth century was Christiaan Huygens, the son of Descartes's old friend Constantijn, who would have been the greatest scientist of his generation if he had not had the misfortune to be active in science at almost exactly the same time as Isaac Newton.

Christiaan Huygens: his work on optics and the wave theory of light

Huygens's father was not the first member of the family to serve the House of Orange, and it was expected that Christiaan, who was born at The Hague on 14 April 1629, would follow the family tradition. As a member of a prominent and wealthy family, Christiaan was educated at home to the highest standards of his day until the age of 16, and this gave him ample opportunity to meet with the important figures who were frequent visitors to the house, including René Descartes. It may well be that it was this contact with Descartes that helped to arouse Huygens's interest in science, but in 1645 he still seemed to be following the path towards a career in diplomacy when he was sent to study mathematics and law at the University of Leiden. He went on to spend a further two years studying law in Breda, from 1647 to 1649, but at the age of 20, he renounced the family tradition and decided to devote himself to the study of science. Far from objecting to this, his father (who was not just a diplomat but also a talented poet who wrote

1. With his characteristic incisiveness, Isaac Newton later pointed out the obvious flaw in this idea. If vision is caused by the pressure of this invisible fluid on the eyes, it would be possible to see in the dark by running fast enough! Had he still been alive to respond, though, Descartes would surely have replied to Newton that the problem was nobody could run fast enough to make the trick work.

in both Dutch and Latin, and also composed music) was broad-minded enough to provide Christiaan with an allowance that left him free to study whatever he pleased. For the next seventeen years he was based at home in The Hague and devoted himself to the scientific study of nature. It was a quiet life, which gave Huygens ample opportunity to do his work, but provides little in the way of anecdotes, and it took some time for Huygens's reputation as a scientist to spread, because he was always extremely reluctant to publish anything until he had thoroughly worked out all the details. He did, though, travel widely, including a visit to London in 1661, and a stay of five months in Paris in 1655, when he met many leading scientists, including Pierre Gassendi.

Huygens's early work was chiefly in the field of mathematics, where he made improvements in existing techniques and developed his own skills, without making any major new breakthroughs. This led him on to mechanics, where he did important work on momentum and studied the nature of centrifugal force, demonstrating its similarity to gravity, and improved on Galileo's treatment of the flight of projectiles. This work so clearly pointed the way ahead that even if a rare genius such as Isaac Newton had not come along at about this time, there can be little doubt that the famous inverse square law of gravity would have been discovered by somebody in the next generation of scientists.[1] Huygens became widely known, however (even outside scientific circles), for his invention of the pendulum clock (apparently entirely independently of Galileo), which he patented in 1657. His motivation for this work came from his interest in astronomy, where the need for accurate timekeeping had long been obvious, but was becoming more pressing as more accurate observing instruments were designed. Unlike Galileo's design, Huygens's proved to be a rugged and practical time-keeping device (although not rugged enough to keep accurate time at sea, which was one of the main unsolved problems of the day), and in 1658 clocks built to Huygens's design began to appear in church

1. Huygens himself had a crucial blind spot which prevented him taking this step – like Descartes, he did not believe that forces could reach out across empty space, but thought that they could only be transmitted by direct contact, if necessary through an intervening fluid.

towers across Holland, and soon spread across Europe. It was from 1658 onwards, and thanks to Christiaan Huygens, that ordinary people began to have access to accurate timepieces, instead of estimating the time of day by the position of the Sun. It is typical of the thoroughness with which Huygens carried out all his work that the investigation of pendulums led him not only to the design of a practical clock, but to a fully worked out theory of the behaviour of oscillating systems in general, not just pendulums. And all because he had needed an accurate timekeeper for his astronomical work.

Today, few people know about Huygens's work on clocks. Rather more people, though, know that he had something to do with the wave theory of light. This theory, like Huygens's theory of oscillating systems, grew out of a piece of practical work connected with astronomy. In 1655, Christiaan Huygens began working with his brother Constantijn (named after their father) on the design and construction of a series of telescopes that became the best astronomical instruments of their time. All refracting telescopes of the day suffered from a problem called chromatic aberration, which occurs because the lenses in the telescope bend light with different colours by slightly different amounts, producing coloured fringes around the edges of images of objects seen through the telescope. This didn't matter too much if you were using the telescope to identify a ship at sea, but it was a major bugbear for the accurate work required in astronomy. The Huygens brothers found a way to reduce chromatic aberration considerably by using a combination of two thin lenses in the eyepiece of a telescope, instead of one fat lens. It wasn't perfect, but it was a great deal better than anything that had gone before. The brothers were also very good at grinding lenses, producing large, accurately shaped lenses that alone would have made their telescopes better than anything else around at the time. With the first telescope built to the new design, Huygens discovered Titan, the largest of the moons of Saturn, in 1655; this was a discovery only marginally less sensational than Galileo's discovery of the moons of Jupiter. By the end of the decade, using a second, larger telescope also constructed with his brother, Huygens had solved the mystery of the peculiar appearance of Saturn itself, which he found to be surrounded by a thin, flat ring of material, which is sometimes seen edge on from Earth (so that it seems to disappear) and is sometimes seen face on (so that with a small telescope like the one Galileo used, Saturn appears to grow a pair of ears). All of this assured Huygens's reputation. In the early 1660s, he spent a lot of time in Paris, although still based in The Hague, and in 1666, when the French Royal Academy of Sciences was founded, he was invited to work permanently in Paris under the auspices of the Academy, as one of its seven founding members.

The foundation of the first Royal Societies (or Academies) of science around this time was itself a significant landmark in the history of science, marking the middle of the seventeenth century as the time when scientific investigation began to become part of the establishment. The first such scientific society to receive an official imprimatur was the Accademia del Cimento (Academy of Experiment), founded in Florence in 1657 by two of Galileo's former pupils, Evangelista Torricelli and Vincenzo Viviani, under the patronage of Grand Duke Ferdinando II and his brother Leopold. This was the spiritual successor of the failed Lyncean Society, which never recovered from the death of Frederico Cesi. But the Accademia del Cimento itself lasted for only ten years, and its demise in 1667 defines as well as anything the date when the Italian leadership in physical sciences that had been inspired by the Renaissance came to an end.

By then, what was to become the longest continuously enduring scientific society in the world had already begun to meet in London. From 1645 onwards, a group of scientifically minded people began to get together regularly in London to discuss new ideas and to communicate new discoveries to one another and by letters to like-minded thinkers across Europe. In 1662, under a charter of Charles II, this became the Royal Society (as the first such society, it needs no other description; it is *the* Royal Society, sometimes simply referred to as 'the Royal'). But although Royal in name, the London-based society was a collection of private individuals with no official source of funding and no obligations to the government. Huygens became one of the first foreign members of the Royal Society on a brief visit to London in 1663. The French equivalent, the Académie des Sciences, founded four years after the Royal received its charter, had the benefits of being a government institution, set up under the patronage of Louis XIV (the grandson of Henry IV), which allowed it to provide financial support and practical facilities for eminent scientists such as Huygens, but which also gave it certain (sometimes onerous) obligations. The success of the two societies in their different ways spawned many imitations (usually patterned after one or the other model), starting with the Akademie der Wissenschaften in Berlin in 1700.

Huygens had always suffered from ill health, and although he was based in Paris for the next fifteen years, he twice had to return to Holland for lengthy periods to recuperate from sickness. This didn't prevent him from carrying out some of his most important work during his time in Paris, and it was there that (apart from a few details) he completed his work in optics in 1678 (typically of Huygens, though, the work was not published in full until 1690). Although it built to some extent on the work of Descartes, unlike those ideas, Huygens's theory of light was securely based on his practical experiences of working with lenses and mirrors, and grappling with the problems he had encountered (such as chromatic aberration) when constructing telescopes. His theory was able to explain how light is reflected by a mirror and how it is refracted when it passes from air into glass or water, all in terms of a pressure wave occurring in a fluid, which became known as the aether. The theory made one particularly important prediction – that light should travel more slowly in a more dense medium (such as glass) than it does in a less dense medium (such as air). This was important in the long term, because in the nineteenth century it would provide a definitive test of whether light does travel as a wave or whether it travels as a stream of particles. In the short term, it was also highly significant since Descartes, and everyone before him, had assumed that light must travel at infinite speed, so that in Descartes's model a disturbance in the Sun (say) affects the eye instantaneously. Huygens was right at the cutting edge of research when he used a finite speed of light in his model in the late 1670s, and he was able to do so because he was on the spot, in Paris, where the crucial discovery was being made.

The enormous conceptual leap required to appreciate that the speed of light, although very large, was not infinite, stemmed from the work of Ole Rømer, a Dane who carried out the work while he was a contemporary of Huygens at the French Academy. Rømer was born at Århus on 25 September 1644, and after studying at the University of Copenhagen he stayed on there as assistant to the physicist and astronomer Erasmus Bartholin. In 1671, Jean Picard (1620–1682) was sent by the French Academy to Denmark to establish the exact position of Tycho's observatory (important for the precise astronomical analysis of his observations), and Rømer assisted him in this work to such good effect that he was invited back to Paris, where he worked at the Academy and became a tutor to the Dauphin. Rømer's greatest piece of work was achieved as a result of his observations of the moons of Jupiter, carried out in conjunction with Giovanni Cassini (who lived from 1625 to 1712 and is best remembered for discovering a gap in the rings of Saturn, still known as the Cassini division). As each moon goes around its parent planet in a regular orbit, just as the Earth follows a regular one-year orbit around the Sun, each moon ought to be eclipsed behind Jupiter at regular intervals. Rømer noticed, though, that the interval between these eclipses was not always the same, and varied in a way that was related to the position of the Earth in its orbit around the Sun, relative to the position of Jupiter. He interpreted this as being a result of the finite speed of light – when the Earth is further away from Jupiter we see the eclipses later, because it has taken the light carrying news of those eclipses longer to travel from Jupiter to our telescopes. Rømer predicted, on the basis of a pattern he had discovered in the way the eclipse times varied, that an eclipse of Jupiter's innermost Galilean moon, Io, due on 9 November 1679, would occur ten minutes later than expected according to all earlier calculations, and he was sensationally proved right. Using the best estimate available of the diameter of the Earth's orbit,[1] Rømer calculated from this time delay that the speed of light must be (in modern units) 225,00 kilometres per second. Using the same calculation but plugging in the best modern estimate of the size of the Earth's orbit,

1. The distance derived from measurements of the parallax of Mars carried out in 1671 by a French team using simultaneous observations by Jean Richer from Cayenne, in French Guiana, and Cassini from Paris. The slight difference in the position of Mars against the background stars seen from the two ends of this baseline made it possible to work out the distance to Mars, and this, combined with Kepler's laws, gave the diameters of the orbits of all the planets. All this, of course, was striking confirmation (like Rømer's work) of the validity of the Copernican model, if anyone outside Rome still needed convincing.

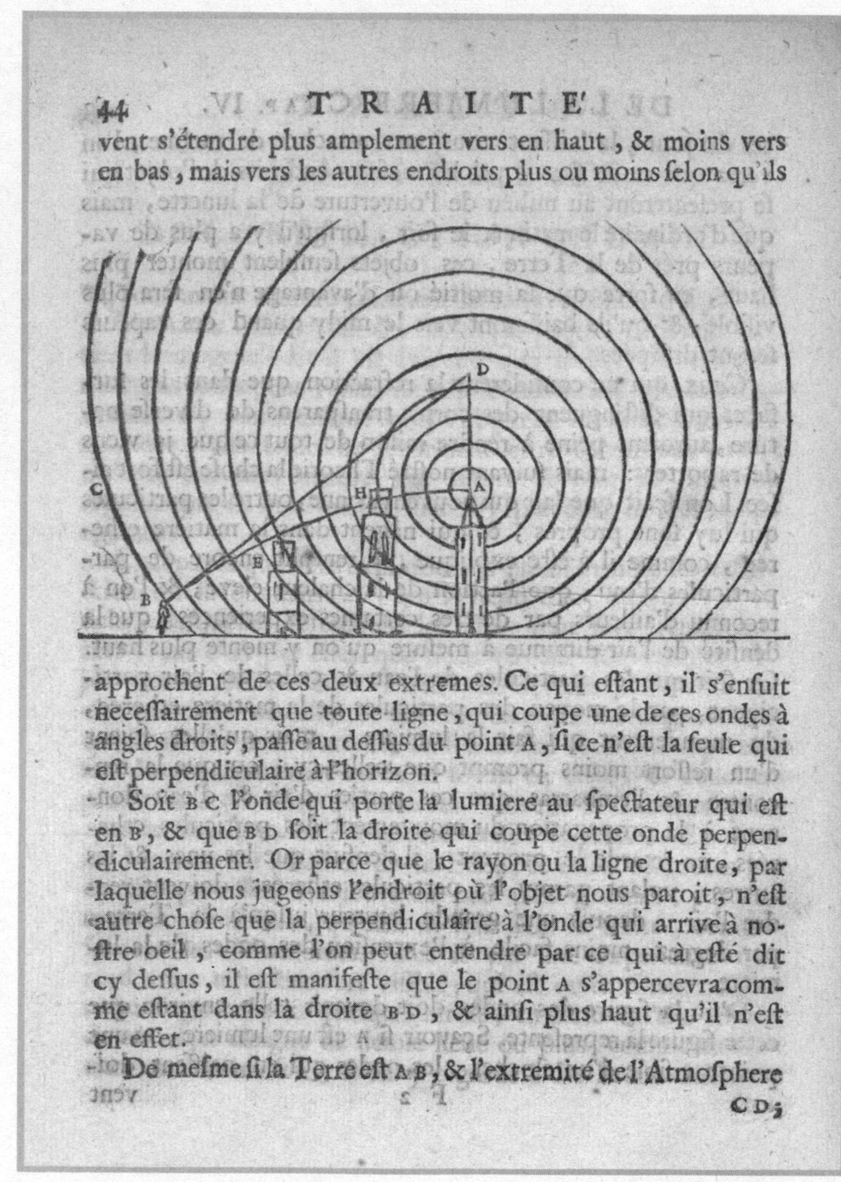
44 TRAITE'

vent s'étendre plus amplement vers en haut, & moins vers en bas, mais vers les autres endroits plus ou moins ſelon qu'ils

approchent de ces deux extremes. Ce qui eſtant, il s'enſuit neceſſairement que toute ligne, qui coupe une de ces ondes à angles droits, paſſe au deſſus du point A, ſi ce n'eſt la ſeule qui eſt perpendiculaire à l'horizon.

Soit B C l'onde qui porte la lumiere au ſpectateur qui eſt en B, & que B D ſoit la droite qui coupe cette onde perpendiculairement. Or parce que le rayon ou la ligne droite, par laquelle nous jugeons l'endroit où l'objet nous paroit, n'eſt autre choſe que la perpendiculaire à l'onde qui arrive à noſtre oeil, comme l'on peut entendre par ce qui à eſté dit cy deſſus, il eſt manifeſte que le point A s'appercevra comme eſtant dans la droite B D, & ainſi plus haut qu'il n'eſt en effet.

De meſme ſi la Terre eſt A B, & l'extremité de l'Atmoſphere

CD;

10. *Depiction of light waves, from Huygens's* Traité de la Lumière, *1690.*

Rømer's own observations give the speed of light as 298,000 kilometres per second. This is stunningly close to the modern value for the speed of light, 299,792 kilometres per second, given that it was the first measurement, ever, of this speed. With his place in history assured (although not everyone was as quick as Huygens to be convinced at the time), Rømer travelled to England, where he met Isaac Newton, Edmond Halley and John Flamsteed, among others. He returned to Denmark in 1681, becoming Astronomer Royal and Director of the Royal Observatory in Copenhagen, where he died on 23 September 1710.

Huygens's work on light, carried out alongside Rømer in Paris, was the crowning achievement of his career, published in 1690 as *Traité de la Lumière* (*Treatise on Light*). The book was completed after Huygens had returned to Holland in 1691, partly because of his deteriorating health, but also because the political climate in France had shifted once again. Bear with me, because the politics are more than a little complicated. Although the independence of the Dutch in the northern part of the Netherlands (the region whose name, Holland, is used in English for the whole country today) had been recognized by Spain in 1648, the Spanish still held the southern part of the Netherlands. In 1660, Louis XIV had married Maria Theresa, The eldest daughter of Philip IV of Spain, and when Philip died in 1665, leaving his infant son Charles II as his successor, Louis used the opportunity to claim the remaining Spanish possessions in The Netherlands (including much of what is now Belgium) and to cast a covetous eye on Holland as well. His ambitions were initially opposed by an alliance of Holland, England and Sweden. But Louis XIV persuaded England to switch sides, offering large financial incentives and a promise of territory on mainland Europe once The Netherlands had been conquered.

This rather unnatural association, which was deeply resented by the English people, came about partly because Charles II of England was a cousin of Louis XIV – Charles I had married Henrietta Maria, the sister of Louis XIII. Charles II was also eager to have a powerful ally, having only recently been restored to the throne following the English Civil War and the Parliamentary interregnum, and to complicate matters further there was a secret clause in the treaty between him and

Louis XIV promising that Charles would become a Catholic. In fact, Charles only converted on his deathbed. Hardly surprisingly, the alliance did not last, and, following defeats of the English Navy by the Dutch, after 1672 France was left alone to invade The Netherlands. Under William of Orange (himself a grandson of Charles I of England and nephew of Charles II, since his mother was the sister of the younger Charles), and with help from several quarters (including Spain, delighted at any opportunity to form an alliance to oppose France even if it meant helping the Dutch), the Dutch forces not only resisted the invasion, but actually made an honourable peace when a treaty was signed at Nijmegen in 1678. It was in the wake of this defeat for French ambition, partly at the hands of the Protestants in Holland, that the situation for Dutch Protestants in Paris became intolerable (they would, of course, have been better tolerated if the French had won!), leading to Huygens returning to his homeland.[1] In spite of his continuing ill health, Huygens made a few more foreign trips, including another visit to London in 1689, when he met Newton. His last illness struck him down in 1694, although he suffered for months until he finally succumbed on 8 July 1695 and died at The Hague.

Robert Boyle: his study of gas pressure

In spite of the war between France and The Netherlands, Huygens's life was mostly without incident outside his scientific work. But the same could hardly be said of his contemporary Robert Boyle, who almost singlehandedly made chemistry respectable, studied the behaviour of gases along the way, promoted the idea of atoms and had a life outside science which could have come straight from the pages of a novel.

If Huygens was born with a silver spoon in his mouth, Boyle was born with a canteen of cutlery in his. Most accounts of Robert Boyle's life mention that he was the fourteenth child (and seventh son, but one had died at birth) of the Earl of Cork, the richest man in the British Isles at the time. However, few of those accounts make it clear that

1. The Peace of Nijmegen was not the end of the story. Louis XIV revoked the Edict of Nantes in 1685, and war broke out again in 1688, lasting for nine years this time, with England now involved on the Dutch side (Charles II had been succeeded in 1685 by his brother, James II, who was a Catholic, as a result of which William of Orange became William III of England, ruling jointly with his wife Mary, the daughter of James II, after her father was forced to abdicate in 1689).

the Earl was not born a member of the aristocracy, but was a self-made man driven by a burning desire to make his fortune and find a respected place in society, an Elizabethan adventurer who made the big time through a combination of luck and skill. He had started life (on 13 October 1566) as plain Richard Boyle, from a gentlemanly but not prominent family. He attended the King's School in Canterbury in the early 1580s, at the same time as Christopher Marlowe, who was two years older than him, then studied at Cambridge. He began to study law in the Middle Temple, but ran out of money and worked as a lawyer's clerk in London before setting off to make his fortune in Ireland (then a colony of England) in 1588, the year of the Spanish Armada and the year he turned 22. Since his father had died long before, and his mother died in 1586, he had to make his own way in life.

According to his own account, Richard Boyle arrived in Dublin carrying £27 3s in cash, a diamond ring and gold bracelet given to him by his mother, and carrying in his bag a new suit and cloak, plus some underwear, to augment the taffeta doublet, black velvet breeches, cloak, dagger and rapier that he wore. He probably also had a hat, although he doesn't mention this. As an intelligent, educated and ruthlessly upwardly mobile young man, Boyle found work in the government department which dealt with land and property that had been seized by the Crown during the reconquest of Ireland that had just about been completed at that time. Huge areas of the country had been seized and given (or sold) to high-ranking Englishmen, while in other cases Irish landowners had to establish proof of their ownership of their own property. Bribes and gifts to officials such as Boyle were routine, while the nature of the work gave him inside information about land that became available at knockdown prices. But even bargains had to be paid for. After seven years of failing to make a fortune, in 1595, Richard married a wealthy widow, who owned land that brought in £500 a year in rent, and set about using this money to make further investments, which eventually prospered beyond even his own wildest dreams, although his wife died giving birth to a stillborn son in 1599.

Before he finally secured his position, Richard Boyle suffered a setback when he lost much of his property in the Munster rebellion of

1598, and had to flee to England; about that time he was also arrested on charges of embezzlement, but acquitted (he was probably guilty, but clever enough to cover his tracks) in a trial presided over by Queen Elizabeth and her Privy Council. Boyle's successful defence of his own case impressed the Queen, and when a new administration was established in Ireland he was appointed Clerk to the Council, the key position in the day-to-day administration of the country. The key purchase that changed his life came in 1602, when he bought derelict estates in Waterford, Tipperary and Cork at a knockdown price from Sir Walter Raleigh, who had so neglected them that they were losing money, and by good management turned them around to make a very healthy profit. Along the way, he built schools and almshouses, new roads and bridges, and even founded whole new towns, establishing himself as one of the most enlightened of the English landlords in Ireland at the time.

By 1603 Richard Boyle had risen so high that he married Catherine Fenton, the 17-year-old daughter of the Secretary of State for Ireland, and received a knighthood on the same day. Catherine produced no less than fifteen children, who were married off, as they became old enough, to provide the most advantageous connections for the family that Sir Richard (who became the first Earl of Cork in 1620, largely thanks to a 'gift' of £4000 in the right quarters) and his money could arrange. The most impressive of these occasions occurred when 15-year-old Francis Boyle married Elizabeth, the daughter of Sir Thomas Stafford, gentleman usher to the Queen, Henrietta Maria (the sister of Louis XIII, remember). King Charles I gave the bride away, the Queen assisted the bride in preparing herself for bed, and both King and Queen stayed to see the young couple into bed together.

Although the various marriages achieved their objective in tying the nouveau riche (not that that was any stigma at the time) Boyles in to established society, they were by no means all successful in personal terms. The only two children to escape this fate were Robert, the Earl's youngest son (born on 25 January 1627, when his mother was 40 and his father 61), and Margaret, Robert's younger sister – and they only did so because the Earl died before they were quite old enough for marriages to be arranged (in Robert's case, the Earl had got as far as choosing a bride for his son but died before the wedding could be organized). Neither of them ever married, not least because of their close-up views of the matrimonial fates of their siblings.

Life as the son of Richard Boyle was never easy physically, even if it ensured total financial security. The father was determined that in spite of their wealth, his male children in particular should not be brought up soft, and to that end sent each of his sons in turn, as soon as they were old enough to leave their mother, off to live with some carefully chosen country family to toughen them up. In Robert's case, this meant that after leaving home as a baby he never saw his mother again, since she died in her mid-forties, when he was 4, a year before he returned home. From the age of 5 until he was 8, Robert lived with his father and those of his siblings who had not yet been married off (a steadily declining number), being taught the basics of reading, writing, Latin and French. He was then deemed ready for the next stage of the toughening process and was sent off to England (with his slightly older brother Francis) to study at Eton, where the Provost was Sir Henry Wotton, former Ambassador to Venice and an old friend of the Earl. Robert took so readily to the academic life that he had to be forced to leave his studies from time to time to take part in the games which were, even then, very much part of the Eton experience, but which he loathed. His studies were also interrupted by recurring illness, which was to plague him all his life.

When Robert was 12, his father bought the manor of Stallbridge, in Dorset, as an English base, and took Francis and Robert to live with him there – Francis actually lived in the manor itself, but Robert, although acknowledged as his father's favourite son (or perhaps because of this?), was boarded out with the parson to encourage him to study rather than idle his hours away. He seemed destined to move on to university, but when Francis was married off to Elizabeth Stafford (later to be known as 'Black Betty' for her beauty and to gain fame (or notoriety) at court, where she became a mistress of Charles II and had a daughter by him), Robert's life changed dramatically. Unwilling as ever to let his sons enjoy anything that might be regarded as idle pleasure, four days after the wedding the Earl sent 15-year-old bridegroom Francis off to France, accompanied by a tutor and his brother Robert. We are told that 'the bridegroom [was] extremely afflicted to be so soon deprived of a joy which he had tasted but just enough to increase his regrets, by the knowledge of what he was forced from'.[1] But there was no arguing with a father like the first Earl of Cork.

After travelling through France via Rouen, Paris and Lyons, the small party settled in Geneva, where Robert finally found a sport he enjoyed (tennis), but continued to study with his old enthusiasm, whatever his surroundings. In 1641, Francis, Robert and their tutor set off on a visit to Italy (funded by the Earl with an incredible budget of £1000 a year), and were actually in Florence when Galileo died.[2] The fuss that was made about this event in Florence roused the young Boyle's curiosity, and he began reading widely about Galileo and his work; this seems to have been a key event in deciding the young man to develop an interest in science.

But back home circumstances were changing dramatically. Although the Earl of Cork was almost a model landlord, most of his English counterparts treated the Irish so harshly that some sort of rebellion was inevitable, and it happened to break out in 1641.[3] Model landlord or not, the Earl could not escape the hostility of the Irish for all things English, and when fighting began (virtually a civil war) all of the Earl's revenue from his huge estates in Ireland was cut off at a stroke. When the Boyle brothers completed their Italian adventure and arrived in Marseilles, they received their first news of the rebellion, a letter telling them that the allowance of £1000 a year was ended and a promise of just £250 (actually still a large sum of money) to pay for their immediate return home. Even the £250 never reached them, though; it seems to have been stolen by the man entrusted by the Earl with delivering it to his sons. In the circumstances, the elder boy, Francis, was sent to make his way back as best he could to assist his

1. Quoted by Pilkington.

2. It was in Florence that the 15-year-old Robert Boyle was introduced to the delights of the brothel (as a spectator) by his tutor, as part of his all-round education. The experience, and another occasion when he was the object of the 'preposterous courtship of two friars, whose lust makes no distinction of sexes', seems to have turned him off sex for life. Boyle's celibacy has led to the inevitable questions about his own sexual orientation, but his description of the two monks as 'gowned sodomites' with 'goatish heats' suggests that he was certainly not that way inclined.

3. No less an authority than Oliver Cromwell is said to have remarked that there would have been no Irish rebellion 'if there had been an Earl of Cork in every province'.

father and brothers (assist them in battle, that is), while the younger Robert stayed with their tutor in Geneva. By the time the fighting ended in 1643, the Earl of Cork, once the richest man in the realm of England, was ruined, and two of his sons had died in battle (Francis distinguished himself in the fighting and survived). The Earl himself soon followed them to the grave, a month short of his seventy-seventh birthday. The following year, Robert returned to England at the age of 17, not only penniless but honour-bound to repay the expenses his tutor had incurred for him in Geneva and assisting with his journey home. As if that weren't enough, although the fighting in Ireland had come to an end, civil war had broken out in England.

The causes of the English Civil War are many and complex, and historians still argue about them. But one of the most important factors in triggering the conflict at the time it actually happened was the rebellion in Ireland which cost the Boyle family so dear. Charles I (who had succeeded his father, James I, in 1625) and his Parliament had long been at loggerheads, and when an army had to be raised to quell the Irish rebellion they disagreed over who should raise the army and who should control it. The outcome was that Parliament established a militia to be under the control of lords lieutenant appointed by Parliament, not by the King. Since the King would not agree to this, the necessary legislation, the Militia Ordinance of 1642, was passed without bothering with the detail of getting the King's signature. On 22 August that year, the King raised his standard at Nottingham, rallying his followers against Parliament. In the battles that followed, Oliver Cromwell came to prominence as a leader of the Parliamentary forces and the first phase of the war came to an end after the defeat of the King's forces at the Battle of Naseby in June 1645 and the fall of Oxford to Parliamentary forces in June 1646. The King himself fell into Parliamentary hands in January 1647.

The peace was short-lived, since Charles escaped from custody on the Isle of Wight in November, rallied his forces and reached a secret agreement with the Scots, offering concessions to the followers of Presbyterianism if he was restored to the throne, before he was recaptured. The Scots tried to carry out their part of the bargain, but in August 1648 their forces were defeated at Preston, and on 30 January the following year Charles I was executed. From 1649 to 1660,

England had no King and was governed by Parliament until 1653 and by Cromwell as Lord Protector from then until his death in 1658. Things then unravelled, like a film of the events of the previous two decades being run backwards at high speed. Having gone to all the trouble of getting rid of a system of hereditary monarchy, England then found Richard Cromwell, the son of Oliver, to be his heir as Lord Protector, but he was ousted by the army in favour of a return of the remnant of the members of the Parliament of 1653, and since nobody else seemed acceptable to act as head of state, Charles II, who had been exiled in France, was restored to the throne in 1660. Although after the Civil War the balance of power in England had clearly tilted more than a little in favour of Parliament and away from the King, it looks from a distance of nearly 350 years like a modest result to have been achieved with so much effort.

At the time Robert Boyle returned, England was more or less partitioned, with parts held by Royalists (with their headquarters in Oxford) and parts (including London and the southeast) by Parliament. But for many people life proceeded without too much disruption, except in the regions where the pitched battles were fought. The youngest son of the first Earl of Cork, though, was not most people. The family were clearly identified as friends of the King, and it might have proved difficult for Robert to follow his natural instinct to keep his head down and avoid involvement in the conflict if it had not been that one of the marriages arranged by his father paid off handsomely. One of Robert's sisters, Katherine (who happened to be his favourite sibling, although she was thirteen years older than him), had been married off to a young man who had now inherited the title Viscount Ranelagh, and although the marriage was a personal disaster and the couple no longer lived together, the Viscount's sister (who remained on good terms with Katherine) was married to a prominent member of the Parliamentary side and Katherine herself was a Parliamentary sympathizer, who often entertained Parliamentarians in her London home. That home provided the first refuge for Robert when he returned to England (where he met, among others, John Milton), and it was thanks in no small measure to Katherine's connections that he was able to keep the manor at Stallbridge, which had been left to him by his father, after the defeat of the King's forces in the Civil War.

In 1645, Boyle did indeed retire to his house in the country, kept his head down as far as politics was concerned and, with a modest (by the family's standards) income trickling in from his estates in spite of the war, was able to read widely (including a thorough study of the Bible), write (on a variety of topics including philosophy, the meaning of life and religion) and carry out his own experiments, which at this time were mainly alchemical. Many letters to Katherine provide a window on his life in Dorset, and in a letter to another friend he mentions seeing an air gun which could use the force of compressed air to fire a lead ball capable of killing a man at a distance of thirty paces – an observation which clearly set him thinking along the lines that were to lead to his discovery of Boyle's law. Katherine herself was an independent, intelligent woman, and her house in London became a meeting place for many of the intellectuals of the day, including members of a group of men interested in science, who began calling themselves the 'invisible college'. This was the forerunner of the Royal Society, and it was through Katherine that Robert began to make the acquaintance of these men on his visits to London. In its early years (around the mid-1640s) the group often met at the not-so-invisible Gresham College in London. The college had been founded in 1596 by Sir Thomas Gresham, a financial adviser to Queen Elizabeth, as the first seat of advanced learning in England outside Oxford and Cambridge. It never quite rivalled those institutions, but this was still a significant step in the spread of learning in England. The focus of the activities of the invisible college shifted to Oxford, however, when several of its prominent members took up posts there in 1648, as the Civil War drew to a close.

In 1652, with the political situation seemingly stabilized, Boyle visited Ireland, accompanied by the physician William Petty, to look into his interests in the family estates there. The prospects for the family had improved politically because one of Robert's brothers, now Lord Broghill, had played a major part in putting down the Irish rebellion, which was bound to gain favour with whoever was ruling England at the time – the last thing Cromwell wanted was trouble in Ireland. But in all the upheavals of the 1640s there had been no opportunity to re-establish the proper flow of income from the estates. Boyle spent the best part of two years in Ireland, benefiting intellectually from a close association with Petty (who taught him anatomy and physiology, and how to dissect, as well as discussing the scientific method with Boyle), and financially, since on his return to England he was guaranteed a share of the income from his late father's estates bringing him in more than £3000 a year for life, enough for him to do anything that he pleased.[1] It pleased him, in 1654 at the age of just 27, to move to Oxford, then the centre of scientific activity in England (possibly in the entire world), where over the next fourteen years he carried out the scientific work which has made him famous. The manor at Stallbridge was passed on to the family of his brother Francis.

Not that Boyle had any need to carry out all of the experiments himself. His huge income enabled him to employ assistants (including a certain Robert Hooke, of whom much more shortly) and run what amounted to a private research institute which would be the envy of many scientists today. The money also meant that Boyle could, unlike many of his contemporaries, get his own books published at his own expense, ensuring that they appeared promptly, in properly printed editions. Since he paid his bills on time, the printers loved him and took extra care with his work.

With his scientific establishment, Boyle was one of the pioneers in the application of the scientific method, following on from practical men such as Galileo and Gilbert, who had done their own experiments, but also drawing inspiration from the more philosophical work of Francis Bacon (1561–1626), who didn't actually do much experimenting himself,[2] but whose writings about the scientific method had a great influence on the generations of British scientists that followed him. Bacon spelled out the need to begin an investigation by collecting as much data as possible, and to proceed by trying to explain the observations – not by dreaming up some wonderful idea and then

1. There is a certain irony in this, since Boyle became an absentee landlord, one of the hated symbols of the English oppression of Ireland. Yet he was, by the standards of the day, a liberal, who once wrote to a friend complaining about the cost of the funeral of a member of the aristocracy that it would have been better to spend the money on the poor.

2. His most famous experiment, rushing out into the cold of a freezing day, at the age of 65, to stuff a chicken with snow to see if this would preserve it, actually killed him, since he developed pneumonia as a result.

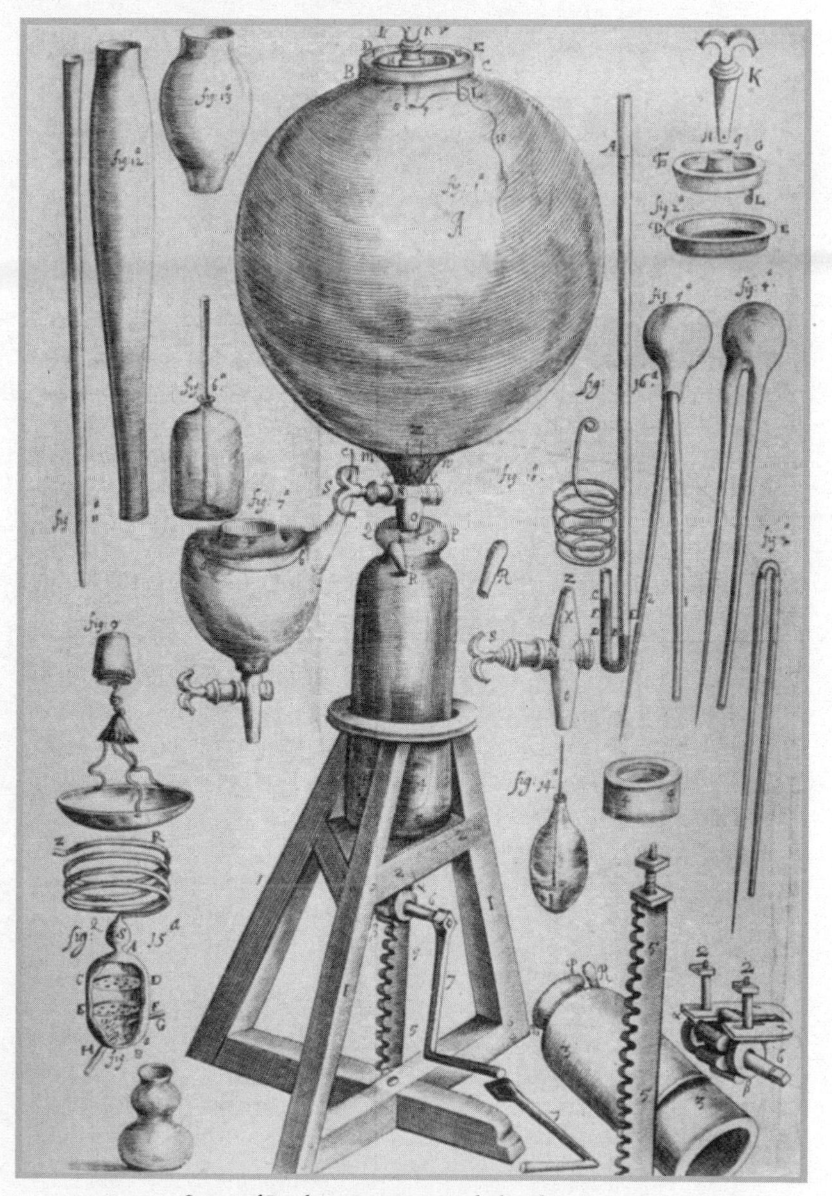

11. *Some of Boyle's apparatus, including his air pump.*

looking for facts to support it. If the Baconian system can be summed up in a sentence, it is that science must be built on the foundations provided by facts – a lesson that Boyle very much took to heart. Writing about Galileo's work on falling bodies, and the discovery that bodies with different weights fall at the same speed, Boyle later held this up as an example of how as scientists 'we assent to experience, even when its information seems contrary to reason'.[1]

It was six years before Boyle published anything on science, but when he did it was well worth the wait. His first significant contribution to science concerned the springiness, or compressibility, of air, and involved the most famous experiment of his illustrious career. In this experiment, he (or his assistants) took a tube of glass shaped like a letter J, with the top open and the short end closed. Mercury was poured into the tube to fill the U bend at the bottom, sealing off the air in the shorter arm of the tube. When the mercury was at the same level on both sides of the tube, the air in the closed end was at atmospheric pressure. But by pouring more mercury into the long arm of the tube, the pressure could be increased, forcing the air in the short arm to contract. Boyle found that if the pressure was doubled, the volume of trapped air was halved; if the pressure was tripled the volume was reduced to a third; and so on. Equally significantly he found that the process was reversible. After being compressed, air would spring back, given an opportunity. All of this can be explained very well in the atomic model of the world, but only with much greater difficulty using Cartesian vortices.

Much of this work (and more, dealing with things like air pumps and the problem of raising water by suction) was published in 1660 in his book *New Experiments Physico-Mechanicall, Touching the Spring of the Air and its Effects*. Usually referred to simply as *The Spring of the Air*, the first edition did not explicitly state what is now known as Boyle's law, that the volume occupied by a gas is inversely proportional to the pressure on it (other things being equal), which was spelled out in a second edition published in 1662. Boyle's work with the vacuum (strictly speaking, very low pressure air) was carried out using an improved air pump, based on the ideas of Otto von Guericke, that he

1. Quoted in Hunter, from Boyle's *Christian Virtuoso*.

12. *Experiment carried out at Magdeburg, in Germany, in 1654. Sixteen horses were unable to separate the two halves of an evacuated sphere, against the pressure of the atmosphere. From von Guericke's* Experimenta Nova, *1672.*

designed and built with Hooke. Whereas von Guericke's pump needed two strong men to operate it, their design could be operated with reasonable ease by one man. Boyle repeated all of von Guericke's experiments and went on to show that water boiled at a lower temperature when the air pressure was reduced (no mean achievement, since it involved setting up a mercury barometer inside the sealed glass vessel so that the fall in pressure could be monitored as air was pumped out). Boyle also came close to discovering oxygen by showing that life, like a flame, depended on the presence of air to sustain it, and specifically pointed out that there were essential similarities between the processes of respiration and combustion. Some of these experiments were not for the squeamish, but they certainly made people sit up and take notice. One of Boyle's colleagues in the invisible college (we are not sure who) composed a 'ballad' about a demonstration of the scientific work of the group, which includes the lines:

> To the Danish Agent late was showne
> That where noe Ayre is, there's no breath.
> A glasse this secret did make knowne
> Wherein a Catt was put to death.
> Out of the glasse the Ayre being screwed,
> Puss dyed and ne'er so much as mewed.
>
> The self same glasse did likewise clear
> Another secret more profound:
> That nought but Ayre unto the Eare
> Can be the Medium of Sound
> For in the glasse emptied of Ayre
> A striking watch you cannot hear.

It may not be great poetry, but it gives you a feel for how impressed the scientific world was with Boyle's discoveries.[1] But the fact that the book was published in English and written in clear, accessible prose was almost as important as what it contained. Like Galileo, Boyle was bringing science to the masses (or at least to the middle classes; in his famous diary, Samuel Pepys writes enthusiastically about the pleasure of delving into one of Boyle's new books). Unlike Galileo, though, he did not have to fear that this might upset the Inquisition.

Boyle's scientific approach to alchemy

In 1661, between the first two editions of *The Spring of the Air*, Boyle published his most famous book, *The Sceptical Chymist*. The extent of Boyle's involvement with alchemy after he left Dorset is still a matter of debate, and Lawrence Principe of Johns Hopkins University has made a persuasive case that Boyle was not so much trying to discard alchemy in favour of what we would now call chemistry, but that he was trying to bring the Baconian method into alchemy – to make alchemy scientific, as it

1. If the fate of Puss seems somewhat cruel, remember that this was still a time when human beings were put to death by burning. But I do think the poet might have regarded the death of the cat as more profound than the muffling of the chime of a watch.

THE
SCEPTICAL CHYMIST:
OR
CHYMICO-PHYSICAL
Doubts & Paradoxes,
Touching the
SPAGYRIST'S PRINCIPLES
Commonly call'd
HYPOSTATICAL;
As they are wont to be Propos'd and
Defended by the Generality of
ALCHYMISTS.
Whereunto is præmis'd Part of another Discourse
relating to the same Subject.

BY
The Honourable *ROBERT BOYLE*, Esq;

LONDON,
Printed by *J. Cadwell* for *J. Crooke*, and are to be
Sold at the *Ship* in St. *Paul's* Church-Yard.
MDCLXI.

13. *Title page of Robert Boyle's* The Sceptical Chymist, *1661.*

were. This certainly fits his place in history as a seventeenth-century man of science (even Isaac Newton, as we shall see, was seriously involved in alchemical work at the end of the seventeenth century), and it would be wrong to claim that Boyle's book transformed alchemy into chemistry overnight. Indeed, it had much less initial influence than *The Spring of the Air*. But as chemistry developed in the eighteenth and nineteenth centuries, people began to look back on Boyle's book as a turning point. The fact is that applying the scientific method to alchemy did, in the end, turn alchemy into chemistry and remove any rational basis for belief in things like the Philosophers' Stone, supposed to turn base metal into gold. And Boyle was a leading light in establishing the scientific method in England.

As an example of how Boyle approached alchemy in a scientific way, he took issue with the idea that gold could be made by removing impurities from other metals. Since gold is more dense than these other metals, he argued, how could it be made by taking something away from them? He did not, you notice, say that transmutation was impossible; but he approached the problem scientifically. He did, though, say that it was impossible to accept the old idea of the world being composed of the four Aristotelian 'elements' – air, earth, fire and water – mixed in different proportions, having carried out experiments that disproved it. Instead, he espoused a form of the atomic hypothesis, that all matter is composed of some form of tiny particles which join together in different ways – an early version of the idea of elements (in the modern meaning of the term) and compounds. 'I now mean by elements,' Boyle wrote, 'certain primitive and simple bodies, which not being made of any other bodies or of one another, are the ingredients of which all those called perfectly mixed bodies are immediately compounded and into which they are ultimately resolved.' This was a theme he developed in his book *Origin of Forms and Qualities*, published in 1666, suggesting that these atoms can move about freely in fluids but are at rest in solids, and that their shapes are important in determining the properties of the material objects they composed. He saw the principal role of chemistry as finding out what things are made of, and coined the term 'chemical analysis' to describe this process.

All of this represents just a small part of Boyle's work, although it is the part that is most relevant to the story of the development of science in the seventeenth century. To pick a few other samples more or less at random, he invented the match, went one better than Bacon by using cold to preserve meat without catching a chill and proved by experiment that water expands on freezing. He was also a major literary figure of the Restoration period, writing on many subjects, including works of fiction. But although he became the most respected scientist of his time, Boyle retained his retiring, modest nature and declined many honours. Like his three surviving brothers, Robert Boyle was offered a peerage after the restoration of Charles II (who, remember, numbered among his lovers the wife of Francis Boyle).[1] But unlike them, he declined. His esteem as a theologian was such that he was asked (by the Lord Chancellor of England) to take holy orders, with the promise of being fast-tracked to a bishopric, but he said no, thank you. He was offered the Provostship of Eton, but turned it down. And when elected President of the Royal Society, in 1680, he expressed his regrets that he could not take up the post because his personal religious beliefs prevented him from swearing the necessary oaths. Throughout his life, he remained The Honourable Robert Boyle, Esquire; and he spread his almost indecently large income widely in charitable donations (he also left most of his property to charity when he died).

When the Royal Society received its charter in 1662, Boyle was not only one of the first members (or Fellows, as they were called) but one of the first members of the council of the society. Partly because the centre of activity in science in England became closely associated with the Royal Society in London in the 1660s, but also to be with his sister, Boyle moved to the capital in 1668 and took up residence with Katherine. His great days of scientific research were behind him (although he carried on experimenting), but he remained at the centre of the scientific scene, and Katherine's house remained a meeting place for intellectuals. One of his contemporaries was John Aubrey, who describes him at this time as:

1. Even the second Earl of Cork received an English title, the Earl of Burlington, to add to his collection; his London residence, Burlington House, is now the home of the Royal Academy and several scientific societies.

> Very tall (about six foot high) and streight, very temperate, and vertuouse, and frugall: a batcheler; keeps a coach; sojournes with his sister, the lady Ranulagh. His greatest delight is chymistrey. He haz at his sister's a noble laboratory, and severall servants (prentices to him) to looke to it. He is charitable to ingeniose men that are in want.

But Boyle's health had never been good, and as the diarist John Evelyn, an old friend, described his appearance in his later years:

> The contexture of his body, during the best of his health, appeared to me so delicate that I have frequently compared him to a chrystal, or Venice glass; which, though wrought never so thin and fine, being carefully set up, would outlast the hardier metals of daily use. And he was withal as clear and candid; not a blemish or spot to tarnish his reputation.

The Venice glass lasted only as long as its companion. Just before Christmas in 1691, Katherine died; Robert Boyle followed her a week later, on 30 December, a month short of his sixty-fifth birthday. After his funeral, on 6 January 1691, Evelyn wrote in his diary: 'It is certainly not onely England, but all the learned world suffred a public losse in this greate & good man, & my particular worthy friend.'

Boyle's experiments which showed that both fire and life depended on something in the air, tie his work in with another main thread of scientific development in the second half of the seventeenth century, the biological investigation of human beings and other living organisms in the wake of the work of Harvey and Descartes. As is so often the case in science, the new developments went hand in hand with new developments in technology. Just as the telescope revolutionized the way people thought about the Universe, the microscope revolutionized the way people thought about themselves; and the first great pioneer of microscopy was the Italian physician Marcello Malpighi, born at Crevalcore, near Bologna, probably on 10 March 1628 (the day he was baptized).

Marcello Malpighi and the circulation of the blood

Malpighi studied philosophy and medicine at the University of Bologna, graduating in 1653, and became a lecturer in logic at Bologna, before moving to the University of Pisa in 1656 as professor of theoretical medicine. But the climate in Pisa did not suit him, and in 1659 he returned to Bologna to teach medicine. In 1662, he moved again, to the University of Messina, but in 1666 he became professor of medicine in Bologna and stayed there for the next 25 years. In 1691, Malpighi moved to Rome, where he retired from teaching but became the personal physician to Pope Innocent XII (apparently reluctantly, but at the Pope's insistence); he died there on 30 November 1694.

From 1667 onwards, a great deal of Malpighi's work was published by the Royal Society in London, a sign of just how significant the Royal had already become (Malpighi was the first Italian to be elected a Fellow of the Society, in 1669). This work was almost entirely concerned with microscopy and dealt with a variety of subjects, including the circulation of blood through the wing membranes of a bat, the structure of insects, the development of chick embryos and the structure of stomata in the leaves of plants. But Malpighi's greatest contribution to science was made as a result of work he carried out in Bologna in 1660 and 1661, and reported in two letters that were published in 1661.

Before that time, following the discovery of the circulation of the blood, it had been widely thought that blood from the heart flowing to the lungs actually emerged from tiny holes in blood vessels into the spaces filled with air inside the lungs, mixed somehow with the air (for reasons that remained unclear) and then somehow found its way back through tiny holes into other blood vessels to return to the heart. Through his microscopic studies of the lungs of frogs, Malpighi found that the inner surface of the lungs is actually covered by tiny capillaries, very close to the surface of the skin, through which the arteries on one side are directly connected to the veins on the other side. He had found the missing link in Harvey's description of blood circulation, a link which Harvey himself had guessed must be there, but could not detect with the instruments at his disposal. 'I could clearly see,' wrote Malpighi, 'that the blood is divided and flows through tortuous vessels, and that it is not poured into spaces but is always driven through tubules and distributed by the manifold bendings of the vessels.' A few years later, the same discovery was made independently by the Dutch microscopist Antoni van Leeuwenhoek (of whom more in Chapter 5), unaware of Malpighi's work.

Soon after Malpighi's discovery, Richard Lower (1631–1691), a member of the Oxford group that was to become the nucleus of the Royal Society, demonstrated in a series of experiments (including the rather simple one of shaking up a glass vessel containing venous blood and watching the dark purplish blood change colour to bright red as it mixed with air) that the red colour of blood flowing from the lungs and heart around the body is produced by something in the air:

> That this red colour is entirely due to the penetration of particles of air into the blood is quite clear from the fact that, while the blood becomes red throughout its mass in the lungs (because the air diffuses in them through all the particles, and hence becomes thoroughly mixed with the blood), when venous blood is collected in a vessel its surface takes on this scarlet colour from exposure to air.[1]

From research such as this (Boyle and Hooke were among the other people who carried out similar experiments), the Oxford group began to regard blood as a kind of mechanical fluid, which carried essential particles from food and from the air around the body. This was very much in line with the Cartesian image of the body as a machine.

Giovanni Borelli and Edward Tyson: the increasing perception of animal (and man) as machine

This theme of the body as a machine was developed in the seventeenth century by another Italian, Giovanni Borelli, who was an older contemporary and friend of Malpighi. Malpighi seems to have stimulated Borelli's interest in living things, while Borelli seems to have stimulated Malpighi to investigate the way living systems work and encouraged his efforts at dissection. Both achieved more than they might have done had they not met.

Born at Castelnuovo, near Naples, on 28 January 1608, Borelli studied mathematics in Rome and became professor of mathematics at Messina some time before 1640, although the exact date is not known. He met Galileo at his house outside Florence in the early 1640s, and became professor of mathematics at the University of Pisa (Galileo's old post) in 1656, where he met Malpighi. Both were founder members of the short-lived Accademia del Cimento, founded in Flor-

1. Lower, quoted in Conrad *et al.*

ence the following year, and Borelli studied anatomy around this time. Borelli returned to Messina in 1668, but in 1674 he was involved (or suspected of involvement) in political intrigues which led to him being exiled to Rome, where he became part of a circle associated with ex-Queen Christina of Sweden (the one who had made Descartes get out of bed at such an unearthly hour). Christina had been forced to abdicate in 1654, having become a Catholic, and was also living in exile in Rome. Borelli died in Rome on 31 December 1679.

Although Borelli was a distinguished mathematician who was the first person to suggest that the trajectory of a comet past the Sun follows a parabolic path, and who tried to explain the movement of the moons of Jupiter by postulating that Jupiter exerts an influence on its satellites similar to the influence the Sun exerts on the planets, his most important scientific work was in the biological field of anatomy. This work was largely carried out while he was in Pisa, but was still only in manuscript form when Borelli died; the resulting book *De Motu Animalium* (*On the Movement of Animals*) was published after he died, in two volumes, in 1680 and 1681. Borelli treated the body as a system of levers acted on by the forces exerted by the muscles, and analysed geometrically how muscles in the human body acted in walking and running. He also described the flight of birds and the swimming motion of fish in mathematical terms. But the crucial point is that he did not seek a special place for humankind, distinct from that of other animals. The *human* body was likened to a machine made up of a series of levers. Borelli still saw a role for God in setting up the system in the first place – as the designer of the machine, if you like. But this was very different from the idea of a human body being operated by some kind of guiding spirit which controlled its activities from minute to minute.

The relationship between man (as they would have put it at the time) and the animals was spelled out by a remarkable (if slightly fortuitous) piece of dissection carried out by Edward Tyson, in London, right at the end of the seventeenth century. Tyson was born at Clevedon, in Somerset, in 1650 (the exact date is not known) and was educated both at the University of Oxford (where he obtained his BA in 1670 and MA in 1673) and in Cambridge, where he obtained his medical degree in 1677. He then moved to London, where he practised as a doctor but carried out anatomical observations and dissections, publishing much of his work in the *Philosophical Transactions* of the Royal Society, having been elected as a Fellow in 1679. As one of the leading physicians of his time (he became a Fellow of the Royal College of Physicians), in 1684 Tyson was appointed physician and governor to the Bethlehem Hospital in London. This was a mental institution that gave us the word 'bedlam', from the common pronunciation of its name – and that word gives you an accurate idea of the kind of place it was when Tyson took up his appointment. Although the first asylum for the insane in Britain (and the second in Europe, after one founded in Granada, in Spain), it was hardly a rest home. The mentally ill were abused in almost every conceivable way and were treated as a kind of entertainment, with 'Bedlam' a place for fashionable people to go to view the curiosities, rather like a zoo. Tyson was the person who began to change all that, introducing women nurses to care for the patients instead of the male nurses who were in reality jailers, establishing a fund to provide clothing for the poorer inmates and carrying through other reforms. In human terms, this was Tyson's greatest achievement. He died in London on 1 August 1708.

In scientific terms, though, Tyson is regarded as the founding father of comparative anatomy, which looks at the physical relationships between different species. One of his most memorable dissections took place in 1680, when an unfortunate porpoise found its way up the river Thames and ended up with a fishmonger who sold it to Tyson for seven shillings and sixpence (a sum recouped by him from the Royal Society). Tyson dissected the supposed 'fish' at Gresham College, with Robert Hooke on hand to make drawings as the dissection progressed, and was astonished to discover that the animal was in fact a mammal, with an internal structure very similar to that of the quadrupeds that lived on land. In his book *Anatomy of a Porpess*, published later that same year, he presented the discovery to an astonished world:

> The structure of the *viscera* and inward parts have so great an Analogy and resemblance to those of Quadrupeds, that we find them here almost the same. The greatest difference from them seems to be in the external shape, and

wanting feet. But here too we observed that when the skin and flesh was taken off, the fore-fins did very well represent an Arm, there being the *Scapula*, an *os Humeri*, the *Ulna*, and *Radius*, and bone of the *Carpus*, the *Metacarp*, and 5 *digiti* curiously jointed . . .

This hinted – more than hinted – at closer relationships between animals than their external appearances might suggest. Tyson carried out many other famous dissections, including a rattlesnake and an ostrich. But his most famous was that of a young chimpanzee (mistakenly described as an Orang-Outang) which was brought to London as a pet by a sailor in 1698. The young chimp had been injured on the voyage from Africa, and was clearly ailing; word soon reached the famous anatomist, who seized the opportunity to study the chimp's appearance and behaviour while it remained alive, and to dissect it (this time with William Cowper[1] to assist with the drawings) as soon as it died. The fruits of their labours appeared in a book with the splendid title *Orang-Outang, sive Homo Sylvestris: or, the Anatomy of a Pygmie Compared with that of a Monkey, an Ape, and a Man.* The heavily illustrated volume, just 165 pages long, presented incontrovertible evidence that humans and chimpanzees were built to the same body plan. At the end of the book, Tyson listed the most significant features of the chimp's anatomy, noting that 48 of them resembled the equivalent human features more than they did those of a monkey, and 27 more closely resembled those of a monkey than those of a human. In other words, the chimpanzee resembled a human being more than it resembled a monkey. He was particularly struck by the way in which the brain of the chimpanzee resembled (apart from its size) the brain of a human being.

The element of luck in Tyson's analysis lies in the fact that the specimen he examined was a young chimp, and human beings resemble infant chimpanzees even more than they do adult chimpanzees. There is a sound reason for this, although it was not understood until quite recently – one of the ways in which evolution can produce variations on old themes is to slow down the process of development, something known as neoteny (meaning holding on to youth). People develop

1. Cowper lived from 1666 to 1709 and was a surgeon and Fellow of the Royal Society.

much more slowly than chimpanzees and other apes do, so that we are born in a relatively undeveloped state – which is one reason why human infants are so helpless, but also why they are capable of learning so many different kinds of things – instead of arriving in the world pre-programmed for a specific role (such as swinging through the trees). But that is getting ahead of my story. In 1699, what mattered was that with the publication of Tyson's book the place of human beings as part of the animal kingdom was clearly established, setting the agenda for the centuries of work that would lead to an understanding of exactly how we fit into that kingdom. That, of course, will be a major theme of the later part of this book. Now, though, the time has come to take stock of the work of the man who did more than anyone else to set the agenda for science in the centuries ahead, Isaac Newton, and his close contemporaries.

5

The 'Newtonian Revolution'

Robert Hooke: the study of microscopy and the publication of Micrographia *– Hooke's study of the wave theory of light – Hooke's law of elasticity – John Flamsteed and Edmond Halley: cataloguing stars by telescope – Newton's early life – The development of calculus – The wrangling of Hooke and Newton – Newton's* Principia Mathematica*: the inverse square law and the three laws of motion – Newton's later life – Hooke's death and the publication of Newton's* Opticks

The three people who between them established both the scientific method itself and the pre-eminence of British science at the end of the seventeenth century were Robert Hooke, Edmond Halley and Isaac Newton. It is some measure of the towering achievements of the other two that Halley clearly ranks third out of the trio in terms of his direct contribution to science; but in spite of the Newton bandwagon that has now been rolling for 300 years (and was given its initial push by Newton himself after Hooke had died), it is impossible for the unbiased historian to say whether Newton or Hooke made the more significant contribution. Newton was a loner who worked in isolation and established the single, profound truth that the Universe works on mathematical principles; Hooke was a gregarious and wide-ranging scientist who came up with a dazzling variety of new ideas, but also did more than anyone else to turn the Royal Society from a gentlemen's gossip shop into the archetypal scientific society. His misfortune was to incur the enmity of Newton, and to die before Newton, giving his old enemy a chance to rewrite history – which he did so effectively that Hooke has only really been rehabilitated in the past few decades. Partly to put Newton in his place, but also because Hooke was the first of the trio to be born, I will start with an account of his life and work, and introduce the other two in the context of their relationships with Hooke.

Robert Hooke was born on the stroke of noon on 18 July 1635, seven years before Galileo Galilei died. His father, John Hooke, was curate of the Church of All Saints at Freshwater, on the Isle of Wight; this was one of the richest livings on the island, but the chief beneficiary

Robert Hooke: the study of microscopy and the publication of Micrographia

was the Rector, George Warburton. As a mere curate, John Hooke was far from wealthy, and he already had two other children, Katherine, born in 1628, and John, born in 1630. Robert Hooke's elder brother became a grocer in Newport, where he served as Mayor at one time, but hanged himself at the age of 46 – we do not know exactly why. His daughter Grace, Robert's niece, would feature strongly in Robert's later life.

Robert Hooke was a sickly child who was not expected to live. We are told that for the first seven years of his life he lived almost entirely on milk and milk products, plus fruit, with 'no Flesh in the least agreeing with his weak Constitution'.[1] But although small and slight and lacking in physical strength, he was an active boy who enjoyed running and jumping. It was only later, at about the age of 16, that he developed a pronounced malformation of his body, some sort of twist, which he later attributed to long hours spent hunched-over working at a lathe and with other tools. He became very skilful at making models, including a model ship about a metre long, complete with working rigging and sails; and after seeing an old brass clock taken to pieces he made a working clock out of wood.

Initially because of his ill health, Hooke's formal education was neglected. When it seemed there was a chance of him surviving after all, his father began to teach him the rudiments of education, intending him to have a career in the church. But Robert's continuing ill health and his father's own growing infirmity meant that little progress was made and Robert continued to be left largely to his own devices. When a professional artist visited Freshwater to undertake a commission, Hooke took one look at how he set about his work and decided that he could do that, so after first making his own paints he set about copying any paintings that he could find, with sufficient skill that it was thought he might become a professional painter himself. When Robert's father John Hooke died in 1648, after a long illness, Robert was 13 years old. With an inheritance of £100 he was sent to London to be apprenticed to the artist Sir Peter Lely. Robert first decided that there was not much point in using the money for an apprenticeship

1. From Richard Waller's preface to *The Posthumous Works of Robert Hooke*, published in 1705.

since he reckoned he could learn painting by himself, then found that the smell of the paints gave him severe headaches. Instead of becoming an artist, he used the money to pay for an education at Westminster School, where as well as his academic studies, he learnt to play the organ.

Although he was too young to be involved directly in the Civil War, the repercussions of the conflict did affect Hooke. In 1653, he secured a place at Christ Church College in Oxford as a chorister – but since the puritanical Parliament had done away with such frippery as church choirs, this meant that he had in effect got a modest income – a scholarship – for nothing. One of his contemporaries at Oxford, a man with a keen interest in science, was Christopher Wren, three years older than Hooke and another product of Westminster School. Like many other poor students at the time, Hooke also helped to make ends meet by working as servant to one of the wealthier undergraduates. At this time, many of the Gresham College group had been moved to Oxford by Oliver Cromwell to replace academics tarred with the brush of Oxford's support for the Royalist side in the war, and Hooke's skill at making things and carrying out experiments made him invaluable as an assistant to this group of scientists. He soon became both Robert Boyle's chief (paid) assistant and his lifelong friend. Hooke was largely responsible for the success of Boyle's air pump, and therefore for the success of the experiments carried out with it, and was also closely involved in the chemical work Boyle carried out in Oxford. But Hooke also carried out astronomical work for Seth Ward, the Savilian professor of astronomy at the time (inventing, among other things, improved telescopic sights), and in the mid-1650s he worked out ways to improve the accuracy of clocks for astronomical timekeeping.

Through this work, Hooke came up with the idea of a new kind of pocket watch regulated by a balance spring. This could have been the forerunner of the kind of chronometer that would have been accurate and dependable enough to determine longitude at sea, and Hooke claimed to have worked out how to achieve this. But when (without revealing all his secrets) he discussed the possibility of patenting such a device, negotiations broke down because Hooke objected to a clause in the patent allowing other people to reap the financial benefits of any improvements to his design. He never did reveal the secret of his idea, which died with him. But the pocket watch itself, although no sea-going chronometer, was a significant improvement on existing designs (Hooke gave one of these watches to Charles II, who was well pleased), which would alone have ensured him a place in the history books.

When the Royal Society was established in London in the early 1660s, it needed a couple of permanent staff members, a Secretary to look after the administrative side and a Curator of Experiments to look after the practical side. On the recommendation of Robert Boyle, the German-born Henry Oldenburg got the first job and Robert Hooke the second. Oldenburg hailed from Bremen, where he had been born in 1617, had been that city's representative in London in 1653 and 1654, where he met Boyle and other members of his circle, and was for a time the tutor to one of Boyle's nephews, Lord Dungarvan. His interest in science stirred, Oldenburg became a tutor in Oxford in 1656 and was very much a member of the circle from which the first Fellows of the Royal Society were drawn. He was fluent in several European languages and acted as a kind of clearing house for scientific information, communicating by letter with scientists all over Europe. He got on well with Boyle, becoming his literary agent and translating Boyle's books, but unfortunately he took a dislike to Hooke. He died in 1677.

Hooke left Oxford to take up his post with the Royal in 1662; he had never completed his degree, because of his involvement as an assistant to Boyle and others, but in 1663 he was appointed MA anyway and also elected as a Fellow of the Royal Society. Two years later, his position as curator of experiments was changed from that of a servant employed by the Society to a position held by him as a Fellow and a member of the Council of the Society, an important distinction which marked his recognition as a gentleman supposedly on equal footing with the other Fellows, but which still (as we shall see) left him with a huge burden of duties. The honours were all very well, but to the impoverished Hooke his salary was just as important; unfortunately, the Royal was, in its early days, both disorganized and short of funds, and for a time Hooke was only kept afloat financially by the generosity of Robert Boyle. In May 1664, Hooke was a candidate for

14. (opposite) *A louse. From Hooke's* Micrographia, *1664.*

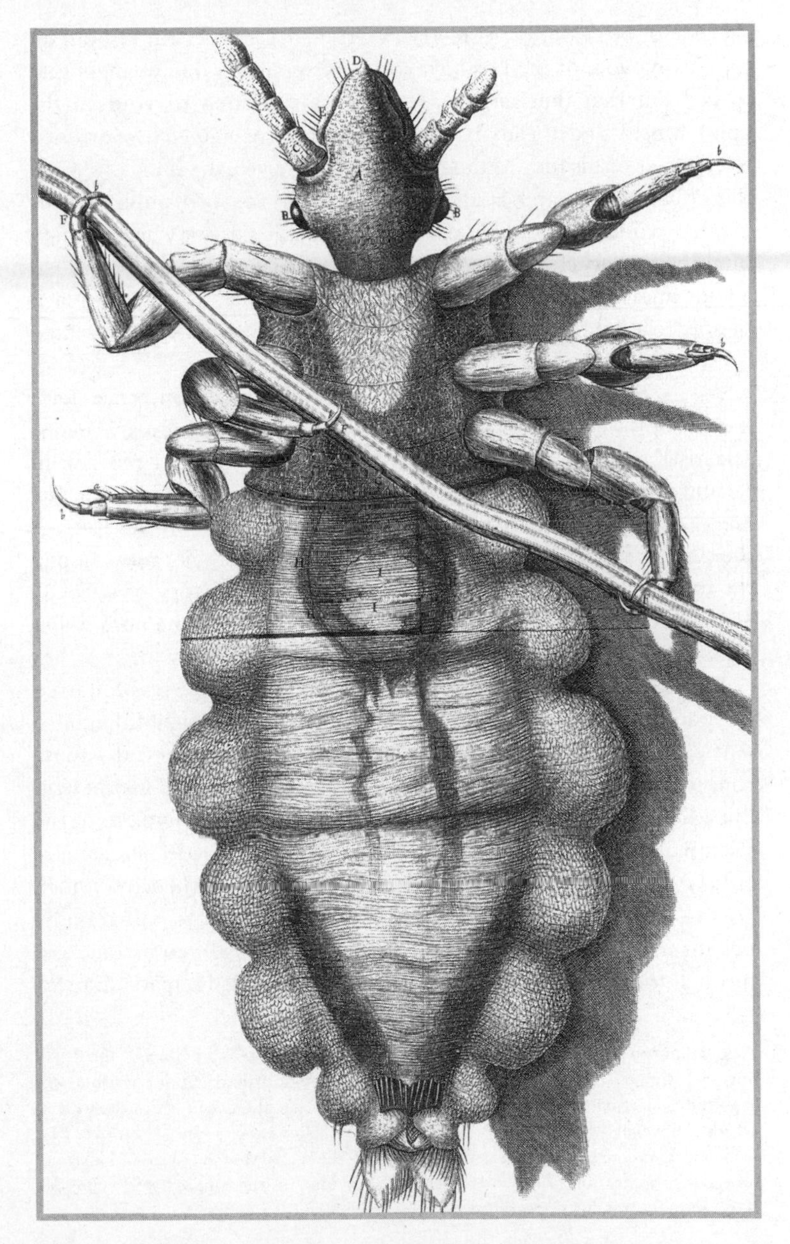

the post of professor of geometry at Gresham College, but lost out on the casting vote of the Lord Mayor. After considerable wrangling, it turned out that the Lord Mayor was not entitled to vote on the appointment, and in 1665 Hooke took over the post, which he retained for the rest of his life. At the beginning of the year he finally took up the appointment, and at the age of 29, Hooke also published his greatest work, *Micrographia*. Unusually for the time, it was written in English, in a very clear and readable style, which ensured that it reached a wide audience, but which may have misled some people into a lack of appreciation of Hooke's scientific skill, since the way he presented his work made it seem almost easy.

As its title suggests, *Micrographia* was largely concerned with microscopy (the first substantial book on microscopy by any major scientist), and it is no exaggeration to say that it was as significant in opening the eyes of people to the small-scale world as Galileo's *The Starry Messenger* was in opening the eyes of people to the nature of the Universe at large. In the words of Geoffrey Keynes, it ranks 'among the most important books ever published in the history of science'. Samuel Pepys records how he sat up reading the volume until 2 am, calling it 'the most ingenious book that ever I read in my life'.[1]

Hooke was not the first microscopist. Several people had followed up Galileo's lead by the 1660s and, as we have seen, Malpighi in particular had already made important discoveries, especially those concerning the circulation of the blood, with the new instrument. But Malpighi's observations had been reported piece by piece to the scientific community more or less as they had been made. The same is largely true of Hooke's contemporary Antoni van Leeuwenhoek (1632–1723), a Dutch draper who had no formal academic training but made a series of astounding discoveries (mostly communicated through the Royal Society) using microscopes that he made himself.

1. 21 January 1665. On 15 February that year, Pepys, who was a Fellow of the Royal Society and enjoyed attending meetings (although he made little scientific contribution), reported a meeting at Gresham College attended by scientific men 'of most eminent worth. Above all, Mr Boyle today was at the meeting, and above him Mr Hooke, who is the most, and promises the least, of any man in the world that ever I saw.' This gives an accurate indication of Hooke's intellectual standing at the time, as well as an idea of his unprepossessing appearance.

These instruments consisted of very small, convex lenses (some the size of pinheads) mounted in strips of metal and held close to the eye – they were really just incredibly powerful magnifying glasses, but some could magnify 200 or 300 times. Van Leeuwenhoek's most important discovery was the existence of tiny moving creatures (which he recognized as forms of life) in droplets of water – microorganisms including varieties now known as protozoa, rotifera and bacteria. He also discovered sperm cells (which he called 'animalcules'), which provided the first hint at how conception works, and independently duplicated some of Malpighi's work (which he was unaware of) on red blood cells and capillaries. These were important studies, and the romance of van Leeuwenhoek's role as a genuine amateur outside the mainstream of science has assured him of prominence in popular accounts of seventeenth-century science (some of which even credit him with inventing the microscope). But he was a one-off, using unconventional techniques and instruments, and Hooke represents the mainstream path along which microscopy developed, following his own invention of improved compound microscopes using two or more lenses to magnify the objects being studied. He also packaged his discoveries in a single, accessible volume, and provided it with beautifully drawn and scientifically accurate drawings of what he had seen through his microscopes (many of the drawings were made by his friend Christopher Wren). *Micrographia* really did mark the moment when microscopy came of age as a scientific discipline.

The most famous of the microscopical discoveries reported by Hooke in his masterpiece was the 'cellular' structure of slices of cork viewed under the microscope. Although the pores that he saw were not cells in the modern biological meaning of the term, he gave them that name, and when what we now call cells were identified in the nineteenth century, biologists took over the name from Hooke. He also described the structure of feathers, the nature of a butterfly's wing and the compound eye of the fly, among many observations of the living world. In a far-sighted section of the book he correctly identifies fossils as the remains of once-living creatures and plants. At that time, there was a widespread belief that these stones that looked like living things were just that – rocks, which, through some mysterious process, mimicked the appearance of life. But Hooke crushingly dismissed the notion that fossils were 'Stones form'd by some extraordinary *Plastick virtue latent* in the Earth itself,' and argued persuasively (referring to the objects now known as ammonites) that they were 'the Shells of certain Shelfishes, which, either by some Deluge, Inundation, Earthquake, or some such other means, came to be thrown to that place, and there to be filled with some kind of mud or clay, or *petrifying* water, or some other substance, which in tract of time has been settled together and hardened.' In lectures given at Gresham College around this time, but not published until after his death, Hooke also specifically recognized that this implied major transformations in the surface of the Earth. 'Parts which have been sea are now land,' he said, and 'mountains have been turned into plains, and plains into mountains, and the like.'

Hooke's study of the wave theory of light

Any of this would have been enough to make Hooke famous and delight readers such as Samuel Pepys. But there was more to *Micrographia* than microscopy. Hooke studied the nature of the coloured patterns produced by thin layers of material (such as the colours of an insect's wing, or the rainbow pattern familiar today from oil or petrol spilled on water), and suggested that it was caused by some form of interference between light reflected from the two sides of the layer. One of the phenomena Hooke investigated in this way concerned coloured rings of light produced where two pieces of glass meet at a slight angle; in the classic form of the experiment, a convex lens rests on a piece of flat glass, so that there is a slight wedge-shaped air gap between the two glass surfaces near the point where they touch. The rings are seen by looking down on the lens from above, and the phenomenon is related to the way a thin layer of oil on water produces a swirling pattern of colours. It is a sign of how successful Newton was in rewriting history that this phenomenon became known as 'Newton's rings'. Hooke's ideas about light were based on a wave theory, which he later developed to include the suggestion that the waves might be a transverse (side to side) oscillation, not the kind of push–pull waves of compression envisaged by Huygens. He described experiments involving combustion, from which he concluded that both burning and respiration involved something from the air being absorbed, coming very close to the discovery of oxygen (a century before it was actually discovered), and he made a clear distinction between heat, which he said arose in a body from 'motion or agitation of its parts' (almost two centuries ahead of his time!), and combustion, which involved two things combining. Hooke experimented on himself, sitting in a chamber from which air was pumped out until he felt pain in his ears, and was involved in the design and testing of an early form of diving bell. He invented the now-familiar 'clock face' type of barometer, a wind gauge, an improved thermometer and a hygroscope for measuring moisture in the air, becoming the first scientific meteorologist and noting the link between changes in atmospheric pressure and changes in the weather. As a bonus, Hooke filled a space at the end of the book with drawings based on some of his astronomical observations. And he spelled out with crystal clarity the philosophy behind all of his work, the value of 'a sincere hand, and a faithful eye, to examine, and to record, the things themselves as they appear', rather than relying on 'work of the brain and the fancy' without any grounding in experiment and observation. 'The truth is,' wrote Hooke, 'the science of Nature has been already too long made only a work of the brain and the fancy: It is now high time that it should return to the plainness and soundness of observations on material and obvious things.'

John Aubrey, who knew Hooke, described him in 1680 as:

> But of middling stature, something crooked, pale faced, and his face but little belowe, but his head is lardge; his eie full and popping, and not quick; a grey eie. He has a delicate head of haire, browne, and of an excellent moist curle. He is and ever was very temperate, and moderate in dyet, etc.
>
> As he is of prodigious inventive head, so is [he] a person of great vertue and goodness.

Several factors conspired to prevent Hooke building on the achievements described in *Micrographia* to the extent that he might have done. The first was his position at the Royal Society, where he kept the whole thing going by carrying out experiments (plural) at every weekly meeting, some of them at the behest of other Fellows, some of his own design. He also read out papers submitted by Fellows who were not present, and described new inventions. Page after page of the minutes of the early years of the Royal contain variations on the themes of 'Mr Hooke produced . . .', 'Mr Hooke was ordered . . .', 'Mr Hooke remarked . . .', 'Mr Hooke made some experiments . . .', and so on and on. As if that were not enough to have on his plate (and remember that Hooke was giving full courses of lectures at Gresham College as well), when Oldenburg died in 1677, Hooke replaced him as one of the Secretaries of the Royal (although at least by that time there was more than one Secretary to share the administrative load), but gave up that post in 1683.

In the short term, soon after the publication of *Micrographia*, plague interrupted the activities of the Royal Society and, like many other people, Hooke retreated from London to the countryside, where he took refuge as a guest at the house of the Earl of Berkeley at Epsom. In the medium term, Hooke was distracted from his scientific work for years following the fire of London in 1666, when he became one of the principal figures (second only to Christopher Wren) in the rebuilding of the city; many of the buildings attributed to Wren are at least partially Hooke's design, and in most cases it is impossible to distinguish between their contributions.

The fire took place in September 1666. In May that year, Hooke had read a paper to the Royal in which he discussed the motion of the planets around the Sun in terms of an attractive force exerted by the Sun to hold them in their orbits (instead of Cartesian whirlpools in the aether), similar to the way a ball on a piece of string can be held 'in orbit' by the force exerted by the string if you whirl it round and round your head. This was a theme that Hooke returned to after his architectural and surveying work on the rebuilding of London, and in a lecture given in 1674 he described his 'System of the World' as follows:

First, that all Coelestial Bodies whatsoever, have an attraction or gravitating power towards their own Centers, whereby they attract not only their own parts, and keep them from flying from them . . . but that they do also attract all the other Coelestial Bodies that are within the sphere of their activity . . . The second supposition is this, That all bodies whatsoever that are put into a direct and simple motion, will so continue to move forward in a streight line, till they are by some other effectual powers deflected and bent into a Motion, describing a Circle, Ellipsis, or some other more compounded Curve Line.

The third supposition is, That these attractive powers are so much the more powerful in operating, by how much the nearer the body wrought upon is to their own Centers.'[1]

The second of Hooke's 'suppositions' is essentially what is now known as Newton's first law of motion; the third supposition wrongly suggests that gravity falls off as one over the distance from an object, not one over the distance squared, but Hooke himself would soon rectify that misconception.

It is almost time to introduce Halley and Newton and their contributions to the debate about gravity. First, though, a quick skim through the rest of Hooke's life.

Hooke's law of elasticity

We know a great deal about Hooke's later life from a diary which he started to keep in 1672. This is no work of literature, like Pepys' diary, but a much more telegraphic record of day to day facts. But it describes almost everything about Hooke's private life in his rooms at Gresham College, with such candour that it was felt unsuitable for publication until the twentieth century (one of the reasons why Hooke's character and achievements did not receive full recognition until recently). Although he never married, Hooke had sexual relationships with several of his maidservants, and by 1676 his niece Grace, who was probably 15 at that time and had been living with him since she was a child, had become his mistress. He was devastated when she died in 1687, and throughout the rest of his life was noticeably melancholic; 1687 was also a key year in the dispute with Newton, which can hardly have helped matters. On the scientific side, apart from the work on gravity, in 1678 Hooke came up with his best-known piece of work, the discovery of the law of elasticity which bears his name. It is typical of the way history has treated Hooke that this rather dull piece of work (a stretched spring resists with a force proportional to its extension) should be known as Hooke's Law, while his many more dazzling achievements (not all of which I have mentioned) are either forgotten or attributed to others. Hooke himself died on 3 March 1703, and his funeral was attended by all the Fellows of the Royal Society then present in London.

1. Quoted by 'Espinasse.

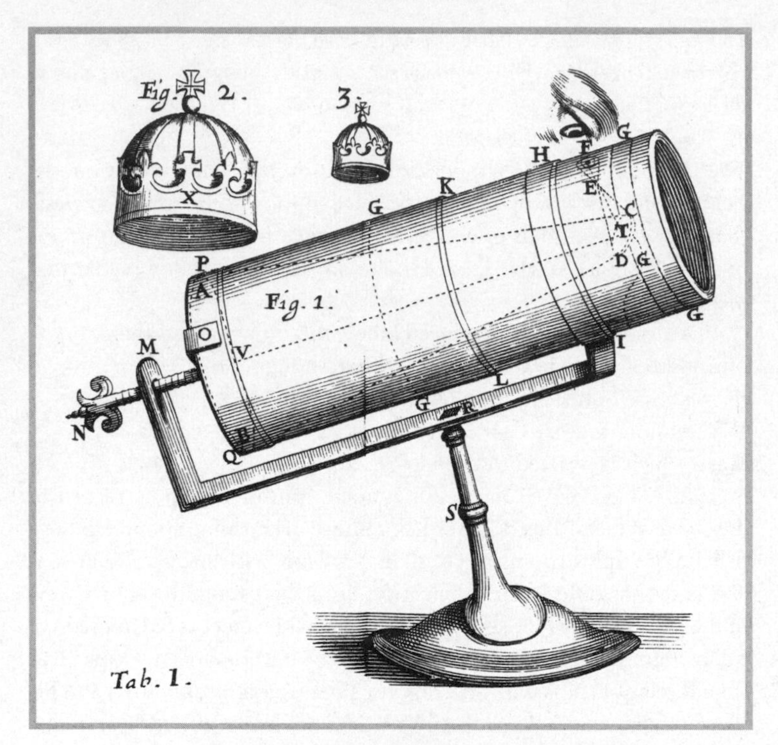

15. *Newton's telescope. From* Philosophical Transactions of the Royal Society, *1672.*

The following year, Isaac Newton published his epic work on light and colour, *Opticks*, having deliberately sat on it for thirty years, waiting for Hooke to die.

Newton's hostility to Hooke (I would have said monomaniacal hostility, but he felt the same way about other people, too) dated back to the early 1670s, when he was a young professor in Cambridge and first came to the attention of the Royal Society. In the 1660s, Newton (who was seven years younger than Hooke) completed his undergraduate studies in Cambridge, then became first a Fellow of Trinity College and, in 1669, the Lucasian professor of mathematics. The chair was in succession to Isaac Barrow, the first Lucasian professor, who resigned allegedly in order to spend more time at religious studies but who soon became Royal Chaplain and then Master of Trinity College, so may have had an ulterior motive. During all this time, Newton had been carrying out experiments and thinking about the world more or less on his own, hardly discussing his ideas with anyone at all. Among other things, he studied the nature of light, using prisms and lenses. In his most important optical work, he split white light (actually sunlight) up into the rainbow colours of the spectrum using a prism, then recombined the colours to make white light again, proving that white light was just a mixture of all the colours of the rainbow.

Previously, other people (including Hooke) had passed white light through a prism and cast the beam on a screen a few inches away, producing a white blob of light with coloured edges. Newton was able to go further because he used a pinhole in a window blind as his source of light and cast the beam from the prism on to a wall on the other side of a large room, many feet away, giving a bigger distance for the colours to spread out. From this work, his interest in colours led him to consider the problem of the coloured fringes produced around the edges of the images seen through telescopes built around lenses, and he designed and built a reflecting telescope (unaware of the earlier work of Leonard Digges) which did not suffer from this problem.

News of all this began to break when Newton described some of his work on light in the lectures he gave as Lucasian professor, and from visitors to Cambridge who either saw or heard about the telescope. The Royal Society asked to see the instrument, and late in 1671 Isaac Barrow took one (Newton probably made at least two) to London and demonstrated it at Gresham College. Newton was immediately elected a Fellow (the actual ceremony took place on 11 January 1672) and was asked what else he had up his sleeve. His response was to submit to the society a comprehensive paper on light and colours. It happened that Newton favoured the corpuscular theory of light as a stream of particles, but the discoveries he spelled out at this time held true regardless of whether this model or the wave model (favoured by people such as Huygens and Hooke) was used to try to explain them.

From several sidelong remarks in the paper, not least a statement that Newton started to experiment in optics in 1666, it seems clear that the trigger for his interest in light had been reading Hooke's

Micrographia, but he played this down by making reference to 'an unexpected experiment, which Mr Hook, somewhere in his *Micrography*, reported to have made with two wedge-like transparent vessels' rather than going into details of Hooke's work (in this example, on what became known as Newton's rings).

Hooke, an older and well-established scientist, was decidedly miffed at receiving less credit from the young whipper-snapper than he thought he was due, and said as much to his friends. Hooke was always touchy about receiving proper recognition for his work, understandably so in view of his own humble origins and his recent past as a servant to the learned gentleman who established the Royal Society.[1] Newton, though, even at this early age, had the highest opinion of his own abilities (largely justified, but still not an appealing characteristic) and regarded other scientists, no matter how respectable and well established, as scarcely fit to lick his boots. This attitude was reinforced over the next few years, when a succession of critics, clearly Newton's intellectual inferiors, raised a host of quibbles about his work, which showed their own ignorance more than anything else. At first Newton tried to answer some of the more reasonable points raised, but eventually he became infuriated at the time-wasting and wrote to Oldenburg: 'I see I have made myself a slave to philosophy . . . I will resolutely bid adieu to it eternally, except what I do for my private satisfaction, or leave it to come after me; for I see a man must either resolve to put out nothing new, or become a slave to defend it.'

When Oldenburg mischievously reported an exaggerated account of Hooke's views to Newton, deliberately trying to stir up trouble, he succeeded better than he could have expected. Newton replied, thanking Oldenburg 'for your candour in acquainting me with Mr Hook's insinuations' and asking for an opportunity to set the record straight. J. G. Crowther has neatly summed up the real root of the problem, which Oldenburg fanned into flame: 'Hooke could not understand what tact had to do with science . . . Newton regarded discoveries

1. Significantly, though, precisely because he was so aware of the importance of giving credit where it was due, Hooke was always careful (even generously so) in acknowledging the work of his colleagues such as Boyle in his own publications. He expected the credit he was due, but there is no evidence that he ever claimed more than his due.

as private property'.[1] At least, he did so when they were his discoveries. After four years, the very public washing of dirty linen resulting from this personality clash had to be brought to an end or the Royal Society would become a laughing stock, and several Fellows got together to insist, through Oldenburg (who must have been unhappy to have his fun at Hooke's expense brought to a close), on a public reconciliation (whatever the two protagonists thought in private), which was achieved through an exchange of letters.

Hooke's letter to Newton seems to bear the genuine stamp of his personality, always ready to argue about science in a friendly way (preferably among a few companions in one of the fashionable coffee houses), but really only interested in the truth:

> I judge you have gone farther in that affair [the study of light] than I did . . . I believe the subject cannot meet with a fitter and more able person to enquire into it than yourself, who are in every way fitted to complete, rectify, and reform what were the sentiments of my younger studies, which I designed to have done somewhat at myself, if my other more troublesome employments would have permitted, though I am sufficiently sensible it would have been with abilities much inferior to yours. Your design and mine, are, I suppose, both at the same thing, which is the discovery of truth, and I suppose we can both endure to hear objections, so as they come not in a manner of open hostility, and have minds equally inclined to yield to the plainest deductions of reason from experiment.

There speaks a true scientist. Newton's reply, although it could be interpreted as conciliatory, was totally out of character, and carries a subtext which is worth highlighting. After saying that 'you defer too much to my ability' (a remark Newton would never have made to anyone except under duress), he goes on with one of the most famous (and arguably most misunderstood) passages in science, usually interpreted as a humble admission of his own minor place in the history of science:

> What Des-Cartes did was a good step. You have added much in several ways, & especially in taking ye colours of thin plates into philosophical

1. *Founders of British Science*, p. 248.

consideration. If I have seen farther, it is by standing on the shoulders of Giants.

John Faulkner, of the Lick Observatory in California, has suggested an interpretation of these remarks which flies in the face of the Newton legend, but closely matches Newton's known character. The reference to Descartes is simply there to put Hooke in his place by suggesting that the priority Hooke claims actually belongs to Descartes. The second sentence patronizingly gives Hooke (the older and more established scientist, remember) a little credit. But the key phrase is the one about 'standing on the shoulders of Giants'. Note the capital letter. Even granted the vagaries of spelling in the seventeenth century, why should Newton choose to emphasize this word? Surely, because Hooke was a little man with a twisted back. The message Newton intends to convey is that although he may have borrowed from the Ancients, he has no need to steal ideas from a little man like Hooke, with the added implication that Hooke is a mental pygmy as well as a small man physically. The fact that the source of this expression predates Newton and was picked up by him for his own purposes only reinforces Faulkner's compelling argument. Newton was a nasty piece of work (for reasons that we shall go on to shortly) and always harboured grudges. True to his word to Oldenburg, he did go into his shell and largely stop reporting his scientific ideas after this exchange in 1676. He only properly emerged on to the scientific stage again, to produce the most influential book in the history of science, after some cajoling by the third member of the triumvirate that transformed science at this time, Edmond Halley.

John Flamsteed and Edmond Halley: cataloguing the stars by telescope

Halley, the youngest of the three, was born on 29 October 1656 (this was on the Old Style Julian calendar still in use in England then; it corresponds to 8 November on the modern Gregorian calendar), during the Parliamentary interregnum. His father, also called Edmond, was a prosperous businessman and landlord who married his mother, Anne Robinson, in a church ceremony only seven weeks before the birth. The most likely explanation for this is that there had been an earlier civil ceremony, of which no record has survived, and that the imminent arrival of their first child encouraged the couple to exchange religious vows as well; it was fairly common practice at the time to have a civil ceremony first and a church ceremony later (if at all). Edmond had a sister Katherine, who was born in 1658 and died in infancy, and a brother Humphrey, whose birth date is not known, but who died in 1684. Little is known about the details of Halley's early life, except that in spite of some financial setbacks resulting from the fire of London in 1666, his father was comfortably able to provide young Edmond with the best available education, first at St Paul's School in London (the family home was in a peaceful village just outside London, in what is now the borough of Hackney) and then at Oxford University. By the time he arrived at Queen's College in July 1673, Halley was already a keen astronomer who had developed his observational skills using instruments paid for by his father; he arrived in Oxford with a set of instruments that included a telescope 24 feet (about 7.3 metres) long and a sextant 2 feet (60 centimetres) in diameter, as good as the equipment used by many of his contemporary professional astronomers.

Several events around that time would have a big impact on Halley's future life. First, his mother had died in 1672. No details are known, but she was buried on 24 October that year. The repercussions for Halley would come later, as a result of his father's second marriage. Then, in 1674, the Royal Society decided that it ought to establish an observatory to match the Observatoire de Paris, which had just been founded by the French Academy. The urgency of the proposal was highlighted by a French claim that the problem of finding longitude at sea had been solved by using the position of the Moon against the background stars as a kind of clock to measure time at sea. The claim proved premature – although the scheme would work in principle, the Moon's orbit is so complicated that by the time the necessary tables of its motion could be compiled, accurate chronometers had provided the solution to the longitude problem. The astronomer John Flamsteed (who lived from 1646 to 1719) was asked to look into the problem, and correctly concluded that it would not work because neither the position of the Moon nor the positions of the reference stars were known accurately enough. When Charles II heard the news, he decided that, as a seafaring nation, Britain had to have the necessary information as an aid to navigation, and the planned observatory became a project of the Crown. Flamsteed was appointed 'Astronomical Observator' by a royal warrant issued on 4 March 1675 (the first Astronomer Royal) and the Royal Observatory was built for him to work in, on Greenwich Hill (a site chosen by Wren). Flamsteed took up residence there in July 1676 and was elected as one of the Fellows of the Royal Society in the same year.

In 1675, the undergraduate Edmond Halley began a correspondence with Flamsteed, initially writing to describe some of his own observations which disagreed with some published tables of astronomical data, suggesting the tables were inaccurate, and asking Flamsteed if he could confirm Halley's results. This was music to Flamsteed's ears, since it confirmed that modern observing techniques could improve on existing stellar catalogues. The two became friends, and Halley was something of Flamsteed's protégé for a time – although, as we shall see, they later fell out. In the summer of that year, Halley visited Flamsteed in London, and he assisted him with observations, including two lunar eclipses on 27 June and 21 December. After the first of these observations, Flamsteed wrote in the *Philosophical Transactions* of the Royal Society that 'Edmond Halley, a talented young man of Oxford was present at these observations and assisted carefully with many of them.' Halley published three scientific papers in 1676, one on planetary orbits, one on an occultation of Mars by the Moon observed on 21 August that year and one on a large sunspot observed in the summer of 1676.[1] He was clearly a rising star of astronomy. But where could he best make a contribution to the subject?

Flamsteed's primary task at the new Royal Observatory was to make an accurate survey of the northern skies, using modern telescopic sights to improve the accuracy of old catalogues which used so-called open sights (the system used by Tycho), where the observer looked along a rod pointing at the chosen star. With telescopic sites, a fine

1. This was a noteworthy event because very few spots were observed on the Sun in the second half of the seventeenth century. This coincided with a period of cold in Europe known as the Little Ice Age. It was so severe that in several winters, notably 1683/4 (as graphically described by John Evelyn), the river Thames froze hard enough for tented cities, known as Frost Fairs, to be set up on its surface. There is almost certainly a link between the quietness of the Sun at this time and the chilling of the Earth.

hair in the focal plane of the telescope provides a far more accurate guide to the precise alignment of a star. Impatient to make a name and a career for himself, Halley hit on the idea of doing something similar to Flamsteed's survey for the southern skies, but concentrating on just the brightest couple of hundred stars to get results reasonably quickly. His father supported the idea, offering Halley an allowance of £300 a year (three times Flamsteed's salary as Astronomer Royal), and promising to cover many of the expenses of the expedition. Flamsteed and Sir Jonas Moore, the Surveyor General of the Ordnance, recommended the proposal to the King, who arranged free passage with the East India Company for Halley, his equipment and a friend, James Clerke, to the Island of St Helena, at that time the most southerly British possession (this was almost a hundred years before James Cook landed in Botany Bay in 1770). They sailed in November 1676, with Halley, now just past his twentieth birthday, abandoning his degree studies.

The expedition was a huge scientific success (in spite of the foul weather Halley and Clerke encountered on St Helena), and also seems to have given Halley an opportunity for some social life. Hints of sexual impropriety surround Halley's early adult life, and in his *Brief Lives*, John Aubrey (who lived from 1626 to 1697, and who not only knew Newton but knew people who had met Shakespeare, but is admittedly not always to be relied on) mentions a long-married but childless couple who travelled out to St Helena in the same ship as Halley. 'Before he came [home] from the Island,' writes Aubrey, 'she was brought to bed of a Child.' Halley seems to have mentioned this event as a curious benefit of the sea voyage or the air in St Helena on the childless couple; Aubrey implies that Halley was the father of the child. This was the kind of rumour that would dog the young man for several years.

Halley came home in the spring of 1678 and his catalogue of southern stars was published in November that year, earning him the sobriquet 'Our Southern Tycho' from Flamsteed himself; he was elected a Fellow of the Royal Society on 30 November. As well as his stellar cataloguing, Halley had observed a transit of Mercury across the face of the Sun while he was in St Helena. In principle, this provided a way to calculate the distance to the Sun by a variation on the theme

16. *Hevelius calculating star positions using a sextant.*
From Hevelius's Machina Coelestis, *1673.*

of parallax, but these early observations were not accurate enough for the results to be definitive. Nevertheless, Halley had planted a seed that would bear fruit before too long. On 3 December, at the 'recommendation' of the King, Halley was awarded his MA (*after* becoming FRS), in spite of failing to complete the formal requirements for his degree. He was now a member, on equal terms, of the group that included Boyle, Hooke, Flamsteed, Wren and Pepys (Newton had by now retreated into his shell in Cambridge). And they had just the job for him.

Because of its potential importance for navigation, the whole business of obtaining accurate star positions was of key importance in the late seventeenth century, both commercially and because of its military implications. But the main observing programme in the 1670s, an attempt to improve on Tycho's catalogue, was being carried out in Danzig (now Gdansk) by a German astronomer who insisted on using the traditional open sights system of Tycho himself, albeit updated at great expense with new instruments, to the frustration of his contemporaries, particularly Hooke and Flamsteed. This backward-looking astronomer had been born in 1611, which perhaps explains his old-fashioned attitude, and christened Johann Höwelcke, but Latinized his name to Johannes Hevelius. In a correspondence beginning in 1668, Hooke implored him to switch to telescopic sights, but Hevelius stubbornly refused, claiming that he could do just as well with open sights. The truth is that Hevelius was simply too set in his ways to change and distrusted the new-fangled methods. He was like someone who persists in using an old-fashioned manual typewriter even though a modern word-processing computer is available.

One of the key features of Halley's southern catalogue (obtained, of course, telescopically) was that it overlapped with parts of the sky surveyed by Tycho, so that by observing some of the same stars as Tycho, Halley had been able to key in his measurements to those of the northern sky, where Hevelius was now (in his own estimation) busily improving on Tycho's data. When Hevelius wrote to Flamsteed at the end of 1678 asking to see Halley's data, the Royal Society saw an opportunity to check up on Hevelius's claims. Halley sent Hevelius a copy of the southern catalogue and said that he would be happy to use the new data from Hevelius instead of the star positions from

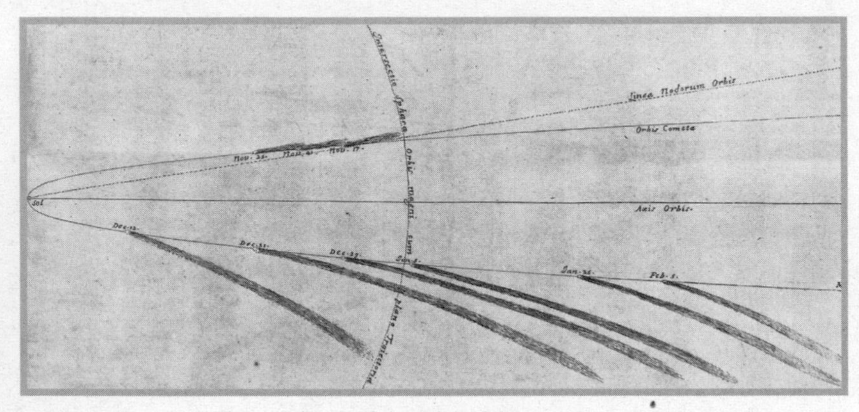

17. *Newton's sketch of the orbit of the comet seen in 1680.*

Tycho to make the link between the northern and southern skies. But, of course, he would like to visit Danzig to confirm the accuracy of the new observations.

So, in the spring of 1679, Halley set out to see if the incredible claims for accuracy that Hevelius, now aged 68, was making could be justified. Initially, Halley supported Hevelius's claims, reporting back that his positions obtained with open sights really were as good as he claimed. But on his return to England, he eventually changed his tune and said that the telescopic observations were much better. Halley later claimed that he had only been tactful to Hevelius's face, not wanting to hasten, the death of 'an old peevish Gentleman'. In fact, Hevelius lived for another nine years, so he cannot have been that frail when Halley saw him. Gossip at the time, though, suggests there was more to the story.

Hevelius's first wife had died in 1662, and in 1663 he had married a 16-year-old beauty, Elisabetha. When Halley visited Danzig, Hevelius was 68 and Elisabetha was 32. Halley was a handsome 22-year-old with a track record of sexual indiscretion. There may have been nothing in the inevitable rumours that followed Halley back to Britain, and there may be a perfectly respectable explanation for the fact that when news reached London later that year falsely reporting Hevelius's death, Halley's immediate reaction was to send the presumed widow a present of an expensive silk dress (at £6 8s 4d the cost of the dress amounted to the equivalent of three weeks' salary for the Astronomer Royal, though merely a week of Halley's allowance). But the kind of behaviour which made such rumours credible helped to create the rift between Flamsteed (a very serious-minded individual) and Halley, while Halley's initial endorsement of Hevelius's unrealistic claims certainly did not meet with his mentor's approval.

Not that Halley seemed to care much about the prospects for his career in astronomy at this time. Having achieved so much so soon, like a pop star after the first wave of success he seems to have been content to sit back on his laurels and make the most of the good thing offered by his money (or rather, his father's money). After returning from Danzig he spent more than a year basically just having a good time, attending meetings of the Royal but not contributing anything, visiting Oxford and hanging out in fashionable coffee houses, the equivalent of today's trendy wine bars (he particularly favoured Jonathan's in Change Alley). However, towards the end of this period, comets entered Halley's life for the first time, although at first only in a minor way.

In the winter of 1680/81, a bright comet became visible. It was first seen in November 1680, moving towards the Sun before being lost in the Sun's glare. It reappeared, going away from the Sun, a little later, and was at first thought to be two separate comets. Flamsteed was one of the first people to suggest that the phenomenon was really a single object, postulating that it had been repelled from the Sun by some sort of magnetic effect. It was a very prominent object in the night sky, clearly visible from the streets of London and Paris, the brightest comet that anyone alive at the time had ever seen. When it first appeared, Halley was about to embark on the typical wealthy young gentleman's Grand Tour of Europe, accompanied by an old schoolfriend, Robert Nelson. They travelled to Paris in December (arriving on Christmas Eve), and saw the second appearance of the comet from the Continent. On his travels through France and Italy, Halley took the opportunity to discuss the comet (and other astronomical matters) with other scientists, including Giovanni Cassini. Robert Nelson stayed in Rome, where he fell in love with (and later married) the second daughter of the Earl of Berkeley (the one who had provided a refuge for Hooke from the plague). Halley made his way back to England via Holland as well as Paris, perhaps deviating from the usual dilettante Grand Tour, since by the time he arrived back in London on 24 January 1682 he had, in the words of Aubrey, 'contracted an acquaintance and friendship with all the eminent mathematicians in France and Italy'.

Halley spent only a little over a year abroad, quite short for a Grand Tour in those days, and it may be that he hurried home because his father remarried around this time, although we don't know the exact date of the wedding. It is an indication of how little we know abut Halley's private life that his own wedding, which took place on 20 April 1682, comes completely out of the blue as far as any historical records are concerned. We know that Halley's wife was Mary Tooke, and that the wedding took place in St James' Church, in Duke Place, London. The couple were together (happily, it seems) for 50 years and had three children. Edmond junior was born in 1698, became a surgeon in the Navy and died two years before his father. Two daughters were born in the same year as each other, 1688, but were not twins. Margaret remained a spinster, but Catherine married twice. There were possibly other children who died in infancy. And that is essentially all we know about Halley's family.

After his marriage Halley lived in Islington, and over the next two years carried out detailed observations of the Moon intended to provide (eventually) the essential reference data for the lunar method of finding longitude. This would require accurate observations over some 18 years, the time it takes for the Moon to complete one cycle of its wanderings against the background stars. But in 1684, Halley's affairs were thrown into confusion by the death of his father and the project was abandoned for many years, as other matters took precedence.

Halley's father walked out of his house on Wednesday 5 March 1684 and never returned. His naked body was found in a river near Rochester five days later. The official verdict was murder, but no murderer was ever found, and the evidence was also consistent with suicide. Halley senior died without leaving a will and with his wealth considerably reduced by the extravagance of his second wife. There was an expensive legal battle between Halley and his stepmother over the estate. Halley was certainly not plunged into abject poverty by these events – he had property of his own and his wife had brought a good dowry with her. But his circumstances were sufficiently altered that in January 1686 he gave up his Fellowship of the Royal Society in order to become the paid Clerk to the society – the rules laid down that paid servants of the Royal could not be Fellows. Only a need for money – albeit a temporary need – could have encouraged this course of action.

Like Halley's private life, the European world was in turmoil in the middle of the 1680s. In France, the Edict of Nantes was revoked in 1685, while further afield Turkish forces reached the gates of Vienna, Buda and Belgrade around this time. In England, Charles II died and was succeeded by his Catholic brother, James II. And while all this was going on, Halley became involved with what is now regarded as the most important publication in the history of science, Isaac Newton's *Principia*.

Back in January 1684, after a meeting of the Royal Society, Halley had fallen into conversation with Christopher Wren and Robert Hooke about planetary orbits. The idea of an inverse square law of attraction being behind the orbits of the planets was not new even then – it went back at least to 1673, when Christiaan Huygens had calculated the outward (centrifugal) force on an object travelling in a circular orbit, and speculation along these lines had been taken up by Hooke, as we shall see, in correspondence with Newton after 1674. Wren had also discussed these ideas with Newton, in 1677. The three Fellows agreed that Kepler's laws of motion implied that the centrifugal force 'pushing' planets outwards from the Sun must be inversely proportional to the squares of their distances from the Sun, and that therefore, in order for the planets to stay in their orbits, they must be attracted by the Sun by an equivalent force which cancelled out the centrifugal force.

But was this an inevitable consequence of an inverse square law? Did such a law uniquely require that planets *must* follow elliptical orbits? Proving this mathematically was horrendously difficult using the traditional techniques available to them and Halley and Wren were willing to admit that the task was beyond them. Hooke, though, told the other two that he could derive all the laws of planetary motion starting out from the assumption of an inverse square law. The others were sceptical, and Wren offered to give Hooke a book to the value of 40 shillings if he could produce the proof in two months.

Hooke failed to produce his proof and the debate lapsed while Halley attended to the complicated business of sorting out his father's affairs after the murder (or suicide). It was probably in connection with this that Halley visited relatives in Peterborough in the summer of 1684, and this in turn may be why he visited Newton in Cambridge in August that year – because he was in the area anyway. There is no evidence that this was any kind of official visit on behalf of the Royal Society (in spite of myths to that effect) and only circumstantial evidence that Halley made the visit because he was in the area on family business. But he had certainly corresponded with Newton about the comet, and may have met him in 1682, so it is natural that he should have taken an opportunity to visit Cambridge when it arose. Whatever, there is no doubt that when Halley did visit Newton they discussed planetary orbits and the inverse square law. Newton later told his mathematician friend Abraham De Moivre (a French Huguenot refugee) exactly what happened:

> In 1684 Dr Halley came to visit him in Cambridge, after they had been some time together the Dr asked him what he thought the Curve would be that would be described by the Planets supposing the force of attraction towards the Sun to be reciprocal to the square of their distance from it. Sr Isaac replied immediately it would be an Ellipsis, the Dr struck with joy & amazement asked him how he knew it, why saith he, I have calculated it, whereupon Dr Halley asked him for the calculation without any further delay, Sr Isaac looked among his papers but could not find it, but he promised him to renew it, & then send it him.[1]

It was this encounter that led Newton to write the *Principia*, establishing his image as the greatest scientist who ever lived. But almost everything he described in that book had been done years before and kept hidden from public view until that happy encounter in Cambridge in 1684. This may seem hard to understand today, when scientists are almost too eager to rush their ideas into print and establish their

1. From a manuscript in the Joseph Halle Schaffner Collection, University of Chicago Library.

priority, but when you look at Newton's background and upbringing, his secretiveness is less surprising.

Newton's early life

Isaac Newton came, on his father's side, from a farming family that had just started to do well for themselves in a material way, but lacked any pretensions to intellectual achievement. His grandfather, Robert Newton, had been born some time around 1570 and had inherited farmland at Woolsthorpe, in Lincolnshire. He prospered from his farming so much that he was able to purchase the manor of Woolsthorpe in 1623, gaining the title of Lord of the Manor. Though not as impressive as it sounds to modern ears, this was a distinct step up the social ladder for the Newton family, and probably an important factor in enabling Robert's son Isaac (born in 1606) to marry Hannah Ayscough, the daughter of James Ayscough, described in contemporary accounts as 'a gentleman'. The betrothal took place in 1639. Robert made Isaac the heir to all his property, including the lordship of the manor, and Hannah brought to the union as her dowry property worth £50 per year. Neither Robert Newton nor his son Isaac ever learned to read or write, but Hannah's brother William was a Cambridge graduate, a clergyman who enjoyed the living at the nearby village of Burton Coggles. The marriage between Isaac and Hannah took place in 1642, six months after the death of Robert Newton; six months after the wedding, Isaac also died, leaving Hannah pregnant with a baby who was born on Christmas Day and christened Isaac after his late father.

Many popular accounts note the coincidence that 'the' Isaac Newton was born in the same year, 1642, that Galileo died. But the coincidence rests on a cheat – using dates from two different calendar systems. Galileo died on 8 January 1642 according to the Gregorian calendar, which had already been introduced in Italy and other Catholic countries; Isaac Newton was born on 25 December 1642 according to the Julian calendar still used in England and other Protestant countries. On the Gregorian calendar, the one we use today, Newton was born on 4 January 1643, while on the Julian calendar Galileo died right at the end of 1641. Either way, the two events did not take place in the same calendar year. But there is an equally noteworthy and genuine coincidence that results from taking Newton's birthday as 4 January 1643, in line with our modern calendar. In that case, he was born exactly 100 years after the publication of *De Revolutionibus*, which highlights how quickly science became established once it became part of the Renaissance.

Although the English Civil War disrupted many lives, as we have seen, over the next few years it largely passed by the quiet backwaters of Lincolnshire, and for three years Isaac Newton enjoyed the devoted attention of his widowed mother. But in 1645, just when he was old enough to appreciate this, she remarried and he was sent to live with his maternal grandparents. Almost literally ripped from his mother's arms and dumped in more austere surroundings at such a tender age, this scarred him mentally for life, although no unkindness was intended. Hannah was just being practical.

Like most marriages among the families of 'gentlemen' at the time (including Hannah's first marriage), this second marriage was a businesslike relationship rather than a love match. Hannah's new husband was a 63-year-old widower, Barnabas Smith, who needed a new partner and essentially chose Hannah from the available candidates (he was the Rector of North Witham, less than 3 kilometres from Woolsthorpe). The business side of the arrangement included the settling of a piece of land on young Isaac by the Rector, on condition that he lived away from the new matrimonial home. So while Hannah went off to North Witham, where she bore two daughters and a son before Barnabas Smith died in 1653, Isaac spent eight formative years as a solitary child in the care of elderly grandparents (they had married back in 1609 and must have been almost as old as Hannah's new husband), who seem to have been dutiful and strict, rather than particularly loving towards him.

The bad side of this is obvious enough, and clearly has a bearing on Isaac's development as a largely solitary individual who kept himself to himself and made few close friends. But the positive side is that he received an education.[1] Had his father lived, Isaac Newton would

1. And in the longer term, he benefited in terms of financial security, inheriting (via his mother) not only his father's property, but also a share of the not inconsiderable property of Barnabas Smith and some of the property Hannah inherited from her parents. Isaac Newton never had to worry about money once he reached the age of 21 and came into the income from the land that had been settled on him by Barnabas Smith as part of his wedding contract with Hannah.

surely have followed in his footsteps as a farmer; but to the Ayscough grandparents it was natural to send the boy to school (one suspects not least because it kept him out of the way). Although Isaac returned to his mother's house in 1653 when he was 10 and she was widowed for the second time, the seed had been sown, and when he was 12 he was sent to study at the grammar school in Grantham, about 8 kilometres from Woolsthorpe. While there, he lodged with the family of an apothecary, Mr Clark, whose wife had a brother, Humphrey Babington. Humphrey Babington was a Fellow of Trinity College in Cambridge, but spent most of his time at Boothby Pagnall, near Grantham, where he was rector.

Although Isaac seems to have been lonely at school, he was a good student and also showed an unusual ability as a model maker (echoing Hooke's skill), constructing such devices (much more than mere toys) as a working model of a windmill and flying a kite at night with a paper lantern attached to it, causing one of the earliest recorded UFO scares. In spite of his decent education (mostly the Classics, Latin and Greek), Newton's mother still expected him to take over the family farm when he became old enough, and in 1659 he was taken away from the school to learn (by practical experience) how to manage the land. This proved disastrous. More interested in books, which he took out to read in the fields, than in his livestock, Newton was fined several times for allowing his animals to damage other farmers' crops, and many stories of his absent-mindedness concerning his agricultural duties have come down to us, doubtless embroidered a little over the years. While Isaac was (perhaps to some extent deliberately) demonstrating his incompetence in this area, Hannah's brother William, the Cambridge graduate, was urging her to let the young man follow his natural inclinations and go up to the university. The combination of her brother's persuasion and the mess Isaac was making of her farm won her grudging acceptance of the situation, and in 1660 (the year of the Restoration) Isaac went back to school to prepare for admission to Cambridge. On the advice of (and, no doubt, partly thanks to the influence of) Humphrey Babington, he took up his place at Trinity College on 8 July 1661. He was then 18 years old – about the age people go to university today, but rather older than most of the young gentlemen entering Cambridge in the 1660s, when it was usual to go up to the university at the age if 14 or 15, accompanied by a servant.

Far from having his own servant, though, Isaac had to act as a servant himself. His mother would not tolerate more than the minimum expenditure on what she still regarded as a wasteful indulgence, and she allowed Isaac just £10 a year, although her own income at this time was in excess of £700 a year. Being a student's servant at this time (called a subsizar) could be extremely unpleasant and involve such duties as emptying the chamber pots of your master. It also had distinctly negative social overtones. But here Newton was lucky (or cunning); he was officially the student of Humphrey Babington, but Babington was seldom up at college and he was a friend who did not stress the master–servant relationship with Isaac. Even so, possibly because of his lowly status and certainly because of his introverted nature, Newton seems to have had a miserable time in Cambridge until early in 1663, when he met and became friendly with Nicholas Wickins. They were both unhappy with their room-mates and decided to share rooms together, which they did in the friendliest way for the next 20 years. It is quite likely that Newton was a homosexual; the only close relationships he had were with men, although there is no evidence that these relationships were consummated physically (equally, there is no evidence that they weren't). This is of no significance to his scientific work, but may provide another clue to his secretive nature.

The scientific life began to take off once Newton decided pretty much to ignore the Cambridge curriculum, such as it was, and read what he wanted (including the works of Galileo and Descartes). In the 1660s, Cambridge was far from being a centre of academic excellence. Compared with Oxford it was a backwater and, unlike Oxford, had not benefited from any intimate contact with the Greshamites. Aristotle was still taught by rote, and the only thing a Cambridge education fitted anyone for was to be a competent priest or a bad doctor. But the first hint of what was to be came in 1663, when Henry Lucas endowed a professorship of mathematics in Cambridge – the first scientific professorship in the university (and the first new chair of any kind since 1540). The first holder of the Lucasian chair of mathematics was Isaac Barrow, previously a professor of Greek (which gives you some

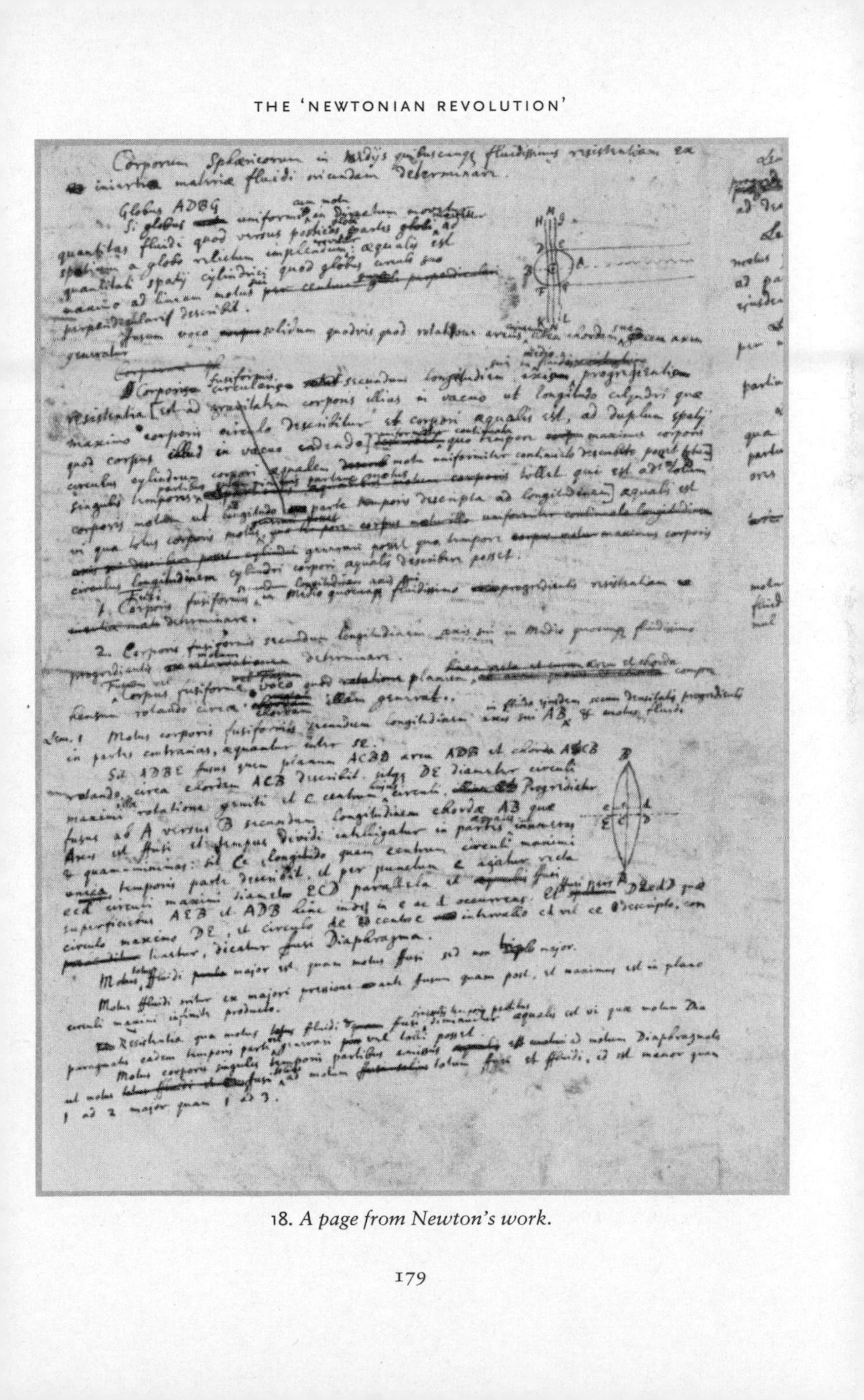

18. *A page from Newton's work.*

idea of where science stood in Cambridge at the time). The appointment was doubly significant – first because Barrow did teach some mathematics (his first course of lectures, in 1664, may well have been influential in stimulating Newton's interest in science) and then, as we have seen, because of what happened when he resigned from the position five years later.

According to Newton's own later account, it was during those five years, from 1663 to 1668, that he carried out most of the work for which he is now famous. I have already discussed his work on light and colour, which led to the famous row with Hooke. But there are two other key pieces of work which need to be put in context – Newton's invention of the mathematical techniques now known as calculus (which he called fluxions) and his work on gravity that led to the *Principia*.

Whatever the exact stimulus, by 1664 Newton was a keen (if unconventional) scholar and eager to extend his time at Cambridge. The way to do this was first to win one of the few scholarships available to an undergraduate and then to gain election, after a few years, to a Fellowship of the college. In April 1664, Newton achieved the essential first step, winning a scholarship in spite of his failure to follow the prescribed course of study, and almost certainly because of the influence of Humphrey Babington, by now a senior member of the college. The scholarship brought in a small income, provided for his keep and removed from him the stigma of being a subsizar; it also meant that after automatically receiving his BA in January 1665 (once you were up at Cambridge in those days it was impossible not to get a degree unless you chose, as many students did, to leave early), he could stay in residence, studying what he liked, until he became MA in 1668.

The development of calculus

Newton was an obsessive character who threw himself body and soul into whatever project was at hand. He would forget to eat or sleep while studying or carrying out experiments, and carried out some truly alarming experiments on himself during his study of optics, gazing at the Sun for so long he nearly went blind and poking about in his eye with a bodkin (a fat, blunt, large-eyed needle) to study the coloured images resulting from this rough treatment. The same obsessiveness would surface in his later life, whether in his duties at the Royal Mint or in his many disputes with people such as Hooke and Gottfried Leibnitz, the other inventor of calculus. Although there is no doubt that Newton had the idea first, in the mid-1660s, there is also no doubt that Leibnitz (who lived from 1646 to 1716) hit on the idea independently a little later (Newton not having bothered to tell anyone of his work at the time) and that Leibnitz also came up with a more easily understood version. I don't intend to go into the mathematical details; the key thing about calculus is that it makes it possible to calculate accurately, from a known starting situation, things that vary as time passes, such as the position of a planet in its orbit. It would be tedious to go into all the details of the Newton–Leibnitz dispute; what matters is that they did develop calculus in the second half of the seventeenth century, providing scientists of the eighteenth and subsequent centuries with the mathematical tools they needed to study processes in which change occurs. Modern physical science simply would not exist without calculus.

Newton's great insights into these mathematical methods, and the beginnings of his investigation of gravity, occurred at a time when the routine of life in Cambridge was interrupted by the threat of plague. Not long after he graduated, the university was closed temporarily and its scholars dispersed to avoid the plague. In the summer of 1665, Newton returned to Lincolnshire, where he stayed until March 1666. It then seemed safe to return to Cambridge, but with the return of warm weather the plague broke out once more, and in June he left again for the country, staying in Lincolnshire until April 1667, when the plague had run its course. While in Lincolnshire, Newton divided his time between Woolsthorpe and Babington's rectory in Boothby Pagnell, so there is no certainty about where the famous apple incident occurred (if indeed it really did occur at this time, as Newton claimed). But what is certain is that, in Newton's own words, written half a century later, 'in those days I was in the prime of my age for invention & minded Mathematicks & Philosophy more than at any time since'. At the end of 1666, in the midst of this inspired spell, Newton enjoyed his twenty-fourth birthday.

The way Newton later told the story, at some time during the plague years he saw an apple fall from a tree and wondered whether, if the influence of the Earth's gravity could extend to the top of the tree, it might extend all the way to the Moon. He then calculated that the force required to hold the Moon in its orbit and the force required to make the apple fall from the tree could both be explained by the Earth's gravity if the force fell off as one over the square of the distance from the centre of the Earth. The implication, carefully cultivated by Newton, is that he had the inverse square law by 1666, long before any discussions between Halley, Hooke and Wren. But Newton was a great one for rewriting history in his own favour, and the inverse square law emerged much more gradually than the story suggests. From the written evidence of Newton's own papers that can be dated, there is nothing about the Moon in the work on gravity he carried out in the plague years. What started him thinking about gravity was the old argument used by opponents of the idea that the Earth could be spinning on its axis, to the effect that if it were spinning it would break up and fly apart because of centrifugal force. Newton calculated the strength of this outward force at the surface of the Earth and compared it with the measured strength of gravity, showing that gravity at the surface of the Earth is hundreds of times stronger than the outward force, so the argument doesn't stand up. Then, in a document written some time after he returned to Cambridge (but certainly before 1670), he compared these forces to the 'endeavour of the Moon to recede from the centre of the Earth' and found that gravity *at the surface of the Earth* is about 4000 times stronger than the outward (centrifugal) force appropriate for the Moon moving in its orbit. This outward force would balance the force of the Earth's gravity if gravity fell off in accordance with an inverse square law, but Newton did not explicitly state this at the time. He also noted, though, from Kepler's laws, that the 'endeavours of receding from the Sun' of the planets in their orbits were inversely proportional to the squares of their distances from the Sun.

To have got this far by 1670 is still impressive, if not as impressive as the myth Newton later cultivated so assiduously. And remember that by this time, still not 30, he had essentially completed his work on light and on calculus. But the investigation of gravity now took a back seat as Newton turned to a new passion – alchemy. Over the next two decades, Newton devoted far more time and effort to alchemy than he had to all of the scientific work that we hold in such esteem today, but since this was a complete dead end there is no place for a detailed discussion of that work here.[1] He also had other distractions revolving around his position at Trinity College and his own unorthodox religious convictions.

In 1667, Newton was elected to a minor Fellowship at Trinity, which would automatically become a major Fellowship when he became an MA in 1668. This gave him a further seven years to do whatever he liked, but involved making a commitment to orthodox religion – specifically, on taking up the Fellowship all new Fellows had to swear an oath that 'I will either set Theology as the object of my studies and will take holy orders when the time prescribed by these statutes arrives, or I will resign from the college.' The snag was that Newton was an Arian.[2] Unlike Bruno, he was not prepared to go to the stake for his beliefs, but nor was he prepared to compromise them by swearing on oath that he believed in the Holy Trinity, which he would be required to do on taking holy orders. To be an Arian in England in the late seventeenth century was not really a burning offence, but if it came out it would exclude Newton from public office and certainly from a college named after the Trinity. Here was yet another reason for him to be secretive and introverted; and in the early 1670s, perhaps searching for a loophole, Newton developed another of his long-term obsessions, carrying out detailed studies of theology (rivalling his studies of alchemy and helping to explain why he did no new science after he was 30). He was not saved by these efforts, though, but by a curious stipulation in the terms laid down by Henry Lucas for his eponymous chair.

Newton had succeeded Barrow as Lucasian professor in 1669, when he was 26. The curious stipulation, which ran counter to all the traditions of the university, was that any holder of the chair was barred from accepting a position in the Church requiring residence outside Cambridge or 'the cure of souls'. In 1675, using this stipulation as an excuse, Newton got permission from Isaac Barrow (by now Master of Trinity) to petition the King for a dispensation for all Lucasian pro-

1. For a popular account which focuses on this aspect of Newton's life, see *Isaac Newton: the last sorcerer*, by Michael White.

2. Newton followed the teaching of the fourth-century Alexandrian Arius. Arianism held that God is a unique being and that therefore Jesus was not truly divine. These were heretical ideas in the eyes of a Church based on the concept of the Holy Trinity.

fessors from the requirement to take holy orders. Charles II, patron of the Royal Society (where, remember, Newton was by now famous for his reflecting telescope and work on light) and enthusiast for science, granted the dispensation in perpetuity, 'to give all just encouragement to learned men who are & shall be elected to the said Professorship'. Newton was safe – on the King's dispensation, he would not have to take holy orders, and the college would waive the rule which said that he had to leave after his initial seven years as a Fellow were up.

The wrangling of Hooke and Newton

In the middle of all the anxiety about his future in Cambridge, Newton was also embroiled in the dispute with Hooke about priority over the theory of light, culminating in the 'shoulders of Giants' letter in 1675. We can now see why Newton regarded this whole business with such irritation – he was far more worried about his future position in Cambridge than about being polite to Hooke. Ironically, though, while Newton was distracted from following up his ideas about gravity, in 1674 Hooke had struck to the heart of the problem of orbital motion. In a treatise published that year, he discarded the idea of a balance of forces, some pushing inwards and some pushing outwards, to hold an object like the Moon in its orbit. He realized that the orbital motion results from the tendency of the Moon to move in a straight line, plus a *single* force pulling it towards the Earth. Newton, Huygens and everyone else still talked about a 'tendency to recede from the centre', or some such phrase, and the implication, even in Newton's work so far, was that something like Descartes's swirling vortices were responsible for pushing things back into their orbits in spite of this tendency to move outwards. Hooke also did away with the vortices, introducing the idea of what we would now call 'action at a distance' – gravity reaching out across *empty* space to tug on the Moon or the planets.

In 1679, after the dust from their initial confrontation had settled, Hooke wrote to Newton asking for his opinion on these ideas (which had already been published). It was Hooke who introduced Newton to the idea of action at a distance (which immediately appears, without comment, in all of Newton's subsequent work on gravity) and to the idea that an orbit is a straight line bent by gravity. But Newton was reluctant to get involved, and wrote to Hooke that:

I had for some years past been endeavouring to bend my self from Philosophy to other studies in so much y^{t} I have long grutched the time spent in y^{t} study . . . I hope it will not be interpreted out of any unkindness to you or ye R. Society that I am backward in engaging myself in these matters.

In spite of this, Newton did suggest a way to test the rotation of the Earth. In the past, it had been proposed that the rotation of the Earth ought to show up if an object was dropped from a sufficiently tall tower, because the object would be left behind by the rotation and fall behind the tower. Newton pointed out that the top of the tower had to be moving faster than its base, because it was further from the centre of the Earth and had correspondingly more circumference to get round in 24 hours. So the dropped object ought to land in front of the tower. Rather carelessly, in a drawing to show what he meant, Newton carried through the trajectory of the falling body as if the Earth were not there, spiralling into the centre of the Earth under the influence of gravity. But he concluded the letter by saying:

But yet my affection to Philosophy being worn out, so that I am almost as little concerned about it as one tradesman uses to be about another man's trade or a country man about learning, I must acknowledge my self avers from spending that time in writing about it W^{ch} I think I can spend otherwise more to my own content.

But drawing that spiral drew Newton into more correspondence on 'Philosophy' whether he liked it or not. Hooke pointed out the error and suggested that the correct orbit followed by the falling object, assuming it could pass through the solid Earth without resistance, would be a kind of shrinking ellipse. Newton in turn corrected Hooke's surmise by showing that the object orbiting inside the Earth would not gradually descend to the centre along any kind of path, but would orbit indefinitely, following a path like an ellipse, but with the whole orbit shifting around as time passed. Hooke in turn replied to the effect that Newton's calculation was based on a force of attraction with 'an aequall power at all Distances from the center . . . But my supposition is that the Attraction always is in a duplicate proportion to the Distance from the Center Reciprocall', in other words, an inverse square law.

Newton never bothered to reply to this letter, but the evidence is that, in spite of his affection to philosophy being worn out, this was the trigger that stimulated him, in 1680, to *prove* (where Hooke and others could only surmise) that an inverse square law of gravity requires the planets to be in elliptical or circular orbits, and implied that comets should follow either elliptical or parabolic paths around the Sun.[1] And that is why he was ready with the answer when Halley turned up on his doorstep in 1684.

Newton's Principia Mathematica*: the inverse square law and the three laws of motion*

It wasn't all plain sailing after that, but Halley's cajoling and encouragement following that encounter in Cambridge led first to the publication of a nine-page paper (in November 1684) spelling out the inverse square law work, and then to the publication in 1687 of Newton's epic three-volume work *Philosophiae Naturalis Principia Mathematica*, in which he laid the foundations for the whole of physics, not only spelling out the implications of his inverse square law of gravity and the three laws of motion, which describe the behaviour of everything in the Universe, but making it clear that the laws of physics are indeed *universal* laws that affect everything. There was still time for one more glimpse of Newton's personality to surface – when Hooke complained that the manuscript (which he saw in his capacity at the Royal) gave him insufficient credit (a justifiable complaint, since he had achieved and passed on important insights even if he didn't have the skill to carry through the mathematical work as Newton could), Newton at first threatened to withdraw the third volume from publication and then went through the text before it was sent to the printers, savagely removing any reference at all to Hooke.

Apart from the mathematical brilliance of the way he fitted everything together, the reason why the *Principia* made such a big impact is

1. In fact, the hardest piece of mathematics in all of this is the proof that it is indeed correct to measure distances used in the inverse square law from the centre of the Earth or the Sun, with gravity acting as if all the mass were concentrated at a single point. Calculus makes this relatively straightforward, but Newton deliberately avoided using calculus in his published proofs, realizing that his peers would not accept the calculations unless they were couched in familiar language. Nobody knows if he did it all by calculus first and then translated into old-fashioned maths, but if he did, that is almost as impressive as if he did it the old-fashioned way to start with.

because it achieved what scientists had been groping towards (without necessarily realizing it) ever since Copernicus – the realization that the world works on essentially mechanical principles that can be understood by human beings, and is not run in accordance with magic or the whims of capricious gods.

For Newton and many (but by no means all) of his contemporaries there was still a role for God as the architect of the whole thing, even a 'hands-on' architect who might interfere from time to time to ensure the smooth running of His creation. But it became increasingly clear to many who followed after Newton that however it started, once the Universe was up and running it ought to need no interference from outside. The analogy that is often used is with a clockwork mechanism. Think of a great church clock of Newton's time, not just with hands marking off the time, but with wooden figures that emerge from the interior on the hour, portraying a little tableau and striking the chimes with a hammer on a bell. A great complexity of surface activity, but all happening as a result of the tick-tock of a simple pendulum. Newton opened the eyes of scientists to the fact that the fundamentals of the Universe might be simple and understandable, in spite of its surface complexity. He also had a clear grasp of the scientific method, and once wrote (to the French Jesuit Gaston Pardies):

> The best and safest method of philosophizing seems to be, first to inquire diligently into the properties of things, and to establish those properties by experiences and then to proceed more slowly to hypotheses for the explanation of them. For hypotheses should be employed only in explaining the properties of things, but not assumed in determining them; unless so far as they may furnish experiments.

In other words, science is about facts, not fancy.

The publication of the *Principia* marked the moment when science came of age as a mature intellectual discipline, putting aside most of the follies of its youth and settling down into grown-up investigation of the world. But it wasn't just because of Newton. He was a man of his times, putting clearly into words (and, crucially, into equations) ideas that were bubbling up all around, expressing more clearly than they could themselves what other scientists were already struggling to express. That is another reason why his book drew such a response – it struck a chord, because the time was ripe for such a summing up and laying of foundations. For almost every scientist who read the *Principia*, it must have been like the moment in C. P. Snow's *The Search*, when 'I saw a medley of haphazard facts fall into line and order . . . "but it's true," I said to myself. "It's very beautiful. And it's true."'

Newton himself became a famous scientist, far beyond the circle of the Royal Society, as a result of the publication of the *Principia*. The philosopher John Locke, a friend of Newton, wrote of the book:

> The incomparable Mr Newton has shewn, how far mathematicks, applied to some Parts of Nature, may upon Principles that Matter of Fact justifies, carry us in the knowledge of some, as I may so call them, particular Provinces of the Incomprehensible Universe.

But in 1687, Newton had stopped being a scientist (he would be 45 at the end of that year and had long since lost his affection for philosophy). True, his *Opticks* would be published at the beginning of the eighteenth century – but that was old work, sat upon until Hooke died and it could be published without him getting a chance to comment on it or claim any credit for his own work on light. But the status that resulted from the *Principia* may have been one of the factors that encouraged Newton to become a public figure in another sense, and although the rest of his life story has little direct bearing on science, it is worth sketching out just how much he achieved apart from his scientific work.

Newton's later life

Newton's first move into the political limelight came early in 1687, after the *Principia* was off his hands and being seen through printing by Halley. James II had succeeded his brother in 1685 and, after a cautious start to his reign, by 1687 he was starting to throw his weight around. Among other things, he tried to extend Catholic influence over the University of Cambridge. Newton, by now a senior Fellow at Trinity (and perhaps influenced by fears of what might happen to him as an Arian under a Catholic regime), was one of the leaders of the opposition to these moves in Cambridge and was one of nine Fellows who had to appear before the notorious Judge

Jeffreys to defend their stand. When James was removed from the throne at the end of 1688 and replaced early in 1689 by William of Orange (a grandson of Charles I) and his wife Mary (a daughter of James II) in the so-called Glorious Revolution,[1] Newton became one of two Members of Parliament sent to London by the University. Although far from active in Parliament, and not offering himself for re-election when the Parliament (having done its job of legalizing the takeover by William and Mary) was dissolved early in 1690, the taste of London life and participation in great events increased Newton's growing dissatisfaction with Cambridge. Although he threw himself into his alchemical work in the early 1690s, in 1693 he seems to have suffered a major nervous breakdown brought on by years of overwork, the strain of concealing his unorthodox religious views and (possibly) the breakup of a close friendship he had had over the previous three years with a young mathematician from Switzerland, Nicholas Fatio de Duillier (usually known as Fatio). When Newton recovered, he sought almost desperately for some way to leave Cambridge, and when in 1696 he was offered the Wardenship of the Royal Mint (by Charles Montague, a former Cambridge student, born in 1661, who knew Newton and was by now Chancellor of the Exchequer, but also found time to serve as President of the Royal Society from 1695 to 1698), he leapt at the chance.

The Wardenship was actually the number two job at the Mint, and could be treated as a sinecure. But since the then Master of the Mint effectively treated his own post as a sinecure, Newton had a chance to get his hands on the levers of power. In his obsessive way he took the place over, seeing through a major recoinage and cracking down on counterfeiters with ferocity and cold-blooded ruthlessness (the punishment was usually hanging, and Newton became a magistrate in order to make sure that the law was on his side). When the Master

1. William and Mary actually gained the throne as a result of a full-scale invasion of Britain, seizing London by force, even if it was largely bloodless and welcomed by many. But history is written by the victors, and Glorious Revolution sounds so much better than Invasion if you want to keep the populace happy. The most significant feature of the 'revolution' (almost justifying the name) was that it tilted the balance of political power in Britain away from the King and in favour of Parliament, without whom William and Mary could not have succeeded.

died in 1699, Newton took over the post – the only time in the long history of the Mint that the Warden moved up in this way. Newton's great success at the Mint encouraged him (probably at the urging of Montague, who was by now Lord Halifax and later became the Earl of Halifax) to stand again for Parliament, which he did successfully in 1701, serving until May 1702, when William II died (Mary had pre-deceased him in 1694) and Parliament was dissolved. William was succeeded by Anne, the second daughter of James II, who, before and during her twelve-year reign, was greatly influenced by Halifax. During the election campaign of 1705, she knighted both Newton, Halifax's protégé, and Halifax's brother, in the hope that the honour would encourage voters to support them.

It did them no good – Halifax's party as a whole lost the election, as did Newton as an individual, and Newton, now in his sixties, never stood again. But the story is worth telling, since many people think that Newton received his knighthood for science and some think that it was a reward for his work at the Mint, but the truth is that it was a rather grubby bit of political opportunism by Halifax as part of his attempt to win the election of 1705.

Hooke's death and the publication of Newton's Opticks

By now, though, Newton was happy to be out of politics, as he had found his last great stage. Hooke had died in March 1703 and Newton, who had largely kept well away from the society as long as Hooke, who had essentially created it, was still around, was elected President of the Royal Society in November that year. *Opticks* was published in 1704, and Newton ran the Royal with his usual meticulous attention to detail for the next two decades. One of the tasks he had to oversee, in 1710, was the move away from the cramped quarters at Gresham College to larger premises at Crane Court. There is no doubt that the move was long overdue – a visitor to Gresham College just before the move took place wrote that 'finally we were shown the room where the Society usually meets. It is very small and wretched and the best things there are the portraits of its members, of which the most noteworthy are those of Boyle and Hoock'.[1] There were many portraits that had to be taken from Gresham College to Crane Court, and this was overseen by the

1. Conrad von Uffenbach, in *London in 1710*.

obsessive stickler for detail Sir Isaac Newton. The only one that got lost, never to be seen again, was the one of Hooke; no portrait of him has survived. If Newton went to such lengths to try to play down Hooke's role in history, Hooke must have been an impressive scientist indeed.

It was Hooke who had made the Royal Society work in the first place, and Newton who moulded it into the form in which it was the leading scientific society in the world for two centuries or more. But Newton didn't mellow with age and fame. As President of the Royal, he was also involved in another unedifying dispute, this time with Flamsteed, the first Astronomer Royal, who was reluctant to release his new star catalogue until everything had been checked and double-checked, while everyone else was desperate to get their hands on the data. In spite of his unpleasantly disputatious streak, Newton's achievements at the Mint, or at the Royal, would have been enough to make him an important historical figure, even if he had never been much of a scientist himself.

In London, Newton's house was looked after (in the sense of telling the servants what to do; he was by now a very rich man) by his niece Catherine Barton. Born in 1679, she was the daughter of his half-sister Hannah Smith, who had married a Robert Barton, who died in 1693 leaving his family destitute. Newton was always generous to his family, and Catherine was a great beauty and excellent housekeeper. There was no hint of the shenanigans that Hooke had got up to with his own niece, but Catherine made a conquest of Halifax, who seems to have met her for the first time around 1703, when he was in his early forties and recently widowed. In 1706, Halifax drew up a will in which, among his other bequests, he left £3000 and all his jewels to Catherine 'as a small Token of the Great Love and Affection I have long had for her'. Later that year, he purchased an annuity for her worth £300 per year (three times the salary of the Astronomer Royal). And in 1713 (a year before he became Prime Minister under George I[1]), he changed

1. Britain had to turn to the Hanovarian George as King because they had run out of Stuarts; although Anne had 17 children only one survived infancy, and he died in 1700. George was a great-grandson of James I, spoke hardly a word of English and lived in Hanover throughout most of his reign (1714–27). By now, though, it didn't much matter who was King, since the country was run by Parliament.

the will to leave her £5000 (50 times Flamsteed's annual income) and a house (which he didn't actually own, but never mind) 'as a Token of the sincere Love, Affection, and esteem I have long had for her Person, and as a small Recompense for the Pleasure and Happiness I have had in her Conversation'.

The choice of phrase caused much mirth when the will became public on the death of Halifax in 1715, not least to Newton's bitter enemy Flamsteed. If there was more to it than conversation we shall never know. But it is easy to see how extraordinarily generous the bequest was – Newton himself, when he died in his eighty-fifth year on 28 March 1727, left a little more than £30,000, shared equally among eight nephews and nieces from his mother's second marriage. So Catherine received considerably more from Halifax for her conversational skills than from her very rich uncle.

While other scientists were still absorbing the implications of the *Principia* in the early eighteenth century (indeed, physical science throughout the century remained, in a sense, in the shadow of Newton), the first person to take up the challenge and opportunity offered by Newton's work was, appropriately, Edmond Halley, who was not only the midwife to the *Principia* but also, scientifically speaking, the first post-Newtonian scientist. Hooke and Newton were both, as far as their science is concerned, entirely seventeenth-century figures, though they did both live into the new century. The younger Halley, however, straddled the transition and produced some of his best work in the new century, after the Newtonian revolution. Along the way, though, he managed, as we shall see, to pack even more into his life than Hooke and Newton combined.

6 Expanding Horizons

Edmond Halley – Transits of Venus – The effort to calculate the size of an atom – Halley travels to sea to study terrestrial magnetism – Predicts return of comet – Proves that stars move independently – Death of Halley – John Ray and Francis Willughby: the first-hand study of flora and fauna – Carl Linnaeus and the naming of species – The Comte de Buffon: Histoire Naturelle *and thoughts on the age of the Earth – Further thoughts on the age of the Earth: Jean Fourier and Fourier analysis – Georges Couvier:* Lectures in comparative anatomy; *speculations on extinction – Jean-Baptiste Lamarck: thoughts on evolution*

In the century after Newton, the most profound change that took place in the understanding of humankind's place in the Universe was the growing realization of the immensity of space and the enormity of the span of past time. In some ways, the century represented a catching up, as science in general came to terms with the way Newton had codified physics and demonstrated the lawful, orderly nature of the Universe (or of the World, as it would have been expressed in those days). These ideas spread from physics itself, the core of science, to obviously related areas such as astronomy and geology; but it also (slowly) spread into the biological sciences, where the patterns and relationships of living things were established as an essential precursor (we can see with hindsight) to discovering the laws on which the living world operates, in particular the law of evolution and the theory of natural selection. Chemistry, too, became more scientific and less mystical as the eighteenth century wore on. But this was all set in the framework of a huge expansion of the realm which science attempted to explain.

Edmond Halley

The most important thing about Newton's (or Hooke's, or Halley's, or Wren's) universal inverse square law of gravity is not that it is an inverse square law (interesting and important though that is), and certainly not who thought of it first, but that it is *universal*. It applies to everything in the Universe, and it does so at all times in the history of the Universe. The person who brought this home to the scientific world (and was also one of the first to extend the boundaries of time) was Edmond Halley, who we last met acting as midwife to the publication of the *Principia*. This was a mammoth task. As well as soothing Newton when he got angry with people like Hooke, dealing with the printers, reading the proofs and so on, Halley ended up paying for the publication of the book, in spite of his own difficult financial circumstances at the time. In May 1686, the Royal Society (then under the Presidency of Samuel Pepys) agreed to publish the book under its own imprint and at its own cost. But it was unable to meet this obligation. It has been suggested that their plea of poverty was a political device resulting from the flare up of the row between Newton and Hooke about priority, with the Royal not wishing to be seen to take sides; but it seems more likely that the Society was genuinely in no financial position to make good its promise to Newton. The Royal Society's finances remained shaky for decades after its foundation (indeed, until Newton himself put the house in order during his time as President), and what meagre funds it had available had recently been used to pay for the publication of Francis Willughby's *History of Fishes*. The book proved almost unsellable (there were still copies in the Society's inventory in 1743), and this left the Royal so hard up that in 1686, instead of receiving his £50 salary, Halley received 50 copies of the book. Happily for Halley, unlike the *History of Fishes* the *Principia* sold moderately well (in spite of being written in Latin and very technical) and he made a modest profit out of it.

Unlike Newton, Halley played no part in the politics associated with the problems of the succession at the end of the seventeenth century. He seems to have been entirely apolitical, and once commented:

> For my part, I am for the King in possession. If I am protected, I am content. I am sure we pay dear enough for our Protection, & why should we not have the Benefit of it?

Keeping out of politics and keeping busy with his scientific work and administrative duties at the Royal, over the next few years Halley came up with a variety of ideas almost on a par with Hooke's achievements in his great days of inventiveness. These included an investigation into the possible causes of the Biblical Flood, which led him to question the accepted date of the Creation, 4004 BC (the date arrived at by Archbishop Ussher in 1620 by counting back all the generations in the Bible). Halley accepted that there had been a catastrophic event like the one described in the Bible, but saw, from comparison with the way features on the surface of the Earth are changed by erosion today, that this must have happened much more than 6000 years ago. He also tried to estimate the age of the Earth by analysing the saltiness of the sea, assuming that this had once been fresh and its saltiness had been increasing steadily as rivers carried minerals down into it off the land, and came up with a similarly long timescale. These views led him to be regarded as something of a heretic by the Church authorities, although by now this meant that he would find difficulty getting an academic post rather than that he would be burnt at the stake. Halley was interested in terrestrial magnetism and had an idea that variations in magnetism from place to place around the globe might, if they were first mapped accurately, be used as a means of navigation. He also studied variations in atmospheric pressure and winds, and published (in 1686) a paper on the trade winds and monsoons, which included the first meteorological chart. But he was also a practical man and carried out experiments for the Admiralty with a diving bell that he developed, allowing men to work on the sea floor at a depth of 10 fathoms (some 18 metres) for up to two hours at a time. Along the way, he also worked out and published the first tables of human mortality, the scientific basis on which life-insurance premiums are calculated.

Transits of Venus

Halley's first contribution to understanding the size of the Universe came in 1691, after he had pushed back the boundaries of time, when he published a paper showing how observations of a transit of Venus across the face of the Sun, seen from different points on Earth, could be used through a variation on the technique of triangulation and parallax to measure the distance to the Sun. These transits of Venus are rare, but predictable, events, and Halley returned to the theme in 1716, when he predicted that the next pair of such transits would occur in 1761 and 1769, and left detailed instructions on how to make the necessary observations. But between 1691 and 1716 Halley's life had undergone enormous upheavals.

In the same year that Halley published his first paper on Venus transits, the Savilian chair of astronomy (in Oxford) fell vacant. Halley was eager (almost desperate) for an academic post and would have been an ideal candidate, except for the objections of the Church authorities to his views on the age of the Earth. Halley applied for the post, but without optimism, writing to a friend that there was 'a caveat entered against me, till I can show that I am not guilty of asserting the eternity of the world'. He was, indeed, rejected in favour of David Gregory, a protégé of Isaac Newton. In all honesty, though, Gregory was a fine candidate, and the appointment seems to have been made largely on merit, not just because of Halley's unconventional religious views.

The effort to calculate the size of an atom

Some indication of just what Oxford was missing out on can be seen if we pick just one piece of work from the many that Halley undertook about this time. He puzzled over the fact that objects the same size but made of different materials have different weights – a lump of gold, for example, has seven times the weight of a lump of glass the same size. One of the implications of Newton's work is that the weight of an object depends on how much matter it contains (its mass), which is why all objects fall with the same acceleration – the mass cancels out of the equations. So, Halley reasoned, gold contains seven times as much matter as glass (size for size), and therefore glass must be at least six-sevenths empty space. This led him to think about the idea of atoms, and to try to find a way to measure their sizes. He did so by finding out how much gold was needed to coat silver wire in order to obtain silver-gilt. The technique used to do this involved drawing out the wire from an ingot of silver, with gold around the circumference. From the known size of the piece of gold used, and the diameter and length of the final wire, the skin of gold around the silver proved to be just 1/134,500th of an inch thick. Assuming this represented a single layer of atoms, Halley calculated that a cube of gold with a side one-hundredth of an inch long would contain more than 2433 million atoms. Since the gold surface of the silver-gilt wire was so perfect that no sign of silver showed through, though, he knew that even this huge number must be a serious underestimate of the actual number of atoms. All this was published (in the *Philosophical Transactions* of the Royal Society) in 1691.

Halley travels to sea to study terrestrial magnetism

Frustrated in his academic ambitions, some time over the next couple of years Halley conceived a new plan with his friend Benjamin Middleton, a wealthy Fellow of the Royal Society who seems to have been the instigator of the scheme. In 1693, they made a proposal to the Admiralty for an expedition to find ways of improving navigation at sea, in particular by studying terrestrial magnetism in different parts of the globe. In Robert Hooke's diary for 11 January that year, he writes that Halley had spoken to him 'of going in Middleton's ship to discover'. To modern eyes this begs the question, 'to discover what?' But, of course, Hooke was using the expression in the same way we would say 'to explore'. The proposal, whatever it was exactly, met with an enthusiastic response from the Admiralty, and on the direct order of the Queen (Mary II) a small ship, a kind known as a pink, was built specially for the task and launched on 1 April 1694. (When William of Orange invaded England in 1688, his fleet had included sixty pinks.) She was known as the *Paramore*, and was just 52 feet long, 18 feet wide at her broadest and drew 9 feet 7 inches, with a displacement of 89 tons. In modern units, some 16 metres long by 5 metres wide, for a voyage to the far South Atlantic (originally the idea was for a voyage around the world!).

Little more is then heard of the project in any surviving documents for a further two years, while the ship was rather slowly fitted out. This may have been particularly fortunate for science, since it was during that time that Halley developed his interest in comets, exchanging a stream of letters on the subject with Newton, and showing that many comets move in elliptical orbits around the Sun in accordance with the inverse square law. Studying the historical records, he began to suspect that the comet seen in 1682 was in such an orbit, and that it had been seen at least three times before, at intervals of 75 or 76 years. None of this was published at the time, largely because Flamsteed had the most accurate observations of the comet of 1682 and wouldn't let anybody see them (especially not Halley, with whom he was no longer on speaking terms). Newton, who was on good terms with Flamsteed at the time, tried to persuade him to pass the data on, but for his pains received a letter from Flamsteed saying that Halley 'had almost ruined himself by his indiscreet behaviour' and alludes to deeds 'too foule and large for [mention in] a letter'. There is no evidence that Halley was any more indiscreet or 'foule' than contemporaries such as Pepys and Hooke, but every indication that Flamsteed was rather prissy by the standards of his age.

In 1696, the *Paramore* expedition seemed ready to go, but experienced an unexplained setback at the last minute. On 19 June, the Navy Board received a letter from Halley listing the ship's company of fifteen men, two boys, himself, Middleton and a servant, and clearly this was an indication that they were all but ready to sail. But nothing more is heard of Middleton, and in August the ship was laid up in wet dock. The best guess is that the delay was caused by Middleton withdrawing from the scheme (he may have died for all we know) or by the latest of many wars with France, but it left Halley at a loose end, which Newton took advantage of. Newton, as Warden of the Royal Mint, was overseeing the reform of the currency, and he appointed Halley as Deputy Comptroller of the Mint at Chester. This was obviously intended as a favour, but Halley found the work tedious, although he stuck it out until 1698, when the currency reform was completed.

In the interim, although Queen Mary had died, official enthusiasm for the voyage 'to discover' had, if anything, increased and the ship was now to sail as a Royal Navy vessel, under the patronage of William III, carrying guns and a Navy crew. But Halley was still to be in charge of the expedition and to that end he was commissioned as a master and commander (a rank one step below that of captain) in the Royal Navy, and given command of the ship (with the courtesy title Captain). He is the only landsman ever to have been given such a working commission as the actual captain of a Royal Navy ship. Although a few other people (including scientists) in the long history of the Navy have been given temporary commissions, either for administrative purposes or as honorary rank, none of them actually commanded a vessel.[1] On 15 October,

1. Halley certainly had previous experience at sea (or at least in boats), and among other things had been involved in surveying the approaches to the Thames in the late 1680s, so he was not a complete landsman. But the record is frustratingly blank about the details of this aspect of his early life.

Halley was issued with detailed orders for the year-long voyage (which came as no surprise, since he had drafted them himself for the Admiralty), but before he left he had a chance to meet Tsar Peter the Great (then in his late twenties), who was visiting England to study ship-building techniques.

Peter was a 'hands-on' student who learned how to build ships by working in the dockyard at Deptford. He stayed at John Evelyn's house, which he more or less wrecked with his wild parties. Halley dined with him there on more than one occasion and may have joined in Peter's favourite game, pushing people at high speed in a wheelbarrow through the ornamental hedges. When Peter left, the Exchequer had to pay Evelyn more than £300 for the cost of the damage – half of what it was costing the King to send *Paramore* on a year-long voyage to the South Atlantic.

The story of that voyage, which began on 20 October 1698, would make a book in itself. Halley's first lieutenant, Edward Harrison,[1] a career Navy officer, understandably (if inexcusably) took exception to being put under the command of a landsman approaching his forty-second birthday, and by the spring of 1699, when the vessel was in the West Indies, matters came to such a head that Harrison retired to his cabin and left Halley to navigate the ship alone, obviously hoping that the captain would make a fool of himself. Halley did no such thing, but sailed the vessel home with great aplomb, arriving on 28 June. After attending to some business affairs and resigning his Clerkship of the Royal, on 16 September he was back at sea, without Harrison, and carried out magnetic observations all the way down to 52° South (nearly level with the tip of South America), returning in triumph to Plymouth on 27 August 1700.

Although now reinstated as a Fellow of the Royal Society, Halley was not finished with the Navy or government business. In 1701, he took the *Paramore* to study the tides of the English Channel, but with the hidden agenda, it now seems, of carrying out clandestine surveys of the approaches to French ports and spying out harbour defences. In 1702, with Queen Anne now on the throne, Halley was sent as an envoy to Austria, ostensibly to advise on harbour fortifications in the

1. No relation to the clock maker.

Adriatic (the Austrian Empire stretched that far south). On this and a follow-up trip, Halley seems to have carried out a little light spying along the way (in January 1704 he was paid a modest sum, for carefully unspecified services, out of the secret service fund), and on the second trip he dined in Hanover with the future George I and his heir.

Predicts return of comet

Just before Halley returned from this second diplomatic mission, the Savilian professor of geometry at Oxford died. This time, with his friends in high places and record of service to the Crown, there was no question that Halley should replace him – even though Flamsteed objected that Halley 'now talks, swears, and drinks brandy like a sea captain'. After all, he *was* a sea captain and took great delight in being referred to in Oxford as Captain Halley, at least until 1710, when he rather belatedly became Dr Halley. He was appointed to the Savilian chair in 1704, and a year later (having given up hope of getting more accurate data from Flamsteed) published his book *A Synopsis of the Astronomy of Comets*, the work for which he is best remembered. In it he predicted that the comet of 1682 would return, in obedience to Newton's laws, 'about the year 1758'. Although Halley remained very active in science after 1705, one piece of work stands out above all the rest of his later achievements. It stemmed from his return to the subject that had first made his name, the study of the positions of the stars.

Proves that stars move independently

During all the time since Halley's first expedition to St Helena, while his career had followed all the twists and turns we have described, Flamsteed had been beavering away at the task the Royal Greenwich Observatory had been set up for – preparing more accurate astronomical star tables as an aid to navigation. But virtually nothing had been published, and Flamsteed claimed that because he was paid only a token sum by the Crown, and had to supply all his own instruments, he owned the data and could sit on them as long as he liked.[1] In 1704, Newton, as President of the Royal Society, persuaded Flamsteed to hand over some of his measurements and the printing of a new star catalogue began. But this was stopped in the face of Flamsteed's objections and his

1. This is reminiscent of the claim by private researchers to 'own' the patent rights to human genes, as ludicrous as claiming that you can patent a tree or the Sun.

claim to own the data. The situation could only be resolved by royal authority, and in 1710 Queen Anne issued a warrant appointing Newton and such other Fellows of the Royal Society as he chose to act as a Board of Visitors to the Observatory, with authority to demand all of Flamsteed's data so far and to be given his annual results within six months of the end of each year.[1] Even Flamsteed couldn't go against the orders of the Queen. Halley was appointed to put all the material in order, and the result was the publication of the first version of Flamsteed's star catalogue in 1712. The arguments didn't stop there, and eventually a definitive version, more or less approved by Flamsteed, appeared in 1725, published by his widow six years after his death. The catalogue gave some 3000 star positions to an accuracy of 10 seconds of arc, and was indeed the best of its kind published up to that time, something any normal person would have been proud to see in print while they were still alive.

Long before then, though, Halley was able to work with Flamsteed's earlier material and he compared Flamsteed's star positions with those in a much more limited catalogue put together by Hipparchus in the second century BC. He found that although most of the star positions obtained by the Greeks matched up closely to the more accurate positions found by Flamsteed, in a few cases the positions measured by Flamsteed were so different from those measured nearly 2000 years earlier that they could not be explained as mistakes on the part of the Ancients (especially not since the other positions clearly were correct, within the errors of the techniques used by the Greeks). Arcturus, for example (a bright and easily observed star), was seen in the eighteenth century twice the width of the full Moon (more than a degree of arc) away from the position recorded by the Greeks. The only conclusion was that these stars had physically moved across the sky since the time of Hipparchus. This was the last nail in the coffin of the crystal sphere idea – the first direct, observational evidence that the notion of stars as little lights attached to a sphere only a little bigger than the orbit of Saturn (remember that Uranus and Neptune had yet to be discovered)

1. It's a sign of how much Flamsteed had dragged his feet that the observatory had been set up by Charles II, who had since been succeeded by James II, followed by William and Mary, before Anne broke the logjam.

was wrong. Evidence that the stars move relative to one another is also evidence that the stars are at different distances from us, spread out in three dimensions through space. It lent credence to the idea that the stars are other suns, at such huge distances from us that they only show up as tiny pinpoints of light – but it was to be more than a hundred years before the distances to the nearest stars were first measured directly.

Death of Halley

When Flamsteed died in 1719, Halley, now 63, succeeded him as Astronomer Royal (he was formally appointed on 9 February 1720). After replacing (this time with the aid of official funds) the instruments that Flamsteed had bought and which had been removed by his widow, in his old age Halley carried out a full programme of observations, including (at last, but too late to solve the navigation problem, because of the advent of portable chronometers) a complete eighteen-year cycle of lunar motions. His old age was eased because as a former Navy officer with more than three years' service he received a pension equal to half his Navy salary. Although his wife died in 1736, and Halley himself suffered a slight stroke around that time, he carried on observing until shortly before his death, on 14 January 1742, soon after his eighty-fifth birthday. But even after Halley died, two of his greatest observations remained to be carried out by others.

The object now known as Halley's comet duly reappeared as predicted, and was first seen again on Christmas Day 1758, although now astronomers date the passage of the comet from its closest approach to the Sun, which occurred on 13 April 1759. This was a triumphant vindication of Newton's theory of gravity and the laws of mechanics spelled out in the *Principia*, and set the seal on Newton's achievement in the same way that, just 160 years later, observations of a total eclipse of the Sun would set the seal on Albert Einstein's general theory of relativity. In 1761, and again in 1769, the transits of Venus predicted by Halley were observed from more than sixty observing stations around the world, and the techniques he had spelled out half a century earlier were indeed used to work out the distance to the Sun, coming up with a figure equivalent to 153 million kilometres impressively close to the best modern measurement, 149.6 million kilometres. So Halley made his last great contribution to science twenty-seven years after he died, ninety-one years after he first burst on to the astronomical scene with his *Catalogue of the Southern Stars* and 113 years after he was born – surely one of the longest periods of 'active' achievement ever recorded. He left the world on the brink of an understanding of the true immensity of space and time, which had been inferred from studies of the physical Universe, but which would soon become (especially in the case of time) of key importance in understanding the origins of the diversity of species in the living world.

Erasmus Darwin, who features in the story of evolution in his own right, as well as being the grandfather of Charles Darwin, was born in 1731, when Halley was still very much an active Astronomer Royal and Newton had been dead for only four years. But to set the scene for a discussion of evolution properly, we need to go back to the seventeenth century, where as convenient a place as any to pick up the story is with the work of Francis Willughby, the naturalist whose book about fishes inadvertently left the Royal Society in such desperate financial straits that Halley had to pay for the publication of the *Principia*.

John Ray and Francis Willughby: the first-hand study of flora and fauna

But there are two peculiarities about Willughby's book. First, he had been dead for fourteen years when the book was published in 1686; second, he didn't write it. The reason the book appeared at all, let alone under Willughby's name, was because of his partnership with the greatest naturalist of the seventeenth century, John Ray, who did more than anyone else to lay the foundations of the scientific study of the natural world. Ray has sometimes been represented as the biological equivalent of Newton, setting the natural world in order as Newton set the physical world in order; but his position is really more like that of Tycho, making the observations which others would later use as the basis for their theories and models of how the biological world works.[1] Willughby's rightful place in the story is as a friend, sponsor and working partner of Ray, and the right place to begin the story is with Ray himself, who was

1. A more appropriate candidate for Newton's biological counterpart would, of course, be Charles Darwin. The fact that Darwin's great work was published nearly 150 years after Newton died accurately reflects how far behind the physical sciences biological sciences were (partly for psychological reasons, with people reluctant to accept themselves as a valid subject for scientific investigation) at the end of the seventeenth century.

born in the village of Black Notley, in Essex, on 29 November 1627. He was one of three children of the village blacksmith, Roger Ray, an important member of the local community but definitely not wealthy; their mother, Elizabeth, was a kind of herbalist and folk healer, who used plants to treat sick villagers. Their family name is variously spelled Ray, Raye and Wray in parish records, and John himself was known as Wray from the time he entered Cambridge University until 1670, when he reverted to Ray – it may be that the 'W' was added by mistake when he enrolled at the university and he was too diffident to point out the error at the time.

The fact that he got to Cambridge at all reflects in some ways the story of Isaac Newton's early years, although without the trauma of separation from his mother or the death of his father. He was clearly a bright boy capable of far more than he could be taught at the village school, and he seems to have benefited from the interest shown in his abilities by two rectors of Black Notley – Thomas Goad, who died in 1638, and his successor Joseph Plume, a Cambridge graduate who was probably responsible for Ray being sent to study at the grammar school in Braintree. The school provided little education except in the classics, and Ray received such a thorough grounding in Latin that almost all his scientific work was written in that language – in many ways he was more fluent in Latin than in English. But in Braintree Ray was noticed by another clergyman, the vicar of Braintree,[1] who was a graduate of Trinity College. It was thanks to him that Ray went up to Cambridge in 1644, when he was 16½.

There was no way that Ray's family could have paid for a university education, and this seems to have caused some problems. Ray was formally admitted to Trinity College as a subsizar on 12 May 1644, apparently on the promise of a scholarship of some kind arranged by Samuel Collins. But this must have fallen through, with the result that some strings were pulled and Ray switched to Catherine Hall on 28 June. He did so because the strings that the vicar of Braintree was able to pull were attached to a bequest in the gift of the holder of that post, providing for the maintenance of 'hopeful poor scholars, students in the University

1. The vicar's name was Samuel Collins, but he was not the more famous Samuel Collins who was at one time Provost of King's College, in Cambridge.

of Cambridge, namely in Catherine Hall and Emmanuel College'. At Trinity, Ray had been admitted as 'Ray, John, Sizar'; at Catherine Hall he was admitted as 'Wray, a scholar'. He was probably too relieved to have landed on his feet to worry about the spelling of his name.

These were not easy times in Cambridge, at the height of the Civil War and associated troubles. The region was firmly in the hands of the Parliamentary (Puritan) faction and Royalists (or even anti-Royalists who weren't Puritan enough) ran the risk of being ejected from their positions at the University. This happened to the Master of Catherine Hall in 1645, and partly as a result of the upheavals caused by this (but also because Catherine Hall was one of the lesser academic lights of the day), Ray transferred back to Trinity as a subsizar in 1646 – he was by then well known as an outstanding scholar and Trinity welcomed him back. There, he became a friend of Isaac Barrow (the future Lucasian professor), a fellow student who had also transferred to Trinity (from Peterhouse) after the Master of his old college had been ejected. Barrow was a Royalist (which is why he only came to prominence in Cambridge after the Restoration) and Ray a Puritan, but the two became good friends and shared a set of rooms.

But although Ray was a Puritan by inclination, he did not blindly follow the official party line, and this would profoundly affect his later life. One of the outward signs of Puritan conformity was to subscribe to a set of ideas known as the Covenant, as a badge of Presbyterianism. The original Covenant, signed by Scottish churchmen in 1638, rejected attempts by Charles I and William Laud, then Archbishop of Canterbury, to impose Church of England practices, seen as too close to Catholicism, on Scotland, and affirmed (or re-affirmed) the Reformed faith and Presbyterian principles of the Scottish Church. Acceptance of the Covenant was the chief condition on which the Scots had supported Parliament in the first phase of the English Civil War, with a promise that Parliament would reform the English Church along Presbyterian lines (in accordance with the Covenant), which meant (among other things) abolishing bishops.[1] Many people formally subscribed to the

1. And in the horribly complicated mess that the Civil War developed into, it was because both Charles I and Charles II later accepted the Covenant that the Scots switched sides, but that has little bearing on the history of science.

Covenant through genuine religious conviction; many others did so as a matter of form, to avoid any conflict with authority. Some, like Ray, never formally signed up to it at all, even though sympathetic to the Puritan cause. And others, like Barrow, refused to sign on principle, even though this ruined their career prospects.

The first personal result of these reformations for Ray was that although he graduated in 1648, and the following year became a Fellow of Trinity (on the same day that Barrow was elected to a Fellowship), he did not take holy orders. Like other institutions, Trinity held the opinion that if there were no bishops, there was no legal way for anyone to be ordained, and so the requirement was waived – even though Ray had always intended to seek ordination and to devote his life to ecclesiastical work. Over the next dozen years or so, Ray held a succession of teaching posts, as a lecturer in Greek, a lecturer in mathematics and a lecturer in humanities, as well as serving in various administrative posts in the college. He was now sufficiently secure that when his father died in 1655 he had been able to have a modest house built for his mother at Black Notley and to support her in her widowhood (his siblings seem to have died young). It was alongside his college duties that, exercising the freedom of Fellows to study what they liked, he began to turn his attention to botany. Fascinated by the differences between plants, and finding nobody who could teach him how to identify different varieties, he set out to provide his own classification scheme, enlisting the help of any of the students who were willing to join in. Which is where Willughby comes on the scene.

Francis Willughby came from a very different background to John Ray. He had been born in 1635, the son of a minor member of the aristocracy, Sir Francis Willughby of Middleton Hall, in Warwickshire; his mother was the daughter of the first Earl of Londonderry, and money would never be a problem for young Francis. With no financial worries but a keen brain and an interest in the natural world, Willughby became one of the archetypal well-connected gentleman amateur scientists of his day – indeed, he would become one of the founding Fellows of the Royal Society at the age of 25. He came up to Cambridge in 1652 and soon became a member of Ray's circle of naturalists, and a firm friend of the older man. The first public fruits of Ray's interest in plants came in 1660 (a year after Willughby became an MA), with the publication of his *Cambridge Catalogue*,[1] which described the plant life of the region around the university. He seemed set for a career as a distinguished Cambridge academic – but everything changed with the Restoration.

Things had started to change in 1658, when the authorities at Trinity decided that Fellows should be ordained after all. Ray stalled, even when he was offered the living of Cheadle in 1659. He regarded it as immoral to swear the required oaths simply as a matter of form, and wanted time to search his conscience to decide whether he really did want to commit himself to God's work in the way that those oaths implied. In the summer of 1660, at the time of the Restoration, Ray, still undecided, was travelling in the north of England and Scotland with Willughby studying the flora and fauna.[2] On his return to Cambridge, he found that many of his Puritan colleagues had been ejected and replaced by Royalists, and that the old Church rituals that he despised as all form without substance had been reinstated, along with the bishops. Expecting to lose his own place at Cambridge, he stayed away, but he was urged to return by the college, where he was a valued Fellow – and, after all, he had never been a signed-up Covenanter. Persuaded that he did belong at the university, he returned and fulfilled the requirement of ordination, being admitted into the priesthood by the Bishop of Lincoln before the end of the year. In 1661, he turned down the offer of a rich living in Kirkby Lonsdale, preferring to make his career in the university.

Then it all went pear-shaped. Although Charles I had sworn allegiance to the Covenant as an act of political expediency during the Civil War, his son had no intention of standing by this oath now he was in power. He also saw no reason why anybody else should be bound by such an oath, and in 1662 the Act of Uniformity was passed, requiring all clerics and all holders of university posts to declare that the oath by which the Covenant had been taken was unlawful and that nobody who had sworn such an oath was bound by it. Most people went

1. His books are usually referred to by their English titles, although most were originally written in Latin.

2. The fruits of this and similar expeditions around the country with various colleagues eventually appeared as the *English Catalogue* in 1670.

through the motions of agreeing to the Act of Uniformity. But Ray had strong feelings about oaths, and could not accept that the King (or anyone else) had the right to set them aside in this way. Although he had not subscribed to the Covenant himself, he refused to declare formally that those who had done so were wrong and that their oaths were unlawful and not binding. He was the only Fellow of Trinity to refuse to obey the King's instructions, and one of only twelve in the entire university (but remember that the die-hard Covenanters had already been expelled in 1660). On 24 August he resigned all his posts and became an unemployed cleric. As a priest, he could not take up secular work; but he could not practise as a priest because of his nonconformity. He went back to his mother's house in Black Notley with no prospects, but was rescued from a life of obscure poverty by his friend Willughby.

In 1662, Ray, Willughby and Philip Skippon, one of Ray's former pupils, had made another long field expedition, this time to the west of England, developing their studies of flora and fauna as they occurred in the wild. They pioneered the idea that first-hand knowledge of the environment and habitat of a living species was essential to any understanding of its physical form and way of life, and that any classification scheme must take account of observed field behaviour and not rely entirely on preserved museum specimens. It was during this trip that they decided that since Ray now had no other obligations, they would make an extended trip to continental Europe, where Willughby would study the birds, beasts, fishes and insects (in those days, the term insect covered just about everything that wasn't a bird, beast or fish), while Ray concentrated on the plant life. The party, augmented by another Trinity man, Nathaniel Bacon, set sail from Dover in April 1663, with Ray's expenses, of course, being met by his companions. Their travels took in northern France, Belgium, Holland, parts of Germany, Switzerland, Austria and Italy, before Willughby and Bacon left the others and returned home in 1664. Willughby gave an account of this first part of the expedition to the Royal Society in 1665. Ray and Skippon, meanwhile, visited Malta and Sicily, travelled through central Italy, stayed a while in Rome (where Ray made astronomical observations, later published by the Royal, of a comet), and then returned home via northern Italy, Switzerland and France, arriving back in England in the spring of 1666. Ray and Skippon produced detailed accounts of their travels and the countries they passed through, but their primary purpose was to study the living world, and this trip provided much of the raw material for Ray's greatest books and his lasting fame. It has been said that the European trip was the equivalent for Ray of the voyage of the *Beagle* for Charles Darwin, and, like Darwin, it took Ray many years to put all his data and specimens in order and to work out the implications. But it was well worth the wait.

When Ray returned to England he had a comprehensive mental picture of the living world, and access to a huge number of specimens, sketches and other observations compiled by himself and his companions. It was a fertile time for science in England, with the first flowering of the Royal Society, and among the things Ray had to catch up on (which he did voraciously) were Hooke's *Micrographia* and Boyle's early works. But he had no base in which to establish himself and organize his material and his ideas. He stayed with various friends over the next few months, and spent the winter of 1666/7 with Willughby at Middleton Hall (where Willughby was now the head of the household, Sir Francis having died), putting their collections in some sort of order. This gradually developed into a permanent relationship. Ray and Willughby travelled in the west of England again in the summer of 1667, and Ray went on other expeditions over the next few years, but he became Willughby's private chaplain at Middleton Hall, formalizing his position in the household. At the end of 1667 he was also elected as Fellow of the Royal Society, but, in recognition of his unusual circumstances, excused payment of the subscription.

At the age of 40, Ray seemed to have found a secure niche for life (after all, Willughby was eight years his junior) with ample opportunity to organize the wealth of material he had collected and to publish, with Willughby, a series of books providing a catalogue of the living world. In 1668, Willughby married an heiress, Emma Barnard, and, as with so many couples in those days, they quickly produced a succession of children – Francis, Cassandra and Thomas (who, after the death of his elder brother at the age of 19, inherited the estates and was later made Lord Middleton by Queen Anne). But in 1669, while on a visit to Chester with Ray, Willughby was taken ill with a severe fever. His health remained poor until well into 1670; in 1671 he seemed much his old self, but in 1672 he became seriously ill again and died on 3 July, in his thirty-seventh year. Ray was one of five executors of Willughby's will, which left him an annuity of £60 a year and gave him the responsibility for the education of Willughby's two sons (it was taken for granted that girls needed no education). Ray took these responsibilities seriously and made no further expeditions, settling at Middleton Hall and devoting himself to writing up the fruits of his and Willughby's past labours.

His position was not as comfortable as it might seem, because Willughby's widow had no fondness for Ray and treated him more as a servant than as her late husband's friend. At first, the tension this caused was moderated by the influence of Willughby's mother, Lady Cassandra, who was much more well disposed towards him. But when she died in 1675, Emma Willughby had a free hand and soon married one Josiah Child, a hugely wealthy man described by Ray as 'sordidly covetous'. His position at Middleton Hall became impossible and he had to move on. He still had the income of £60 a year left to him by Willughby, but for the foreseeable future he had no access to Willughby's collections at Middleton Hall and could not easily complete their planned programme of publications.

Perhaps with an eye to the future, Ray's personal circumstances had already changed in 1673, when he married a girl in the Middleton household, Margaret Oakley – not a serving girl, but a 'gentlewoman' responsible in some way, perhaps as governess, for the children. She was 24 years younger than him and the relationship was clearly more of a practical arrangement than a love match (like the second marriage of Isaac Newton's mother), but it seems to have been a happy one, although it produced no children until Ray was 55, when Margaret gave birth to twin daughters; two other girls quickly followed.

After being ejected from Middleton Hall, the Rays lived first at Sutton Coldfield, then near Black Notley, until in 1679 Ray's mother died and they took over the house he had provided for her, living on the £60 a year plus about £40 a year in rents from some land near by (we don't know exactly how the land came into the family). It was just enough to support the family and leave Ray (who turned down several offers of employment over the next few years) free to spend the next quarter of a century with uninterrupted freedom to work at what he liked, completing a series of great books setting the biological world in order. We shall only mention the most important titles, although Ray wrote many others (including books on English language and dialect).

Ray always felt, with genuine modesty, that without Willughby's help (both intellectual and financial) he would have achieved nothing, and his first priority was to get into print the books on birds and fishes which, according to their friendly division of the living world, Willughby would have been responsible for had he lived. The book on birds was essentially complete by the time Ray had to leave Middleton Hall and was published, as Francis Willughby's *Ornithology*, in 1677. But although Ray had also done a great deal of work on fishes (both on his own and with Willughby), much remained to be done on this topic in 1675, and it was only after he settled at Black Notley in 1679 that he was able to get back to work on the project. He completed it in spite of the difficulties with access to the material at Middleton Hall. The *Ornithology* might just be regarded as being, in reality, a joint publication of Ray and Willughby; the *History of Fishes* actually owes very little to Willughby (except his collecting) and ought really to be regarded as a Ray book. No matter; it was under Francis Willughby's name that the *History of Fishes* was published in a splendidly illustrated edition, in 1686, at a cost of £400 to the Royal Society.

In the gaps between working on Willughby's part of the living world, Ray had also been working on his first love, botany, and the first volume of his immense *History of Plants* was also published in 1686, with the second volume following in 1688 and a third in 1704. The book covered more than 18,000 plants, classifying them in terms of their family relationships, morphology, distribution and habitats, as well as listing their pharmacological uses and describing general features of plant life such as the process of seed germination. Most important of all, he established the species as the basic unit of taxonomy – it was Ray who established the very concept of a species in the modern sense of the term, so that, in Ray's own words, members of one species are 'never born from the seed of another species'. In more Biblical language, dogs beget dogs, cats beget cats and so on; dogs and cats are therefore separate species.

Ray died the year after the third volume of his *History of Plants* was published, on 17 January 1705, at the age of 77. He left unpublished a rough draft of his last great work, a *History of Insects*, which appeared posthumously in 1710. In spite of Willughby's early death, his own financial problems and severe ill health in later years, and in addition to an outpouring of other work, he had single-handedly completed the task that he and Willughby had set themselves so long ago, of setting the biological world in order.

It was Ray who, more than anybody else, made the study of botany and zoology a scientific pursuit, bringing order and logic to the investigation of the living world out of the chaos that existed before.[1] He invented a clear taxonomic system based on physiology, morphology and anatomy, thereby paving the way for the work of the much more famous Linnaeus, who drew heavily (without always acknowledging the debt) on the work published by Ray, both under his own name and under Willughby's name. Although deeply religious, Ray also found it hard to reconcile the Biblical account of the Creation with the evidence of his own eyes, not just in the variety of the living world (where he came close to suggesting that species are not fixed, but can change as generations pass) but also from his studies of fossils, which he was among the first to recognize as the remains of once-living creatures and plants. As early as 1661 he was making notes on what were then called 'serpent stones', and following the pioneering work of Hooke and Steno (of whom more shortly) in the 1660s, he returned repeatedly to this theme in his writings, puzzling over the idea that the absence of living forms of fossilized species today seems to imply that whole species have been wiped from the face of the Earth, and struggling with the idea (which he rejected) that the presence of fossil fishes in the rocks from high mountains implies that the mountains have been uplifted over immense lengths of time.[2] As early as 1663, writing in Bruges, he describes a buried forest found 'in places which 500 years ago were sea' and writes:

1. Before Ray, plants and animals were 'classified' in alphabetical order of their names, and included mythical beasts such as the unicorn. This may not, strictly speaking, amount to chaos, but was hardly scientific!

2. To appreciate just how confused people still were in the 1660s about the nature and origin of life, note that it was only in 1668 that Francesco Redi (1626–1698) proved by careful experiments with pieces of meat kept sealed away from egg-laying flies that maggots did not arise spontaneously from the rotten meat itself.

Many years ago before all records of antiquity these places were part of the firm land and covered with wood; afterwards being overwhelmed by the violence of the sea they continued so long under water till the rivers brought down earth and mud enough to cover the trees, fill up these shallows and restore them to firm land again . . . that of old time the bottom of the sea lay so deep and that that hundred foot thickness of earth arose from the sediment of those great rivers which there emptied themselves into the sea . . . is a strange thing considering the novity of the world, the age whereof, according to the usual account, is not yet 5600 years.'[1]

Ray's puzzlement accurately reflects the way people struggled towards an understanding of the true immensity of the span of geological time, seeing the evidence with their own eyes, but not being able to bring themselves to accept the implications at first. But before we pick up the threads of the history of geology, we ought to see how Ray's work led, through Linnaeus, to the development of the scientific understanding of the living world that was the essential precursor to a satisfactory theory of evolution.

Carl Linnaeus and the naming of species

Carl Linnaeus is unique among the scientists whose work we describe in this book in changing his name from a Latinized version to a vernacular version. But that happened only because the family name had been Latinized in the first place by his father, a clergyman who was originally known as Nils Ingemarsson but invented the family name Linnaeus after a large linden tree growing on his land. And the only reason Carl (sometimes known as Carolus), a vain man with an inflated idea of his own importance, changed this splendid name is that in 1761 he was granted a patent of nobility (backdated to 1757) and became Carl von Linné. But it is simply as Linnaeus that he is known to posterity.[2]

Linnaeus was born at South Râshult, in southern Sweden, on 23 May 1707. The family was not wealthy, and intended Linnaeus to

1. Quoted by Raven, from Ray's *Observations*.

2. It's a measure of the high opinion Linnaeus had of himself that he wrote no less than four autobiographical memoirs, which were published after his death and established his image for immediate posterity. The gloss put on some of his achievements should be taken with a pinch of salt, but the status of his scientific work is clear without any such polishing.

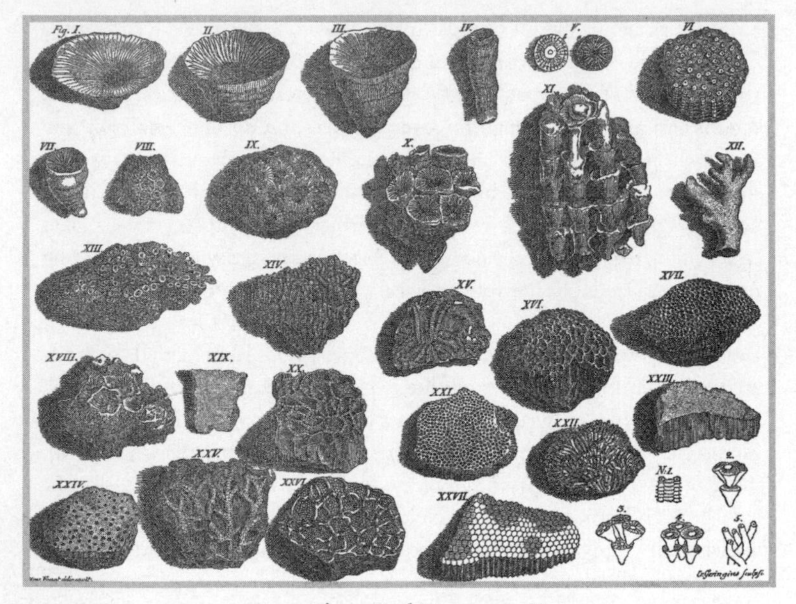

19. *A page from Carl Linnaeus's* Såsom Naturforskare Och Läkare, *1746.*

follow in his father's footsteps as a clergyman. But he showed so little interest or aptitude for this that his father was about to have him apprenticed to a shoemaker when one of the young man's teachers intervened and suggested that he might make a career in medicine. With the aid of sponsorship from several patrons, Linnaeus was able to complete his medical studies, commencing at the University of Lund in 1727 and continuing in Uppsala from 1728 to 1733. Linnaeus had been interested in flowering plants since he was a child, and at university his reading on botany went far beyond the curriculum required for medical students. He was particularly taken with the new idea, put forward by the French botanist Sébastian Vaillant (1669–1722) in 1717, that plants reproduced sexually, and had male and female parts which corresponded to the reproductive organs of animals. The novelty and boldness of this realization in the eighteenth century can perhaps be appreciated from the fact that Linnaeus himself never fully understood the role of insects in pollination, although he was one of the first to accept and make use of the idea of sexual reproduction in plants.

Alongside his medical studies, Linnaeus developed the idea of using the differences between the reproductive parts of flowering plants as a means of classifying and cataloguing the plants. This was a natural step for him to take because he was an obsessive cataloguer who made lists of everything (the archetypal anorak-ish stamp collector). When he became a professor, his botanical outings with the students were organized with military precision; the students even had to wear special light clothing, referred to as the 'botanical uniform'. They always set out at precisely 7 am, with a meal break at 2 pm and a short rest at 4 pm, while the professor gave demonstrations precisely every half-hour. In a letter to a friend, Linnaeus once commented that he could not 'understand anything that is not systematically ordered'.[1] In many people, this would be regarded as an aberration, something to worry about rather than to be proud of; but Linnaeus found exactly the right channel in which to direct his abnormally obsessive behaviour. His talents were soon recognized – at Uppsala from 1730, he stood in for one of the professors, Olaf Rudbeck, carrying out demonstrations at the university's botanical gardens, and in 1732 the Uppsala Society of Science sent him on a major expedition to Lapland to gather botanical specimens and to investigate the local customs of what was then still a mysterious northern land.

In 1734, Linnaeus undertook another botanical expedition, this time to central Sweden, before completing his medical degree at the University of Hardewijk, in Holland, in 1735. He then moved on to the University of Leiden, returning to Sweden in 1738 and marrying Sara Moraea (a doctor's daughter) in 1739. He practised as a physician in Stockholm until 1741, when he was appointed to the chair of medicine at Uppsala. In 1742, he switched to the chair of botany, which he held for the rest of his life. He died in Uppsala on 10 January 1778. For all his faults, Linnaeus was a charming man and popular teacher, whose many students spread his ideas about taxonomy far and wide both during his own lifetime and after his death. But the most remarkable thing about those ideas is that they were essentially completed while he was still a student and published in a mature form, as *Systema Naturae* as early as 1735, soon after Linnaeus arrived in

1. Quoted by Sten Lindroth, in Frängsmyr.

CAROLI LINNÆI
Naturæ Curiosorum *Dioscoridis Secundi*

SYSTEMA
NATURÆ

IN QUO
NATURÆ REGNA TRIA,
SECUNDUM
CLASSES, ORDINES, GENERA, SPECIES,
SYSTEMATICE PROPONUNTUR.

Editio Secunda, Auctior.

STOCKHOLMIÆ
Apud GOTTFR. KIESEWETTER.
1740.

20. *Title page of the* Systema Naturae, *1740.*

Holland. The work went through many revisions and editions, and the innovation for which Linnaeus is now best remembered, the binomial system of classifying every species with a two-word name, was spelled out in volume 1 of the tenth edition, which appeared in 1758 (the year of the return of Halley's comet), after having been introduced in his book *Species Plantarum* in 1753. It was that tenth edition that introduced to biology, and defined, the terms *Mammalia*, *Primates* and *Homo sapiens*, among others.

The idea of giving species two-word names was not in itself new, and occurs in vernacular descriptions going back to ancient times; but what Linnaeus did was to turn this into a systematic method of identification, with precise ground rules. But none of this would have had much point without all the effort that went into identifying and classifying species in terms of their characteristics – the field research carried out by Linnaeus himself, his students and predecessors such as Ray. In his various publications, Linnaeus provided descriptions of some 7700 species of plants and 4400 species of animals (just about everything known in Europe at the time), eventually giving them all names in the binomial system. Everything in the living world was arranged by Linnaeus in a hierarchy of family relationships, from the broad classifications of their Kingdom and Class down through the subdivisions Order and Genus to the Species itself. Although some of the names have been changed down the years, and some species rearranged in the light of later evidence, the point is that from the time of Linnaeus onwards, whenever a biologist refers to a species (as, say, *Canis lupus*, the wolf) their colleagues know precisely which species they are referring to. And if they don't, they can go and look up all the particulars of that species in standard texts, and even see preserved type specimens of that species in the vaults of natural history museums.[1] The influence of this system can be seen in the way in which it preserves the last vestige of Latin, formerly the universal language of science, in scientific work right up to the present day. As future generations of

1. One of the most important of these collections is built around the material gathered by Linnaeus himself. After his death, this was purchased by a wealthy English botanist, James Smith, who helped found the London-based Linnean Society in 1788. When Smith died in 1828, the society bought the collection, which it still has, for the then vast sum of £3150, incurring a debt that took 33 years to pay off.

botanists and zoologists explored the world beyond Europe, the new species they discovered could be classified and fitted in to the naming system in the same way, providing the raw material from which the relationships between species and the laws of evolution would begin to become clear in the nineteenth century.

All of his cataloguing work could, if you were being uncharitable, be dismissed as mere stamp collecting. But Linnaeus took one bold step which changed humankind's view of our place in nature for ever. He was the first person to include 'man' (as they referred to humankind in those days) in a system of biological classification. Just how man fitted in to the biological scheme of things took him some time to decide, and the whole idea of classifying man in the same way as the animals was, of course, controversial in the eighteenth century. The ultimate modern version of the classification (going beyond the original Linnaean system) gives our precise position in the living world:

Kingdom	Animalia
Phylum	Chordata
Subphylum	Vertebrata
Class	Mammalia
Order	Primates
Family	Hominidae
Genus	Homo
Species	sapiens

As things stood in Linnaeus' classification, *Homo sapiens,* our own species, was unique in being the only member of a genus – the genus *Homo* has but a single member. Linnaeus saw things differently, however, and should not be accused of displacing *Homo sapiens* too far from the other animals – he included several other species of 'man' in the genus *Homo,* based on folk legends and myths of tailed men, 'troglodytes' and so on. He also agonized about whether there ought to be a separate genus for *Homo* at all. In the foreword to his *Fauna Svecica,* published in 1746, he said 'the fact is that as a natural historian I have yet to find any characteristics which enable man to be distinguished on scientific principles from an ape', and in response to criticism of this position he wrote to a colleague, Johann Gmelin, in 1747:

I ask you and the whole world for a generic differentia between man and ape which conforms to the principles of natural history. I certainly know of none ... If I were to call man ape or vice versa, I should bring down all the theologians on my head. But perhaps I should still do it according to the rules of science.'[1]

In other words, it was Linnaeus's own belief that man belonged in the same genus as the apes, a belief thoroughly born out by modern studies of the similarities between the DNA of humans, chimpanzees and gorillas. If the classification were being made from scratch today, using the DNA evidence, man would actually be classified as a chimpanzee – *Pan sapiens*, perhaps. It is only through a historical accident and Linnaeus's fear of arousing the wrath of the theologians that *Homo sapiens* sits in unique and isolated splendour as the sole member of a genus.

Linnaeus was religious, and certainly believed in God. Like so many of his contemporaries, he saw himself as uncovering God's handiwork in his classification of nature, and he said on more than one occasion that the number of species existing on Earth in his day was the same as the number created by God in the beginning.[2] But this did not stop him having doubts about the standard eighteenth-century interpretation of the Bible, particularly when it came to the question of the age of the Earth.

Linnaeus was drawn into this debate by a controversy which raged in Sweden in the 1740s, following the discovery that the level of water in the Baltic Sea seemed to be going down.[3] One of the first people to investigate this phenomenon properly, and to present convincing evidence that there really was a change in sea level, was Anders Celsius

1. Quoted by Gunnar Broberg, in Frängsmyr.

2. More than once, though, Linnaeus writes 'there are as many varieties as there are plants produced by the seed of the same species', implying that no two individuals are identical and coming close to finding one of the keys that Darwin would use to unlock the secrets of evolution. Quoted by Gunnar Eriksson, in Frängsmyr.

3. We now know that in fact the land there is rising. During the most recent Ice Age, the weight of ice made the solid crust of the region sink down into the fluid layers below the surface of the Earth, and it is still rebounding from the release of that weight some 10,000 years ago.

(1701–1744), known today for the temperature scale which bears his name. One of the possible explanations for the 'diminution of the waters' put forward by Celsius built on an idea discussed by Newton in the third volume of the *Principia*, to the effect that water is turned into solid matter by the action of plants. The idea was that plant material is chiefly made of fluids,[1] taken up from their surroundings, and when plants decay they form solid matter which is carried into the seas and lakes by rivers, settles on the bottom and builds up new rocks. Linnaeus developed the idea with an elaborate model in which the role of surface weeds in keeping the waters still (as in the Sargasso Sea) and encouraging sedimentation played a big part; but the details of the model need not concern us, since it was wrong in almost every respect. What is significant, though, is that these investigations led Linnaeus to consider the age of the Earth.

By the 1740s, the existence of fossils far from any present-day seas was well known, and it was widely accepted that these were the remains of once-living creatures. The idea had gained currency after the work of the Dane Niels Steensen (1638–1686), who Latinized his name to Nicolaus Steno and is usually known simply as Steno. In the mid-1660s, he made the connection between the distinctive characteristics of sharks' teeth and fossil remains, which he realized were those of sharks, found in rock strata far inland today. Steno argued that different rock strata had been laid down under water at different times in the history of the Earth, and many of his successors in the eighteenth century (and even into the nineteenth) identified this process with the Biblical Flood (or a series of floods). Linnaeus accepted the Biblical account of the Flood, but reasoned that such a short-lived event (a flood lasting less than 200 days) could not have moved living things far inland and covered them in sediment in the time available – 'he who attributes all this to the Flood, which suddenly came and as suddenly passed, is verily a stranger to science,' he wrote, 'and himself blind, seeing only through the eyes of others, as far as he sees anything at all'.[2] Instead, he argued that the whole Earth was initially covered by water, which had been receding ever since, continually being turned

1. Actually not so; plants are mainly made of carbon dioxide, from the air.
2. Quoted by Frängsmyr.

into dry land and leaving behind fossils as evidence of the waters that once covered the Earth. All of this clearly required much longer than the 6000 years of history allowed by Bible scholars of the time, but Linnaeus never quite brought himself to say as much.

In the eighteenth century, there were already grounds to question the date of 4004 BC calculated for the Creation by Archbishop James Ussher in 1620, not only from science but from history. At that time, information about China was beginning to filter out to Europe, mainly as a result of the work of French Jesuit missionaries, and it was becoming well known that the first recorded Emperor was on the throne of China some 3000 years before the birth of Christ, with the implication that Chinese history went back significantly further still. Although some theologians simply attempted to make Ussher's chronology match the Chinese records, Linnaeus wrote that he 'would gladly have believed that the earth was older than the Chinese had claimed, had the Holy Scriptures suffered it', and that 'in former times, as now, nature built up the land, tore it away, and built it up again'.[1] Linnaeus couldn't quite bring himself to say, in so many words, that the standard theological interpretation of the Bible was wrong. But in France, his exact contemporary Georges Louis Leclerc, known to posterity as the Comte de Buffon, was going that vital step further and carrying out the first truly scientific experiments aimed at determining the age of the Earth.

The Comte de Buffon: Histoire Naturelle *and thoughts on the age of the Earth*

Buffon (as I shall refer to him for consistency) was born on 7 September 1707, in Montbard, just to the northwest of Dijon (then, as now, the region's capital) in the Burgundy country. His paternal family had been peasants just a couple of generations earlier, but Buffon's father, Benjamin-François Leclerc, was a minor civil servant involved in the local administration of the salt tax. Then, in 1714, Buffon's maternal uncle died, leaving a huge fortune to his sister, Buffon's mother. With the money, Leclerc purchased the entire village of Buffon, close to Montbard, extensive lands, property in Montbard and Dijon, and obtained an appointment for himself as a councillor in the local parliament in Dijon. The term 'nouveau riche' might have been invented for him, and

1. *ibid.*

Buffon himself (perhaps sensitive about his humble origins) remained a vain social climber throughout his life. The family took up residence in Dijon and Buffon became a student at the Jesuit College, from where he graduated with qualifications in law (although he had also studied mathematics and astronomy) in 1726.

Only the barest outline of Buffon's life is known for the next few years. It seems that he spent some time in Angers, where he studied medicine and probably botany, but left without obtaining any formal qualifications (he later said this was as a result of a duel, but that is almost certainly a story he made up to impress people with). At some point, he met up with two travellers from England, the second Duke of Kingston (then in his late teens) and his tutor/companion Nathan Hickman. Buffon joined them on the Grand Tour. This really was grand – the Duke travelled with an entourage of servants and possessions in several coaches, and stayed in magnificent lodgings for weeks or months on end. It was a lifestyle that young Buffon found it easy to aspire to, and he soon had a chance to put his wishes into practice. In the summer of 1731, Buffon left his companions and returned to Dijon, where his mother was seriously ill. She died on 1 August and he rejoined the English party in Lyons, from where they made their way through Switzerland and into Italy. By August 1732 Buffon was back in Paris, but then his life changed dramatically, and after the end of that year he never travelled again, except for regular trips between Montbard and Paris.

The turning point in Buffon's life came when his father remarried, on 30 December 1732, and attempted to appropriate the entire family fortune, including the share left to Buffon by his mother. Although Leclerc had a total of five children by his first marriage, two had died (in their early twenties) in 1731, the same year that their mother died; one surviving son had become a monk, the only surviving daughter had become a nun. Which left Buffon, now 25, and his father to fight it out over the inheritance. The outcome was that Buffon acquired a substantial fortune in his own right, plus the house and lands in Montbard, and the village of Buffon. The latter was particularly important because he had already started signing himself Georges-Louis Leclerc de Buffon during the Grand Tour, perhaps feeling that his real name was not impressive enough for a friend of the Duke of Kingston.

But he never spoke to his father again, and after 1734 he dropped the 'Leclerc' and simply signed himself Buffon.

He could have led a life of indolent ease. To put his wealth in perspective, Buffon's income eventually amounted to about 80,000 livres a year, at a time when the minimum required for a gentleman to maintain himself in a manner befitting his position (if not quite at the level of the Duke of Kingston) was about 10,000 livres a year. But Buffon did not rest on his inherited wealth. He managed his estates successfully and profitably, developed a tree nursery to provide trees to line the roads of Burgundy, established an iron foundry in Buffon and developed other businesses. Alongside all this, he developed his interest in natural history into what would have been a demanding full-time career on its own for most people. To achieve all this and overcome what he felt to be his natural laziness, Buffon hired a peasant to physically drag him out of bed at 5 am each morning and make sure he was awake. For the next half-century he would start work as soon as he was dressed, stop for breakfast (always two glasses of wine and a bread roll) at 9 am, work until 2 pm before taking a leisurely lunch and entertaining any guests or casual visitors,[1] take a nap and a long walk, then a final burst of work from 5 pm to 7 pm, with bed at 9 pm and no supper.

This dedication to hard work explains, physically, how Buffon was able to produce one of the most monumental, and influential, works in the history of science, his *Histoire Naturelle*, which appeared in 44 volumes between 1749 and 1804 (the last eight published, using Buffon's material, after his death in 1788). This was the first work to cover the whole of natural history, and it was written in a clear and accessible style which made the books popular bestsellers. It ran into many editions and translations, and both added to Buffon's wealth and spread interest in science widely in the second half of the eighteenth century. Buffon did not make any great original contributions to the

1. One of those visitors, Thomas Jefferson, recalled 'It was Buffon's practice to remain in his study till dinner time, and receive no visitors under any pretense; but his house was open and his grounds, and a servant showed them very civilly, and invited all strangers and friends to remain to dine . . . we dined with him, and he proved himself then, as he always did, a man of extraordinary powers in conversation.' (Quoted by Fellows and Milliken.)

understanding of the natural world (in some ways he may have hindered progress, particularly with his opposition to Linnaeus's ideas about species), but he did bring together a wealth of material and put it into a coherent shape, providing a jumping-off point for other researchers and a stimulus to encourage people to become naturalists. Even this isn't the end of the story, though, since on top of (or alongside) everything else, from 1739 onwards Buffon was the superintendent (or keeper) of the Jardin du Roi, the King's botanical gardens in Paris.

The way Buffon got this post was typical of the way the *ancien régime* worked. He had aristocratic contacts (not least thanks to the time he had spent in Paris in the company of the Duke of Kingston), he was at the very least a gentleman (and had the airs of an aristocrat), he had independent means (a key factor, since the government was nearly bankrupt and, far from being able to draw his full salary, he actually had to put his own money in from time to time to cover expenses and keep the gardens operating effectively) and (almost as a bonus) he was good at the job.

During the 1730s, Buffon had become noticed in scientific circles through his publications on mathematics (in the *Mémoires* of the Académie des Sciences) and his experiments in silviculture, aimed at encouraging reforestation and providing wood of a higher standard for the ships of the French navy. He joined the Académie in 1734, and rose through its hierarchical structure to become an Associate Member in June 1739, when he was 31. Just a month later, the superintendent of the Jardin died unexpectedly, and with the other main candidate for the post away in England, Buffon's contacts quickly managed to slip him in to the post that he would hold for the next 41 years. Although he was hugely important and influential as an administrator and popularizer of science,[1] it is Buffon's contribution to the development of original ideas in science that concerns us here, and these can be dealt with fairly quickly.

1. His skill as a popularizer was amply shown in 1747, when he demonstrated in public that by focusing the Sun's rays with an array of mirrors it was possible to set fire to wood at a distance of 200 feet, just as Archimedes was said to have done when the Greeks defeated the Roman fleet at Syracuse.

His personal life can be dealt with even more quickly. In 1752, at the age of 44, Buffon married a girl of 20, Marie-Françoise, who bore him one daughter, who died in infancy, and a son on 22 May 1764. After the birth of the second child, Marie-Françoise suffered more or less continual ill health until she died in 1769. The son, nicknamed Buffonet, was a sad disappointment to his father, becoming a wastrel and reckless horseman, and possessing just about enough intellectual capacity for a career as an officer in the army (where, of course, commissions were bought, not earned on merit). In spite of all this, Buffon made arrangements for his son to succeed him at the Jardin; but when Buffon became seriously ill, the authorities quickly took steps to change the arrangements for the succession. Buffon was given the title Comte in July 1772. Buffonet inherited the title, but lived to regret it; he eventually ended up a victim of the Terror, after the French Revolution, guillotined in 1794.

Buffon's most unconventional (though not entirely original) contribution to science was the speculation that the Earth had formed out of material thrown out of the Sun as the result of the impact of a comet (an idea which he based on a comment by Newton that 'comets occasionally fall upon the Sun'[1]). This idea suggested that the Earth formed in a molten state and had gradually cooled to the point where life could exist. But it seemed clear that this must have taken much longer than the 6000 years or so allowed by the theologians' estimate of the time since the Creation. Newton himself said as much, in the *Principia*:

> A globe of red-hot iron equal to our earth, that is about 40000000 feet in diameter, would scarcely cool in an equal number of days, or in above 50000 years.

But he didn't try to calculate how long it would take for such a globe to cool, and contented himself by pointing the way for posterity:

> I suspect that the duration of heat may, on account of some latent causes, increase in a yet less ratio than that of the diameter; and I should be glad that the true ratio was investigated by experiments.

1. Quoted by Fellows and Milliken.

Buffon took the hint. The experiment he devised involved heating balls of iron of different sizes until they were red hot, then timing how long it took them to cool until they could just be touched without burning the skin. Then, using a rather rough and ready technique, Buffon extrapolated how long it would take to cool a similar ball the size of the Earth. The experiment was not really very accurate, but it was a serious scientific attempt to estimate the age of the Earth, owing nothing at all to the Bible but extrapolating from actual measurements. This makes it a landmark event in science. In the *Histoire*, Buffon writes:

> Instead of the 50,000 years which [Newton] assigns for the time required to cool the earth to its present temperature, it would require 42,964 years and 221 days to cool it just to the point at which it ceased to burn.

Ignore the spurious accuracy of the quoted number. Buffon went on to calculate that (rounding his figures off) the Earth must be at least 75,000 years old. Don't be fooled by the fact that this is so much less than the best current estimate, 4.5 *billion* years. What matters, in context, is that the number is more than ten times the age inferred by Bible scholars, bringing science into conflict direct with theology in the second half of the eighteenth century.[1] But Buffon's estimate of the age of the Earth pales into insignificance compared with one made by a member of the next generation of French scientists, Jean Fourier. Unfortunately, though, Fourier seems to have been so stunned by his own calculation that he never published it.

Further thoughts on the age of the Earth: Jean Fourier and Fourier analysis

Fourier, who lived from 1768 to 1830, is best remembered today for his work in mathematics. He is one of the people whose life and work we do not have space to elaborate on here – except to mention that he served as scientific adviser to Napoleon in Egypt, where he ended up running half the country from 1798 to 1801, held positions

1. Buffon avoided conflict with the Church authorities using the time-honoured formula of presenting his ideas as mere 'philosophical speculations'. Well, if it was good enough for Galileo, why not? Outwardly at least, he remained a practising Catholic to the end of his life, and took the last rites.

in the civil administration of France under Napoleon, and was made first a Baron and then a Count for his services to the Empire. He survived the upheaval of the restoration of Louis XVII to achieve a position of prominence in French science before he died as a result of an illness contracted while in Egypt. The essence of his work is that he developed mathematical techniques for dealing with time-varying phenomena – to break down the complicated pattern of pressure variations in a blast of sound, for example, into a collection of simple sine waves that can be added together to reproduce the original sound. His techniques of Fourier analysis are still used at the forefront of scientific research today, for example by astronomers trying to measure the variability of stars or quasars. Fourier developed those techniques not out of a love of mathematics for its own sake, but because he needed them to describe mathematically a phenomenon that he was really intrigued by – the way heat flows from a hotter object to a cooler object. Fourier then went one better than Buffon by developing sets of mathematical equations to describe heat flow (based on many experimental observations, of course), and he used these equations to calculate how long it would have taken the Earth to cool. He also made allowance for a factor that Buffon had overlooked. The solid crust of the Earth acts like an insulating blanket around its molten interior, restricting the flow of heat so that, indeed, the core is still molten today even though the surface is cool.[1] Although Fourier must have written down the number that came out of his calculation, he seems to have destroyed the paper he wrote it on. What he did leave for posterity was a formula for calculating the age of the Earth (written down in 1820). Using this, any interested scientist could put the appropriate heat flow numbers in and get the age of the Earth out. When you do this, Fourier's formula gives an age of 100 *million* years – more than a thousand times longer than Buffon's estimate and only 50 times shorter than the best modern estimate. By 1820, science was well on the way to measuring the true timescale of history.

Buffon's other contributions, though, emphasize just how much science at the end of the nineteenth century was struggling to come to

1. The main reason the core is still so hot today, though, is because of the energy released by radioactivity; more of this later.

terms with the growing weight of evidence from fossil bones of the antiquity of life on Earth. He argued that heat alone had been responsible for creating life, and took the seemingly logical step of arguing that since the Earth was hotter in the past, it was easier to make living things, which is why the ancient bones (now known to be those of mammoths and dinosaurs) were so big. From place to place in the *Histoire*, Buffon also hinted at an early version of the idea of evolution – such ideas were discussed long before the work of Charles Darwin, whose key contribution, as we shall see, was to find the *mechanism* of evolution (natural selection). But it is still rather startling (even allowing for the now outmoded idea of some species being 'higher' or 'lower' than others) to find Buffon saying as early as 1753, in volume IV of the *Histoire*:

> If we once admit that there are families of plants and animals, so that the ass may be of the family of the horse, and that the one may only differ from the other through degeneration from a common ancestor, we might be driven to admit that the ape is of the family of man, that he is but a degenerate man, and that he and man have had a common ancestor, even as the ass and horse have had. It would follow then that every family whether animal or vegetable, had sprung from a single stock, which after a succession of generations, had become higher in the case of some of its descendants and lower in that of others.

He also presented one of the clearest arguments against living creatures having been individually designed by an intelligent Creator (although he didn't express this conclusion in so many words). The pig, he pointed out:

> Does not appear to have been formed upon an original, special, and perfect plan, since it is a compound of other animals; it has evidently useless parts, or rather parts of which it cannot make any use, toes all the bones of which are perfectly formed, and which, nevertheless, are of no service to it. Nature is far from subjecting herself to final causes in the formation of these creatures.

Even in translation, these passages give you an idea of why Buffon's writings were so popular; in France, he is regarded as a major literary figure for his fine style, independently of what he was writing about.

When it came to present-day life, Buffon was also involved in the debate about just how sexual reproduction worked. There were three schools of thought. One held that the seed of future generations was stored inside the female, and that the only contribution of the male partner was to trigger it into life. One held that the seed came from the male, and that the female's role was only to nurture it. And a few held that contributions from both partners were essential, explaining why a baby might have 'its father's eyes' and 'its mother's nose'. Buffon subscribed to this third view, but in the form of a horribly complicated model that there is no point in describing here.

Buffon, very much a man (and a scientist) of his time, died in Paris, after a long and painful illness involving kidney stones, on 16 April 1788. The society he knew was about to be turned upside down by revolution; but in science the revolution had already happened and its impact would gather momentum into the nineteenth century, even in the midst of the political upheavals (as Fourier's work typifies). In terms of the understanding of life on Earth, the next great leap was also taken in Paris, by Georges Cuvier, who picked up in the 1790s where Buffon had left off in the 1780s.

Georges Cuvier: Lectures in Comparative Anatomy; *speculations on extinction*

Cuvier was born on 23 August 1769 at Montbéliard, then the capital of an independent principality but now part of France, on the border with Switzerland. Although the Montbéliards spoke French, they were largely Lutheran, and had many cultural links with the German-speaking states to the north, together with a deep antipathy towards the French, who had repeatedly tried to absorb their tiny neighbour. At the time Cuvier was born, Montbéliard had been politically linked to the Grand Duchy of Württemberg for a couple of hundred years and was ruled by a junior branch of the Grand Duke's family on his behalf. Instead of Montbéliard being a parochial backwater, this meant that there was a clear and well-trodden path for able young men to follow out of the principality and into the wider European world. Cuvier's father had been a soldier, serving as a mercenary officer in a French regiment, but was retired on half-pay at the time the boy was born. Although the family was hard up (Cuvier's mother, who was some 20 years younger than her husband, had no money of her own), the French connection gave Cuvier a powerful potential patron in the form of the Comte de Waldner, who had been the commanding officer of his father's regiment and who became the boy's godfather. This was more than just a nominal relationship and young Cuvier often visited the Waldner household when he was a child.

Earlier in 1769, the first child of the ex-soldier and his wife, a boy called Georges, had died at the age of four; when the new baby was born, he was christened Jean-Leopold-Nicholas-Frédéric, and soon after the name Dagobert (one of his godfather's names) was formally added to the list. But the boy was always known by his dead brother's name and he signed himself Georges throughout his adult life. It is as Georges Cuvier that he has gone down in history. The boy became a channel for all his parents' hopes and ambitions, receiving the best education they could provide; when another boy, Frédéric, was born four years later, he received far less attention.

From about the age of 12 onwards, Cuvier often visited the house of his paternal uncle, Jean-Nicholas, who was a pastor and had a complete collection of all the volumes of Buffon's *Histoire Naturelle* published to that time. Georges was fascinated by the book and spent hours absorbed in it, as well as going out into the countryside to collect his own specimens; but there was no suggestion at this time that it might be feasible for him to earn a living as a naturalist. The route Cuvier's parents planned for him to follow to a respectable and secure career was the traditional one of becoming a Lutheran pastor, but he was turned down for a free place at the University of Tübingen and the family was too poor to pay his fees. But although the family was poor, it had connections at court through the Comte de Waldner, and at about this time the Grand Duke of Württemberg, Charles-Eugène, paid a visit to the governor of Montbéliard, Prince Frédéric. The Duke was informed of the young man's plight and offered him a free place at the new Academy in Stuttgart, which the Duke himself had founded in 1770 and which was awarded the status of a university by the Emperor Joseph II in 1781. Cuvier took up his place at the new university in 1784, when he was 15.

The Academy had been established as a training ground for civil servants, to prepare young men to become administrators in the many states of the fragmented Germany of the period. It was run like a military establishment, with uniforms, rigid codes of conduct and strictly enforced rules; but it provided both a superb education and, initially at least, a guaranteed job for life at the end of that education. Not that everyone appreciated this. Friedrich Schiller had graduated from the Academy in 1782 and promptly clashed with the authorities because he did not want a job for life – he wanted to be a poet and playwright, but was forced to take up a career as a regimental surgeon[1] until he escaped from the region under the influence of Charles-Eugène in 1784, just when Cuvier was getting to grips with his new life. But when Cuvier himself graduated in 1788, the situation was reversed. The Academy (and similar institutions across Germany) had been so successful that by then it had produced more potential administrators than there were jobs for them, and, like many of his contemporaries, instead of having a job for life Cuvier was left entirely to his own resources. Unfortunately, he had no financial resources at all, and as a short-term expedient to earn a living while he took stock of the situation, he took up a post as tutor to a family in Normandy, at Caen, following in the footsteps of another young man from Montbéliard, Georg-Friedrich Parrot, who was moving on to a better position and recommended his compatriot as his successor.

This was, as the apocryphal Chinese curse would say, an interesting time to be living in France. Fortunately, Normandy was at first remote from the political upheavals in Paris and Cuvier was able to re-establish his botanical and zoological interests (which had flourished in Stuttgart and which he wrote about to friends he had made at the university), gaining access to the botanical gardens in Caen and to the library of the university. Cuvier worked for the Marquis d'Héricy and his family, as tutor to their son Achille. The family owned a house in Caen and two modest châteaux, although they mainly used just one of them, at Fiquainville, as their summer residence.

Although the anniversary of the French Revolution is officially

1. This conjures up an irresistible Pythonesque image of the surgeon leaning over his patient while clutching a scalpel and confiding, 'I really wanted to be a poet, you know', before setting to work.

marked on 14 July, commemorating the storming of the Bastille in 1789, the limited reforms made by the National Assembly in response to the upheavals of that year more or less kept a lid on things until 1791, when the royal family set off the next wave of change by their unsuccessful attempt to flee. That year, Normandy was caught up in the turmoil, with the university closed and rioting in the streets, triggered by hunger. The Marquise d'Héricy, and to a lesser extent her husband, were not unsympathetic to some of the demands of the reformers, but as members of the aristocracy they were clearly threatened, and the Marquise, Achille and Cuvier moved permanently to the summer residence at Fiquainville for safety. They were visited occasionally by the Marquis, but the Marquis and Marquise officially separated from one another at this time (which may have been a ploy to preserve some of the family estates in her name, whatever the fate of the Marquis). France became a republic and Cuvier had the opportunity, living quietly in the countryside, to become a real field naturalist, deliberately following in the footsteps of Linnaeus in identifying and describing hundreds of species. This encouraged him to develop his own ideas about how species ought to be classified, and the relationships between different kinds of animals and plants. He began to publish in the leading French journals and established contact by correspondence with the leading natural historians in Paris. But just as he was starting to make a name for himself, France entered the most vicious stage of the Revolution, the period known as the Terror, which began with the execution of Louis XVI and Marie Antoinette in 1793. The Terror lasted for more than a year and reached every corner of France. More than 40,000 opponents (or imagined opponents) of the Jacobin regime were executed in that time; you were either with the Jacobins or against them, and in the commune of Bec-aux-Cauchois, which included Fiquainville, Cuvier wisely chose to be with them. From November 1793 to February 1795 he worked as secretary (on a salary of 30 livres a year) for the commune, giving him a significant amount of influence, which he used to save the d'Héricy family from the worst excesses of the period. Already known to the scientific world in Paris, Cuvier now became known, if only in a minor way, as an able administrator with impeccable political credentials. As the Terror eased[1] early in 1795, Cuvier visited Paris with Achille d'Héricy (who was now nearly 18 and would not need a tutor much longer). It is not clear exactly what the purpose of the visit was, since it seems to have been deliberately concealed, but the most likely explanation is that Cuvier was lobbying on behalf of the family for the return of some of their property, seized during the Revolution, and that he also took the opportunity of sounding out his scientific contacts in Paris about the possibility of a job at the Museum of Natural History (which incorporated the Jardin des Plantes, formerly the Jardin du Roi). The outcome of these talks must have been promising, because Cuvier went back to Normandy, resigned his post as secretary to the commune at Bec-aux-Cauchois and returned to Paris, still a few months short of his twenty-sixth birthday.

Cuvier joined the staff of the Museum of Natural History as assistant to the professor of comparative anatomy, and remained associated with the museum (rising to more senior positions and also holding outside posts) for the rest of his life. After the upheavals of his youth, he put down roots in Paris, turning down (among other things) an offer to accompany Napoleon's expedition to Egypt in 1798. A year later he was appointed professor of natural history at the Collège de France, and a year after that he began publishing what ended up as his five-volume masterwork, *Lectures in Comparative Anatomy*. He was often short of money, though, and took on a variety of jobs in government and education, at the same time or overlapping one another, to establish some security. Among other things, Cuvier played a major role in organizing the new Sorbonne, and from about 1810 until his death in Paris (a victim of a cholera outbreak) on 13 May 1832 was probably the most influential biologist in the world, so well established that his positions were not seriously threatened by the restoration of the Bourbons in 1815. Along the way he had married (in 1804) a

1. The Jacobins were removed from power and replaced by the Directory in 1795, which lasted until 1799 before being overthrown by the coup which brought Napoleon to power; along the way, Montbéliard had been swallowed up by revolutionary France in 1793.

widow, Anne-Marie Duvaucel, who brought with her four children; there is some evidence[1] that before this he had at least two children by a long-term mistress whose name is not recorded. In 1831, he was made a Baron; it was very rare for a Protestant to be elevated to the peerage in France at that time.

Cuvier set new standards in comparative anatomy, providing insights into the way that different parts of a living animal work together that soon proved invaluable in interpreting and classifying fossil remains. He highlighted this approach by comparing the body plans of meat-eating and plant-eating animals. A meat eater must have the right kind of legs to run fast and catch its prey, the right kind of teeth for tearing flesh, claws to hold on to its victim with and so on. The plant eater, by contrast, has flat, grinding teeth, hoofs instead of paws and other distinctive features. Exaggerating only slightly, Cuvier claimed in his *Lectures* that an expert could reconstruct a whole animal by looking at a single bone; it is certainly true that, even if you are far from being an expert, identifying a single tooth as an incisor can tell you unequivocally that, for example, the animal the tooth belonged to had paws and claws, not hoofs.[2]

As far as the living world was concerned, Cuvier's comparative studies led him to the realization that it was not possible to represent all forms of animal life on Earth as members of a single linear system, linking the so-called lower forms of life with so-called higher forms of life (with man, of course, at the top of the supposed ladder of creation). Instead, he arranged all animals into four major groups (vertebrates, molluscs, articulates and radiates), which each had its own kind of anatomy. The actual classification that Cuvier came up with is no longer used; but the fact that he made any such classification was a significant break with past thinking about zoology and pointed the way ahead.

Applying these ideas to the study of fossil remains, Cuvier reconstructed extinct species and almost single-handedly invented the sci-

1. See Outram.

2. It is perhaps worth putting all this in an historical context by mentioning that Cuvier's great work was carried out almost exactly 200 years after Galileo's great work; the interval in time from Galileo to Cuvier is almost the same as the interval from Cuvier to ourselves.

ence of paleontology (along the way, he was the first person to identify the pterodactyl, which he named). One of the most important practical results of this kind of work was that it began to be possible to place the strata in which fossils were found in order – not to date them in any absolute sense, but to say which ones were older and which ones were younger. Working with Alexandre Brongniart (1770–1847), the professor of mineralogy at the Museum of Natural History, Cuvier spent four years probing the rocks of the Paris Basin, identifying which fossils occurred in which strata, so that once the original comparison had been provided, the discovery of known types of fossils elsewhere could be used to place strata in their correct geological and chronological sequence. It was even possible to see where life began. In later editions of his *Discours sur la Théorie de la Terre* (published in 1825 but based on material published in 1812), Cuvier wrote:

> What is even more surprising is that life itself has not always existed on the globe, and that it is easy for the observer to recognize the precise point where it has first left traces.

With clear evidence from these studies that many species that had formerly lived on Earth were now extinct, Cuvier subscribed to the idea that there had been a series of catastrophes, of which the Biblical Flood was just the most recent, during which many species went extinct. Some people took this idea further and argued that after each catastrophe God had been involved in a special Creation to repopulate the Earth; but after flirting with this idea Cuvier followed most of his colleagues in accepting that there had been only one Creation and that events since then had been worked out in accordance with the plans (or laws) laid down by God in the beginning. He saw no problem in repopulating the Earth after each catastrophe, arguing that seemingly 'new' species in the fossil record were actually immigrants from parts of the world which had not yet been explored at the beginning of the nineteenth century. Either way, Cuvier saw that the history of life on Earth went back for at least hundreds of thousands of years, far beyond Ussher's estimate – but even a timescale of hundreds of thousands of years meant that there had to have been repeated, major catastrophes on the scale of the Flood to explain the amount of change that Cuvier was finding in the fossil record. His ideas about the fixity of species, though, brought him into conflict with some of his French contemporaries and set back the study of evolution in France at a crucial time.

Jean-Baptiste Lamarck: thoughts on evolution

The ideas that Cuvier opposed were essentially those of Jean-Baptiste Lamarck, who was born in 1744 and who will feature more strongly in Chapter 8. A protégé of Buffon, Lamarck worked at the Museum of Natural History in Paris before Cuvier arrived there. From 1809 onwards, he developed a model of how evolution works based on the idea that characteristics can be acquired by an individual during its lifetime and then passed on to succeeding generations. In the classic example, it is supposed (wrongly) that by stretching to reach the topmost leaves on a tree, the neck of a giraffe gets longer during its lifetime; when that giraffe has offspring, they are therefore born with longer necks than if the parent had never tried to eat leaves. The particular bone of contention between Lamarck and Cuvier, however, was that Lamarck thought that no species ever went extinct, but developed into another form, while Cuvier thought that no species ever changed, but that entire species could be wiped out by catastrophes.

Lamarck's ideas were taken up and promoted by Étienne Geoffroy Saint-Hillaire (usually referred to as Geoffroy), a close contemporary of Cuvier (Geoffroy was born in 1772 and died in 1844) who had already established himself at the Jardin des Plantes before Cuvier arrived in Paris (unlike Cuvier, Geoffroy did go to Egypt with Napoleon). In the second decade of the nineteenth century, Geoffroy began to develop a variation on the evolutionary theme which went beyond Lamarck's ideas and suggested that there might be a direct role of the environment in evolution. He suggested that the environment might produce changes in living organisms (more or less following Lamarck's incorrect lead), but then went on to raise the suggestion of a process that might well be called natural selection:

> If these modifications lead to injurious effects, the animals which exhibit them perish and are replaced by others of a somewhat different form, a form changed so as to be adapted to the new environment.[1]

1. Quoted by David Young in *The Discovery of Evolution*.

This is tantalizingly close to Darwinism; but in no small measure because of the influence of Cuvier, the next step was not taken at the time.

Although they were initially firm friends, after the turn of the century a certain professional antipathy developed between Cuvier and Geoffroy, and in 1818 Cuvier went ballistic when Geoffroy published work that claimed to prove that *all* animals are built on the same body plan, and elaborated with descriptions not only of the way that different parts of an insect's body correspond to various parts of the vertebrate design, but also to relate both body plans (vertebrate and insect) to the structure of a mollusc. In 1830, a year after Lamarck died, Cuvier launched a blistering attack on Geoffroy, which laid into not only these fanciful notions about the relationships between vertebrates, insects and molluscs, but also the far more respectable (given the state of knowledge at the time) idea of Lamarckian evolution. Cuvier held firmly to the idea that species, once created, stayed fixed in the same form for ever, or at least until they went extinct. He urged younger naturalists to confine themselves to *describing* the natural world, without wasting time on theories purporting to *explain* the natural world. Under the weight of his authority, Lamarckism (which might otherwise have been developed by the next generation into something more like Darwinian evolution) was buried and remained essentially forgotten until after Darwin himself published the theory of evolution by natural selection.

That makes this an opportune moment to leave, temporarily, our account of the development of the life sciences and catch up on progress in the eighteenth century in the physical sciences. While human horizons were expanding in both time and space during the eighteenth century and into the beginning of the nineteenth century, largely thanks to the astronomers and the biologists, the hands-on investigation of the physical world (by physicists themselves and by chemists, at last shaking off the dead hand of alchemy) had also been taking huge strides. There was no single breakthrough to rank with the achievements of Newton and his contemporaries, but the steady increase in knowledge that occurred during this period (which is, appropriately, often referred to as the Enlightenment) can now be seen to have been the essential precursor to the spectacular way in which science took off in the nineteenth century.

Book Three

THE ENLIGHTENMENT

Enlightened Science I: Chemistry catches up

The Enlightenment – Joseph Black and the discovery of carbon dioxide – Black on temperature – The steam engine: Thomas Newcomen, James Watt and the Industrial Revolution – Experiments in electricity: Joseph Priestley – Priestley's experiments with gases – The discovery of oxygen – The chemical studies of Henry Cavendish: publication in the Philosophical Transactions *– Water is not an element – The Cavendish experiment: weighing the Earth – Antoine-Laurent Lavoisier: study of air; study of the system of respiration – The first table of elements; Lavoisier renames elements; He publishes* Elements of Chemistry *– Lavoisier's execution*

The Enlightenment

Historians often refer to the period more or less following the Renaissance as the Enlightenment, a name also given to the philosophical movement that reached its peak in the second half of the eighteenth century. The basic feature of the Enlightenment was a belief in the superiority of reason over superstition. This incorporated the idea that humankind was in the process of progressing socially, so that the future would be an improvement on the past; and one of those improvements was a challenge to orthodox religion with its overtones of superstition. Both the American and the French revolutions were justified intellectually, in part, on the basis of human rights, a guiding principle of Enlightenment philosophers such as Voltaire and activists such as Thomas Paine. Although it was just one of many factors in the Enlightenment, the success of Newtonian physics in providing a mathematical description of an ordered world clearly played a big part in the flowering of this movement in the eighteenth century, encouraging philosophers of a rationalist persuasion, and also encouraging chemists and biologists to think that their parts of the natural world might be explained on the basis of simple laws. It isn't so much that Linnaeus, say, consciously modelled his approach on Newton, but rather that the idea of order and rationality as a way to investigate the world had taken root by the early eighteenth century and seemed the obvious way forward.

It is probably not entirely a coincidence that the Industrial Revolution took place first in England (in round terms, during the period from 1740 to 1780) before spreading to the rest of Europe. There are many factors which contributed to this revolution occurring when and where it did, including the geographical and geological circumstances of Britain (an 'island of coal'), the early flowering of what might be called democracy (while France was still ruled by the conservative, aristocratic *ancien régime*, and Germany was a fragmented cluster of statelets) and perhaps an element of pure chance. But one of the factors was that the Newtonian mechanistic world view became firmly established most quickly, naturally enough, in Newton's homeland. Once the Industrial Revolution got under way, it gave a huge boost to science, both by stimulating interest in topics such as heat and thermodynamics (of great practical and commercial importance in the steam age), and in providing new tools for scientists to use in their investigations of the world.

Nowhere is this seen more clearly than in the case of chemistry. It isn't because chemists were particularly stupid or superstitious that they lagged behind the other sciences, particularly physics, until well into the eighteenth century. They simply lacked the tools for the job. Astronomy could be carried out to some extent without any tools at all, using only the human eye; physics, in the seventeenth century, involved studying easily manipulated objects, such as balls rolling down inclined planes, or the swing of a pendulum; even botanists and zoologists could make progress with the aid of the simplest magnifying glasses and microscopes. But what chemists needed, above all, was a reliable and controllable source of heat to encourage chemical reactions. If your source of heat was basically a blacksmith's forge, and you couldn't measure temperatures, any chemical experimenting was bound to be a bit rough and ready. Even into the nineteenth century, to provide more controllable heat and make more subtle experiments, chemists were forced to use different numbers of candles or spirit lamps with several different wicks that could be lit or extinguished individually; to obtain a localized source of intense heat, they had to use a burning glass to concentrate the rays of the Sun. As for making accurate measurements of what was going on, Gabriel Fahrenheit (1686–1736) invented the alcohol thermometer only in 1709, and did not come up with the mercury thermometer until 1714, when he also developed the temperature scale now named in his honour.[1] This was just two years after Thomas Newcomen (1663–1729) completed the first practical steam engine to pump water from mines. As we shall see, what was wrong with Newcomen's design was even more important than what was right with it in stimulating the progress of science in the next generation.

All of these factors help to explain why there was such a gap from Robert Boyle, who laid the ground rules for chemistry to become a science, to the people who actually did make chemistry scientific at the time of the Industrial Revolution. From the 1740s onwards, progress was rapid (if sometimes confused) and can be understood in terms of the working lives of a handful of men, most of whom were contemporaries and knew each other. Pride of place goes to Joseph Black, who pioneered the application of accurate quantitative techniques to chemistry, measuring (as far as possible) everything that went into a reaction and everything that came out.

Joseph Black and the discovery of carbon dioxide

Black was born in Bordeaux on 16 April 1728 (just a year after Newton died). This fact alone gives some insight into the cultural links between different parts of Europe at that time – Black's father, John, had himself been born in Belfast, but was of Scottish descent, and settled in Bordeaux as a wine merchant. With the state of the roads between Scotland and the south of England in the seventeenth and eighteenth centuries, the easiest way to travel from, say, Glasgow or Edinburgh to London (and the only way to travel from Belfast to London!) was by sea, and once you were in a boat it was almost as easy to go to Bordeaux. And, of course, there was the recent historical connection between Scotland and France, the Auld Alliance, dating from the time when Scotland was an independent country regarding England as a natural enemy. A Scottish gentleman, or merchant like John Black, was as much at home in France as in Britain. There, he married Margaret Gordon, the daughter of another expatriate Scot, and together they raised thirteen children – eight sons and five daughters, all of whom, unusually for the time, grew to adulthood.

1. Anders Celsius (1701–1744) came up with his eponymous scale only in 1742.

21. *The Newcomen engine.*

As well as a town house in the Chartron suburb of Bordeaux, the Black family owned a farm and a country house with a vineyard. Joseph grew up in these comfortable surroundings, and was largely educated by his mother, until he was 12, when he was sent to Belfast to live with relations and attend school in preparation for his admission to the University of Glasgow, which he entered in 1746. At first, Black studied languages and philosophy, but since his father required him to join one of the professions, in 1748 he switched to medicine and anatomy, which he studied for three years under William Cullen (1710–1790), the professor of medicine. Cullen's classes included chemistry, and as well as being an excellent teacher with an up-to-date knowledge of the science of the time, he made an important contribution in his own right when he proved that very low temperatures could be achieved when water, or other fluids, evaporated. By using an air pump to produce cold by encouraging liquids to evaporate at low pressure, Cullen (assisted by one his pupils, a Dr Dobson) invented what was, in effect, the first refrigerator. After passing his medical examinations in Glasgow, Black moved on (in 1751 or 1752) to Edinburgh to carry out research leading to the award of his doctor's degree. It was this research that led to his own most famous contribution to science.

At that time, there was a great deal of concern in the medical profession about the use of quack remedies to relieve the symptoms of 'stones' in the urinary system (urinary calculus). These remedies, intended to dissolve the offending stones, involved drinking what to modern eyes seem astonishingly powerful concoctions such as caustic potash and other strong alkalis; but they were very much in vogue following the endorsement of one particular such remedy a few years earlier by Robert Walpole, the first 'Prime Minister' of Britain, who was convinced that it had cured him. When Black was a medical student, a milder alkali, known as white magnesia, had recently been introduced into medicine as a treatment for 'acid' stomach. He decided that for his thesis work he would investigate the properties of white magnesia, in the hope of finding that it would be an acceptable treatment for stones. This hope was dashed; but it was the way Black carried out his investigations that pointed the way to a genuinely scientific study of chemistry, and which led him to discover what we now know to be carbon dioxide, showing for the first time that air is a mixture of gases and not a single substance.

To put all this in perspective, the chemists of Black's day recognized two forms of alkali substance, mild and caustic. Mild alkalis could be converted into caustic alkalis by boiling them with slaked lime, and the slaked lime itself was produced by slaking quicklime with water. Quicklime was made by heating limestone (chalk, essentially) in a kiln – and this is the key, because the 'caustic' properties of the materials were thought to be the result of some form of fire-stuff from the kiln getting into the lime and being passed along through the various processes to make caustic alkalis. Black's first discovery was that when white magnesia was heated it lost weight. Since no liquid was produced, this could only mean that 'air' had escaped from the material. He then found that all mild alkalis effervesce when treated with acids, but the caustic alkalis did not. So the cause of the difference between the two alkalis was that mild alkalis contain 'fixed air', which can be liberated by the action of heat or of acid, while caustic alkalis do not. In other words, the caustic properties are not a result of the presence of fire-stuff.

This led to a series of experiments in which the balance was a key tool, with everything being weighed at every step. For example, Black weighed an amount of limestone before heating it to produce quicklime, which was weighed. A weighed amount of water was added to the quicklime to make slaked lime, which was weighed. Then a weighed amount of mild alkali was added, converting the slaked lime back into the original amount of limestone. From the changes in weight at different stages of the experiment, Black could even calculate the weight of 'fixed air' gained or lost in the various reactions.

In a further series of experiments on the 'air' released by the mild alkalis, such as using it to extinguish a lighted candle, Black showed that it is different from ordinary air, but must be present in the atmosphere, dispersed through it. In other words, as we would now put it, air is a mixture of gases. This was a dramatic discovery at the time. All this work formed the basis of Black's thesis, submitted in 1754 and published in expanded form in 1756. It not only obtained Black his doctorate, but made his name, immediately in Scotland and soon throughout the scientific world, as a leading chemist. After completing his medical studies, Black started to practise medicine in Edinburgh, but the following year the chair of chemistry in Edinburgh became vacant and Black's former teacher, William Cullen, was appointed to the post. This left a vacancy in Glasgow, where Cullen recommended his former pupil, who became professor of medicine and lecturer in chemistry there in 1756, with a private medical practice on the side, at the age of 28. Black was a conscientious professor who gave absorbing lectures that drew students to Glasgow (and later to Edinburgh) from all over Britain, Europe and even from America,[1] and were a major influence on the next generation of scientists (one of his pupils kept detailed notes of the lectures, which were published in 1803, continuing to inspire students into the nineteenth century). But although he continued to do research, he hardly published any of his results, instead presenting them in his lectures to undergraduates or to learned societies. So the young men really were given a front row seat at which they could see the new science unfolding. For the next few years, Black developed the investigations of the subjects of his thesis, showing, among other things, that 'fixed air' is produced by the respiration of animals, in the process of fermentation and by burning charcoal. But he never made any other major breakthrough in chemistry, and by the 1760s his attention had largely turned to physics.

Black on temperature

Black's other major contribution to science concerned the nature of heat. Heat fascinated people like Cullen, Black and their contemporaries, not just because of its intrinsic importance in laboratory chemistry, but because of its role in the burgeoning Industrial Revolution. The development of the steam engine (of which more shortly) is an obvious example; but think also of the flourishing whisky industry in Scotland, which used huge amounts of fuel turning liquids into vapours and then had to remove equally large amounts of heat from the vapours as they condensed back into liquids. There were eminently practical reasons why Black investigated such problems in the early 1760s, although it is also likely that his interest in what happens when liquids evaporate was stimulated by his close association with Cullen. Black investigated the well-known phenomenon that

1. One of Black's pupils, Benjamin Rush (1746–1813), became the first professor of chemistry in the US, at Philadelphia College in 1769.

when ice melts, it stays at the same temperature while the solid is turning into liquid. Applying his usual careful, quantitative approach, he carried out measurements which showed that the heat needed to melt a given quantity of ice into water at the same temperature was the same as the amount of heat needed to raise the temperature of that much water from the melting point to 140° Fahrenheit. He described the heat which the solid absorbed while melting to liquid at the same temperature as latent heat, and realized that it was the presence of this heat that made the water liquid instead of solid – making a crucial distinction between the concepts of heat and temperature. In a similar way, there is a latent heat associated with the transition of liquid water into vapour (or any other liquid into its vapour state), which Black also investigated quantitatively. He also gave the name 'specific heat' to the amount of heat required to raise the temperature of a certain amount of a chosen substance by a certain amount (in modern usage, this might be the amount of heat needed to raise the temperature of 1 gram of a substance by 1° C). Because all water has the same specific heat, if a pound weight (say) of water at freezing point (32° F) is added to a pound weight of water at the boiling point (212° F), the result is two pounds of water at 122° F, the temperature halfway between the two extremes. One pound of water has its temperature increased by 90° F, the other has its temperature decreased by 90° F. But because iron (say) has a smaller specific heat than water, if a pound of water at 212° F were poured on to a pound of iron at 32° F the temperature of the iron would be increased by much more than 90° F. All of these discoveries were described by Black to the University Philosophical Club on 23 April 1762, but never published in a formal, written sense. In his experiments on steam, Black was assisted by a young instrument maker at the university, a certain James Watt, who made apparatus for Black as well as carrying out his own investigations of steam. The two became firm friends, and nobody was more gratified than Black when Watt's work on steam engines brought him riches and fame.

Black himself left Glasgow in 1766, when he was appointed professor of chemistry in Edinburgh, in succession to William Cullen. He was both physician and friend to Adam Smith, David Hume and the pioneering geologist James Hutton (among others). He never married, and left the development of the techniques of analytical chemistry he had invented to be taken up by others (notably Antoine Lavoisier), but he remained a figure of high standing in the Scottish Enlightenment. Although he retained his professorship until his death, Black became increasingly frail in later life and gave his last series of lectures in the academic year 1796/7. He died quietly on 10 November 1799, at the age of 71.

The steam engine: Thomas Newcomen, James Watt and the Industrial Revolution

Although this is not a history of technology, any more than it is a history of medicine, it's worth taking a look at the achievements of Black's friend James Watt, because these marked a particularly significant step towards the science-based society we live in today. The special thing about Watt's achievement is that he was the first person to take a set of ideas from the cutting edge of then-current research in science and apply them to make a major technological advance; and the fact that he was working at a university, in direct contact with the researchers making the scientific breakthroughs,[1] presages the way in which modern high-tech industries have laboratories with close research connections. In the second half of the eighteenth century, Watt's improvements to the steam engine were very high-tech indeed; and it was the whole style of Watt's approach that pointed the way for the development of technology in the nineteenth and twentieth centuries.

Watt was born in Greenock, on Clydeside, on 19 January 1736. His father (also James) was a shipwright who had diversified his activities to become also a ship's chandler, builder, ship owner and merchant (so he could build a ship, fit it out, provide a cargo and send it off for the cargo to be sold in a foreign port). Young James's mother, Agnes, had borne three children before James, but all died young; a fifth child, John, was born three years after James and survived infancy but was lost at sea (from one of his father's ships) as a young man.

The younger James Watt had a comfortable upbringing and a good basic education at the local grammar school, although he suffered from migraines and was regarded as physically delicate; he was more interested in his father's workshop than school, and made working

1. His friends also included, as well as Black, James Hutton, with whom he went on geological expeditions.

models of different machines and other devices, including a barrel organ. Because his father intended James to take over the ship-based family businesses, he was not sent to university. But as a result of a succession of business failures involving the web of James senior's interests, this prospect disappeared and, now in his late teens, young James suddenly had to face the prospect of making his own living. In 1754, he went to Glasgow to begin learning the trade of mathematical instrument maker and then, in 1755, to London, where for a fee of 20 guineas and his labour he was given a one-year crash course, a kind of compressed apprenticeship, by one of the best instrument makers in the country. He returned to Scotland in 1756 and wanted to set up in business in Glasgow, but was prevented from doing so by the powerful craftsmen's guilds, since he had not served a traditional apprenticeship; the following year, however, he was provided with a workshop and accommodation within the university precincts, where he became mathematical instrument maker to the university, and could also undertake private work. The university had the power to do what it liked on its own premises, and one thing people like Adam Smith (then a professor at Glasgow) certainly disliked was the way the guilds exercised their power.

Watt just about made a living in his new position and had time to indulge in some experiments with steam power, stimulated by one of the students at Glasgow, John Robison, who suggested to Watt, in 1759, the possibility that steam power might be used to drive a moving carriage. Although nothing came of these experiments, it meant that Watt already had some understanding of steam engines when he was asked, in the winter of 1763/4, to repair a working model of a Newcomen engine which the university had acquired (the problem with the working model being that it did not work).

Thomas Newcomen (1664–1729) and his assistant John Calley had erected the first successful steam engine in 1712, at a coal mine near Dudley Castle, in the English midlands. Although other people had experimented with steam power before, this was the first practicable engine that did a useful job – pumping water out of the mine. The key feature of the Newcomen design involves a vertical cylinder with a piston, set up so that the piston is attached by a beam to a counterweight. Other things being equal, the weight falls and lifts the piston to the top of the cylinder. To make the engine work, the cylinder below the piston is filled with steam. Then, cold water is sprayed into the cylinder, making the steam condense, creating a partial vacuum. Atmospheric pressure drives the piston down into the vacuum in spite of the counterweight. As the piston reaches the bottom of the cylinder, steam is allowed back in below the piston, equalizing the pressure (or even raising it slightly above atmospheric pressure, although this is not necessary) so that the counterweight can lift the piston back to the top of the cylinder. Then, the cycle repeats.[1]

But after Watt had repaired the mechanics of the model Newcomen engine, he found that when it was fired up and its little boiler was full of steam, it ran for only a very few strokes before all the steam was exhausted, even though it was supposed to be a perfect scale model of an engine that would run for much longer. Watt realized that this was because of what is known as the scale effect – Isaac Newton had pointed out in his *Opticks* that a small object loses heat more rapidly than a large object of the same shape (the reason is that the small object has a larger surface area, across which heat escapes, in proportion to its volume, which stores heat). But instead of just shrugging his shoulders and accepting that the scale model could never run as well as the real thing, Watt took a detailed look at the scientific principles on which the engine operated to see if it could be made more efficient – with the implication that the same improvements would make full-size steam engines much more efficient than Newcomen engines.

Watt identified the major loss of heat in the Newcomen engine as being a result of the need to cool the entire cylinder (which could achieve a lot of heat, being very massive) at every stroke of the piston, and then heat it all the way back above the boiling point of water every time to allow it to fill with steam. He realized that the solution was to use two cylinders, one which was kept hot all the time (in which the piston moved) and one which was kept cold all the time (in the early models, by immersing it in a tank of water). When the piston

1. Because it is atmospheric pressure that drives the piston down, a Newcomen engine is sometimes called an atmospheric engine. It was Watt who introduced steam as the working fluid in his designs, which is why he is often referred to as the inventor of the steam engine, even though Newcomen's engines did use steam.

22. *Watt's steam engine.*

was at the top of its stroke, a valve opened to let steam flow from the hot cylinder to the cold cylinder, where it condensed, creating the required partial vacuum. At the bottom of the stroke, this valve closed and another one opened, allowing fresh steam to enter the still-hot working cylinder. There were many other improvements, including the use of hot steam at atmospheric pressure to drive the piston down from above, helping to keep the working cylinder hot; but the key development was the separate condenser.

In the course of these experiments, Watt independently came across the phenomenon of latent heat, a few years later than Black. It seems he had not been aware of Black's work (hardly surprising, since Black never published anything), but discussed his discoveries with Black, who brought him up to date, helping to further improve his engine. What Watt noticed was that if one part of boiling water is added to thirty parts of cold water, the rise in temperature of the cold water is barely perceptible; but if a comparably small amount of steam (at the same temperature as boiling water, of course) is bubbled through the cold water, it soon makes the water boil (we now know, because the latent heat released as steam[1] condenses to water).

Watt patented his steam engine in 1769, but it was not an immediate commercial success, and from 1767 to 1774 his main work was as a surveyor for the Scottish canals, including the Caledonian Canal. He had married in 1763, but his first wife, Margaret, died in 1773 (leaving two sons), and in 1774 he moved to Birmingham, where he became part of a scientifically inclined group known as the Lunar Society (because they met once a month) which included Joseph Priestley, Josiah Wedgwood and Erasmus Darwin (the latter two the grandfathers of Charles Darwin). It was here that Watt went into partnership with Matthew Boulton (1728–1809), who he had met through Erasmus Darwin, which led to the commercial success of his steam engines. He also invented and patented many detailed features of the machines, including the automatic regulator, or governor, to shut off the steam if the machine was running too fast. Watt married again in 1775 and had a son and a daughter by his second wife, Ann. He retired from his steam-engine business in 1800, at the age of 64, but continued inventing until he died, in Birmingham, on 25 August 1819.

Experiments in electricity: Joseph Priestley

Just as the development of the steam engine by Watt rested on scientific principles, so the importance of steam in the Industrial Revolution would encourage the further development of the study of the connection between heat and motion (thermodynamics) in the nineteenth century. This in turn led to the development of more efficient machines, in a classic example of the feedback between science and technology. But while Boulton and Watt were playing their part in the Industrial Revolution in the last quarter of the eighteenth century, one of their friends in the Lunar Society, Joseph Priestley, was taking the next great step forward in chemistry

1. Strictly speaking, as water vapour condenses. 'Steam' is a mixture of water vapour and droplets of very hot water.

– even though science was far from being the most important thing in his life.

Priestley was born in Fieldhead, near Leeds, on 13 March 1733. His father, James, was a weaver and cloth-dresser, working at his looms in the cottage in which he lived.[1] He was also a Calvinist. James Priestley's wife, Mary, bore six children in six years, then died in the unusually severe winter of 1739/40. Joseph was the first of these children, and as his siblings arrived in quick succession he was sent to live with his maternal grandfather and saw little of his mother. When she died he returned home, but James found it impossible to cope with all the children and his work, and Joseph then (at about the age of 8) went to live with an aunt (whose own husband died shortly after he joined their family) who had no children of her own. Also an enthusiastic Calvinist, she made sure that the boy had a good education at the local schools (still mostly Latin and Greek) and encouraged him towards a career as a minister of the cloth.

In spite of a bad stammer, Priestley achieved this aim. In 1752, he went to study at a nonconformist academy. Such academies (which were not necessarily as grand as the name now implies – some consisted of just a couple of teachers and a handful or two of pupils) had their origins in the same Act of Uniformity of 1662 that had led to John Ray leaving Cambridge. When some 2000 nonconformists were ejected from their parishes as a result of this Act, most of them became private teachers (as, indeed, did Ray) – the only real opportunity they had to earn a living. In 1689, after the Glorious Revolution, an Act of Toleration allowed nonconformists to play a fuller part in society and they founded about forty academies for the training of nonconformist ministers. For obvious religious reasons these academies generally had good relations with the Scottish universities, and many of their pupils went on to study in Glasgow or Edinburgh. They flourished in the middle of the eighteenth century (among others, Daniel Defoe, Thomas Malthus and William Hazlitt attended such academies), but later fell into decline as nonconformists were more fully integrated into society and once again allowed to teach in mainstream schools and colleges.

After completing his courses, in 1755, Priestley became a minister

1. The classic cottage industry that would soon be replaced by steam-powered mills.

at Needham Market in Suffolk, where he alienated many of his congregation by being an Arian – he was, so to speak, a nonconformist nonconformist. He had started out with more or less orthodox Trinitarian views, but it was while at Needham Market, after making his own careful study of the Bible, that he became convinced that the notion of the Holy Trinity was absurd and became an Arian. From Needham Market, Priestley moved on to Nantwich, in Cheshire, and then to teach at Warrington Academy, midway between Liverpool and Manchester. It was there, in 1762, that he married Mary Wilkinson, the sister of John Wilkinson, an ironmaster who made a fortune out of armaments; the couple had three sons and a daughter. Largely thanks to Priestley, the Warrington Academy was one of the first educational institutions in England to replace the traditional study of classics with lessons in history, science and English literature.

Priestley's intellectual interests were wide-ranging, and among his early writings were an English grammar and a biographical chart, laying out the chronological relationship between important figures from history, covering the period from 1200 BC to the eighteenth century. This was such an impressive piece of work that he was awarded the degree of Doctor of Laws (LLD) by the University of Edinburgh in 1765. On a visit to London that same year (he took to spending a month in London every year), Priestley met Benjamin Franklin and other scientists interested in electricity (then known as electricians), which encouraged him to carry out his own experiments, in one of which he showed that there is no electric force inside a hollow charged sphere. He suggested (among other things) that electricity obeys an inverse square law, and this work saw him elected as a Fellow of the Royal Society in 1766. He wrote a history of electricity, published in 1767, which ran to some 250,000 words (he had already written six books on non-scientific subjects) and established him as a teacher and historian of science. He was now 34, but all his achievements so far were merely a taste of what was to come.

Although science was, relatively speaking, only an incidental part of Priestley's full and active life, we do not have space here to set it in its proper context, and can only sketch out his role as a theologian and radical dissenter in the turbulent decades of the late eighteenth century. In 1767, Priestley returned to work as a minister at a chapel in Leeds.

Alongside his developing interest in chemistry, he wrote pamphlets criticizing the British government's treatment of the American colonies[1] and continued his search for religious truth, now leaning towards the views of the Unitarians (a strongly Arian sect founded in 1774). Priestley's fame spread, and at the end of his term in Leeds in 1773, he was invited by the Whig politician Lord Shelburne to become his 'Librarian' at a salary of £250 a year, with free accommodation in a house on Shelburne's estate during his employment and a pension for life at the end of his service. The library work occupied little of Priestley's time and his actual role was as a political adviser and intellectual sounding board to Shelburne, and part-time tutor to Shelburne's two sons, giving him plenty of time for his own scientific work (largely subsidized by Shelburne) and other interests. As Secretary of State from 1766 (when he was just 29) to 1768, Shelburne had tried to push through a policy of conciliation towards the Americans, but was dismissed for his pains by the King, George III. In 1782, after the King's disastrous policy had resulted in the defeat of Britain in the American War of Independence, he was forced to recall Shelburne as the only credible statesman who could carry out the difficult task of establishing peace with the former colonies. But by then Priestley had moved on. By 1780, his outspokenness as a dissenter had become a political embarrassment even to Lord Shelburne, who retired his 'Librarian' with the promised pension (£150 a year). Priestley moved to Birmingham, where he lived in a house provided by his wealthy brother-in-law, worked as a minister and was, one way and another, quite comfortably off. It was during this period of his life that he was an active member of the Lunar Society.

In Birmingham, Priestley continued to speak out against the established Church of England, and just as he had been sympathetic to the cause of the American colonists, he was outspoken in his support of the French Revolution (which was initially a popular democratic movement). Things came to a head on 14 July 1791, when Priestley and other supporters of the new government in France organized a dinner in Birmingham to celebrate the second anniversary

1. Some of the fine phrases from Priestley's writings on liberty were taken over by Thomas Jefferson and incorporated into the American Declaration of Independence.

of the storming of the Bastille. Their opponents (some political, some business competitors eager for a chance to damage their rivals) organized a mob, which first went to the hotel where the dinner had been held but then, finding the diners had left, went on the rampage, burning and looting the houses and chapels of the dissenters. Priestley escaped in time, but his house was destroyed, along with his library, manuscripts and scientific equipment. Priestley moved to London, intending at first to stay and (verbally) fight his case, but his position became untenable as the French Revolution turned into a bloody mess and hostility towards France was whipped up by war (his position in England was hardly helped by the offer of French citizenship from the revolutionaries in Paris). In 1794, when he was 61, Priestley and his wife emigrated to North America (following his sons, who had emigrated the previous year), where he lived quietly (by his standards – he still managed to publish thirty works after 1791) in Northumberland, Pennsylvania, until his death on 6 February 1804.

Priestley's experiments with gases

As a chemist, Priestley was a great experimenter and a lousy theorist. When he began his work, only two gases (or 'airs') were known – air itself (which was not yet, in spite of Black's work, widely recognized as being a mixture of gases) and carbon dioxide ('fixed air'); hydrogen ('inflammable air') was discovered by Henry Cavendish in 1776. Priestley identified another ten gases, including (to give them their modern names) ammonia, hydrogen chloride, nitrous oxide (laughing gas) and sulphur dioxide. His greatest discovery, of course, was oxygen – but even though he carried out the experiments which revealed the existence of oxygen as a separate gas, he explained those experiments in terms of the phlogiston model, which had been promoted by the German chemist George Stahl (1660–1734). This model 'explained' burning as a result of a substance (phlogiston) leaving the thing that is being burnt. For example, in modern terminology, when a metal burns it combines with oxygen to form a metallic oxide, a substance known in Priestley's day as a calx. According to the phlogiston model, what is happening is that phlogiston escapes from the metal and leaves the calx behind. When the calx is heated (on this picture), phlogiston recombines with it (or rather, re-enters it) to form the metal. The reason why things do not burn in the absence of air, said Stahl, was because air is necessary to absorb the phlogiston.

The phlogiston model worked, after a fashion, as long as chemistry was a vague, qualitative science. But as soon as Black and his successors started making accurate measurements of what was going on, the phlogiston doctrine was doomed, since it could only be a matter of time before someone noticed that things get heavier when they burn, not lighter – suggesting that something is entering into (combining with) them, not escaping. The surprise is that Priestley did not see this (but remember that he was only a part-time chemist with many other things on his mind), so that it was left to the Frenchman Antoine Lavoisier to make the connection between burning and oxygen, and to pull the rug from under the phlogiston model.

Priestley began his experiments involving 'airs' during his time in Leeds, where he lived close to a brewery. The air immediately above the surface of the brew fermenting in the vats had recently been identified as Black's fixed air, and Priestley saw that he had a ready-made laboratory in which he could experiment with large quantities of this gas. He found that it formed a layer roughly nine inches to a foot (23–30 centimetres) deep above the fermenting liquor, and that although a burning candle placed in this layer was extinguished, the smoke stayed there. By adding smoke to the carbon dioxide layer, Priestley made it visible, so that waves on its surface (the boundary between the carbon dioxide and ordinary air) could be observed, and it could be seen flowing over the side of the vessel and falling to the floor. Priestley experimented with dissolving fixed air from the vats in water, and found that by sloshing water backwards and forwards from one vessel to another in the fixed air for a few minutes, he could produce a pleasant sparkling drink. In the early 1770s, partly as a result of an (unsuccessful) attempt to find a convenient preventative for scurvy,[1] Priestley refined this technique by obtaining carbon dioxide

1. Through this work, Priestley was considered for the post of naturalist on the second voyage of James Cook around the world, but was rejected on religious grounds, leading him to respond, 'I thought that this had been a business of *philosophy* and not of *divinity*.'

from chalk using sulphuric acid, and then dissolving the gas in water under pressure. This led to a craze for 'soda water', which spread across Europe. Although Priestley sought no financial reward for his innovation, justice was done, since it was through his invention of soda water that Lord Shelburne first heard of Priestley, while travelling in Italy in 1772. Since it was Shelburne who provided the time for Priestley to concentrate more on chemistry over the next few years, the place (his estate at Calne, in Wiltshire) and the money for the experiments which soon led to the discovery of oxygen, you could say that this discovery owed much to the brewing industry.

The discovery of oxygen

Priestley also began to suspect that air was not a simple substance while he was at Leeds. He found, using experiments with mice, that the ability of air to sustain life could somehow be 'used up' in respiration, so that it was no longer fit to breathe, but that the respirability of the air could be restored by the presence of plants – the first hints of the process of photosynthesis in which carbon dioxide is broken down and oxygen is released. But his discovery of oxygen, the gas that is used up during respiration, was made at Calne on 1 August 1774, when he heated the red calx of mercury (mercuric oxide) by focusing the rays of the Sun through a 12 inch (about 30 centimetres) diameter lens on to a sample in a glass vessel. Gas (what Priestley and his contemporaries called an 'air') was released as the calx reverted to the metallic form of mercury. It was some time before Priestley discovered that the new 'air' he had made was better than ordinary air for respiration. In the course of a long series of experiments, he first discovered that a lighted candle plunged into the gas flared up with unusual brightness, and eventually, on 8 March 1775, he placed a full-grown mouse in a sealed vessel filled with the new air. A mouse that size, he knew from experience, could live for about a quarter of an hour in that amount of ordinary air; but this mouse ran about for a full half-hour and then, taken out of the vessel seemingly dead, revived when warmed by the fire. Cautiously, recognizing the possibility that he might have picked an unusually hardy mouse for his experiment, Priestley wrote in his notes on the experiment only that the new air was at least as good as ordinary air; but further experiments showed that in the sense of supporting respiration it was between four and five times as good as common air. This corresponds to the fact that only about 20 per cent of the air that we breathe is actually oxygen.

Priestley's discovery had actually been pre-empted by the Swedish chemist Carl Scheele (1742–1786), whose surviving laboratory notes show that by 1772 he had realized that air is a mixture of two substances, one of which prevents burning while the other promotes combustion. He had prepared samples of the gas that encourages burning by heating mercuric oxide, and by other techniques, but he made no attempt to publicize these discoveries immediately – he wrote about them in a book which was prepared in 1773, but not published until 1777. News of the work only began to spread in the scientific community shortly before Priestley carried out his experiments in August 1774. It seems that Priestley was unaware of Scheele's work at that time, but in September 1774, while Priestley was still carrying out his experiments, Scheele wrote about his own discovery in a letter to Lavoisier. Scheele made many other discoveries of great importance to chemistry, but he worked as a pharmacist, only published the one book and turned down several offers of academic posts. He also died young, at the age of 43. This combination of circumstances has meant that he is sometimes overlooked in historical accounts of eighteenth-century chemistry. The real relevance of the near-simultaneous discovery of oxygen by Scheele and by Priestley, though, is not who did it first, but that it reminds us that in most cases science progresses incrementally, building on what has already been discovered and making use of the technology of the day, so that it is largely a matter of luck which individual makes which discovery first, and who gets their name in the history book. For better or worse, it is Priestley whose name is associated with the discovery of oxygen, even though Scheele undoubtedly discovered it first, and even though Priestley continued to try to explain his discoveries in terms of the phlogiston model.

Just occasionally, though, a discoverer fails (if that is the right word) to get his name attached to a particular discovery, or law, in the history books because he never bothered to tell anyone about his work and gained his scientific satisfaction from carrying out his own experiments to assuage his own curiosity. The archetypal example of this rare breed of scientist is Henry Cavendish, a contemporary of Priestley, who published enough to make him an important figure in the development of chemistry in the second half of the eighteenth century, but who *didn't* publish a wealth of results (notably in physics) that were rediscovered independently by others (whose names are duly attached to the discoveries in the history books) in the following century. But there were unusual family reasons (essentially, vast wealth) why, whatever his inclinations, Cavendish was in a position to follow those inclinations and pick and choose when it came to deciding what he would publish.

Cavendish came from not one but two of the most wealthy and influential aristocratic families in England at the time. His paternal grandfather was William Cavendish, the second Duke of Devonshire, and his mother, Anne de Grey, was a daughter of Henry de Grey, Duke of Kent (and twelfth Earl – he was elevated to Duke in 1710). As the fourth out of five brothers (there were also six sisters), Henry's father, Charles Cavendish (1704–1783), had no grand title of his own, but such was the status of the Cavendish family that he was known as Lord Charles Cavendish throughout his life. If he really had been a Lord, his son Henry would have been the Honourable Henry Cavendish, Esquire, just as Robert Boyle, the son of an Earl, had been 'The Honourable'. Henry Cavendish was indeed addressed in this way during his father's lifetime, but as soon as his father died he let it be known that he preferred plain Henry Cavendish, Esquire.

Both sides of the family had scientific interests. From 1736 onwards, for ten or more years, the Duke of Kent and his family encouraged work in physics and astronomy, not least by employing the astronomer Thomas Wright (whose work features in Chapter 8) as a tutor for the Duchess and two of the Duke's daughters, Sophia and Mary (but not Anne, who had not only left home but died young, of tuberculosis, in 1733). Wright also did surveying work on the estate and carried out astronomical observations there, which were reported to the Royal Society in the 1730s. The teaching work continued even after the death of the Duke in 1740. Through the family connection, both 'Lord' Charles and Henry Cavendish visited the Duke of Kent's estate while Wright was there (he was certainly there until Henry was at least 15), and must have met him and discussed astronomy with him.

This is all the more certain since Charles Cavendish himself was so keenly interested in science that around this time, in the middle of his life, he gave up the traditional role of junior members of the aristocracy – politics – for science. As was more or less de rigueur for someone in his position, Charles had been elected to the House of Commons (a misnomer, if ever there was one, in those days) in 1725, where he served alongside one of his brothers, an uncle, two brothers-in-law and a first cousin. Charles Cavendish was a diligent and capable Member of Parliament, who turned out to be an able administrator and was closely involved with the work related to building the first bridge at Westminster (the first new bridge across the Thames in London since the London Bridge commemorated in the nursery rhyme). But after sixteen years (during all of which time Horace Walpole was Prime Minister[1]) he decided that he had done his duty for the country, and in 1741, when he was 37 years old and young Henry was just ten, he retired from politics to follow his interest in science. As a scientist, he was a keen amateur, rather in the tradition of the first members of the Royal, and very skilful at experimental work (his skills were praised by Benjamin Franklin). One of his most intriguing pieces of work was the invention, in 1857, of thermometers that show the greatest and least temperature recorded during the absence of the observer – what we now know as 'maximum–minimum' thermometers. But although not a scientist of the first rank, Charles Cavendish soon turned his administrative skills to good use for both the Royal Society (where he had been elected a Fellow just three months after Newton died) and the Royal Greenwich Observatory, as well as providing encouragement for his son Henry.

Charles Cavendish had married Anne de Grey in 1729, when he was not quite 25 and she was two years younger. Their fathers had been friends for years, and were undoubtedly pleased at the match, but we have no information about the romantic side of the relationship (except that it seems that love was involved, since in those days younger

1. Walpole headed a Whig administration, closely identified at the time with the success of the Glorious Revolution; the opposition Tories, right up to the 1740s, were still largely Jacobite, and it is conceivable that a change of government during those sixteen years could have led to the restoration of the Stuarts; this possibility only really receded after the defeat of Bonnie Prince Charlie at Culloden, following his 1745 rebellion.

sons of the aristocracy did not normally marry until their thirties). What we do know is how wealthy the young couple were, since the details were all laid out in the marriage settlement. Charles had land and income settled on him by his father, and Anne brought with her income, securities and the promise of a substantial inheritance. Christa Jungnickel and Russell McCormmach[1] have calculated that at the time of his marriage, as well as considerable property, Charles Cavendish had a disposable annual income of at least £2000, which grew as time passed. This at a time when £50 a year was enough to live on and £500 enough for a gentleman to live comfortably.

Anne Cavendish, as she now was (although usually referred to as Lady Anne), had already shown signs of what was to become her fatal illness, prone to what were described as severe colds but which ominously included spitting blood. In the winter of 1730/31, which was very severe, the couple travelled on the Continent, first visiting Paris and then moving on to Nice, which was regarded as a good place for people convalescing from lung diseases, with plenty of sunshine and fresh air. It was there, on 31 October 1731, that Anne gave birth to their first son, Henry, named after his maternal grandfather. After further travels on the Continent (partly to seek medical advice about Anne's condition), the family returned to England, where Henry's brother Frederick (named after the then Prince of Wales) was born on 24 June 1733.[2] Less than three months later, on 20 September 1733, Anne died. Charles Cavendish never remarried, and for all practical purposes Henry Cavendish never had a mother, which may help to account for some of his peculiarities as an adult. Five years later, in 1738, Charles Cavendish sold his country estate and settled with his two young sons in a house in Great Marlborough Street, in London, convenient for his public and scientific committee work.

Although Charles Cavendish had been educated at Eton, both his sons were sent to a private school in Hackney and then on to Peterhouse, Cambridge, with Frederick always following the trail blazed by his brother. Henry went up to Cambridge in November 1749, when

1. *Cavendish: the experimental life.*

2. Born two years after Henry, Frederick would also die two years after Henry, in 1812. Charles, Henry and Frederick Cavendish each lived to their seventy-ninth year.

he was 18, and stayed there for three years and three months; he left without taking his degree, as many aristocratic young gentlemen did, but not without taking full advantage of what Cambridge could offer in the way of education (which wasn't all that much, even in the 1760s). It was after Henry had left Peterhouse that Frederick fell from the window of his room, some time in the summer of 1754, and suffered head injuries which left him with permanent brain damage. Partly thanks to the family wealth, which meant that he always had reliable servants or companions to keep an eye on him, this did not stop him leading an independent life, but meant that he could never follow in his father's footsteps on either the political or scientific stages.

Henry Cavendish simply had no interest in politics, but was fascinated by science. After the two brothers completed their Grand Tour of Europe together, Henry settled in the house at Great Marlborough Street and devoted his life to science, initially in collaboration with his father. Some members of the family disapproved of this as self-indulgence, and also felt that it wasn't quite decent for a Cavendish to be involved in laboratory experiments, but Charles Cavendish can hardly have objected to his son following his own passion for science. There are several stories about Charles's alleged financial meanness towards Henry, but in so far as there is any truth in these stories they only reflect the elder Cavendish's well-known caution with money. Charles Cavendish was always on the lookout for ways to increase his wealth and careful to spend no more than was necessary – but his idea of 'necessary' was what was appropriate for the son of a Duke. Some accounts say that while his father was alive Henry received an allowance of only £120 a year (which would have been more than adequate, given that he lived at home with all found); others, more plausibly, say that his allowance was £500 a year, the same sum that Charles Cavendish had received from his own father at the time of his marriage. What is undoubtedly true is that Henry Cavendish had no interest in money at all (the way that only very rich people can have no interest in money). For example, he owned only one suit at a time, which he wore every day until it was worn out, when he would purchase another in the same old-fashioned style. He was also very set in his eating habits, almost always dining off leg of mutton when he was at home. On one occasion, when several scientific friends were coming to dine, the housekeeper asked what to prepare. 'A leg of mutton,' said Cavendish. Told that this would not be enough, he replied, 'Then get two.'

But Henry Cavendish's attitude to money is best highlighted by a well-documented account of a visit paid to him by his banker, long after Charles Cavendish had died. The banker was concerned that Henry had accumulated some £80,000 in his current account and urged him to do something with the money. Cavendish was furious that he had been 'plagued' with this unwelcome enquiry, telling the banker that it was his job to look after the money and that if he bothered him with such trivia, he would take his account elsewhere. The banker continued, somewhat nervously, to suggest that maybe half of the money should be invested. Cavendish agreed, telling the banker to do what he thought best with the money, but not to 'plague' him about it any more, or he really would close the account. Fortunately, the bankers were honest, and his money was invested safely, along with all the rest. By the time he died, Henry Cavendish had investments with a nominal value of over a million pounds, although their actual value on the market at that time was just under a million.

The basis for that wealth had come partly from Charles Cavendish's success in building up his own wealth, but also from an inheritance which came to Charles shortly before he died, and formed part of the estate inherited by Henry (Frederick was provided with enough for a comfortable life as a gentleman, but apart from being the younger son his mental problems made him unfit to have control of a fortune). Charles Cavendish had a first cousin, Elizabeth, who was the daughter of his uncle James. Elizabeth married Richard Chandler, who was the son of the Bishop of Durham, and a politician; her brother (and only sibling), William, married another member of the Chandler family, Barbara. In 1751, James Cavendish and William Cavendish both died, William without leaving an heir, so that Elizabeth and Richard (who now took the name Cavendish) were left to continue that line of the family and inherited accordingly. But Richard and Elizabeth were also childless, and Richard died before her, leaving Elizabeth as heiress to a huge fortune, both land and securities. It was this wealth that she left to Charles, her only surviving male Cavendish cousin and her closest relative in the Cavendish family, when she died in 1779. When Charles Cavendish died in 1783, at the age of 79, the accumulated wealth passed to Henry, then in his fifty-second year. It was after this that he was referred to as 'the richest of the wise, and the wisest of the rich'.

Following the family tradition, when Henry died in 1810 he left his fortune to close relations, the chief beneficiary being George Cavendish, a son of the fourth Duke of Devonshire (himself a first cousin of Henry Cavendish) and brother to the fifth Duke (George's mother was Charlotte Boyle, daughter of the third Earl of Burlington). One of George's descendants, his grandson William, became the seventh Duke of Devonshire when the sixth Duke, who never married, died in 1858. It was this William Cavendish who, after increasing the family wealth even more through his iron and steel interests, and (among many other things) serving for nine years as Chancellor of the University of Cambridge, provided the endowment for the construction of the Cavendish Laboratory in Cambridge in the 1870s. William Cavendish never officially recorded whether he intended the laboratory to be a monument to his ancestor, but it ensured, as we shall see, that the name Cavendish was at the forefront of research in physics during the revolutionary developments that took place in the late nineteenth century and throughout the twentieth century.

It is possible that Henry Cavendish himself might have had doubts about the wisdom of letting some of the family wealth go outside the family. But although hardly a spendthrift, he had no qualms about spending money when there was a good reason to do so. He employed assistants to help with his scientific work, of course; but he also ensured that he had suitable premises in which to carry out that work. So in all likelihood, had he been around in the 1870s, he would have seen the need for an institution like the Cavendish Laboratory and approved of the expenditure. Shortly before his father died, Henry rented a country house in Hampstead, which he used for about three years. After 1784, he rented out the house in Great Marlborough Street, but bought another town house in nearby Bedford Square (the house still stands), and after leaving Hampstead he purchased another country house in Clapham Common, south of the river Thames. At all these locations, his life revolved around his scientific work and he had no social life at all, except to meet other scientists.

Henry was painfully shy and hardly ever went out except to scientific gatherings – even at these, latecomers sometimes found him standing outside the door trying to pluck up enough courage to enter, long after he was a respected scientist in his own right. He communicated with his servants by writing them notes, wherever possible; and there are several stories about how on unexpectedly encountering a woman he did not know, he would shield his eyes with his hand and literally run away. But he often travelled around Britain by coach in the summer months, with one of his assistants, carrying out scientific investigations (he was interested in geology) and visiting other scientists.

With his social life revolving around science, Henry was first taken to a meeting of the Royal Society as the guest of his father in 1758. He was elected a Fellow in his own right in 1760, and the same year became a member of the Royal Society Club, a dining society made up from members of the Royal, but a separate institution. He attended nearly every one of the Club's dinners (which were held at weekly intervals throughout most of the year) for the next fifty years.[1] To give you some idea of the value of money in those days, for three shillings (15 pence today) on one occasion the meal included a choice of nine dishes of meat, poultry or fish, two fruit pies, plum pudding, butter and cheese, and wine, porter or lemonade.[2]

The published work which established Cavendish's reputation as 'the wisest of the rich' was just the tip of the iceberg of his research activities, the fruits of most of which were never published in his lifetime. The totality of his work was wide ranging, and he would also have had a profound influence on physics (especially the study of electricity) if his results had been known to his contemporaries; but the published work was mostly in chemistry, where it was no less influential and sat right in the mainstream of developments in the second half of the eighteenth century. The first of Cavendish's chemical researches that we know about were carried out around 1764, and involved a study of arsenic; but these results were not published, and we don't know why Cavendish chose this particular substance for

1. There is no evidence at all for the myth that on these occasions Charles Cavendish would send Henry out of the house with only the few shillings in his pocket required to pay for the meal, and not a penny more.

2. Minute book of the Royal Society Club, quoted by Jungnickel and McCormmach.

investigation. To put his skills in perspective, though, his notes show that at this time he developed a method for preparing arsenic acid (still in use today), which was independently invented by Scheele in 1775 and is usually credited to him (quite rightly, in view of Cavendish's reticence). But when Cavendish did publish for the first time, in the *Philosophical Transactions* in 1766, he did so with a bang.

The chemical studies of Henry Cavendish: publication in the Philosophical Transactions

Cavendish, who was by then 35 years old, actually prepared a linked set of four papers describing his experiments with different gases ('airs'). For some unknown reason, only the first three of these papers were actually submitted for publication, but they contained what was probably his single most significant discovery, that the 'air' given off when metals react with acids is a distinct entity in its own right, different from anything in the air that we breathe. This gas is now known as hydrogen; Cavendish, for obvious reasons, called it 'inflammable air'. Following Black's lead, Cavendish carried out many careful, quantitative tests, including comparing the kind of explosions produced when different quantities of inflammable air were mixed with ordinary air and ignited, and determining the density of inflammable air. He thought that the gas was released by the metals involved in the reaction (we now know that it comes from the acids), and he identified the gas with phlogiston – to Cavendish, although not to all of his contemporaries, hydrogen *was* phlogiston. Cavendish also investigated the properties of Priestley's 'fixed air' (carbon dioxide), always making accurate measurements, but never claiming greater accuracy for results than was justified by the accuracy of his measuring instruments. In 1767, he published a study of the composition of mineral water, but then (perhaps stimulated by the publication that year of Priestley's *History of Electricity*) he seems to have turned his attention to electrical research and published a theoretical model, based on the idea of electricity as a fluid, in the *Philosophical Transactions* in 1771.

The paper seems to have been totally ignored, and although Cavendish continued to experiment with electricity, he published nothing else on the subject. This was a great loss to science at the time, but all of Cavendish's results (for example, 'Ohm's law') were independently reproduced by later generations of scientists (in this case, Ohm) and will be discussed in their context later. It's worth mentioning, though, that in one set of beautifully accurate experiments involving one conducting sphere mounted concentrically inside another (charged) conducting sphere, Cavendish established that the electric force obeys an inverse square law ('Coulomb's law') to within an accuracy of ± 1 per cent.

In the early 1780s, Cavendish went back to the study of gases. As he put it in the great paper that came out of this work,[1] the experiments were made 'principally with a view to find out the cause of the diminution which common air is well known to suffer by all the various ways in which it is phlogisticated, and to discover what becomes of the air thus lost'. In modern terminology, the reason why air is 'diminished' when something burns in it is that oxygen from the air is combining with the material that is burning, so up to 20 per cent of the common air can be locked up in a solid or liquid compound. But although Priestley had already discovered oxygen and found that it makes up roughly one-fifth of the common air, by the time Cavendish carried out these new experiments the process of combustion was far from being properly understood, and Cavendish, like many others, thought that it involved *adding* phlogiston to the air, not taking oxygen *out* of the air.

Since Cavendish thought that his inflammable air *was* phlogiston, it was natural to use the gas we now call hydrogen in these experiments. The technique Cavendish used was one that had been invented by the pioneering electrician Allessandro Volta and then used in experiments by John Warltire, a friend of Priestley (who later also carried out similar experiments). In this technique, a mixture of hydrogen and oxygen in a sealed copper or glass vessel was exploded using an electric spark. Because the vessel was sealed, this made it possible to weigh everything before and after the explosion, with only light and heat escaping from the vessel, and avoid any contagion by other materials which would have been impossible if it had been ignited by, say, a candle. The technology may not seem very sophisticated by modern standards, but it is another example of the way in which progress in science depended completely on that improved technology.

1. *Philosophical Transactions*, volume 74, p. 119, 1784.

Warltire noticed that the inside of the glass vessel became covered in dew after the explosion, but neither he nor Priestley, who reported Warltire's results, realized the significance of this. They were more interested in the possibility that heat had weight, which was lost from the vessel as heat escaped during the explosion.[1] Warltire's experiments seemed to show this loss of weight, but early in 1781, much more careful experiments carried out by Cavendish (who was, incidentally, an early subscriber to the idea that heat is associated with motion) showed that this was not the case. It's worth quoting what he said about these experiments in his 1784 paper:

> In Dr Priestley's last volume of experiments is related an experiment of Mr Warltire's in which it is said that, on firing a mixture of common and inflammable air by electricity in a closed copper vessel holding about three pints, a loss of weight was always perceived, on an average about two grains . . . It is also related that on repeating the experiment in glass vessels, the inside of the glass, though clean and dry before, immediately becomes dewy; which confirmed an opinion he had long entertained, that common air deposits its moisture by phlogistication. As the latter experiment seemed likely to throw great light on the subject I had in view, I thought it well worth examining more closely. The first experiment also, if there was no mistake in it, would be very extraordinary and curious: but it did not succeed with me; for although the vessel I used held more than Mr Warltire's, namely 24,000 grains of water, and though the experiment was repeated several times with different proportions of common and inflammable air, I could never perceive a loss of weight of more than one-fifth of a grain, and commonly none at all.

In a footnote, Cavendish mentioned that since his experiments had been carried out, Priestley had also found that attempts to reproduce Warltire's results did not succeed. In modern terminology, Cavendish showed that the weight of water formed in the explosion was equal to the combined weight of hydrogen and oxygen used up in the explosion. But he didn't put it like that.

Cavendish took so long to get these results into print because they

1. Einstein's famous equation $E = mc^2$ tells us that the energy lost does correspond to a loss of weight, but far too small, of course, to be measured in such experiments.

were just the beginning of a careful series of experiments in which he investigated the results of exploding different proportions of hydrogen with air and analysed carefully the dewy liquid deposited on the glass. He was particularly cautious about this because in some of his first experiments the liquid turned out to be slightly acidic – we now know that if there is insufficient hydrogen to use up all of the oxygen in the sealed vessel, the heat of the explosion makes the remaining oxygen combine with nitrogen from the air to make nitrogen oxides, which form the basis of nitric acid. But eventually Cavendish found that with a sufficient amount of 'inflammable air' there was always the same proportion of common air lost and the liquid produced by the explosion was pure water. He found that '423 measures of inflammable air are nearly[1] sufficient to phlogisticate 1000 of common air; and that the bulk of the air remaining after the explosion is then very little more than four-fifths of the common air employed'. In earlier experiments, he had found that 20.8 per cent of common air, by volume, is (in modern terminology) oxygen. So the ratio of hydrogen gas to oxygen gas by volume needed to convert all of the mixture of gases to water was, from his numbers, 423:208, within 2 per cent of the ratio (2:1) in which these gases are now known to combine.

Water is not an element

Cavendish, of course, described his results in terms of the phlogiston model (he even explained the production of nitric acid from 'phlogisticated air', our nitrogen, in this way, although the explanation is horribly complicated), and he did not think of hydrogen and oxygen as elements which combined physically to make water. But he did show that water itself was not an element and is somehow formed from a mixture of two other substances. This was a key step in the later stages of the transformation of alchemy into chemistry. Unfortunately, because Cavendish was so painstaking and thorough in following up every possibility before publishing his results, by the time he did publish, other people were working along similar lines, and for a time this led to a confusion about priorities. In England, partly on the basis of the kind of experiments pioneered by Volta and Warltire, James Watt arrived at the idea of the compound nature of water some time around 1782 or 1783, and his speculations (nowhere

1. He uses 'nearly' to mean 'very closely'.

near as complete and accurate as Cavendish's work) were also published by the Royal in 1784. In France, Lavoisier heard of Cavendish's early results from Charles Blagden, a scientific associate of Cavendish (who also became Secretary of the Royal Society), when he visited Paris in 1783.[1] Lavoisier promptly investigated the phenomenon (also using a more sloppy experimental technique than Cavendish, although Lavoisier was usually a careful experimenter) and wrote his results up without giving full credit to Cavendish's prior work. But this is all water that has long since flowed under the bridge, and nobody now doubts Cavendish's role in identifying water as a compound substance – which was to be a key component in Lavoisier's demolition of the phlogiston model and development of a better understanding of combustion.

Before we go on to Lavoisier's own work, though, there are two more of Cavendish's achievements that are too important to omit from our story, even though they are not part of the development of chemistry in the eighteenth century. The first is an example of just how incredibly accurate Cavendish was as an experimenter, and how far ahead of his time in many respects. In a paper published in 1785, Cavendish described experiments on air which involved prolonged sparking of nitrogen (phlogisticated air), and oxygen (dephlogisticated air) over alkali. This used up all the nitrogen and produced a variety of oxides of nitrogen. Entirely as a by-product of this work, Cavendish noted that it had proved impossible to remove all of the gas from his sample of air, and that even after all the oxygen and nitrogen had been removed there remained a small bubble, 'certainly not more than 1/120 of the bulk of the phlogisticted air'. He put this down to experimental error, but noted it anyway, for completeness. More than a century later, this work was brought to the attention of William Ramsay, working at University College in London, and Lord Rayleigh, working at the Cavendish Laboratory in Cambridge (accounts differ as to exactly how it was drawn to their attention). They decided to follow up Cavendish's mysterious bubble, and in 1894 found a

1. It was Blagden who, at Cavendish's suggestion, carried out measurements of sea temperature on a voyage to America in the mid-1770s, and discovered the warmth of the Gulf Stream.

previously unknown gas, argon, present in tiny traces (0.93 per cent, or 1/107th) in the atmosphere. The work led to the award of one of the first Nobel prizes, in 1904 (actually *two* Nobel prizes, since Rayleigh received the one for physics and Ramsay received the prize for chemistry). Nobel prizes are never awarded posthumously, but if they were, Cavendish would surely have been included in this honour, for work carried out 120 years earlier.

The Cavendish experiment: weighing the Earth

The last contribution of Cavendish to mention is also the last major piece of work he carried out, his most famous and the subject of his last important publication, which was read to the Royal Society on 21 June 1798, four months before Cavendish's sixty-seventh birthday. At an age when most scientists have long stopped making any significant contribution to their trade, Cavendish, in an outbuilding of his house on Clapham Common, had just weighed the Earth.

What became known as 'the Cavendish experiment' was actually devised by Cavendish's long-time friend John Michell, who features in the next chapter. Michell thought up the experiment and got as far as building the apparatus needed to carry it out, but died in 1793, before he could do the experiment himself. All of Michell's scientific apparatus was left to his old college in Cambridge, Queens, but since there was nobody there competent to follow up Michell's idea for weighing the Earth, Francis Wollaston, one of the professors in Cambridge, passed it on to Cavendish (one of Wollaston's sons was a neighbour of Cavendish at Clapham Common, which may have been a factor, although in any case Cavendish was the obvious man for the job, in spite of his age). The experiment was very simple in principle, but required great skill in practice because of the tiny forces that had to be measured. The apparatus (largely rebuilt by Cavendish) used a strong, light rod (six feet long and made of wood) with a small lead ball, about two inches across, on each end. The rod was suspended by a wire from its mid-point. Two much heavier lead balls, each weighing about 350 pounds, were suspended so that they could be swung into position a precise distance from the small balls, and the whole apparatus was located inside a wooden case to prevent it being disturbed by air currents. Because of the gravitational attraction between the large weights and the small balls, the rod was twisted slightly in the horizontal plane until it was stopped by the torsion of the wire; and to measure the force corresponding to this amount of twist, Cavendish carried out experiments with the large weights absent and with the horizontal bar rotating to and fro as a horizontal pendulum. The whole setup is called a torsion balance.

From all this, Cavendish found the force of attraction between a 350 pound weight and each small ball. But he already knew the force of attraction exerted by the Earth on the small ball – its weight – so he could work out the mass of the Earth from the ratio of these two forces. It is also possible to use such experiments to measure the strength of the gravitational force, in terms of a number known as the gravitational constant, written as *G*; this is what such experiments are still used for today. But Cavendish did not think in those terms and did not himself measure *G*, although a value for this constant can be inferred from his data. Indeed, Cavendish didn't present a value for the mass of the Earth as such, in his results, but rather gave a figure for its density. He carried out a series of eight experiments in August and September 1797, and nine more in April and May 1798. The results, published in the *Philosophical Transactions*[1] and taking account of many possible sources of error as well as using two different wires and comparing the two sets of results, gave a value for the density of the Earth which he quoted as 5.48 times the density of water.

This was slightly higher than an estimate which had been derived by geologists a little earlier, based on measurements of how much a pendulum was deflected out of the vertical towards a large mountain. But those studies depended on guessing the density of the rocks making up the mountain, and in a letter to Cavendish written in 1798, James Hutton, one of the geologists involved in that work, said that he now thought that this figure had been underestimated and that the true value for the density of the Earth derived from this method was 'between 5 and 6'.[2] Many years later, it was noticed that in spite of all his care, Cavendish had made a tiny arithmetical slip in his calculations, and that the figure for the density of the Earth should, using his own numbers, be given as 5.45 times the density of water. The modern value

1. Volume 88, p. 526, 1798.
2. Quoted by Jungnickel and McCormmach.

for the mean density of the Earth, derived from several techniques, is 5.52 times the density of water, just over 1 per cent bigger than Cavendish's corrected value. The best analogy we have seen that gives a feel for the accuracy involved in these experiments comes from a book written by one of the people involved in them at the end of the nineteenth century, the English physicist John Poynting (1852–1914).[1] Writing about experiments that used the common vertical balance to measure the tiny forces produced by placing a large mass under one pan of the balance, he said:

> Imagine a balance large enough to contain in one pan the whole population of the British Islands, and that all the population had been placed there but one medium-sized boy. Then the increase in weight which had to be measured was equivalent to measuring the increase due to putting that boy on with the rest. The accuracy of the measurement was equivalent to observing whether or no he had taken off one of his boots before stepping on to the pan.

Cavendish had been just as accurate, almost a hundred years earlier.

During the first decade of the nineteenth century, into his own seventh decade, Cavendish's life continued much as it always had done, with scientific experiments (although nothing worth mentioning here), dinners with the Royal Society Club and attendance at scientific meetings (he was an early subscriber to the Royal Institution, where he served on its governing body and took an active interest in the work of Humphry Davy). He died quietly, at home, on 24 February 1810, after a short illness and was buried in the family vault at All Saints' Church in Derby (now Derby Cathedral). Scientists such as Black, Priestley, Scheele and Cavendish made the discoveries that laid the foundations of chemistry as a science. Cavendish lived to see these discoveries pulled together in a synthesis that made chemistry truly scientific by the man regarded as the greatest chemist of them all, Antoine Lavoisier. Indeed, Cavendish outlived Lovoisier, who, as we shall see, fell victim to the Terror during the French Revolution.

Antoine-Laurent Lavoisier was born into a Catholic family in Paris on 26 August 1743, in the district known as the Marais. Both his

1. *The Earth* (CUP, Cambridge, 1913).

Antoine-Laurent Lavoisier: study of air; study of the system of respiration

grandfather (another Antoine) and his father (Jean-Antoine) were successful lawyers, and young Antoine was brought up in middle-class comfort. He had one sibling, a sister born in 1745 and christened Marie-Marguerite-Emilie, but their mother, Emilie, died in 1748 and the family went to live with his widowed maternal grandmother near what is now Les Halles. There, the unmarried sister of Emilie, another Marie, became a surrogate mother who devoted herself to the children. Antoine attended the Collège Mazarin (founded under the will of Cardinal Mazarin, who lived from 1602 to 1661 and ruled France during the minority of Louis XIV), where he shone at the classics and literature, but also began to learn about science. In 1760, Lavoisier's sister Marie died, at the age of 15; a year later, he entered the School of Law at the University of Paris, intending to follow the traditional family career, where he graduated as a Bachelor of Law in 1763 and qualified as Licentiate in 1764. But his legal studies still gave him time to develop his interests in science, and he attended courses in astronomy and mathematics, botany, geology and chemistry alongside his official work. After completing his education, it was science, not the law, that called, and Lavoisier spent three years working as an assistant to Jean-Etienne Guettard (1715–1786) on a project to complete a geological map of France, surveying and collecting specimens.

By the time this field work was over, Lavoisier was free to choose whatever career he liked. His grandmother had died in 1766 and left most of her wealth (ample enough to live on) to him. The same year, Lavoisier first became widely known in scientific circles, when he was awarded a gold medal, presented by the President of the Royal Academy of Sciences on behalf of the King, for an essay on the best way to light the streets of a large town at night. In 1767, with Guettard's geological survey now officially sponsored by the government, he set out with Guettard (as his equal collaborator, not as an assistant) to survey Alsace-Lorraine. This work established his reputation so effectively that he was elected as a member of the Royal Academy of Sciences in 1678, at the remarkably young age of 25.

The French Academy operated quite differently from the English Royal Society. The Royal was, strictly speaking, merely a gentlemen's club, with no official status. But the Academy was funded by the

French government, and its members were paid salaries and expected to carry out work of a scientific nature for the government, even if they had other posts as well. Lavoisier, an able administrator, played a full part in the activities of the Academy, and during his time as a member worked on very many reports covering topics as diverse as the adulteration of cider, the Montgolfier balloon, meteorites, cultivation of cabbages, the mineralogy of the Pyrenees and the nature of the gas arising from cesspools. But Lavoisier's undoubted abilities did not include precognition, and in 1768 he took what turned out to be the worst decision of his life, when he bought a one-third share in a 'Tax Farm'.

The French taxation system at the time managed to be unfair, incompetent and corrupt. Many of the problems arose from the stability of the French political system over the seventeenth and eighteenth centuries, when Louis XIV reigned for 72 years (initially with the help of Cardinal Mazarin), covering the period when England *twice* disposed of monarchs that they took exception to, and then Louis XV reigned for a further 59 years, from 1715 to 1774. Neither of them took much notice of the will of the people. The result was that practices (such as exempting nobles from taxes) which might have been acceptable as 'the natural order of things' in the early seventeenth century were still in force, fossilized relics of a bygone age, in the last quarter of the eighteenth century. This was a major factor in the discontent that led to the French Revolution.

At the time Lavoisier became, in effect, a tax collector, the right to collect taxes (such as taxes on salt and customs duties on alcoholic drink) was farmed out to groups of financiers known as Tax Farmers, who paid the government (usually using borrowed capital) for the privilege. The King's Farmers-General ('Fermiers'), as they were known, then recouped their investment plus a reasonable profit from the taxes, and if they managed to collect more than they had paid the King, they could keep it.[1] To make things worse, even honest and efficient Fermiers (and there were some) could only obtain the tax-

1. Not that people like Lavoisier actually *collected* the taxes; they were at the top of the pyramid, and the Tax Farms employed administrators and heavies to do the actual work.

collecting rights by providing sinecures for ministers, their families or members of the royal family, who would draw an income (known as a pension) from the proceeds of the 'farm' without doing any work. It requires little imagination to see how unpopular this system was with all the people who paid taxes (which meant everyone except the rich); from the perspective of the early twenty-first century, however, it requires rather more imagination to appreciate that Lavoisier was not a bad man out to oppress the poor, but simply made what he considered to be a sound investment. He certainly worked hard on behalf of his 'farm', and there is no evidence that he was particularly harsh as a tax gatherer. But he was, indeed, a tax gatherer, and the system itself was harsh, even if administered within the law. But if the association would end unhappily, it certainly started out well, and not just financially. On 16 December 1771, Lavoisier entered into an arranged marriage with Marie-Anne-Pierette Paulze, the 13-year-old daughter of his fellow Fermier, the lawyer Jacques Paulze. To mark the occasion, Lavoisier's father bought him a minor title; although he seldom used it, from now on he was officially 'de Lavoisier'. Although childless, the marriage seems to have been a happy one, and Marie became keenly interested in science, working as Lavoisier's assistant and helping to keep notes on his experiments.

At the end of the 1760s, Lavoisier had carried out a series of experiments building from Black's meticulous approach to chemistry, which finally proved that water could not be transmuted into earth. But the work for which he is now famous began in the 1770s, after his marriage. By heating diamonds using sunlight concentrated by a huge lens (four feet in diameter and six inches thick) he proved, in 1772, that diamond is combustible, and by the end of the year he had established that when sulphur burns it gains weight, instead of losing it. It was his first, independent, step towards the modern understanding of combustion as a process involving oxygen from the air combining with the substance that is being burnt. This was the prelude to a huge series of experiments, including a careful study of Black's 'fixed air' and the production of what we now know to be oxygen by heating the red calx of lead (red lead) with the great lens. In 1774, not long after his discovery of oxygen, Priestley had set out with Lord Shelburne on a continental trip, and in October that year Priestley had visited

23. *Lavoisier's experiment on human respiration. After a drawing by Marie-Anne Lavoisier.*

Lavoisier in Paris, where he gave him the news of his early results. Starting in November 1774, Lavoisier carried out his own experiments along these lines, and in May 1775 he published a paper in which he said that the 'principle' that combined with metals during the process of calcination (forming a calx) came from the atmosphere and was the 'pure air' discovered by Priestley.

It was about this time that Lavoisier began to be more involved in government work. When Louis XVI succeeded to the throne in 1774, he attempted to reform some of the corrupt administrative practices he had inherited. These included the way gunpowder was supplied (or, as often as not, not supplied) to the army and navy. Like the tax system, this was privatized, corrupt and inefficient. In 1775, Louis XVI effectively nationalized the gunpowder industry, appointing four commissioners (one of them Lavoisier) to run it. Lavoisier moved into the Paris Arsenal to facilitate this work (which, as with everything he touched, he carried out diligently and effectively), and set up his laboratory there. It was there that he established the superiority of the modern model of combustion to the phlogiston model and eventually, in 1779, gave oxygen its name.[1]

Like other chemists of his time, Lavoisier was deeply interested in the nature of heat, which he called 'matter of fire'. After carrying out experiments which showed that oxygen from the air is converted into fixed air by animals (including human animals) during respiration, he concluded that an animal maintains its body temperature by the conversion of oxygen into fixed air in the same way that charcoal gives off heat when it burns (he was correct in principle; although, of course, the processes which generate body heat are a little more complicated than simple burning). But how could he test this hypothesis? In order to prove it, in the early 1780s he carried out some ingenious experiments involving a guinea pig, in collaboration with his fellow academician Pierre Laplace, whose story is told in the next chapter.

The guinea pig was placed in a container surrounded by ice, all inside a larger container (the whole thing is known as an ice calorimeter), and after ten hours in the cold environment the warmth of the animal's body had melted 13 ounces of ice. By burning small pieces of charcoal in an ice calorimeter, Lavoisier and Laplace then found out how much charcoal was needed to melt that much ice. Then, in a separate series of experiments, they measured both how much fixed air the guinea pig breathed out over ten hours while resting, and how much fixed air was produced by burning different amounts of charcoal. Their conclusion was that the amount of fixed air breathed out by the guinea pig would have been the same as the amount produced by charcoal burning that produced enough heat to melt 10½ ounces of ice. The agreement was not exact, but Lavoisier and Laplace were well aware that the experiment was imperfect and regarded this as confirmation of the idea that animals keep warm by converting what we would now call carbon (from their food) into what we would now call carbon dioxide (which they breathe out) by combining it with oxygen from the air (which they breathe in). Lavoisier and Laplace said that respiration 'is therefore a combustion, admittedly very slow, but otherwise exactly similar to that of charcoal'. This was a key step in setting human beings

1. From the Greek for 'acid giver'. Lavoisier mistakenly thought that oxygen was present in all acids.

in their context as systems (albeit complicated systems) obeying the same laws as falling stones or burning candles. By the end of the eighteenth century, science had shown that there was no need to invoke anything outside the known world of science to produce the life-associated warmth of the human body – no need for Harvey's 'natural heat'.

Overlapping the work on respiration, Lavoisier developed the theory of combustion further, and he published his definitive demolition of the phlogiston model in the *Mémoires* of the Academy in 1786, although it took some time for the last supporters of the phlogiston model to die out. It's worth giving Lavoisier's summing up from that paper in his own words, remembering that he used the term 'air' where we would say 'gas':

1. There is true combustion, evolution of flame and light, only in so far as the combustible body is surrounded by and in contact with oxygen; combustion cannot take place in any other kind of air or in a vacuum, and burning bodies plunged into either of these are extinguished as if they had been plunged into water.
2. In every combustion there is an absorption of the air in which the combustion takes place; if this air is pure oxygen, it can be completely absorbed, if proper precautions are taken.
3. In every combustion there is an increase in weight in the body that is being burnt, and this increase is exactly equal to the weight of the air that has been absorbed.
4. In every combustion there is an evolution of heat and light.

We have already mentioned how Charles Blagden brought news of Henry Cavendish's work on the composition of water to Lavoisier in Paris in June 1783; now we can see how naturally this fitted in to Lavoisier's model of combustion, even though his initial experiments with the combustion of hydrogen were less accurate than those of Cavendish. It reflects badly on Lavoisier that he did not give proper credit to Cavendish when he first published his own results, but the important point is that Lavoisier, rather than Cavendish, was the first person to appreciate that water is a compound substance formed from a combination of 'inflammable air' and oxygen in the same sort of way that 'fixed air' is formed from a combination of carbon and oxygen.

TRAITÉ ÉLÉMENTAIRE DE CHIMIE.

PREMIERE PARTIE.

De la formation des fluides aériformes & de leur décomposition ; de la combustion des corps simples & de la formation des acides.

CHAPITRE PREMIER.

Des combinaisons du calorique & de la formation des fluides élastiques aériformes.

C'EST un phénomène constant dans la nature, & dont la généralité a été bien établie par Boerhaave, que lorsqu'on échauffe un corps

Tome I. A

24. *Title page of Lavoisier's* Traité Élémentaire de Chimie, *1789.*

The first table of elements; Lavoisier renames elements; he publishes Elements of Chemistry

Lavoisier summed up his life's work in chemistry in a book, *Traité Elementaire de Chimie* (*Elements of Chemistry*), published in 1789, the year of the storming of the Bastille. There were many translations and new editions of the book, which laid the foundations for chemistry as a genuinely scientific discipline and is sometimes regarded by chemists as the equivalent to chemistry of Newton's *Principia* to physics. As well as providing extensive descriptions of the techniques of chemistry, including the kind of apparatus used and the kind of experiments carried out, Lavoisier provided the clearest definition yet of what was meant by a chemical element, at last putting into practice Robert Boyle's insight from the 1660s, finally consigning the four 'elements' of the Ancient Greeks to the dustbin of history, and giving the first table of the elements that, although very incomplete, is recognizable as the basis from which the modern table of the elements has grown.[1] He stated clearly the law of conservation of mass, got rid of old names (such as phlogisticated air, inflammable air and oil of vitriol) and replaced them with names based on a logical system of nomenclature (such as oxygen, hydrogen and sulphuric acid), introducing a logical way to name compounds, such as nitrates. By giving chemistry a logical language, he greatly eased the task of chemists trying to communicate their discoveries to one another.

In truth, Lavoisier's masterwork, although one of the most important scientific books ever published, isn't quite in the same league as Newton's masterwork; nothing is. But its publication does mark, as far as it is possible to pinpoint such things, the moment when chemistry got rid of the last vestiges of alchemy and became recognizable as the ancestral form of the scientific discipline we know by that name today. Lavoisier was 46 in the year his great book was published, and quite possibly he would not have made any major new contributions to science after that, even if he had lived in less turbulent times. But by the beginning of the 1790s, political developments in France meant that Lavoisier had less and less time to devote to science.

Lavoisier was already actively involved in government, originally

1. But Lavoisier had one blind spot here, including 'caloric', the 'element of heat', in his table along with things like oxygen, hydrogen, charcoal, sulphur, gold and lead.

Lavoisier's execution

as a result of purchasing an estate (including a château and a farm, which he used to experiment with scientific methods of agriculture) at Fréchines, in the Province of Orléans. Although technically a minor member of the nobility, in 1787 (as if he didn't have enough on his plate already) he was elected to the Provincial Assembly of Orléans as a representative of the Third Estate (the other two being the Clergy and the Nobles). Politically, he was what would now be called a liberal, and a reformer, who tried unsuccessfully to achieve a fairer tax system in the Province. In May 1789, Lavoisier wrote to his colleagues in the Provincial Assembly that 'inequality of taxation cannot be tolerated except at the expense of the rich'.

The first stage of the development of the French Revolution, with a democratic majority seeming to be in control of the National Assembly and the King pushed to one side, very much fitted in with Lavoisier's ideas. But things soon started to turn nasty. Although Lavoisier continued to serve the government on the Gunpowder Commission and in other ways, he came under unjustified suspicion that the Commission had been lining its pockets at the public expense – in fact, it was largely thanks to the reforms carried out by the Commission that France had sufficient good-quality gunpowder to wage the Napoleonic wars. More seriously (and more justifiably) he was also tainted with the widespread hatred of the Fermiers. In spite of all these difficulties, Lavoisier continued to work diligently on behalf of the government, even as the nature of the government was changing; he played an important part in planning the reform of the French educational system and was a member of the commission, appointed in 1790, which (eventually) introduced the metric system. But none of this did him any good when elements of the Jacobin administration decided to make an example of the former Fermiers, and Lavoisier was among 28 'Tax Farmers' (some honest, some corrupt) who were guillotined on 8 May 1794. He was fourth in the list for execution that day, and his father-in-law, Jacques Paulze, was third. At the time of the executions, Joseph Priestley was at sea on his way to exile in America. It was truly the end of an era in chemistry.

8

Enlightened Science II: Progress on all fronts

The study of electricity: Stephen Gray, Charles Du Fay, Benjamin Franklin and Charles Coulomb – Luigi Galvani, Alessandro Volta and the invention of the electric battery – Pierre-Louis de Maupertuis: the principle of least action – Leonhard Euler: mathematical description of the refraction of light – Thomas Wright: speculations on the Milky Way – The discoveries of William and Caroline Herschel – John Michell – Pierre Simon Laplace, 'The French Newton': his Exposition *– Benjamin Thompson (Count Rumford): his life – Thompson's thoughts on convection – His thoughts on heat and motion – James Hutton: the uniformitarian theory of geology*

Many accounts of the history of science describe the eighteenth century as a period when, apart from the dramatic progress in chemistry we have just described, nothing much happened. It is regarded as an interregnum, somehow in the shadow of Newton, marking time until the major advances of the nineteenth century. Such an interpretation is very wide of the mark. In fact, progress in the physical sciences in the eighteenth century proceeded on a broad front – not with any single great breakthrough to rank with the achievements of Newton, to be sure; but with a host of lesser achievements as the lessons of Newtonianism – that the world is comprehensible and can be explained in accordance with simple physical laws – were absorbed and applied. Indeed, the lesson was so widely absorbed that from now on, with a few notable exceptions,[1] it will no longer be possible to go into so much biographical detail about the scientists themselves. This is not because they become intrinsically less interesting in more recent times, but simply because there are so many of them and there is so much to describe. It is after the death of Newton that, first in the physical sciences and then in other disciplines, the story of science itself, rather than of the individuals who contributed to the story, becomes the central theme in the history of science, and it becomes harder and harder to know the dancer from the dance.

1. See, for example, Chapter 9.

25. *Demonstration of the way electricity passes through living people and corpses. From Watson's* Experiments and Observations, *1748.*

It was in the decade following the death of Newton that the term 'physics' started to be used, rather than 'natural philosophy', to describe this kind of investigation of the world. Strictly speaking, this was a revival of an old terminology, since the word had been used by Aristotle, and probably even earlier; but it marked the beginning of what we now mean by physics, and one of the first books using the term in its modern sense, *Essai de physique*, written by Pieter van Musschenbroek (1692–1761), was published in 1737. In the same decade, physicists started to come to grips with the mysterious phenomenon of electricity. Musschenbroek himself, who worked in Leiden, later (in the mid-1740s) invented a device that could store large quantities of electricity. It was simply a glass vessel (a jar) coated with metal on the inside and outside – an early form of what is now called a capacitor. This Leiden jar, as such devices came to be called, could be charged up, storing electricity to be used in later experiments, and if several of them were wired together they could produce a very large discharge, sufficient to kill an animal.

The study of electricity: Stephen Gray, Charles Du Fay, Benjamin Franklin and Charles Coulomb

But the first steps towards an understanding of static electricity were carried out without the aid of Leiden jars. Stephen Gray, an English experimenter (born around 1670; died in 1736) published a series of papers in the *Philosophical Transactions* in which he described how a cork in the end of a glass tube gained electrical characteristics (became charged, we would now say) when the glass was rubbed,[1] how a pine stick stuck in the cork would carry the electrical influence right to the end of the stick and how the influence could be extended for considerable distances along fine threads. Gray and his contemporaries made their electricity as and when they needed it, by friction, from simple machines in which a globe of sulphur was rotated while being rubbed (later, glass spheres or cylinders were substituted for the sulphur). Partly influenced by Gray's work, the Frenchman Charles Du Fay (1698–1739) discovered in the mid-1730s that there are two kinds of electricity (what we now call positive and negative charge) and that similar kinds repel one another while opposite kinds attract. Between them, the work of Gray and Du Fay also demonstrated the importance of insulating material in preventing electricity draining away from charged objects, and showed that anything could be charged with electricity if it was insulated – Du Fay even electrified a man suspended by insulating silk cords and was able to draw sparks from the body of his subject. As a result of his work, Du Fay came up with a model of electricity which described it in terms of two different fluids.

This model was refuted by the model developed by Benjamin Franklin (1706–1790), whose interest in electricity extended far beyond the famous, and hazardous, kite experiment (which was, incidentally, carried out in 1752 to charge up a Leiden jar and thereby prove the connection between lightning and electricity). In spite of all his other interests and activities, Franklin found time, from the mid–1740s into the early 1750s, to carry out important experiments

1. This method of 'making' static electricity by friction is the reason why a child's balloon rubbed on a woollen sweater can be made to stick to the ceiling, and why your hair can become electrified by brushing it.

(making use of the recently invented Leiden jars) which led him to a one-fluid model of electricity based on the idea that a physical transfer of the single fluid occurred when an object became electrically charged, leaving one surface with 'negative' charge and one with 'positive' charge (terms that he introduced). This led him naturally to the idea that charge is conserved – there is always the same amount of electricity, but it can be moved around, and overall the amount of negative charge must balance the amount of positive charge. Franklin also showed that electricity can magnetize and demagnetize iron needles, echoing work carried out a little earlier by John Michell (1724–1793), Henry Cavendish's friend who devised the 'Cavendish experiment'. By 1750, Michell had also discovered that the force of repulsion between two like magnetic poles obeys an inverse square law, but although he published all these results that year in *A Treatise on Artificial Magnets*, nobody took much notice, just as nobody took much notice of the various contributions of Franklin, Priestley and Cavendish in determining that the electric force obeys an inverse square law. It was not until 1780 that Charles Coulomb (1736–1806), building on the work of Priestley, carried out the definitive experiments on both electric and magnetic forces, using a torsion balance, which finally convinced everyone that both forces obey an inverse square law. This has therefore gone down in history as Coulomb's law.

Once again, these examples highlight the interplay between science and technology. The study of electricity only began to gather momentum after machines that could manufacture electricity were available, and then when devices to store electricity were developed. The inverse square law itself was only developed through the aid of the technology of the torsion balance. But the biggest technological breakthrough in the eighteenth-century study of electricity came right at the end of the century, paving the way for the work of Michael Faraday and James Clerk Maxwell in the nineteenth century. It was the invention of the electrical battery, and it resulted from an accidental scientific discovery.

The discovery was made by Luigi Galvani (1737–1798), a lecturer in anatomy and professor of obstetrics at the University of Bologna. Galvani carried out a long series of experiments on animal electricity, which he described in a paper published in 1791. In it, he recounted how he had first become interested in the subject after noticing twitch-

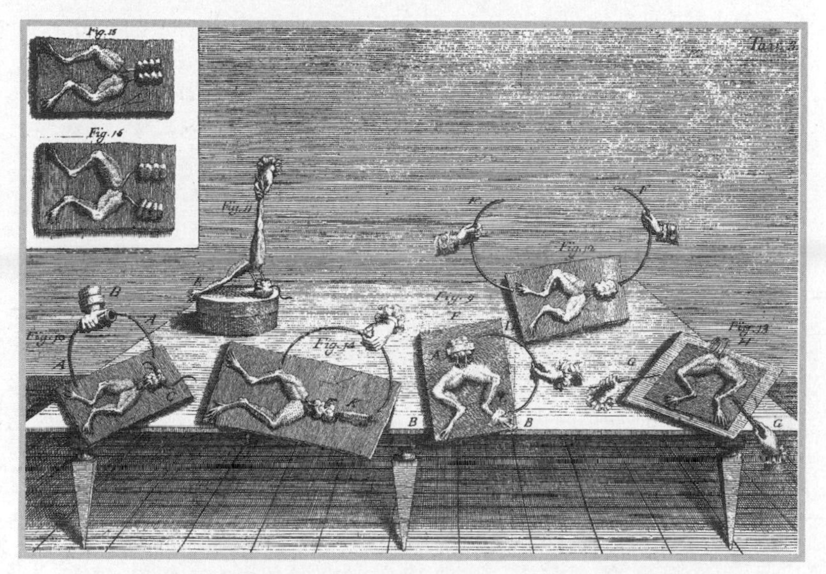

26. *Luigi Galvani's experiments with electricity and frogs' legs. From* De Viribies Electricitatis in Motu Mascalari, *1791.*

Luigi Galvani, Alessandro Volta and the invention of the electric battery

ing in the muscle of a frog laid out for dissection on a table where there was also an electrical machine. Galvani showed that the twitching could be induced by connecting the muscles of the dead frog directly to such a machine or if the frog were laid out on a metal surface during a thunderstorm. But his key observation came when he noticed that frogs' legs being hung out to dry twitched when the brass hook they were suspended on came into contact with an iron fence. Repeating this experiment indoors, with no outside source of electricity around, he concluded that the convulsions were caused by electricity stored, or manufactured, in the muscles of the frog.

Not everyone agreed with him. In particular, Alessandro Volta (1745–1827), the professor of experimental physics at Pavia University, in Lombardy, argued in papers published in 1792 and 1793 that electricity was the stimulus for the contraction of the muscles, but that it came from an outside source – in this case, from an interaction between the two metals (brass and iron) coming into contact. The difficulty was proving it. But Volta was a first-class experimenter who had already carried out important work on electricity (including the design of a better frictional machine to make static electricity and a device to measure electric charge) and had also worked on gases, measuring the amount of oxygen in the air by exploding it with hydrogen. He was well able to rise to this new challenge.

Volta first tested his ideas by placing different pairs of metals in contact and touching them with his tongue, which proved sensitive to tiny electrical currents that could not be detected by any of the instruments available at the time. While he was carrying out these experiments and trying to find a way to magnify the effect he could feel with his tongue into something more dramatic, his work was hampered by the political upheavals that affected Lombardy as a result of the French Revolution and subsequent conflict between France and Austria for control of the region. But by 1799 Volta had come up with the device that would do the trick. He described it in a letter to Joseph Banks, then President of the Royal Society, which was read to a meeting of the Royal in 1800.

His key invention was literally a pile of silver and zinc discs, alternating with one another and separated by cardboard discs soaked in brine. Known as a Voltaic pile, this was the forerunner of the modern battery, and produced a flow of electric current when the top and bottom of the pile were connected by a wire. The battery provided, for the first time, a more or less steady flow of electric current, unlike the Leiden jar, which was an all or nothing device that discharged its store of electricity in one go. Before Volta's invention, the study of electricity was essentially confined to the investigation of static electricity; after 1800, physicists could work with electric currents, which they could turn on and off at will. They could also strengthen the current by adding more discs to the pile, or reduce it by taking discs away. Almost immediately, other researchers found that the electric current from such a pile could be used to decompose water into hydrogen and oxygen, the first hint of what a powerful tool for science the invention would become. Although we shall have to wait until Chapter 11 to follow up the implications, the importance of Volta's work was obvious immediately, and after the French won control of Lombardy in 1800, Napoleon made Volta a Count.

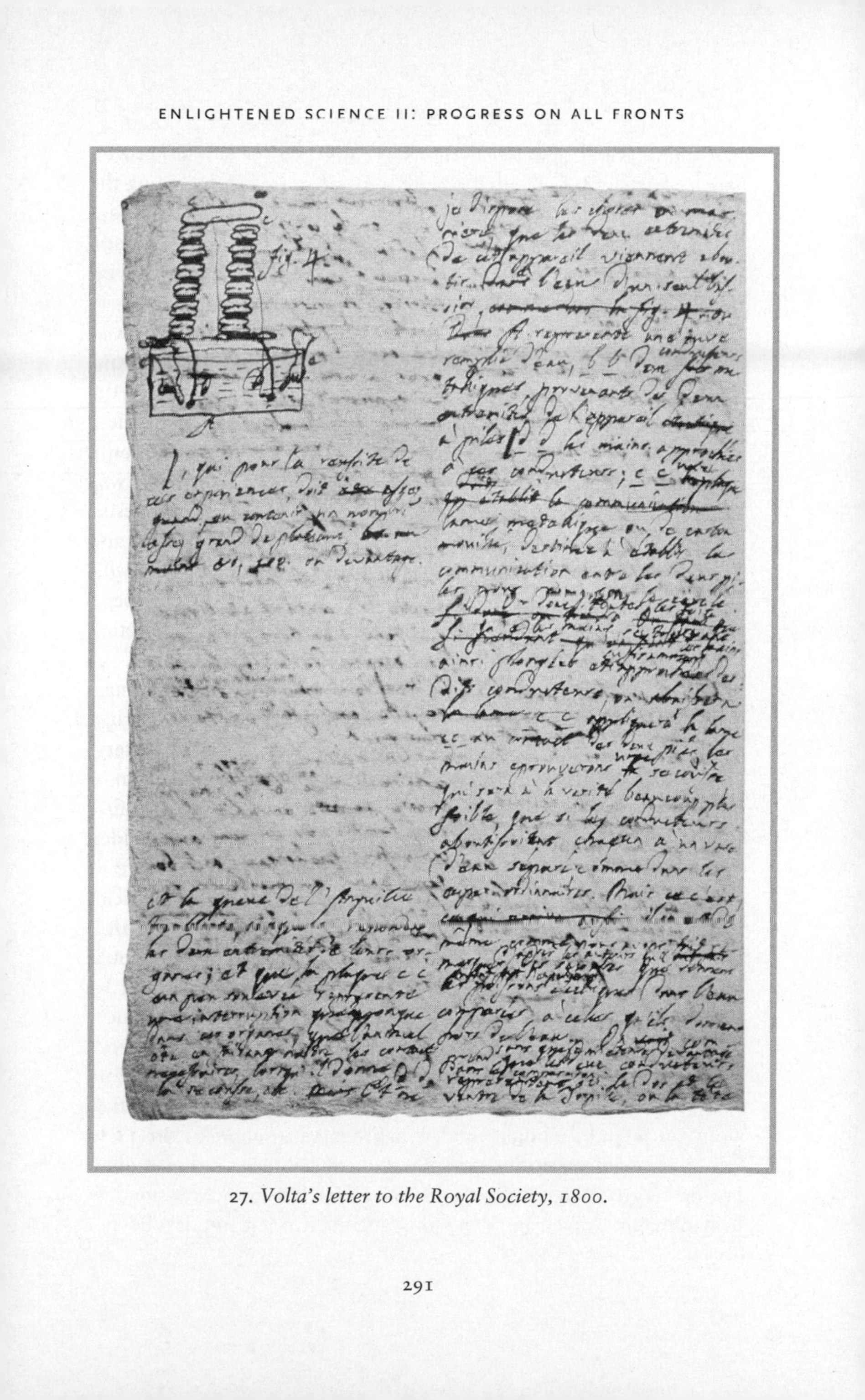

27. *Volta's letter to the Royal Society, 1800.*

If it took some time for physicists to come to grips with electricity, many of their ideas from the eighteenth century seem surprisingly modern, even if they were not always fully developed or widely appreciated at the time. For example, it was as early as 1738 that the Dutch-born mathematician Daniel Bernoulli (1700–1782) published a book on hydrodynamics which described the behaviour of liquids and gases in terms of the impact of atoms on the walls of their container – very similar to the kinetic theory of gases developed more fully in the nineteenth century and, of course, very much a development of the ideas of Newton concerning the laws of motion. Such ideas were also spreading geographically. In 1743, just five years after Bernoulli published his great book, we find Benjamin Franklin among the leading lights in establishing the American Philosophical Society in Philadelphia – the first scientific society in what is now the United States and the tiny seed of what would become a great flowering of science in the second half of the twentieth century.

Pierre-Louis de Maupertuis: the principle of least action

One of the most important insights in the whole of science, whose value only really became apparent during that twentieth century flowering, was formulated by Pierre-Louis de Maupertuis (1698–1759) just a year later, in 1744. De Maupertuis had been a soldier before turning to science; his big idea is known as the principle of least action. 'Action' is the name given by physicists to a property of a body which is measured in terms of the changing position of an object and its momentum (that is, it relates mass, velocity and distance travelled by a particle). The principle of least action says that nature always operates to keep this quantity to a minimum (in other words, nature is lazy). This turned out to be hugely important in quantum mechanics, but the simplest example of the principle of least action at work is that light always travels in straight lines.

Leonhard Euler: mathematical description of the refraction of light

Speaking of light, in 1746 Leonhard Euler (1707–1783), a Swiss regarded as the most prolific mathematician of all time, and the man who introduced the use of the letters *e* and *i* in their modern mathematical context, described mathematically the refraction of light, by assuming (following Huygens) that light is a wave, with each colour corresponding to a different wavelength; but this anti-Newtonian model did not take hold at the time.[1] The wave model of light languished because Newton was regarded in such awe; other ideas languished because they came from obscure scientists in remote parts of the world. A classic example is provided by Mikhail Vasil'evich Lomonosov (1711–1765), a Russian polymath who developed the Newtonian ideas of atoms, came up with a kinetic theory similar to that of Bernoulli and, around 1748, formulated the laws of conservation of mass and energy. But his work was virtually unknown outside Russia until long after his death.

Thomas Wright: speculations on the Milky Way

The discoveries of William and Caroline Herschel

John Michell

There were also ideas that were ahead of their time in astronomy. The Durham astronomer Thomas Wright (1711–1786) published (in 1750) *An Original Theory and new Hypothesis of the Universe*, in which he explained the appearance of the Milky Way by suggesting that the Sun is part of a disc of stars which he likened to a mill wheel. This was the same Wright whose 'day job' as a surveyor brought him into contact with Charles and Henry Cavendish. In 1781, William (1738–1822) and Caroline (1750–1848) Herschel discovered the planet Uranus, a sensation at the time as the first planet that had not been known to the Ancients, but barely hinting at the discoveries to be made beyond the old boundaries of the Solar System. And Henry Cavendish's good friend John Michell (1724–1793) was, as is now well known, the first person to come up with the idea of what are now known as black holes, in a paper read to the Royal Society on Michell's behalf by Cavendish in 1783.[2] Michell's idea was simply based on the (by then well-established) fact that light has a finite speed and the understanding that the more massive an object is, the faster you have to move to escape from its gravitational grip. It's worth quoting from that paper, if only

1. Incidentally, Euler really did go blind through looking at the Sun, just as all books of popular astronomy warn you can happen. The warning is based on fact!

2. Michell, something of a polymath, first made his name by his investigation of the large earthquake that struck Lisbon in 1755, showing that the disturbance originated from beneath the Earth's crust, out under the Atlantic Ocean, and established that earthquakes had nothing to do with atmospheric disturbances, as had previously been thought. He would probably have achieved even more in science, but in 1764 he gave up his post as professor of geology in Cambridge and became rector of a parish at Thornhill, in Yorkshire.

for the pleasure of imagining the otherwise shy Henry Cavendish in his element, reading it out to the packed throng at the Royal:

> If there should really exist in nature any bodies whose density is not less than that of the sun, and whose diameters are more than 500 times the diameter of the sun, since their light could not arrive at us . . . we could have no information from sight; yet, if any other luminiferous bodies should happen to revolve around them we might still perhaps from the motions of these revolving bodies infer the existence of the central ones.

Indeed, that is just how astronomers infer the existence of black holes today – by studying the motion of bright material in orbit around the black hole.

It's surprising enough that one person came up with the idea of black holes (which we are used to thinking of as a quintessential example of twentieth-century theorizing) before the end of the eighteenth century – even more surprising that a second person independently came up with the same idea before the century closed. That person was Pierre Simon Laplace, and it's worth slowing down the pace of our story to take stock of where physics stood at the end of the eighteenth century by taking a slightly more leisurely look at the career of the man sometimes referred to as 'the French Newton'.

Pierre Simon Laplace, 'The French Newton'; his Exposition

Laplace was born at Beaumont-en-Auge, near Caen, in the Calvados region of Normandy, on 28 March 1749. Little is known of his early life – indeed, not much is known about his private life at all. Some accounts refer to him coming from a poor farming family, but although his parents were not rich, they were certainly comfortably off. His father, also Pierre, was in the cider business, if not in a big way, and also served as a local magistrate, giving a clear indication of his status in the community. His mother, Marie-Anne, came from a family of well-off farmers at Tourgéville. Just as Laplace was named after his father, so his only sibling, a sister born in 1745, was named Marie-Anne, after her mother. Laplace went to school as a day boy at a local college run by the Benedictines, and was probably intended by his father for the priesthood. From 1766 to 1768 he studied at the University of Caen, and it seems to have been at this time that he was found to have a talent for mathematics. He helped to support himself at college by working as a private tutor, and there is some evidence that he briefly worked in this capacity for the Marquis d'Héricy, who was to play such an important part in Georges Couvier's life. Laplace left Caen without taking a degree and went to Paris with a letter from one of his professors recommending him to Jean d'Alembert (1717–1783), one of the top mathematicians in France at the time and a high-ranking member of the Academy. D'Alembert was sufficiently impressed by the young man's abilities that he found him a post with the grand title of professor of mathematics at the École Militaire, but which really just consisted of trying to drum the basics of the subject into reluctant officer cadets. He stayed in the post from 1769 to 1776, making a reputation for himself with a series of mathematical papers (as ever with the maths, we won't go into the details here) and being elected to the Academy in 1773.

Laplace was particularly interested in probability, and it was through this mathematical interest that he was led to investigate problems in the Solar System, such as the detailed nature of the orbits of the planets and of the Moon around the Earth. Could these have arisen by chance? Or must there be some physical reason why they have the properties they do? One example, which Laplace discussed in 1776, concerns the nature and orbits of comets. All the planets move around the Sun in the same direction and in the same plane (the plane of the ecliptic). This is a powerful indication (we shall soon see how powerful) that they all formed together by the same physical process. But comets orbit the Sun in all directions and at all angles (at least judging from the evidence of the few dozen comets whose orbits were known at the time). This suggests that they have a different origin, and mathematicians before Laplace had already reached this conclusion. But as a mathematician, Laplace was not so much concerned with the conclusion but with how it had been arrived at; he developed a more sophisticated analysis which showed probabilistically that it was highly unlikely that some force existed that was trying to make comets move in the plane of the ecliptic. In the mid-1770s, Laplace also looked for the first time at the behaviour of the orbits of Jupiter and Saturn.[1] These orbits showed a slight, long-term shift which did not seem to fit the predictions of Newtonian gravitational theory, and Newton himself had suggested that after a long enough time (only a few hundred years!) divine intervention would be required to put the planets back in their proper orbits and prevent the Solar System falling apart. Laplace's first stab at the puzzle didn't find the answer, but returning to the problem in the 1780s, he showed conclusively that these secular variations, as they are called, can be explained within the framework of Newtonian theory and are caused by the disturbing influences of the two planets on one another. The variations follow a cycle 929 years long, which brings everything back to where it started, so the Solar System is stable after all (at least on all but the longest timescales). According to legend, when asked by Napoleon why God did not appear in his discussion of the secular variations, he replied, 'I have no need of that hypothesis.'

Laplace also worked on the theory of tides, explaining why the two tides each day reach roughly the same height (more naive calculations 'predicted' that one high tide should be much higher than the other), developed his ideas on probability to deal with practical problems such as estimating the total population of France from a sample of birth statistics and, as we have seen, worked with Lavoisier (almost six years older than Laplace and then at the height of his reputation) on the study of heat. It is an interesting insight into the state of science in the 1780s that although Lavoisier and Laplace, undoubtedly two of the greatest scientists of the time, discussed their experimental results both in terms of the old caloric model of heat (more of this shortly) and the new kinetic theory, they carefully avoided choosing between them, and even suggested that both might be at work at the same time.

In 1788, we get a chink of an insight into Laplace's private life. On 15 May, by now well established as a leading member of the Academy, he married Marie-Charlotte de Courty de Romanges. They had two children. A son, Charles-Emile, was born in 1789, became a general

1. In this, as on many other occasions, Laplace was stimulated by discussing the problem with Joseph Lagrange (1736–1813), a mathematician whose work on group theory and development of a mathematical function (the Lagrangian) which characterizes the path (or trajectory) of a particle proved immensely valuable to twentieth-century physicists.

and died (childless) in 1874; a daughter, Sophie-Suzanne, died giving birth to her own daughter (who survived) in 1813. It was around the time of his marriage that Laplace made his definitive study of planetary motions. As well as explaining the secular variations of the orbits of Jupiter and Saturn, he solved a long-standing puzzle about similar changes in the orbit of the Moon around the Earth, showing that they are produced by a complicated interaction between the Sun and the Earth–Moon system *and* the gravitational influence of the other planets on the Earth's orbit. In April 1788, he was able to state (using the word 'world' where we would say 'Solar System'):

> The system of the world only oscillates around a mean state from which it never departs except by a very small quantity. By virtue of its constitution and the law of gravity, it enjoys a stability that can be destroyed only by foreign causes, and we are certain that their action is undetectable from the time of the most ancient observations until our own day.[1]

Although we know little about his private life, Laplace was clearly a great survivor, and one reason why we know so little is that he never openly criticized any government or got involved in politics. He survived the various forms of government following the French Revolution, most of which were eager to be associated with him as a symbol of French prestige. At the only time when he might have been at risk, during the Terror, Laplace had already seen which way the wind was blowing and had prudently removed himself and his family to Melun, some 50 kms southeast of Paris. There he kept his head down until after the fall of the Jacobins, when he was called back to Paris to work on the reorganization of science under the Directory.

Laplace had already worked on the metric system before the Jacobin interlude; now, his work on reforming the educational system of France to include proper teaching of science led him to write one of the most influential books about science ever published, the *Exposition du système du monde*, which appeared in two volumes in 1796. Laplace's prestige and his ability to bend with the wind saw him serve in government under Napoleon, who made him a Count in 1806, but

1. English translation from Gillispie.

remain in favour with the restored monarchy, with Louis XVIII making him the Marquis de Laplace in 1817. But in spite of continuing to work in mathematics and all the honours heaped on him in his long life (he died in Paris on 5 March 1827), in terms of the development of science, the *Exposition* remains Laplace's most important achievement, still valuable today as a summing up of where physics stood at the end of the eighteenth century. And it was appreciated as such at the time – on the flyleaf of a copy given to the College of New Jersey (now Princeton University) in 1798, the donor has written:

> This treatise, considering its object and extent, unites (in a much higher degree than any other work on the same subject that we ever saw) clearness, order and accuracy. It is familiar without being vague; it is precise but not abstruse; its matter seems drawn from a vast stock deposited in the mind of the author; and this matter is impregnated with the true spirit of philosophy.[1]

The fundamental basis of that philosophy was spelled out by Laplace himself in his great book, and, if anything, rings more true today than at any time in the past two centuries:

> The simplicity of nature is not to be measured by that of our conceptions. Infinitely varied in its effects, nature is simple only in its causes, and its economy consists in producing a great number of phenomena, often very complicated, by means of a small number of general laws.

There speaks the voice of experience, from the man who explained the complexities of the Solar System in terms of Newton's simple law of gravity.

Laplace's summing up of physics ranges from planetary astronomy, orbital motion and gravity through mechanics and hydrostatics, and right at the end he introduces a couple of new (or newish) ideas. One of these is the so-called 'nebular hypothesis' for the origin of the Solar System, which was also thought up by Immanuel Kant (1724–1804) in 1755, although there is no indication that Laplace knew about Kant's then rather obscure work. This is the idea that the planets

1. Quoted by Gillispie.

formed from a cloud of material around the young Sun, shrinking down into a plane as the cloud, or nebula, contracted. At the time, there were seven known planets and fourteen satellites all orbiting the Sun in the same direction. Eight of these systems were also known to be rotating on their axes in the same sense that they were orbiting around the Sun – if you look down on the North pole of the Earth, for example, you see the Earth rotating anticlockwise on its axis, while the planet is moving anticlockwise around the Sun. Laplace calculated that, since there is a 1 in 2 chance of each orbit or rotation being 'forward' rather than 'backward', the total odds against this happening by chance were $(1 - (1/2)^{29})$, a number so close to 1 that it was certain that these bodies had formed together, and the nebular hypothesis seemed the best way to account for this. It is, indeed, the model still favoured today.

The other new idea was, of course, Laplace's version of black holes. Intriguingly, this discussion (which was along very similar lines to that of Michell, but much more brief) appeared only in the first edition of the *Exposition*; but there is no record of why Laplace removed it from the later editions. His version of the hypothesis of dark stars pointed out that a body with a diameter 250 times that of the Sun and with the same density as the Earth would have such a strong gravitational attraction that even light could not escape from it.[1] In truth, this is merely a historical curiosity, and the speculation had no influence on the development of science in the nineteenth century. But the book as a whole did, not only for its content but for its clarity and easy style, which is typified by the opening sentences with which Laplace sucks his readers in.

> If on a clear night, and in a place where the whole horizon is in view, you follow the spectacle of the heavens, you will see it changing at every moment. The stars rise or set. Some begin to show themselves in the east, others disappear in the west. Several, such as the Pole Star and the Great Bear, never touch the horizon in our climate . . .

1. It's because the Earth is more dense than the Sun that Laplace came up with a figure of 250 times the diameter of the Sun, whereas Michell came up with twice that number.

How could anyone possibly fail to read on!

Laplace's story is almost all science and very little personality, although some of this does seem to shine through in the *Exposition*. But don't run away with the idea that by the late eighteenth century physics (let alone the rest of science) had settled down into some kind of dull routine. There were still plenty of 'characters' around, and of all the eighteenth-century physicists, the most colourful career was that of Benjamin Thompson (later Count Rumford). He made important contributions to science, particularly in the study of heat, and no less important contributions as a social reformer, although these were not motivated by politics but by practicality. Indeed, Thompson seems to have been an opportunist largely motivated by self-interest, and it is rather ironic that the best way he found to promote his own wealth and status turned out to be to do good for others. But since his career gives a sideways look at both the American Revolution and the upheavals in Europe at the end of the eighteenth century (as well as being an entertaining story in its own right) it's worth going into in a little more detail than his purely scientific contribution really justifies.

Benjamin Thompson (Count Rumford): his life

Thompson was born on 26 March 1753, the son of a farmer in Woburn, Massachusetts. His father died not long after he was born, and his mother soon remarried and had several more children. Although Benjamin was an inquisitive and intelligent boy, the family were poor and his position offered no chance of anything other than the most rudimentary education. From the age of 13, he had to work to help to support the large family, first as a clerk to an importer of dry goods in the port of Salem, and later (from October 1769) as a shop assistant in Boston. Part of the appeal of Boston was that it offered an opportunity for the young man to attend evening classes, and it was also a hotbed of political unrest, which carried its own excitement for a teenager; but he neglected his job (which bored him rigid) and soon lost it – one story has it that he was fired, another that he left voluntarily. Either way, he spent most of 1770 back home in Woburn, unemployed, dividing his time between the usual teenage interests and attempting self-education with help from a slightly older friend, Loammi Baldwin. Partly through his charm, and partly because he was clearly interested in what was then still (in that part of the world) known as natural philosophy, the local physician, a Dr John Hay, agreed to take Thompson on as his apprentice, and he used the opportunity to combine a personal programme of study with his duties – he even seems to have attended a few lectures at Harvard, although he had no official connection with the university (however, since this is based only on Thompson's own account, it should, as we shall see, be taken with a pinch of salt).

The trouble with being an apprentice was that it cost money, and in order to pay his way, Thompson took on a variety of part-time teaching jobs – since all that he was expected to teach was reading, writing and a little reckoning, he needed no formal qualifications for this. By the summer of 1792, either Thompson had had enough of being an apprentice or the doctor had had enough of him, and he decided to try his hand at full-time schoolmastering. He found a post in the town of Concord, New Hampshire. The town actually lay right on the border of Massachusetts and New Hampshire, and had previously been known as Rumford, Massachusetts; the name was changed at the end of 1762 as a conciliatory gesture after a bitter wrangle about which state the town belonged in and who it should be paying taxes to. Thompson's patron in Concord was the Reverend Timothy Walker. Walker's daughter, Sarah, had recently married (at the rather advanced age of 30) the richest man in town, Benjamin Rolfe, who had promptly died at the age of 60, leaving her very well off. Schoolmastering lasted an even shorter time than any of Thompson's other jobs to date, and in November 1772 he married Sarah Walker Rolfe and settled down to manage his wife's estate and turn himself into a gentleman. He was still only 19, tall and good-looking, and always said that it had been Sarah who made the running in their relationship. They had one child, Sarah, born on 18 October 1774. But by then, Thompson's life had already begun to take another twist.

The trouble with Thompson was that he was never satisfied with what he had got and always wanted more (at least, right up until the last months of his life). Thompson lost no time in ingratiating himself with the local governor, John Wentworth. He proposed a scientific expedition to survey in the nearby White Mountains (although these plans came to nothing) and commenced a programme of scientific agriculture. But all this was against the background of the turmoil leading up to the American Revolution. Although this is no place to go into the details, it's worth remembering that in its early stages this was very much a dispute between two schools of thought which both regarded themselves as loyal English. Thompson threw in his lot with the ruling authorities, and it is in this connection that his otherwise surprising appointment as a major in the New Hampshire militia occurred in 1773, only a few months after his marriage. In preparation for the fighting that most people knew was inevitable, the colonials (for want of a better term) were encouraging (indeed, bribing) deserters from the British army to join the ranks and train them in organized warfare; Major Thompson, as a landowner with many contacts among the farmers of the region, was in an ideal position to keep an eye on this activity. Since Thompson was also quite outspoken in his belief that true patriotism meant obeying the rule of law and working within the law to make changes, it didn't take too long for those plotting the overthrow of the old regime to become aware of his activities. Shortly before Christmas 1774, just a couple of months after the birth of his daughter, hearing that a mob was gathering with the objective of tarring and feathering him, Thompson headed out of town on horseback never to return. He never saw his wife again, although, as we shall see, his daughter Sarah did eventually re-enter his life.

Thompson now headed for Boston, where he offered his services to the governor of Massachusetts, General Thomas Gage. Officially, these were rejected and Thompson returned to Woburn. In fact, he had now become a spy for the British authorities, passing back information about rebel activities to the headquarters in Boston. Before very long his position in Woburn also became untenable, and in October 1775 he rejoined the British in Boston. When they were thrown out by the rebels in March 1776, most of the garrison and loyalists sailed for Halifax, Nova Scotia, while the official dispatches from Boston bearing the unwelcome news of this setback for the British forces were sent to London in the care of Judge William Brown. Somehow, Major Benjamin Thompson managed to wangle a place in Judge Brown's entourage, and he arrived in London in the summer of 1776 as an expert with first-hand information about the fighting abilities of the American rebels and as an eye-witness to the fall of Boston from the American side. Furthermore, he presented himself as a gentleman who had lost his large estates through his loyalty to the British cause. With these credentials and his own impressive organizational ability, he quickly became the right-hand man of Lord George Germain, the Secretary of State for the Colonies.

Thompson was good at his job and very successful, becoming, by 1780, Under-secretary of State for the Northern Department. But his work as a civil servant lies outside the scope of this book. Alongside that work, however, he also returned to scientific interests, and in the late 1770s he carried out experiments to measure the explosive force of gunpowder (obviously both topical and relevant to his day job), which led to him being elected as a Fellow of the Royal Society in 1779. These experiments also provided the excuse for Thompson to spend three months on manoeuvres with part of the British Navy in the summer of 1779; but although ostensibly studying gunnery, in fact Thompson was once again working as a spy, this time for Lord Germain, reporting back incredible (but true) tales of inefficiency and corruption in the Navy which Germain could use to further his own political career. Thompson was well aware, though, that under the system of patronage that existed at the time, his star was firmly tied to Germain's and if his patron fell from favour he would be out in the cold. So he set about preparing a fall-back position for himself.

This involved a standard ploy for someone of his rank – forming his own regiment. In order to boost the strength of the army at times of need, the King could issue a Royal Charter which allowed an individual to raise a regiment at his own expense and to become a senior officer in that regiment. It was an expensive process (although Thompson could by now afford it), but carried a huge bonus – at the end of hostilities, when the regiment disbanded, the officers kept both their rank and an entitlement to half-pay for life. So Thompson became a lieutenant-colonel in the King's American Dragoons, which were actually recruited in New York for Thompson by a Major David Murray. In 1781, though, Thompson's pretend soldiering suddenly became real. A French spy was caught with details of British naval operations and his information clearly came from someone in high office with detailed knowledge of the fleet. Thompson was suspected and there was much gossip, but no charges were laid. We shall never know the truth, but it is a fact that he abruptly gave up his position in London and headed for New York to take an active part with his regiment. Thompson's role in the fighting was neither glorious nor successful, and in 1783, after the defeat of the British, he was back in London, where his friends were still influential enough to arrange for him to be promoted to full colonel, thereby considerably boosting his income, before being retired on half-pay. Just after his promotion, he had his portrait, in full uniform, painted by Thomas Gainsborough.

Colonel Thompson, as he now was, decided next to try his fortune on the mainland of Europe, and after a few months touring the continent sizing up possibilities, with a combination of charm, luck and rather exaggerated stories of his military service, he was offered a post as military aide in Munich by the Elector of Bavaria, Carl Theodor. At least he was almost offered a post. It was intimated that in order to avoid offending other members of the Bavarian court, it would be helpful if Thompson could be seen to be well in with the British King, George III. In any case, as a colonel in the British army, he would have to return to London to obtain permission to serve a foreign power. While he was there, Thompson persuaded the King that it would be helpful if he received a knighthood, and the favour was duly granted. Thompson almost always got what he wanted, but this impressive piece of cheek owed much of its success to the fact that Britain was keenly interested in improving relations with Bavaria in view of the way the political situation was now (in 1784) developing in France. It's also clear, and hardly surprising, given his track record, that Sir Benjamin offered to spy on the Bavarians for the British, reporting secretly to Sir Robert Keith, the British Ambassador in Vienna.

Thompson's thoughts on convection

Thompson was a phenomenal success in Bavaria, where he applied scientific principles to turning a poorly equipped army with low morale, barely more than a rabble, into an efficient, happy machine (although not a fighting machine). His job, seemingly impossible, was to achieve this while saving the Elector money, and he did so by the application of science. The soldiers needed uniforms, so Thompson studied the way heat was transmitted by different kinds of material to find the most cost-effective option for their clothing. Along the way, he accidentally discovered convection currents,[1] when he noticed the liquid (alcohol) in a large thermometer being used in his experiments rising up the middle of the tube and falling down the sides. The soldiers also needed feeding, so he studied nutrition and worked out how to feed them economically but healthily. In order to make the uniforms, he swept the streets of Munich clear of beggars, and put them to work in (by the standards of the day) well-equipped, clean workshops, where they were also offered the rudiments of education, and where the younger children were obliged to attend a form of school. To provide maximum nutrition for the troops at minimum cost, he fed them on (among other things) his nutritious soup – based on the potato, a vegetable scarcely used in that part of Europe at the time – which required every barracks to have a kitchen garden and grow its own vegetables. This provided useful work and skills that the soldiers would be able to employ when they left the army, and did much to raise morale. In Munich itself, the military vegetable garden was incorporated into a grand public park, which became known as the English Garden, carved out of what had been the Elector's private deer park, which helped to make Thompson very popular with the populace.

Among his many inventions, Thompson designed the first enclosed kitchen ranges, replacing inefficient open fires; portable stoves for use in the field; improved lamps; and (later) efficient coffee pots (a lifelong teetotaller who hated tea, Thompson was eager to promote coffee as a healthy alternative to alcohol). His official position at court made him the most powerful man in Bavaria after the Elector, and before too long he held (simultaneously) the posts of Minister of War, Minister of Police, State Councillor and Chamberlain to the Court, and the rank of major-general. In 1792, the Elector found another way to honour his most trusted aide. At that time, the last vestiges of the Holy Roman Empire still existed, as a very loose coalition of states of central Europe and with an 'Emperor' whose role was no more than ceremonial. That year, the Emperor Leopold II died, and while the various crowned heads were assembling to select a successor, Carl Theodor became, under the system of Buggins' turn operating at the time, caretaker Holy Roman Emperor. He held the post from 1 March to 14 July

1. The term was actually introduced in 1834, by William Prout (1785–1850).

1792, long enough to ennoble a few of his favourites, including Major-General Sir Benjamin Thompson, who became Count Rumford (in German, Graf von Rumford, an unlikely title for an American-English scientist).[1]

Although, as this example indicates, Rumford (as we shall now refer to him) was still very much the Elector's blue-eyed boy, his position as a foreigner who had risen so far so soon earned him many enemies at court. Carl Theodor was old and childless, and there was already jockeying for position among the factions for when the inevitable happened. Rumford had achieved so much that there seemed little more scope for him to move further up the social ladder. He was now 39, and his thoughts were turning to the possibility of returning to America, when he received, out of the blue, a letter from his daughter Sarah, usually known as Sally. Rumford's wife had just died and Sally had been given his address by Loammi Baldwin.

When the French invaded the Rhineland and took over Belgium in November 1792, with war threatening to engulf Bavaria, Rumford, who was genuinely exhausted, had had enough and left for Italy, officially on grounds of ill health, although with an element of political expediency. The Italian sojourn was part holiday, partly an opportunity to revive his interests in science (he saw Volta demonstrate the way frogs' legs twitched under the influence of electricity and met up with Sir Charles Blagden, Secretary of the Royal Society and friend of Henry Cavendish), partly an opportunity for romantic dalliance (Rumford was never short of female companionship; he had several mistresses, including two sisters, each a Countess, one of whom he 'shared' with the Elector, and he fathered at least two illegitimate children).

Rumford spent sixteen months in Italy, but did return to Bavaria in the summer of 1794, now with an ambition to make a name for himself in science – he was no Henry Cavendish, content with the discoveries for themselves without seeking public acclaim. The political situation was still the same, and in any case for Rumford's work to be noticed

1. The Holy Roman Empire finally came to an end in August 1806, as a side effect of the Napoleonic wars, when the last Emperor, Francis II (the one who got the job in July 1792), abdicated and became Francis I of Austria alone. Rumford was left as Count of an Empire that didn't exist.

it would have to be published in England, preferably by the Royal Society. In the autumn of 1795, the Elector granted him a six-months leave of absence to travel to London for this purpose. There, now famous as a scientist and statesman, and with a noble title to boot, Rumford was in his element and stretched the six months out to almost a year. As ever, he combined business, self-promotion, science and pleasure. Shocked by the pall of smoke which hung over London in winter, he used his understanding of convection to design a better fireplace, with a ledge or shelf at the back of the chimney so that cold air falling down the chimney struck this ledge and was deflected to join the hot air rising from the fire without sending clouds of billowing smoke into the room (he later worked on central heating systems, using steam). In 1796, partly out of egotism, to perpetuate his own name (but, to be fair, using his own money) Rumford endowed two prize medals to be awarded for outstanding work in the fields of heat and light, one for America and one for Britain. The same year, he brought Sally over from America to join him, and although he was initially shocked by her colonial country-bumpkin ways, a social embarrassment to the sophisticated Count Rumford, they spent much time together over the rest of his life.

Rumford was called back to Munich in August 1796, partly because the political situation had changed in his favour (the latest heir presumptive to Carl Theodor was a supporter of Rumford) and partly because of military threats to Bavaria (indeed, to Munich itself), which seemed set to be caught between opposing Austrian and French armies.[1] It wasn't so much that Rumford was genuinely thought to be a great military leader, but more that he was a convenient scapegoat – just about everybody of importance fled Munich, leaving the foreigner as Town Commandant, meaning that he would carry the can when the city was invaded. Soon, the Austrians arrived and set up camp on one side of the town. Then the French arrived and set up camp on the other side of the town. Each army was determined to occupy Munich rather than let their opponents have it, but Rumford, shuttling between the camps and playing for time, managed to avoid triggering any conflict until the French were pulled out following the defeat of another

1. Austria was the major central European power at this time, of course.

of their armies on the lower Rhine. Rumford came out of it all smelling of roses, as ever. When the Elector returned, as a reward he appointed Rumford as Commandant of the Bavarian police, and made Sally, who had accompanied her father, a Countess in her own right, although no extra income resulted from this and the pension Rumford was entitled to as a Count was to be divided equally between the two of them. He was also promoted to general.

The unexpected success, however, made Rumford even more unpopular with the opposition and he became eager to move on, neglecting his administrative duties and carrying out his most important scientific work around this time. Even the Elector realized that he was weakening his own position by continuing to favour Rumford – but what could he do with him? A face-saving resolution to the problem seemed to have been found in 1798, when Carl Theodor appointed Count Rumford Minister Plenipotentiary to the Court of St James (that is, Ambassador to Britain). Rumford packed up his belongings and headed back to London, only to discover that George III had no intention whatever of accepting his credentials, using the excuse that as a British national he could not represent a foreign government, but probably actually because George III's ministers disliked Rumford, regarded him as an upstart and had long memories of his previous double-dealings in the spying trade.

Whatever the reason, the effect turned out well for science. Rumford considered, once again, returning to America, but in the end stayed in London and came up with a scheme to establish a combined museum (giving prominence to his own work, of course), research and educational establishment, which came to fruition as the Royal Institution (RI). Raising the necessary money through public subscription (that is, persuading the rich, with his usual charm, to dig deep into their pockets), he saw the RI open its doors in 1800, with a series of lectures by Thomas Garnett, a physician, lately from Glasgow, who was given the title professor of natural philosophy at the RI. But Garnett did not last long in the job – Rumford was unimpressed by his abilities and in 1801 replaced him with a rising young man, Humphry Davy, who was to make the RI a huge success in promoting the public understanding of science.

Soon after appointing Davy, Rumford went back to Munich to pay his respects to the new Elector, Maximilian Joseph, who had recently succeeded Carl Theodor. He was, after all, still being paid by the Bavarian government and Maximilian had expressed an interest in establishing a similar institution to the RI in Munich. After a couple of weeks there, Rumford set off back to London via Paris, where he was greeted with all the acclaim that he thought he deserved and, fatefully, made the acquaintance of the widow Lavoisier, now in her early forties (Rumford, of course was now in his late forties).[1] After all this, London palled. Rumford sorted out his affairs, packed up and left permanently for the Continent on 9 May 1802. There were other visits to Munich, but these ended in 1805, when Austria took over the territory and the Elector fled. Rumford had had the foresight to wind up his affairs there before the storm broke; his heart was now in Paris with Madame Lavoisier. She joined him on an extended tour of Bavaria and Switzerland, and in the spring of 1804 the couple had settled in a house in Paris. They decided to marry, but ran into the technical difficulty that Rumford had to obtain papers from America proving that his first wife was dead; not so easy with war raging and France blockaded by the British. This delayed things until 24 October 1805, when they finally did marry and almost immediately (after some four years of pre-marital bliss together!) found that they were incompatible. Rumford was ready for a quiet life of semi-retirement and science; his wife wanted parties and a full social life. They parted after a couple of years, and Rumford spent his last years at a house on the outskirts of Paris, in Auteuil, consoled by another mistress, Victoire Lefèvre. They had a son, Charles, who was born in October 1813, less than a year before Rumford died on 21 August 1814, at the age of 61. Sally Rumford lived until 1852, but never married, and left a substantial bequest to Charles Lefèvre's son, Amédé, on condition that he changed his name to Rumford. His descendants still carry that name.

His thoughts on heat and motion

Fascinating though the story of Benjamin Thompson/Count Rumford is (and I have only scratched the surface of it here), it would have no place in a history of science if Rumford had

1. He also met Joseph Guillotin, inventor of the guillotine, but not, presumably at the same social gathering as Mme Lavoisier. Rumford described M Guillotin as 'a very mild, polite humane man'; remember that he invented his machine as a more merciful alternative to hanging.

not made one really important contribution to our understanding of the nature of heat. This came about through his work in Munich in 1797, where, following his 'defence' of the city, among his many responsibilities he was in charge of the Munich Arsenal, where cannon were made by boring out metal cylinders. Rumford was throughout his life an intensely practical man, an inventor and engineer more in the mould of a James Watt than a theorist like Newton, and his main scientific interest concerned the nature of heat, which was still very much a puzzle in the second half of the eighteenth century. The model which still held sway in many quarters was the idea that heat was associated with a fluid called caloric. Every body was thought to possess caloric, and when caloric flowed out of a body it made its presence known by raising the temperature.

Rumford became interested in the caloric model while he was carrying out his experiments with gunpowder in the late 1770s. He noticed that the barrel of a cannon became hotter if it was fired without a cannon ball being loaded than it did when there was a cannon ball being fired, even though the same amount of gunpowder was used. If the rise in temperature was simply due to the release of caloric, then it should always be the same if the same amount of powder was burnt, so there must be something wrong with the caloric model.[1] There were rival models. As a young man, Rumford had read the work of Herman Boerhaave (1668–1738), a Dutchman best remembered for his work in chemistry, in which he suggested that heat was a form of vibration, like sound. Rumford found this model more appealing, but it wasn't until he became involved in cannon-boring almost twenty years later that he found a way to convince people of the deficiencies of the caloric model.

It was, of course, very easy for the caloric model to explain, superficially, the familiar fact that friction produces heat – according to this model, the pressure of two surfaces rubbing together squeezes caloric out of them. In the cannon-boring process, the metal cylinders were mounted horizontally against a non-rotating drill bit. The whole cylin-

1. The modern explanation is that when the ball is fired, energy from the explosion goes into making the ball move, so less energy is available to be dissipated as heat in the cannon.

der was rotated (literally, by horsepower) and the drill moved down the cannon as the boring proceeded. When Rumford observed this process, he was impressed by two facts. First, the sheer quantity of heat generated, and second, that the source of this heat seemed to be inexhaustible. As long as the horses kept working and the drill bit was in contact with the metal of the cannon, heat could be generated. If the caloric model were correct, then surely at some point all the caloric would have been squeezed from the rotating cylinder and there would be none left to make it hot.

Rumford made an analogy with a sponge, soaked in water and hung from a thread in the middle of a room. It would gradually give up its moisture to the air, and eventually become dry and free from moisture. That would be equivalent to the caloric model. But heat was more like the ringing of a church bell. The sound produced by a bell is not 'used up' when the bell is struck, and as long as you keep striking it it will continue to make its characteristic sound. Frugally, using surplus metal cast as an extension to the barrel of the cannon and intended to be cut off before the boring, Rumford set out to measure just how much heat was produced, using a dull drill bit to make the experiment more impressive. By enclosing the metal cylinder in a wooden box full of water, he could measure the heat released by seeing how long it took for the water to boil, and he delighted in the astonishment expressed by visitors in seeing large quantities of cold water quickly brought to the boil in this way without the use of any fire. But, as he also pointed out, this was not an efficient way to heat water. His horses had to be fed, and if you really wanted to boil water, a more efficient way to do so would be to dispense with the horses and burn their hay directly under the water. With this almost throwaway remark, he was on the edge of understanding the way in which energy is conserved but can be converted from one form to another.

Repeating the experiment over and over (emptying the hot water away and replacing it with cold), Rumford found that it always took the same time to boil the same amount of water using the heat generated by friction in this way. There was no sign at all of the 'caloric' being used up like water from a sponge. Strictly speaking, these experiments are not absolute proof that an inexhaustible supply of heat can be generated in this way, because they did not literally go on for ever –

but they were very suggestive, and were perceived at the time as a major blow to the caloric model. He also carried out a series of experiments which involved weighing sealed bottles containing various fluids at different temperatures and establishing that there was no connection between the 'amount of heat' in a body and its mass, so nothing material could be flowing in or out as the body was cooled or heated. Rumford himself did not claim to understand what heat *is*, although he did claim to have shown what it *is not*. But he did write:

> It appears to me to be extremely difficult, if not quite impossible, to form any distinct idea of any thing, capable of being excited and communicated in the manner the Heat was excited and communicated in these experiments, except it be MOTION.[1]

This sentence exactly fits the modern understanding of the association between heat and the motion of individual atoms and molecules in a substance; but, of course, Rumford had no idea what the kind of motion associated with heat was, so the statement is not quite as prescient as it seems. Rather, it was evidence from experiments like this that helped to establish the idea of atoms in the nineteenth century. And one reason why science did progress so rapidly in the nineteenth century was that by the end of the 1790s it was obvious to all but the most blinkered of the old school that the ideas of phlogiston and caloric were both dead and buried.

James Hutton: the uniformitarian theory of geology

In terms of understanding the place of humankind in space and time, however, the most significant development in the last decades of the eighteenth century was the growing understanding of the geological processes that have shaped the Earth. The first version of the story was largely pieced together by one man, the Scot James Hutton, whose lead was then followed up in the nineteenth century by Charles Lyell. Hutton was born in Edinburgh on 3 June 1726. He was the son of William Hutton, a merchant who served as City Treasurer for Edinburgh and also owned a modest farm in Berwickshire. He died while James was very young, so the boy was raised by his mother alone. James attended the High School in

1. Quoted by Brown.

Edinburgh and took arts courses at the university there before being apprenticed to a lawyer at the age of 17. But he showed no aptitude for the law and became so deeply interested in chemistry that within a year he was back at the university studying medicine (the nearest thing then available to chemistry, as typified by the work of people like Joseph Black). After three more years in Edinburgh, Hutton moved on to Paris and then to Leiden, where he received his MD in September 1749; but he never practised medicine (and probably had never intended practising medicine, which was for him just a means to study chemistry).

Hutton's inheritance included the farm in Berwickshire, so on his return to Britain he decided he ought to learn about modern farming practices, and in the early 1750s, he went off first to Norfolk and then to the Low Countries to bring himself up to date, before heading back to Scotland, where he applied the techniques he had learned to turn a rather unprepossessing farm well supplied with rocks into an efficient, productive unit. All of this outdoor activity had triggered an interest in geology, and Hutton had also kept up his chemical interests. Chemistry came up trumps when a technique he had invented years before in collaboration with a friend, John Davie, was developed by Davie into a successful industrial process for the manufacture of the important chemical sal ammoniac (ammonium chloride, used in, among other things, preparing cotton for dying and printing) from ordinary soot. With money coming in from his share of the proceeds from the sal ammoniac process, in 1768, at the age of 42, Hutton, who never married, rented out his farm and moved to Edinburgh to devote himself to science. He was a particular friend of Joseph Black (just two years younger than Hutton), and was a founder member of the Royal Society of Edinburgh, established in 1783. But what he is best remembered for is his suggestion that the Earth had been around for much longer than the theologians suggested – perhaps for ever.

From his study of the visible world, Hutton concluded that no great acts of violence (such as the Biblical Flood) were needed to explain the present-day appearance of the globe, but that if enough time were available everything we see could be explained in terms of the same processes that we see around us today, with mountains being worn away by erosion and sediments being laid down on the sea floor before being uplifted to form new mountains by repeated earthquake and volcanic activity of the kind we see today, *not* by huge earthquakes which threw up new mountain ranges overnight. This became known as the principle of uniformitarianism – the same uniform processes are at work all the time and mould the surface of the Earth continually. The idea that occasional great acts of violence are needed to explain the observed features of the Earth became known as catastrophism.[1] Hutton's ideas flew in the face of the received geological wisdom of his time, which was a combination of catastrophism and Neptunism, the idea that the Earth had once been completely covered by water, particularly promoted by the Prussian geologist Abraham Werner (1749–1817). Hutton marshalled his arguments with care, and presented an impressive case for uniformitarianism in two papers read to the Royal Society of Edinburgh in 1785 and published in 1788 in that society's *Transactions* (the first paper was presented by Black to the March 1785 meeting of the Society; Hutton himself read the second paper in May, a few weeks before his fifty-ninth birthday).

Hutton's proposals brought severe (but ill-founded) criticism from the Neptunists in the early 1790s, and in response to these criticisms Hutton, though now in his sixties and unwell, developed his arguments in the form of a book, *Theory of the Earth*, published in two volumes in 1795. He was still working on a third volume when he died on 26 March 1797, in his seventy-first year. Unfortunately, although Hutton made a carefully argued case supported by a wealth of observational facts, his writing style was largely impenetrable, although the book did contain a few striking examples. One of the best of these concerns the Roman roads still visible in Europe some 2000 years after they were laid down, in spite of the natural processes of erosion going on all that time. Clearly, Hutton pointed out, the time required for natural

1. Both these terms are still misused by people trying to discredit rival ideas; the most important point of confusion is that because the history of the Earth is so long, events that seem rare and dramatic on a human timescale (such as large meteors hitting the Earth), and which are certainly catastrophic in the everyday meaning of the word, are both normal and, in geological terminology, uniformitarian as far as the history of the planet is concerned. It's all a matter of perspective. To a butterfly that lives only for one day, nightfall is a catastrophe; to us, it is routine. To us, a new Ice Age would be a catastrophe; to Planet Earth, it is routine.

processes to carve the face of the Earth into its modern appearance must be vastly longer – certainly much longer than the 6000 years or so allowed by the then-standard interpretation of the Bible. Hutton regarded the age of the Earth as beyond comprehension, and in his most telling line wrote 'we find no vestige of a beginning – no prospect of an end'.

Such flashes of clarity were rare in the book, though, and with Hutton dead and no longer around to promote his ideas, which came under renewed and vigorous attack from the Neptunists and Wernerians, they might have languished if it had not been for his friend John Playfair (1748–1819), then professor of mathematics at Edinburgh University (and later professor of natural philosophy there). Picking up the baton, Playfair wrote a masterly, clear summary of Hutton's work, which was published in 1802 as *Illustrations of the Huttonian Theory of the Earth*. It was through this book that the principle of uniformitarianism first reached a wide audience, convincing all those with wit to see the evidence that here was an idea that had to be taken seriously. But it literally took a generation for the seed planted by Hutton and Playfair to flower, since the person who picked up the baton of uniformitarianism from Playfair was born just eight months after Hutton died.

Book Four

THE BIG PICTURE

The 'Darwinian Revolution'

Charles Lyell: his life – His travels in Europe and study of geology – He publishes the Principles of Geology *– Lyell's thoughts on species – Theories of evolution: Erasmus Darwin and* Zoonomia *– Jean-Baptiste Lamarck: the Lamarckian theory of evolution – Charles Darwin: his life – The voyage of the* Beagle *– Darwin develops his theory of evolution by natural selection – Alfred Russel Wallace – The publication of Darwin's* Origin of Species.

There were many dramatic developments in science in the nineteenth century, but undoubtedly the most important of these in terms of understanding the place of humankind in the Universe (and arguably the most important idea in the whole of science) was the theory of natural selection, which, for the first time, offered a scientific explanation of the fact of evolution. The name of Charles Darwin is forever linked with the idea of natural selection, and rightly so; but two other names, Charles Lyell and Alfred Russel Wallace, deserve to stand either side of his at the centre of the evolutionary stage.

Charles Lyell: his life

Charles Lyell came from a well-off family, but the wealth was scarcely two generations old. It originated with his grandfather, also Charles Lyell, who had been born in Forfarshire, Scotland, in 1734. This Charles Lyell was the son of a farmer, but after his father died he was apprenticed as a book-keeper before joining the Royal Navy in 1756 as an able-bodied seaman. His former training helped him to become successively a captain's clerk, gunner's mate and then midshipman, the first step on the road to becoming an officer. But he was not to be another Nelson, and in 1766 he became purser of HMS *Romney*. Fans of Horatio Hornblower and the novels of Patrick O'Brien will appreciate that the job of purser gave opportunities for even an honest man to line his pockets – the purser was responsible for purchasing supplies for his ship, which he sold at a profit to the Navy; grandfather Lyell went even further than most by joining a business partnership to supply Navy ships in the ports of North America. In 1767, he married Mary Beale, a Cornish girl, and in 1769 she gave birth (in London) to another Charles Lyell, who was to become the father of the geologist. By 1778, the elder Charles Lyell was secretary to Admiral John Byron and purser of his flagship, HMS *Princess Royal*. As a result of the action which Byron's fleet saw against the French during the American War of Independence (the French navy's assistance to the rebel cause was instrumental in ensuring that the British lost that war), Lyell received so much prize money[1] that, combining it with his other earnings, in 1782, three years after retiring from the Navy, he was able to buy estates in Scotland running to 5000 acres and including a fine house at Kinnordy, in Forfarshire (now Angus). His son had been educated in a manner fitting the elder Lyell's growing status and spent just over a year at St Andrews University before moving to Peterhouse, Cambridge, in 1787.

The second Charles Lyell was well educated (he graduated in 1791 and then studied law in London) and well travelled, including a long tour of Europe carried out in 1792, visiting Paris when France was in the turmoil of Revolution. In 1794, he became a Fellow of Peterhouse, a useful connection for an aspiring lawyer, but he remained based in London until his father died in January 1796, in his sixty-second year. With no need now to practise law, the second Charles Lyell married a Miss Frances Smith later that year, and moved to Kinnordy, where Charles Lyell the geologist was born on 14 November 1797.

Charles and Frances Lyell never settled in Scotland, however, and before baby Charles was a year old they had moved to the south of England,[2] renting a large house and some land in the New Forest, not far from Southampton. It was there that young Charles grew up, surrounded by younger siblings (eventually, two brothers and no less than seven sisters). The New Forest provided a backdrop for the boy to develop an interest in botany and insects while attending school

1. When enemy ships or booty were captured they were purchased by the Crown (or sold on the open market) and the proceeds shared among the men involved in the action in accordance with strictly laid down rules (most for the Admiral, of course, and least for the men); this was the prime motive that persuaded men to serve in the Royal Navy in spite of the extreme hardships and poor pay. Most men never received significant prize money, if any at all, but were sustained by the example of the few that hit the jackpot.

2. Leaving the Scottish property in the hands of agents.

locally, but in 1810 he moved on to a minor public school at Midhurst, together with his younger brother Tom. Tom left in 1813 to become a midshipman, but Charles, as the eldest son, was groomed to follow in his father's footsteps.

After visiting Scotland in 1815 with his parents and sister Fanny (an extended tour, but taking in the family estates he would one day inherit), Charles went up to Oxford in February 1816, joining Exeter College as a gentleman commoner, the most prestigious (and expensive) 'rank' of undergraduate. He took with him a reputation for academic excellence in the traditional arts-oriented subjects, and arrived at a university that was just (only just) beginning to shake off its well-deserved reputation as an institution only fit for the education of country parsons.[1] Lyell found in himself an unsuspected mathematical ability and became interested in geology after he read a book in his father's library, Robert Bakewell's *Introduction to Geology*, either late in 1816 or early in 1817. Bakewell was an advocate of Hutton's ideas, and it was through reading Bakewell that Lyell was introduced to Hutton's work and went on to read Playfair's book. This was the first time he had any clue that a subject like geology existed, and he then attended some lectures on mineralogy given by William Buckland (1784–1856) at Oxford in the summer term of 1817. Buckland in his turn had been inspired by the pioneering work of William Smith (1769–1839), a surveyor whose work on canals in the late eighteenth and early nineteenth centuries made him familiar with the rock strata of England and an expert in the use of fossils to indicate the relative ages of different strata (which were older and which were younger), even though there was no way at that time to tell their absolute ages. It was Smith, now regarded as the 'father of English geology', who produced the first geological map of England, which was published in 1815, although much of his material had already been circulated to colleagues such as Buckland. Buckland himself had been on a long geological expedition around Europe in 1816, so must have had exciting first-hand news to impart to his students, rather in the way that a university lecturer today might have returned recently from a trip

1. *Exactly* the kind of country parsons you find in the pages of Jane Austen's novels – she died in 1817, the year that Lyell became interested in geology.

abroad visiting one of the large telescopes to carry out observations of the Universe.

Lyell's growing interest in geology did not entirely please his father, who felt that it might distract him from his study of the classics, but as well as attending Buckland's lectures, Lyell was now drinking in the geology, on his travels around Britain (including further visits to Scotland and to East Anglia), not just admiring the fine views. In the summer of 1818, Charles Lyell senior treated the family, including Charles Lyell junior, to an extended tour of Europe. The younger Charles was able to visit the Jardin des Plantes (as it now was) in Paris, see some of Cuvier's specimens and read Cuvier's works on fossils in the library (Cuvier himself was in England at the time). The tour took in Switzerland and northern Italy, giving the young man ample opportunity to take in the geological delights, as well as taking in the cultural delights of cities such as Florence and Bologna. In 1819, Lyell graduated from Oxford at the age of 21, and was also elected a Fellow of the Geological Society of London (no great honour, since in those days any gentleman with an amateur interest in geology could become a Fellow, but a clear indication of where his interests lay). The next step in following in his father's footsteps would be to study law – but the first hint of a problem that would help to change those plans came while Charles was studying hard for his final examinations, and was bothered by problems with his eyesight and severe headaches.

After another tour of England and Scotland (partly in the company of his father and sisters Marianne and Caroline), Lyell took up his legal studies in London in February 1820, but immediately he suffered more problems with his eyes, raising doubts about his ability to make his career in a profession which required minute attention to detail in handwritten documents (in an age, remember, when there was no electric light). To give his eyes a chance to recover, Charles Lyell senior took his son to Rome, via Belgium, Germany and Austria. They were away from August to November, and for a time the rest seemed to have done the trick. Lyell returned to his legal studies but continued to be troubled by his eyes, and in the autumn of 1821 paid an extended visit to the family home, Bartley, in the New Forest. In October that year he went on a leisurely tour along the South Downs, visiting his old school in Midhurst and making the acquaintance of Gideon Mantell

(1790–1852) in Lewes, Sussex; Mantell was a surgeon and amateur (but very good) geologist who discovered several types of dinosaur. Lyell did go back to London and his legal studies from late October to mid-December 1821, but the combination of his eyesight problems and his love of geology meant that without any formal break with his chosen career being made, in 1822 he virtually ceased being a lawyer and began a serious investigation of the geology of the southeast of England, stimulated by his conversations and correspondence with his new friend Mantell.

Thanks to the work of people like William Smith, at that time the geological structure of England and Wales was becoming fairly well known (while the extension of related geological features into France was also being mapped), and it was obvious that rock layers had been twisted and bent, after they were laid down, by immense forces. It was natural to suppose that these forces, and the forces which had lifted what were clearly once sea beds high above sea level, were associated with earthquakes. But in spite of Hutton's insight, the widely held opinion, championed by geologists such as William Conybeare (1787–1857), was that the changes had been brought about by short-lived, violent convulsions, and that the kind of processes now seen at work on the surface of the Earth were inadequate for the task. In the early 1820s, Lyell was intrigued by these arguments, though still more impressed by Hutton's ideas, and learned a great deal about cutting-edge geology from Conybeare's writings.

Lyell did actually keep up his legal studies just enough to be called to the Bar in May 1822, and later he did practise (for a short time and in a rather desultory fashion) as a barrister. But in 1823 he not only visited Paris once again (this time meeting Cuvier, still a confirmed catastrophist) but became involved in the running of the Geological Society, first as Secretary and later as Foreign Secretary; much later, he also served two terms as President. Apart from its scientific importance (Lyell attended several lectures at the Jardin, as well as meeting French scientists), the 1823 trip is significant historically because it was the first time that Lyell crossed the English channel in a steamship, the packet *Earl of Liverpool*, which took him direct from London to Calais in just eleven hours, with no need to wait for a fair wind. A small technological step, to be sure, but one of the first signs of the speeding up of global communications that was about to change the world.

Lyell's own world began to change in 1825, the year that he began his practice as a barrister. He was asked to write for the *Quarterly Review*, a magazine published by John Murray, and began to contribute essays and book reviews (themselves really an excuse for an essay) on scientific topics and issues such as the proposal for a new university in London. He turned out to have a talent for writing, and, even better, the *Quarterly* paid for its contributions. Lyell's legal work brought him very little income (it is not clear whether he actually earned enough to cover his expenses in the profession) and writing enabled him, for the first time, to achieve a degree of financial independence from his father – not that there was any pressure on him from his father to do so, but still a significant step for the young man. The *Quarterly* also brought Lyell's name to the attention of a wider circle of educated people, which opened up other prospects. Having discovered his talent as a writer, early in 1827 he decided to write a book about geology and began to gather material for this project. So the idea for the book already existed, and Lyell had already proved his worth as a writer, before he set out on his most important, and famous, geological expedition, in 1828.

His travels in Europe and study of geology

The expedition has echoes of John Ray's great botanical expedition of the previous century, showing how little things had yet changed in spite of the steam packet. In May 1828, Lyell travelled first to Paris, where he had arranged to meet the geologist Roderick Murchison (1792–1871), and together they then travelled south through the Auvergne and along the Mediterranean coast to northern Italy, with Lyell making extensive notes on the geological features that they encountered. Murchison (who was accompanied by his wife) set off back to England from Padua at the end of September, while Lyell pressed on towards Sicily, the nearest location of volcanic and earthquake activity to mainland Europe. It was what Lyell saw in Sicily, in particular, that convinced him that the Earth had indeed been formed by the same processes that are at work today, operating over immense spans of time. It was Lyell's field work that put flesh on the bones of the idea outlined by Hutton. On Etna, among other things, he found raised sea beds '700 feet & more' above sea level, separated with lava flows, and in one place:

A very strong illustration of the length of the intervals which occasionally separated the flows of distinct lava currents. A bed of [fossilized] oysters, perfectly identifiable with our common eatable species, no less than *twenty feet in thickness*, is there seen resting on a current of basaltic lava; upon the oyster bed again is superimposed a second mass of lava, together with tuff or peperino.

... we cannot fail to form the most exalted conception of the antiquity of this mountain [Etna], when we consider that its base is about ninety miles in circumference; so that it would require ninety flows of lava, each a mile in breadth at their termination, to raise the present foot of the volcano as much as the average height of one lava-current.[1]

•

It was this kind of clear writing, as well as the weight of evidence he gathered in support of his case, that made Lyell's book such an eye-opener, both to geologists and to the educated public. Lyell also realized that since Etna (and, indeed, the whole of Sicily) was relatively young, the plants and animals found there must be species that had migrated from Africa or Europe and adapted to the conditions they found. By adapting to the changing environments of our planet, life itself must be moulded in some way by geological forces, although he was unable to say just how this happened.

He publishes the Principles of Geology

By February 1829, Lyell was back in London and, with his eyesight as good as it had ever been after his long journey away from his legal documents and enjoying a great deal of physical activity, he lost no time in getting down to work on his book. As well as his own field studies, Lyell drew extensively on the work of geologists from across continental Europe, producing by far the most thorough overview of the subject that anyone had yet written. John Murray, the publisher of the *Quarterly*, was the obvious choice to put the material before the public, and although Lyell kept rewriting his work even after it had been sent to the printers, the first volume of the *Principles of Geology* (a name deliberately chosen to echo Newton's *Principia*) appeared in July 1830, and was an immediate success.[2]

1. Quotes from *Principles of Geology*, Lyell's italics.

2. The subtitle on the title page of the volume reads, 'Being an attempt to explain the former changes of the Earth's surface, by reference to causes now in operation'. No scope there for any doubt in the mind of a prospective purchaser as to Lyell's intentions!

28. *Sketch of Santorini, from Lyell's* Principles of Geology, Volume 2, *1868.*

Although Lyell often bickered with Murray about the financial side, the publisher actually treated his author well, by the standards of the day, and it was Lyell's income from the book that eventually made him financially independent, although his father continued to provide an allowance. After more field work (this time chiefly in Spain), the second volume of the *Principles* appeared in January 1832 and was not only a success in its own right, but revived the sales of the first volume.

The delay between the publication of the two volumes wasn't just caused by the field work. In 1831, a chair of geology was established at King's College in London, and Lyell successfully sought the appointment (in spite of some opposition by Church representatives concerned about his views on the age of the Earth), giving a series of highly successful lectures (in a daring innovation, women were allowed to attend some of them), but resigning in 1833 to devote himself to his writing, which he found more profitable and in which he was his own master, with no time-consuming duties. He became the first person to make his living as a science writer (although admittedly with a little help from the family wealth).

There were other distractions. In 1831, Lyell became engaged to Mary Horner, the daughter of a geologist, Leonard Horner (1786–1864); she shared his geological interests, which made for an unusually close and happy relationship between her and Charles. The couple married in 1832, when Lyell's allowance from his father was increased from £400 to £500 per year, while Mary brought with her investments worth £120 a year. All this, in addition to Lyell's increasing income from writing (and the fact that they remained childless), made the couple comfortably off and made the chair at King's an irksome distraction rather than an important source of income. Then there was politics. At the end of 1830, half a century of Tory rule came to an end in Britain and a Whig government pledged to reform Parliament came to power. These were turbulent times across Europe, and earlier in 1830 agricultural workers in England rioted in protest at the loss of work caused by the introduction of new machinery on farms. There was a distinct whiff of revolution in the air, and memories of the French Revolution were still fresh. The reforms proposed by the Whigs, which were at first popular in the country at large, included the abolition of rotten boroughs, where a few voters returned an MP to the House of Commons; but the required legislation was blocked by the House of Lords. In spite of the rottenness of the boroughs, then as now by-elections were seen as important indicators of the will of the people, and in September 1831 (when Charles Lyell happened to be at Kinnordy on holiday) a crucial by-election was held in Forfarshire. There were fewer than ninety voters in the whole of the constituency (landowners, including Charles Lyell senior and his sons) and there was no secret ballot. Every vote counted and everybody knew who had voted which way. Charles Lyell senior voted for the Tory candidate, who won by a narrow margin, while 'our' Charles Lyell abstained. This was a key factor in delaying the reform of Parliament, and it also had an adverse effect on the promotion prospects of Tom Lyell, now a lieutenant in the Navy, dependent on Whig patronage for promotion (since, of course, the Admiralty was run by appointees of the Whig government) but marked out as the son of a man who voted Tory at a crucial moment.

Lyell's thoughts on species

It was in the second volume of the *Principles*, when it eventually did appear, that Lyell turned his attention to the species puzzle, concluding that:

> Each species may have had its origin in a single pair, or individual, where an individual was sufficient, and species may have been created in succession at such times and in such places as to enable them to multiply and endure for an appointed period, and occupy an appointed place on the globe.

The work, in Lyell's view at the time, of a very 'hands-on' God, and not so different from the story of Noah's Ark. Note that this hypothesis explicitly includes the idea, obvious from the fossil record by the 1830s, that many species that once lived on Earth have gone extinct and been replaced by other species. In keeping with the spirit of his times, though, Lyell reserved a special place for humankind, regarding our species as unique and distinct from the animal kingdom. But he *did* suggest that the reason why species went extinct was because of competition for resources, such as food, from other species.

The third volume of the *Principles* appeared in April 1833. Lyell's work for the rest of his life revolved around keeping the massive book up to date, rewriting it and bringing out new editions hot on the heels of one another – the twelfth and final edition appeared posthumously in 1875, Lyell having died in London on 22 February that year (less than two years after the death of his wife), while working on what turned out to be his last revisions for the book. His *Elements of Geology*, which appeared in 1838 and is regarded as the first modern textbook of geology, was based on the *Principles*, and itself underwent refinements. This eagerness for revision wasn't just because geology really was a fast-moving subject at the time;[1] Lyell's obsession with keeping the book bang up to date derived from the fact that it was his main source of income (certainly until his father died in 1849, the year of the California Gold Rush), both from its own sales and in terms of maintaining his profile as a science writer and, by general acclaim, the

1. Although it certainly was; the nearest analogy, both with the drama of the science and the level of popular interest in the subject, is with cosmology in the late twentieth century.

leading geologist of his time. Lyell was knighted in 1848 and became a baronet (a kind of hereditary knight) in 1864. Although he by no means ceased to be an active field geologist after 1833, he was then in his mid-thirties, and it was with the *Principles* and the *Elements* that he made his mark on science; there is no need to say much here about his later life, except (as we shall see) in the context of his relationship with Charles Darwin. But it is worth mentioning one of Lyell's later geological field trips, which shows how the world was changing in the nineteenth century. In the summer of 1841, he went on a year-long visit to North America (by steamship, of course), where he not only encountered new geological evidence for the antiquity of the Earth and saw the forces of nature at work in such places as Niagara Falls, but was pleasantly surprised by the ease with which the new railways made it possible to travel across what had until very recently been unknown territory. He also gave hugely popular public lectures and boosted the sales of his books in the New World. Lyell enjoyed the experience so much that he returned for three later visits, and as a result of his first-hand knowledge of the United States became an outspoken supporter of the Union during the American Civil War (when most people of his social position in Britain supported the Confederates). But everything that Lyell did in later life was overshadowed by the *Principles*, and even the *Principles* has tended to be overshadowed in the eyes of many people by a book which, its author acknowledged, owed an enormous debt to Lyell's book – Charles Darwin's *Origin of Species*. Darwin was the right man, in the right place, at the right time to gain the maximum benefit from the *Principles*. But, as we shall see, this was not entirely the lucky fluke that it is sometimes made out to be.

There was nothing new about the idea of evolution by the time Charles Darwin came on the scene. Evolutionary ideas of a sort can be traced back to the Ancient Greeks, and even within the time frame covered by this book, there were notable discussions about the way species change by Francis Bacon in 1620 and a little later by the mathematician Gottfried Wilhelm Leibnitz; while in the eighteenth century, Buffon, puzzling over the way similar but subtly different species occur in different regions of the globe, speculated that North American bison might be descended from an ancestral form of

European ox that migrated there, where 'they received the impression of the climate and in time became bisons'. What was different about Charles Darwin (and Alfred Russel Wallace) was that he came up with a sound scientific theory to explain why evolution occurred, instead of resorting to vague suggestions such as the idea that it might be due to 'the impression of the climate'. Before Darwin and Wallace, the best idea about how evolution might work (and it really was a good idea, given the state of knowledge at the time, although it has sometimes been ridiculed by those who have the benefit of hindsight) was thought up by Charles Darwin's grandfather, Erasmus, at the end of the eighteenth century, and (independently) by the Frenchman Jean-Baptiste Lamarck at the beginning of the nineteenth century.

Theories of evolution: Erasmus Darwin and Zoonomia

The association between the Darwin family and the mystery of life on Earth actually goes back one generation further still, to the time of Isaac Newton. Robert Darwin, the father of Erasmus, lived from 1682 to 1754 and was a barrister who retired from his profession and settled in the family home at Elston, in the English midlands, at the age of 42. He married the same year and Erasmus, the youngest of seven children, was born on 12 December 1731. Several years before settling into domestic bliss, however, in 1718, Robert had noticed an unusual fossil embedded in a stone slab in the village of Elston. The find is now known to be part of a plesiosaur from the Jurassic period; thanks to Robert Darwin, the fossil was presented to the Royal Society, and as a thank you, Robert was invited to attend a meeting of the Royal on 18 December that year, where he met Newton, then President of the Royal Society. Little is known about Robert Darwin's life, but his children (three girls and four boys) were clearly brought up in a household where there was more than average curiosity about science and the natural world.

Erasmus was educated at Chesterfield School (where one of his friends was Lord George Cavendish, second son of the then Duke of Devonshire) before moving on to St John's College, Cambridge, in 1750, partly financed by a scholarship which brought in £16 per year. In spite of the dire state of the university at the time, Erasmus did well, initially in classics, and also gained a reputation as a poet. But his father was not a rich man and Erasmus had to choose a profession where he could make a living. After his first year in Cambridge he began to study medicine; he also became a friend of John Michell, who was then a tutor at Queen's College. His medical studies continued in Edinburgh in 1753 and 1754 (the year his father died), then he went back to Cambridge to obtain his MB in 1755. He may have spent more time in Edinburgh after that, but there is no record of him ever receiving an MD there, although this didn't stop him from adding those letters to his list of qualifications.

Whatever his paper qualifications, Erasmus Darwin was a successful doctor who soon established a flourishing practice at Lichfield, 24 kilometres north of Birmingham. He also began to publish scientific papers (he was especially interested at the time in steam, the possibilities of steam engines and the way clouds form), and on 30 December 1757, a few weeks after his twenty-sixth birthday, he married Mary Howard (known as Polly), who was herself a few weeks short of her eighteenth birthday. All of this activity, on several fronts simultaneously, is typical of Erasmus Darwin, who certainly lived life to the full. The couple had three children who survived to adulthood (Charles, Erasmus and Robert) and two who died in infancy (Elizabeth and William). The only one who married was Robert (1766–1848), the father of Charles Robert Darwin, of evolution fame. The Charles Darwin who was Erasmus's son was his eldest child, a brilliant student who was the apple of his father's eye and seemed to have a glittering career ahead of him in medicine when, at the age of 20, as a medical student in Edinburgh he cut his finger during a dissection and acquired an infection (septicaemia) from which he died. By then, in 1778, Erasmus junior was already set on the path to becoming a lawyer, but young Robert was still at school and was strongly influenced by his father to become a doctor, which he did successfully, even though he lacked the brilliance of his brother and hated the sight of blood. Erasmus junior also died relatively young, drowned at the age of 40 in what may have been an accident or may have been suicide.

Polly herself had died, after a long and painful illness, in 1770. Although there is no doubt that Erasmus loved his first wife and was deeply affected by her death, when 17-year-old Mary Parker moved into the household to help look after young Robert, the inevitable happened and she produced two daughters, fathered by Erasmus. The girls were openly acknowledged as his and comfortably looked after by him in the Darwin household, even after their mother moved out and married, and everyone involved remained on friendly terms. Erasmus Darwin himself later fell for a married lady, Elizabeth Pole, and succeeded in winning her hand after her husband died; they married in 1781 and produced another seven children together, only one of whom died in infancy.

With all of this and his medical practice to tend to, you might think that Erasmus Darwin had little time for science. But he had become a Fellow of the Royal Society in 1761, was the moving force behind the establishment of the Lunar Society, and mingled with scientists such as James Watt, Benjamin Franklin (who he met through John Michell) and Joseph Priestley. He published scientific papers and kept up to date with many of the new developments in science, and was one of the first people in England to accept Lavoisier's ideas about oxygen. He also translated Linnaeus into English (introducing the terms 'stamen' and 'pistil' to the language of botany). Along the way, he dabbled in canal investments, was one of the backers of an iron works and became firm friends with Josiah Wedgwood, who made a fortune from his pottery, and with whom Erasmus campaigned against slavery. Both men were delighted when Robert Darwin, the son of Erasmus, and Susannah Wedgwood, the daughter of Josiah, became romantically involved; but Josiah died in 1795, the year before they married. Susannah inherited £25,000 from her father, equivalent to about £2 million today, and among other things this would mean that her son Charles Robert Darwin would never have to worry about earning his living in one of the professions.

By the time Robert and Susannah married, Erasmus Darwin had achieved widespread fame for the work which justifies his place in the history of science, but this started with a poetical work, based on the ideas of Linnaeus, designed to introduce new readers to the delights of botany. It was called *The Loves of the Plants*, and initially published anonymously in 1789 (when Erasmus was 57), although it had had a long gestation. Erasmus literally made plants sexy, enchanted a wide audience, and seems to have been an influence on poets such as Shelley,

Coleridge, Keats and Wordsworth.[1] This success was followed in 1792 by *The Economy of Vegetation* (usually referred to as *The Botanic Garden*, which is strictly speaking the title of a collected edition including both *The Economy of Vegetation* and *The Loves of the Plants*), in which 2440 lines of verse are supported (if that is the word) by some 80,000 words of notes, which amount to a book about the natural world. Then, in 1794, Erasmus published the first volume of his prose work *Zoonomia*, running to more than 200,000 words and to be followed in 1796 by a second volume some 50 per cent longer still. It is in Volume I of *Zoonomia* that he at last sets out fully his ideas on evolution, alluded to in the earlier poetical writings, although these form just one chapter out of the forty in Volume I, many of which are devoted to medicine and biology.

Erasmus Darwin's thoughts on evolution go far beyond mere speculation and generalities, although he was, of course, handicapped by the limited state of knowledge at the time. He details the evidence that species have changed in the past, and draws particular attention to the way in which changes have been produced in both plants and animals by deliberate human intervention, for example breeding faster racehorses or developing more productive crops by the process of artificial selection – something that was to be a key feature in the theory developed by his grandson. He also points out the way in which characteristics are inherited by offspring from their parents, drawing attention to, among other things, 'a breed of cats with an additional claw on every foot' that he has come across. He elaborates on the way different adaptations enable different species to obtain food, mentioning (in another pre-echo of Charles Darwin) that 'some birds have acquired harder beaks to crack nuts, as the parrot. Others have acquired beaks adapted to break the harder seeds, as sparrows. Others for the softer seeds . . .'. Most dramatically of all, Erasmus (clearly a Huttonian!) comes out with his belief that all of life on Earth (by implication including humankind) may be descended from a common source:

1. For evidence in support of these startling claims, see the biography of Erasmus Darwin by Desmond King-Hele. Coleridge visited Erasmus in 1796.

Would it be too bold to imagine, that in the great length of time since the earth began to exist, perhaps millions of ages[1] before the commencement of the history of mankind, would it be too bold to imagine, that all warm-blooded animals have arisen from one living filament, which THE GREAT FIRST CAUSE endued with animality, with the power of acquiring new parts, attended with new propensities . . .

God still exists for Erasmus, but only as the first cause who set the processes of life on Earth working; there is no place here for a God who intervenes to create new species from time to time, but a clear sense that whatever the origins of life itself, once life existed it evolved and adapted in accordance with natural laws, with no outside intervention.[2] But Erasmus did not know what those natural laws that govern evolution were. His speculation was that changes were brought about in the bodies of living animals and plants by their striving for something they needed (food, say) or to escape from predators. This would be rather like the way in which a weight lifter puts on muscle. But Erasmus thought that these acquired characteristics would then be passed on to the offspring of the individual that acquired them, leading to evolutionary change. A wading bird that didn't like getting its feathers wet, for example, would constantly be stretching up as high as possible to avoid contact with the water, and thereby stretch its legs a tiny bit. The slightly longer legs would be inherited by its offspring, and over many generations this repeating process could turn a bird with legs like a swan into one with legs like a flamingo.

Although this idea was wrong, it was not crazy, given the state of knowledge at the end of the eighteenth century, and Erasmus Darwin deserves credit for at least trying to come up with a scientific explanation for the fact of evolution. He continued (along with many other activities) to develop his ideas for the rest of his life, and 1803 saw the publication of *The Temple of Nature*, which told in verse of the evolution of life from a microscopic speck to the diversity of the present

1. By an 'age' Erasmus Darwin probably means about a hundred years, so his ideas on the timescale of evolution were way ahead of his time.

2. At this time, remember, the Church still taught that species were created individually by God, and once created were fixed and immutable.

day. Once again, the verse is accompanied by copious notes that amount to a book in their own right. But this time Erasmus did not meet with publishing success; his near atheism and evolutionary ideas were condemned, and were clearly out of step with a society at war with Napoleonic France and longing for stability and security rather than revolution and evolution. Besides, Erasmus himself was no longer around to argue his cause, having died quietly at home on 18 April 1802, at the age of 70. Perhaps appropriately, though, given the political situation, it was indeed in Napoleonic France that similar evolutionary ideas to those of Erasmus Darwin were taken up and developed, in some ways more fully.

Jean-Baptiste Lamarck: the Lamarckian theory of evolution

Jean-Baptiste Pierre Antoine de Monet de Lamarck, to give him his full entitlement of names, was a member of the minor French nobility (it is a reasonable rule of thumb that the longer the list of names, the more minor the branch of the nobility), born at Bazentin, in Picardy, on 1 August 1744. He was educated at the Jesuit College in Amiens from about the age of 11 to 15 (details of his early life are rather vague), and was probably intended for the priesthood. But when his father died in 1760 he set off to become a soldier, joining the army fighting in the Low Countries during the Seven Years War. The war ended in 1763, and Lamarck seems to have become interested in botany as a result of the wildlife he saw on subsequent postings to the Mediterranean and eastern France. In 1768, he received an injury which forced him to give up his military career, and he settled in Paris, where he worked in a bank and attended lectures on medicine and botany. Ten years later, he established his reputation as a botanist with the publication of his *Flore française* (*French Flora*), which became the standard text on the classification of French plants. On the strength of the book (and with the patronage of Buffon, who had helped with the book's publication), Lamarck was elected to the Académie and was soon able to turn his back on the bank.

Buffon's patronage came at a price. In 1781, Lamarck had the unenviable task of acting as tutor and companion to Buffon's useless son Georges during a European tour; but at least this gave Lamarck an opportunity to see more of the natural world. After his travels, Lamarck held a series of minor botanical posts connected with the

Jardin du Roi, although his interests extended far beyond botany (or even biology) and included meteorology, physics and chemistry. He was involved in the reorganization of the Jardin after the French Revolution and was assigned responsibility, as a professor, for the study of what were then called 'insects and worms' in the new French Natural History Museum in 1793; it was Lamarck who gave this rag-bag collection of species the overall name 'invertebrates'. As a reformer, and untainted by any odious connections with tax farming, Lamarck seems to have survived the Revolution without ever being personally threatened. As a professor, Lamarck was required to give an annual series of lectures at the museum, and these lectures show how his ideas on evolution themselves gradually evolved, with the first mention of the idea that species are not immutable coming in 1800. Describing animals from the most complex forms downwards to the simplest, classified by what he (rather confusingly) called their 'degradation', he said that the invertebrates:

> show us still better than the others that astounding degradation in organization, and that progressive diminution in animal faculties which must greatly interest the philosophical Naturalist. Finally they take us gradually to the ultimate stage of animalization, that is to say to the most imperfect animals, the most simply organized, those indeed which are hardly to be suspected of animality. These are, perhaps, the ones with which nature began, while it formed all the others with the help of much time and of favourable circumstances.[1]

In other words, although his argument is presented upside down, Lamarck is saying that the simplest animals evolved into the more complex – and note that reference to 'much time' being required for the process.

Lamarck's biographer L. J. Jordanova says that there is 'no evidence' that he was aware of the ideas of Erasmus Darwin; Darwin's biographer Desmond King-Hele says that Lamarck's ideas were 'almost certainly' influenced by *Zoonomia*. We shall never know the truth, but in one respect Lamarck's behaviour was very similar to that of Darwin. Although little is known of his private life, we do know that

1. Translation quoted by Jordanova.

he had six children by a woman he lived with, and married her only when she was dying. He then married at least twice more (there is some suggestion of a fourth marriage) and produced at least two more children. But unlike Erasmus Darwin (or, indeed, Charles Darwin) he had an abysmal literary style and seems (as the above example, which is far from being the worst, shows) to have been unable to present his ideas clearly in print.

Those ideas on evolution were summed up in his epic *Histoire naturelle des animaux sans vertèbres*, published in seven volumes between 1815 and 1822, when Lamarck was 78 and blind (he died in Paris on 18 December 1829). For our purposes, Lamarck's ideas on evolution can be summed up by the four 'laws' which he presented in Volume 1 of that book, published in 1815:

> First Law: By virtue of life's own powers there is a constant tendency for the volume of all organic bodies to increase and for the dimensions of their parts to extend up to a limit determined by life itself.

(This is more or less true; there does seem to be some evolutionary advantage in having a bigger body, and most multi-celled animal species have got bigger in the course of evolution.)

> Second Law: The production of new organs in animals results from newly experienced needs which persist, and from new movements which the needs give rise to and maintain.

(At the very least this is not completely wrong; if the environmental circumstances change, there are pressures which favour certain evolutionary developments. *But* Lamarck means, wrongly, that the 'new organs' develop *within individuals*, not by tiny changes from one generation to the next.)

> Third Law: The development of organs and their faculties bears a constant relationship to the use of the organs in question.

(This is the idea that the flamingo's legs get longer because the flamingo is always stretching up to avoid contact with the water. Definitely wrong.)

Fourth Law: Everything which has been acquired . . . or changed in the organization of an individual during its lifetime is preserved in the reproductive process and is transmitted to the next generation by those who experienced the alterations.

(This is the heart of Lamarckism – the inheritance of acquired characteristics. Definitely wrong.)

Perhaps the most telling point made by Lamarck, though, was the one which stuck in the throat of Charles Lyell and led him to reject the idea of evolution when writing the *Principles* – he specifically included humankind in the process.

Lamarck's ideas were strongly opposed by the influential Georges Cuvier, who firmly believed in the fixity of species, and were promoted by Isidore Geoffroy Saint-Hilaire (1772–1844), who worked with Lamarck in Paris. Unfortunately, Saint-Hilaire's support did at least as much harm to the Lamarckian cause as it did good. He built from Lamarck's ideas and came very close to the idea of natural selection, suggesting that the kind of 'new organs' described by Lamarck might not always be beneficial and writing (in the 1820s) that:

If these modifications lead to injurious effects, the animals which exhibit them perish and are replaced by others of a somewhat different form, a form changed so as to be adapted to the new environment.[1]

This includes elements of Lamarckism, but also the germ of the idea of survival of the fittest. But Saint-Hilaire also espoused wild ideas about the relationships between species, and although he did a great deal of sound comparative anatomy he went too far when he claimed to have identified the same basic body plan in vertebrates and molluscs, drawing further fire from Cuvier and discrediting all his work, including his ideas on evolution. By the end of the 1820s, with Lamarck dead and his main supporter to a large extent discredited, the way was clear for Charles Darwin to pick up the threads. But it took him a long time to weave those threads into a coherent theory of evolution, and even longer to get up the nerve to publish his ideas.

1. Quoted by Henry Osborn's, *From the Greeks to Darwin.*

Charles Darwin: his life

There are two popular myths about Charles Darwin, neither of which resembles the truth. The first, already alluded to, is that he was a dilettante young gentleman who was lucky enough to go on a voyage around the world, where he saw the rather obvious evidence for evolution at work and came up with an explanation that any reasonably intelligent contemporary might have thought of in the same circumstances. The second is that he was a rare genius whose unique flash of insight advanced the cause of science by a generation or more. In fact, both Charles Darwin and the idea of natural selection were very much products of their time, but he was unusually hard working, painstaking and persistent in his search for scientific truth across a wide range of disciplines.

By the time Erasmus Darwin died, his son Robert was well established in a successful medical practice near Shrewsbury, and had recently moved into a fine house that he had had built, called The Mount and completed in 1800. Robert resembled his father physically, over six feet tall and running to fat as he grew older; in the Darwin tradition, he fathered a healthy brood of children (although not quite on the scale that his father had), but Erasmus did not live to see the birth of his grandson Charles, who was the second-youngest of that brood. His sisters Marianne, Caroline and Susan were born in 1798, 1800 and 1803, elder brother Erasmus in 1804, Charles Robert Darwin on 12 February 1809 and finally Emily Catherine (known to the family as Catty) arrived in 1810, when her mother Susannah was 44. Charles seems to have had an idyllic childhood, spoiled by three older sisters, allowed to roam the grounds of the house and nearby countryside, taught the basics of reading and writing at home by Caroline until he was eight years old, and with an elder brother to look up to. Things changed dramatically in 1817. In the spring of that year, Charles began attending a local day school, prior to becoming a boarder at Shrewsbury School (where his brother Erasmus was already established) in 1818. And in July 1817, after a life troubled by illness of one kind or another, his mother died, taken by a sudden and painful intestinal complaint at the age of 52. Robert Darwin never came to terms with his loss, and far from following his father's example of a happy second marriage, he forbade any discussion of his lost wife and sank into frequent bouts of depression for the rest of his life. His edict must have carried weight, since in later life Charles Darwin wrote that he could recall very little of his mother.

As far as running the household was concerned, Marianne and Caroline were old enough to take over, and the younger daughters later played their part. Some historians (and psychologists) argue that his mother's death, and especially his father's response to it, must have had a profound impact on young Charles and shaped his future personality; others suggest that in a large household with several sisters and servants, his mother was a more remote figure than would be the case for an 8-year-old today, and that her death probably left no lasting scars. But the fact that Charles was sent away to boarding school just a year after his mother died, cutting him off from that supportive family environment (but bringing him closer to his brother Erasmus), suggests that the combination of factors in 1817 and 1818 really did have a profound impact on him. Shrewsbury School was close enough to The Mount – just 15 minutes away across the fields, making relatively frequent visits home quite feasible – but to a 9-year-old boy living away from home for the first time it made little difference whether home was 15 minutes' or 15 days' journey away.

Darwin developed a strong interest in natural history during his time at Shrewsbury School, taking long walks to observe his natural surroundings,[1] collecting specimens and poring over books in his father's library. In 1822, when Erasmus was in his final year at the school and Charles was 13, the elder brother developed a short-lived but passionate interest in chemistry (a very fashionable subject at the time) and easily persuaded Charles to act as his assistant in setting up a laboratory of their own at The Mount, funded to the tune of £50 with the aid of their indulgent father. When Erasmus duly left to go up to Cambridge later that year, Charles had the run of the lab to himself whenever he was at home.

Erasmus was following in the family tradition, training to become a doctor, but had no vocation for the profession and found the academic routine at Cambridge boring, but the extra-curricular activities much

1. Possibly the long walks were the cause of the developing interest in natural history, rather than resulting from it; this would fit in with the idea that Darwin really was deeply affected by the events of 1817 and 1818.

more to his taste. Charles found life at Shrewsbury School without Erasmus equally dull, but made up for it when he was allowed to visit Erasmus in the summer of 1823, having what could only be described as a high old time, which had a distinctly bad influence on the 14-year-old. Back home he developed a passion for shooting game birds, preferred sports to academic work at school and was so clearly showing signs of becoming a wastrel younger son that in 1825 Robert Darwin took him out of school and made him his own assistant for a few months, trying to instil in him something of the Darwin medical tradition. He was then packed off to Edinburgh as a medical student. Although Charles was only 16, Erasmus had just completed his three years in Cambridge and was himself about to spend a year in Edinburgh to complete his medical training; the idea was that Charles could be looked after by Erasmus and attend medical courses during that year, after which he would be settled enough and old enough (hopefully, mature enough) to work formally for his medical qualifications on his own. But it didn't work out like that.

In many ways, the year in Edinburgh was a rerun of the high old time in Cambridge, although Erasmus managed to scrape through his courses and the two young men managed to avoid any detailed reports of their extra-curricular activities getting back to Dr Robert. Any possibility that Charles might himself become a doctor disappeared, however, not through neglect of his studies but through his own squeamishness. Although made physically sick by dissection of a corpse, Charles did stick to certain aspects of his studies. But the turning point came when he watched two operations, one on a child, being carried out, as was the only way possible then, without anaesthetic. The image of the screaming child, in particular, made a deep impression on him, and he later wrote in his *Autobiography*:

I rushed away before they were completed. Nor did I ever attend again, for hardly any inducement would have been strong enough to make me do so; this being long before the blessed days of chloroform.

The two cases fairly haunted me for many a long year.[1]

1. The edition edited by Nora Barlow is the best source for such insights into Darwin's early life.

Unable to bring himself to admit this failing to his father, Darwin returned to Edinburgh in October 1826, ostensibly to continue his medical studies, but enrolled in classes in natural history, attended lectures on geology and in particular came under the influence of Robert Grant (1793–1874), a Scottish comparative anatomist and expert on marine life who was fascinated by sea slugs. Grant was an evolutionist who favoured Lamarckism and also shared some of Saint-Hilaire's views about the universal body plan; he passed these ideas on to young Darwin (who had already read *Zoonomia* for its medical insights, although, according to his autobiography, without the evolutionary ideas in it making any impact on him at the time) and encouraged him to do his own studies of the creatures they found on the sea shore. In geology, Darwin learned about the argument between the Neptunists, who thought that the Earth's features had been shaped by water, and the Vulcanists, who saw heat as the driving force (he preferred the latter explanation). But by April 1827, although Darwin (still only 18) had found something he was deeply interested in and prepared to work hard at, it was clear that the sham of his medical studies could not be sustained and he left Edinburgh for good, with no formal qualifications. Perhaps in order to delay the inevitable confrontation with his father, he took his time getting back to The Mount. After a short tour of Scotland he went on his first visit to London, where he met up with his sister Caroline and was shown around by his cousin Harry Wedgwood, newly qualified as a barrister. He then moved on to Paris, meeting up with Josiah Wedgwood II (Harry's father, and the son of the close friend of Charles's grandfather Erasmus) and his daughters Fanny and Emma, on their way back to England from Switzerland.

In August, however, it was time to face the music, and the upshot was that Robert Darwin insisted that the only prospect was for Charles to go up to Cambridge and obtain his degree, so that Robert could set him up as a country clergyman, the standard respectable way of disposing of rapscallion younger sons at the time. After a summer divided between the country pursuits of the rich (hunting and partying) and cramming rather desperately to bring his knowledge of the classics up to scratch, Charles Darwin was formally accepted by Christ's College in the autumn of 1827, and took up residence, after more

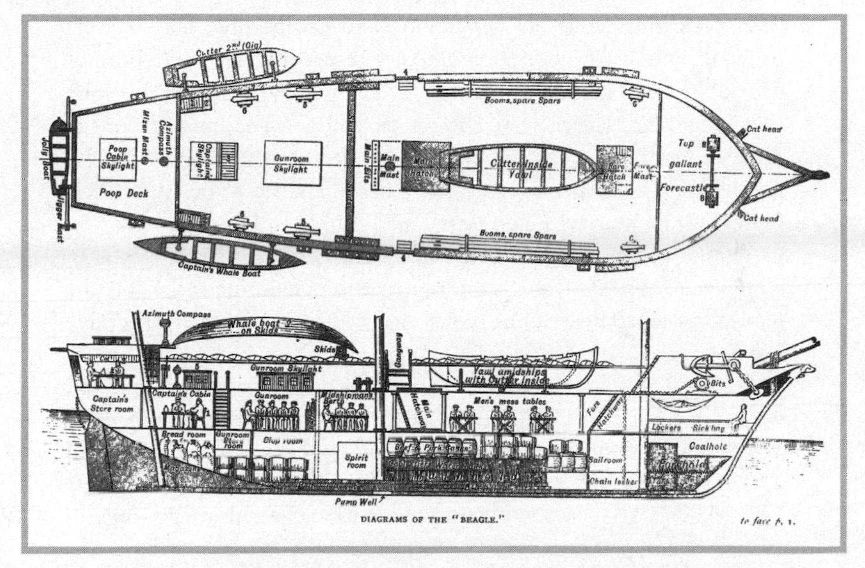

29. *Drawing of HMS* Beagle, *from Darwin's* Journal of Researches, *1845.*

swotting, early in 1828. Once again, he was in the company of Erasmus, now finishing his Bachelor of Medicine degree before setting off on a Grand Tour of Europe as a reward. The contrast for Charles, who faced four years' study and a life as a country parson, must have been hard to swallow.

Darwin's time as an undergraduate in Cambridge followed the pattern he had established in his later months in Edinburgh; he neglected his official studies but threw himself into the study of what really interested him – the natural world. This time, he came under the wing of John Henslow (1795–1861), professor of botany in Cambridge, who became a friend as well as a teacher. He also studied geology under Adam Sedgwick (1785–1873), the Woodwardian professor of geology, who was outstanding at field work, although he rejected the uniformitarian ideas of Hutton and Lyell. Both men regarded Darwin as an outstanding pupil, and his intellectual capacity and ability for hard work were demonstrated when, after a desperate burst of last-minute cramming to catch up on all the things he had been neglecting while out botanizing and geologizing, Darwin surprised even himself by obtaining a very respectable degree (tenth out of 178) in the examinations held at the beginning of 1831. But in spite of the scientific ability he had shown, the route to a country parsonage now seemed more clear-cut than ever, not least since while Charles had been up at Cambridge, Erasmus had managed to persuade their father that he was not suited to the medical life and had been allowed to abandon his career at the age of 25 and settle in London with an allowance from Dr Robert. Indulgent the doctor might be, but he naturally wanted at least one of his sons to settle down at a respectable profession.

Charles spent the summer of 1831 on what he must have thought would be his last great geological expedition, studying the rocks of Wales, before returning to The Mount on 29 August. There, he found a totally unexpected letter from one of his Cambridge tutors, George Peacock. Peacock was passing on an invitation from his friend Captain Francis Beaufort (1774–1857), of the Admiralty (and now famous for the wind scale that bears his name), inviting Darwin to join a surveying expedition to be carried out by HMS *Beagle*, under the command of Captain Robert FitzRoy, who was looking for a suitable gentleman to accompany him on the long voyage and take advantage of the opportunity to study the natural history and geology of, in particular, South America. Darwin's name had been suggested by Henslow, who also sent a letter urging him to seize the opportunity. Darwin was not actually the first choice for the position – Henslow thought briefly of taking the opportunity himself, and another of his protégés turned it down in favour of becoming vicar of Bottisham, a village just outside Cambridge. But he absolutely had the right credentials – FitzRoy wanted a gentleman, a member of his own class, who he could treat on equal terms during the long voyage, when he would otherwise be isolated from social contact by his God-like position in command. The gentleman had (of course) to pay his own way; and the Admiralty were keen that he should be an accomplished naturalist to take advantage of the opportunities offered by the expedition to South America and (possibly) around the world. When Henslow suggested Darwin (via Peacock) to Beaufort, though, the name struck an additional chord. One of grandfather Erasmus Darwin's close friends had been

Richard Edgeworth, a man after his own heart who had four happy marriages and produced 22 children. Twelve years younger than Erasmus, Edgeworth had married for the fourth and last time in 1798, to a Miss Frances Beaufort, the 29-year-old sister of the Francis Beaufort, who by 1831 was Hydrographer to the Royal Navy. So when Beaufort wrote to FitzRoy recommending young Charles for the role of companion and naturalist on the voyage, he was happy to describe him as 'a Mr Darwin grandson of the well known philosopher and poet – full of zeal and enterprize' even though the two had never met.[1]

The voyage of the Beagle

There were some hurdles to overcome before Darwin's role on the *Beagle* was finalized. At first, his father (who would have to fund young Charles on the trip) objected to what seemed to be another madcap scheme, but was won over by Josiah Wedgwood II, Charles Darwin's uncle. Then, FitzRoy (a temperamental man) took exception to the way Darwin seemed to be being foisted on him, sight unseen, and darkly suggested that he might have already found his own companion; things were smoothed out when Darwin and FitzRoy met and hit it off with one another. Eventually all was settled, and the *Beagle*, a three-masted vessel just 90 feet (27 metres) long, set sail on 27 December 1831, when Charles Darwin was not quite 23 years old. There is no need to go into details of the five-year-long voyage (which did indeed go right around the world) here, but there are a few points worth mentioning. First, Darwin was not cooped up on the ship for all that time, but went on long expeditions through South America, in particular, while the ship was busy at the official surveying work. Second, he made his name in scientific circles as a geologist, not as a biologist, through the fossils and other samples that he sent back to England during the voyage. And finally, there is one particular detail worth mentioning – Darwin experienced a large earthquake in Chile and saw for himself how much the disturbance had raised the land, with shellfish beds stranded high and dry several feet (about a metre) above the shoreline. This was first-hand confirmation of the ideas spelled out by Lyell in his *Principles of Geology*. Darwin had taken the first volume with him on the voyage, the second caught up with

1. Quoted by Browne.

him during the expedition and the third was waiting for him on his return to England in October 1836. Seeing the world through Lyell's eyes, he became a confirmed uniformitarian, and this had a profound influence on the development of his ideas on evolution – as Darwin put it late in his life:

I always feel as if my books came half out of Lyell's brain, and that I have never acknowledged this sufficiently . . . I have always thought that the great merit of the Principles was that it altered the whole tone of one's mind.[1]

Darwin came home to a reception he can scarcely have dreamed of, and which must have both puzzled and gratified his father. He soon met Lyell himself, and was introduced as an equal to the geological luminaries of the land. In January 1837 he read a paper to the Geological Society of London on the coastal uplift in Chile (the hottest discovery from his voyage) and was almost immediately elected a Fellow of the Society (significantly, he did not become a Fellow of the Zoological Society until 1839, the same year that he was elected a Fellow of the Royal Society). As well as his fame as a geologist, Darwin soon also received acclaim as a writer, in the mould of Lyell. The first project was a *Journal of Researches*,[2] in which Darwin wrote about his activities on the voyage, while FitzRoy wrote about the more naval aspects. Darwin soon completed his share of the work, drawing on his diaries, but publication was delayed until 1839 because of FitzRoy's naval commitments, which left him little time for writing – and also, to be frank, because FitzRoy wasn't very good at writing. To FitzRoy's chagrin, it soon became clear that Darwin's part of the book was of much wider interest than his own, and it was quickly republished on its own as the *Voyage of the Beagle*.

1839 was a big year in Darwin's life – the year he turned 30, saw the publication of the *Journal*, became a Fellow of the Royal Society and married his cousin Emma Wedgwood. It was also smack in the

1. Letter cited by Jonathan Howard, *Darwin*.

2. In full, *Journal of Researches into the Geology and Natural History of the Various Countries Visited by* HMS 'Beagle,' *under the Command of Captain FitzRoy, R.N., from 1832 to 1836*.

middle of what he later described as his most creative period intellectually, from the return of the *Beagle* in 1836 to the time he left London and settled with his new family in Kent, in 1842. But it was also during this period that he began to suffer a series of debilitating illnesses, the exact cause of which has never been ascertained, but which in all probability resulted from a disease picked up in the tropics. The move out of London, where Darwin had initially settled on his return to England, resulted in no small measure from the political turmoil of the time, with reformers such as the Chartists demonstrating on the streets of the capital and kept in check by the army. The Darwins moved to Down House, in the village of Down, in Kent (the village later changed its name to Downe, but the house kept the old spelling).

Charles and Emma had a long and happy marriage, blighted only by his recurring illness and the early deaths of several of their children. But they also produced many survivors, some of whom went on to achieve eminence in their own right. William, the first-born, lived from 1839 to 1914; then came Anne (1841–1851), Mary (died aged three weeks, 1842), Henrietta (1843–1930), George (1845–1912), Elizabeth (1847–1926), Francis (1848–1925), Leonard (1850–1943), Horace (1851–1928) and Charles (1856–1858). It's worth looking again at those dates for Leonard; born well before the publication of the *Origin*, he lived until well after the atom had been split, which gives some idea of the pace of change of science in the hundred years from 1850 to 1950. But the family life, except as a stable background for Charles Darwin's work, is not what matters here. What we are interested in is Darwin's work, and especially the theory of evolution by natural selection.

Darwin develops his theory of evolution by natural selection

There was no question in Darwin's mind by the time he returned from his voyage (if not before he set out) that evolution was a fact. The puzzle was to find a natural mechanism that would explain that fact – a model, or theory, of how evolution worked. Darwin started his first notebook on *The Transformation of Species* in 1837, and developed his evolutionary ideas privately while publishing geological papers that proved crucial in deciding the uniformitarian/Catastrophist debate in favour of the Uniformitarians. A key step came in the autumn of 1838, not long before his marriage, when Darwin read the famous *Essay on the*

Principle of Population by Thomas Malthus (1766–1834).[1] Originally published anonymously in 1798, the essay was in a sixth (by now signed) edition by the time Darwin read it. Malthus himself, who studied at Cambridge and was ordained in 1788, wrote the first version of the essay while working as a curate, but later became a famous economist and Britain's first professor of political economy. He pointed out in his essay that populations, including human populations, have the power to grow geometrically, doubling in a certain interval of time, then doubling again in the next interval the same size, and so on. At the time he was writing, the human population of North America really was doubling every twenty-five years, and all that is required to achieve this is that, on average, each couple should, by the time they are 25, have produced four children who in turn survive to be 25. The fecundity of the Darwin family must have immediately brought home to Charles how modest a requirement this is.

Indeed, if each pair of even the slowest-breeding mammals, elephants, left just four offspring that survived and bred in their turn, then in 750 years each original pair would have 19 million living descendants. Yet clearly, as Malthus pointed out, there were about the same number of elephants around at the end of the eighteenth century as there had been in 1050. He reasoned that populations are held in check by pestilence, predators and especially by the limited amount of food available (as well as by war, in the case of humans), so that on average each pair leaves just two surviving offspring, except in special cases like the opening up of new land for colonization in North America. Most offspring die without reproducing, if nature takes its course.

Malthusian arguments were actually used by nineteenth-century politicians to argue that efforts to improve the lot of the working classes were doomed to failure, since any improvements in living conditions would result in more children surviving and the resulting increase in population would swallow up the improved resources to leave even more people in the same abject state of poverty.[2] But

1. The essay can still be found in print in an edition edited by Antony Flew.

2. The flaw in this argument can be summed up in one word, almost taboo in Victorian times – contraception.

Darwin, in the autumn of 1838, leaped to a different conclusion. Here were the ingredients of a theory of how evolution could work – pressure of population, struggle for survival *among members of the same species* (more accurately, of course, it is a struggle to reproduce) and survival (reproduction) only of the best-adapted individuals (the 'fittest', in the sense of the fit of a key in a lock, or a piece in a jigsaw puzzle, rather than the sporting sense of the word).

Darwin sketched out these ideas in a document dated by historians to 1839, and more fully in a 35-page outline dated by himself to 1842. The theory of evolution by natural selection was essentially complete before he even moved to Down House, and he discussed it with a few trusted colleagues, including Lyell (who, to Darwin's disappointment, was not convinced). Afraid of the public reaction to the theory, and worried about upsetting Emma, a very conventional Christian, Darwin then sat on the idea for two decades, although in 1844 he did develop his outline into a manuscript about 50,000 words long, running to 189 pages, which he had copied out neatly by a local schoolmaster and left among his papers, with a note to Emma requesting that it be published after his death.

Or rather, he didn't quite sit on it. In the second edition of the *Voyage of the Beagle*, which he worked on in 1845, Darwin added a lot of new material, scattered here and there through the pages. Howard Gruber has pointed out that it is easy to identify these paragraphs by comparing the two editions, and that if you take all the new material out and string it together, it forms 'an essay which gives almost the whole of his thought' on evolution by natural selection.[1] The only explanation is that Darwin was concerned about posterity, and about his priority. If anyone else came up with the idea, he could point to this 'ghost' essay and reveal that he had thought of it first. Meanwhile, in order to make it more likely that his theory would be accepted when he did eventually get around to publishing it, he decided that he ought to make a name for himself as a biologist. Starting in 1846 (ten years after the *Beagle* had brought him home), he began an exhaustive study of barnacles, drawing in part on his South American materials, that eventually formed a definitive, three-volume work completed in 1854.

1. Howard Gruber, *Darwin on Man.*

It was a stunning achievement for a man with no previous reputation in the field, often wracked by illness, during a period which also saw the death of his father in 1848 and of his favourite daughter, Annie, in 1851. It earned him the Royal Medal of the Royal Society, their highest award for a naturalist. He was, for the first time, established as a biologist of the first rank, with a thorough understanding of the subtle differences between closely related species. But he still hesitated about publishing his ideas on evolution, although, at the urging of the few close confidants with whom he discussed the idea, he began, in the mid-1850s, to collect his material together and organize it into what he planned would be a big, fat book that would present such a weight of evidence that it would overwhelm any opposition. 'From September 1854 onwards,' he wrote in his *Autobiography*, 'I devoted all my time to arranging my huge pile of notes, to observing, and experimenting, in relation to the transmutation of species.' It is doubtful if such a book would ever have seen the light of day in Darwin's lifetime, but he was finally forced to go public when another naturalist did indeed come up with the same idea.

Alfred Russel Wallace

The 'other man' was Alfred Russel Wallace, a naturalist based in the Far East, who in 1858 was 35 years old, the same age that Darwin had been in 1844, when he developed the extended outline of his theory. The contrast between Darwin's privileged life and Wallace's own struggle for survival is striking, and worth highlighting as an example of how science at this time was ceasing to be the prerogative of the wealthy gentleman amateur. Wallace was born at Usk, in Monmouthshire (now Gwent), on 8 January 1823. He was the eighth out of nine children of an ordinary family; his father was a rather unsuccessful solicitor and the children were given a basic education at home by him. In 1828, the family moved briefly to Dulwich, then settled in Hertford, the home town of Alfred's mother. There, Alfred and one of his brothers, John, attended the local grammar school, but Alfred had to leave at about the age of 14 to earn a living. In his autobiography *My Life*, published in 1905, Wallace said that school made little impact on him, but that he read voraciously from his father's large collection of books and from the books that became available to him when his father ran a small library in Hertford.

In 1837 (with Darwin already returned from his famous voyage), Wallace went to work with his eldest brother William, a surveyor. He revelled in the open-air life, fascinated by the different kinds of rock strata being uncovered by canal- and road-building work, and intrigued by the fossils revealed in the process. But the money and prospects in surveying were poor at that time and Wallace briefly became apprenticed to a clockmaker, only giving up the trade because the clockmaker moved to London and he didn't want to follow. So it was back to surveying with William, this time as part of the programme of land enclosure in mid-Wales – Wallace didn't appreciate the political implications at the time, but later railed about this 'land-robbery'.[1] The brothers also turned their hands to building, designing the structures themselves, seemingly successfully, even though they had no training in architecture and relied on what they could learn from books. But all the while, Alfred Wallace was becoming more interested in the study of the natural world, reading appropriate books and beginning a scientific collection of wild flowers.

These relatively good times came to an end in 1843, when Alfred's father died and the surveying work dried up as the country was gripped by an economic recession (by now, Darwin had settled in Down House and had already written down at least two outlines of the theory of natural selection). Alfred lived for a few months in London with his brother John (a builder), surviving on a small inheritance. When that ran out in 1844, he managed to get a job at a school in Leicester, teaching the basics of reading, writing and arithmetic to the youngest boys, and surveying (which was probably the key to him getting the job, since anyone could teach the 'three Rs') to older boys. His salary was £30 a year, which puts the £50 spent by young Charles and Erasmus Darwin on setting up their home chemical laboratory in perspective. Wallace was now 21, just a year younger than Darwin had been when he graduated from Cambridge, in a dead-end job, with no prospects. But two significant events occurred during his time in Leicester. He read Malthus's *Essay* for the first time (although it did not initially have a dramatic impact on his thinking) and he met another keen amateur naturalist, Henry Bates (1825–1892), whose

1. *My Life.*

interest in entomology neatly complemented Wallace's interest in flowers.

Wallace was rescued from a life as a (by his own admission) second-rate schoolmaster by a family tragedy. In February 1845, his brother William died from pneumonia, and after settling William's affairs, Alfred decided to take over his surveying work, based in the town of Neath, in South Wales. This time he was lucky; there was plenty of work connected with the railway boom of the time and Alfred was quickly able to build up a small capital fund, for the first time in his life. He brought his mother and brother John to live with him in Neath, and with John's help once again branched out into architecture and building work. His interest in natural history also flourished and was boosted by correspondence with Bates. But Wallace became increasingly frustrated and disillusioned by the business side of the surveying and building work, finding it difficult to cope with businesses which owed him money but delayed payment, and getting depressed when sometimes confronted by smaller creditors who genuinely could not afford to pay. After a visit to Paris in September 1847, when he visited the Jardin des Plantes, Wallace hatched a scheme to change his life once and for all, and proposed to Bates that they should use the small amount of money that Wallace had accumulated to finance a two-man expedition to South America. Once there, they could fund their natural history work by sending back specimens to Britain to be sold to museums and to the wealthy private collectors who were then (partly thanks to Darwin's account of the voyage of the *Beagle*) always on the lookout for curiosities from the tropics. Already a firm believer in evolution, Wallace said in his autobiography that even before setting out on this expedition, 'the great problem of the origin of species was already distinctly formulated in my mind . . . I firmly believed that a full and careful study of the facts of nature would ultimately lead to a solution of the mystery'.

Some four years spent exploring and collecting in the jungles of Brazil, often under conditions of extreme hardship, gave Wallace the same sort of first-hand experience of the living world that Darwin had gained during the voyage of the *Beagle*, and helped him to establish a reputation as a naturalist through papers published as a result of his work in the field, as well as through the specimens he collected. But the expedition was far from being a triumph. Alfred's younger brother Herbert, who travelled out to Brazil to join him in 1849, died of yellow fever in 1851, and Alfred always blamed himself for his brother's death, on the grounds that Herbert would never have gone to Brazil if Alfred had not been there. Alfred Wallace himself nearly died as a result of his South American adventure. On the way home, the ship he was travelling on, the brig *Helen*, carrying a cargo of rubber, caught fire and went to the bottom, taking Wallace's best specimens with her. The crew and passengers spent ten days at sea in open boats before being rescued, and Wallace returned to England late in 1852 almost penniless (although he had had the foresight to insure his collection for £150), with nothing to sell, but with notes which he used as the basis for several scientific papers and a book, *Narrative of Travels in the Amazon and Rio Negro*, which was modestly successful. Bates had stayed in South America and returned three years later, with his specimens intact; but by then Wallace was on the other side of the world.

During the next sixteen months, Wallace attended scientific meetings, studied insects at the British Museum, found time for a short holiday in Switzerland and planned his next expedition. He also met Darwin at a scientific gathering early in 1854, but neither of them could later recall any details of the occasion. More significantly, the two began a correspondence, resulting from Darwin's interest in a paper by Wallace on the variability of species of butterfly in the Amazon basin; this led to Darwin becoming one of Wallace's customers, buying specimens that he sent back from the Far East, and sometimes complaining (gently) in his notes about the cost of shipping them back to England. Wallace went to the Far East because he decided that the best way to pursue his interest in the species problem would be to visit a region of the globe which had not already been explored by other naturalists, so that the specimens he sent home would be more valuable (both scientifically and financially), and the income from them could support him adequately. His studies at the British Museum, and conversations with other naturalists, convinced him that the Malay Archipelago fitted the bill, and he scraped together enough money to set out[1] in the spring of 1854, some six months before Darwin

1. Delayed by the outbreak of the Crimean War.

started arranging his 'huge pile of notes', this time accompanied by a 16-year-old assistant, Charles Allen.

This time, Wallace's expedition was an unqualified success, although once again he endured the hardships of travelling in tropical regions where few Westerners had yet been. He was away for eight years, during which time he published more than forty scientific papers, sent back to the journals in England, and he returned with his specimen collections intact. Apart from his ideas on evolution, his work was extremely important in establishing the geographical ranges of different species, which showed how they had spread from one island to another (later, such work would tie in with the idea of continental drift). But, of course, it is evolution that concerns us here. Influenced, like Darwin, by the work of Lyell, which established the great age of the Earth (what Darwin once called 'the gift of time') and the way an accumulation of small changes could add up to produce large changes, Wallace developed the idea of evolution as like the branching of a huge tree, with different branches growing from a single trunk, and continually dividing and splitting down to the little twigs, still growing, which represent the diversity of living species (all derived from a common stock) in the world today. He presented these ideas in a paper published in 1855, without, at that point, offering an explanation for how or why speciation (the splitting of the branches into two or more closely related growing twigs) occurred.

Darwin and his friends welcomed the paper, but several, including Lyell, soon became concerned that Darwin might be pre-empted, by Wallace or somebody else, if he did not publish soon (Lyell still wasn't convinced about natural selection, but as a friend and a good scientist he wanted the idea in print, to establish Darwin's priority and stir a wider debate). The climate of opinion was much more favourable to open debate about evolution than it had been twenty years before, but Darwin still failed to see any urgency and carried on sorting his vast weight of evidence in support of the idea of natural selection. He did drop hints in his correspondence to Wallace that he was preparing such a work for publication, but gave no details of the theory; they were intended to warn Wallace that Darwin was ahead of him in this particular game. But they had the effect of encouraging Wallace and stimulating him to develop his own ideas further.

The breakthrough came in February 1858, when Wallace was ill with a fever in Ternate, in the Moluccas. Lying in bed all day, thinking about the species problem, he recalled the work of Thomas Malthus. Wondering why some individuals in each generation survive, while most die, he realized that this was not due to chance; those that lived and reproduced in their turn must be the ones best suited to the environmental conditions prevailing at the time. The ones that were most resistant to disease survived any illness they experienced; the fastest escaped predators; and so on. 'Then it suddenly flashed upon me that this self-acting process would necessarily *improve the race*, because in every generation the inferior would inevitably be killed off and the superior would remain – that is, *the fittest would survive*.'[1]

This is the nub of the theory of evolution by natural selection. First, offspring resemble their parents, but in each generation there are slight differences between individuals. Only the individuals best suited to the environment survive to reproduce, so the slight differences which make them successful are selectively passed on to the next generation and become the norm. When conditions change, or when species colonize new territory (as Darwin saw with the birds of the Galapagos Islands and Wallace saw in the Malay Archipelago), species change to match the new conditions and new species arise as a result. What neither Darwin nor Wallace knew, and which would not become clear until well into the twentieth century, was how heritability occurred or where the variations came from (see Chapter 14). But, given the observed fact of heritability with small variations, natural selection explained how, given enough time, evolution could produce an antelope adapted to a grazing lifestyle, the grass itself, a lion adapted to eat antelope, a bird that depends on a certain kind of seed for its food, or any other species on Earth today, including humankind, from a single, simple common ancestor.

It was the insight experienced on his sickbed in February 1858 that led Wallace to write a paper, 'On the Tendency of Varieties to Depart Indefinitely from the Original Type', which he sent to Darwin with a

1. *My Life*. This was written long after the event, which explains Wallace's use of the term 'survival of the fittest', which did not occur in the original formulations of the theory by Darwin or by himself.

covering letter asking his opinion of the contents. The package arrived at Down House on 18 June 1858. The shock to Darwin at seeing his ideas pre-empted, as Lyell and others had warned might happen, came almost simultaneously with another, more personal, one – just ten days later, his infant son Charles Waring Darwin died of scarlet fever. In spite of his family problems, Darwin immediately tried to do the decent thing by Wallace, sending the paper on to Lyell with the comment:

> Your words have come true with a vengeance – that I should be forestalled ... I never saw a more striking coincidence; if Wallace had my MS sketch written out in 1842, he could not have made a better short abstract! ... I shall, of course, at once write and offer to send it to any journal.[1]

But Lyell, in conjunction with the naturalist Joseph Hooker (1817–1911), another member of Darwin's inner circle, found an alternative plan. They took the matter out of the hands of Darwin (who was happy to leave them to it while he came to terms with the loss of little Charles, consoled Emma and made the funeral arrangements) and came up with the idea of adding Darwin's 1844 outline of his theory to Wallace's paper and offering it to the Linnean Society as a joint publication. The paper was read to the Society on 1 July, without causing any great stir at the time,[2] and duly published under the impressive title '*On the tendency of species to form varieties; and on the perpetuation of varieties and species by natural means of selection* by Charles Darwin Esq., FRS, FLS, & FGS and Alfred Wallace Esq., communicated by Sir Charles Lyell, FRS, FLS, and J. D. Hooker Esq., MD, VPRS, FLS, &c.' That '&c.' is irresistible!

You might have expected Wallace to be more than a little upset at this cavalier treatment of his paper without him even being consulted, but in fact he was delighted, and always afterwards referred to the theory of natural selection as Darwinism, even writing a book under

1. See the *Autobiography*, edited by Francis Darwin.

2. In his autobiography, Darwin commented that 'our joint productions excited very little attention, and the only published notice of them which I can remember was by Professor Haughton of Dublin, whose verdict was that all that was new in them was false, and what was true was old'.

The publication of Darwin's Origin of Species

that title. Much later, he wrote, 'the one great result which I claim for my paper of 1858 is that it compelled Darwin to write and publish his *Origin of Species* without further delay'.[1] That he did; *On the Origin of Species by Means of Natural Selection, or the Preservation of favoured races in the struggle for life* was published by John Murray on 24 November 1859 and certainly did make a big impression, both on the scientific community and the world at large. Darwin went on to write other important books, accumulate more wealth and enjoy old age surrounded by his family in Down House, where he died on 19 April 1882; by and large, though, he kept out of the public debate about evolution and natural selection. Wallace also wrote more books, prospered in a more modest way for a time, but became an enthusiast for spiritualism, which tainted his scientific reputation. His spiritualist views also coloured his ideas about human beings, which he saw as specially touched by God, and not subject to the same evolutionary laws as other species. At the age of 43, in 1866 he married Annie Mitten, then just 18 years old, and the couple had a daughter and a son. But they were beset by financial worries, which only eased in 1880, when as a result of a petition which was primarily the idea of Darwin and Thomas Henry Huxley,[2] and was signed by several prominent scientists, Queen Victoria granted Wallace a pension of £200 a year for life. He was elected a Fellow of the Royal Society in 1893, received the Order of Merit in 1910 and died at Broadstone, in Dorset, on 7 November 1913. Charles

1. Quoted by Wilma George.

2. Huxley (1825–1895) deserves far more space than we can give him here, not so much for his own scientific work, which was important but not groundbreaking, or even his role as 'Darwin's bulldog' in promoting the theory of natural selection. His real importance in the history of science is that by dint of sheer ability and hard work he clawed his way up from humble origins to become a leading scientific figure, fought for better education for the working classes and was instrumental in seeing new seats of learning, with entry not restricted to gentlemen, opened up in London, Birmingham and Manchester, as well as Johns Hopkins University in Baltimore. He helped to establish science as a profession that people were paid to do, rather than a hobby indulged in by the rich. It is one of life's little ironies that in 1858 he effectively championed the cause of the gentleman amateur Darwin (who stood for everything Huxley hated, except that he was a brilliant scientist) and not the working-class Wallace. Our only consolation for relegating Huxley to a footnote is that you can find out all about him in Adrian Desmond's superb biography.

Darwin was the first scientist we have encountered in these pages who was born after 1800; Alfred Wallace was the first who died after 1900. In spite of all the other achievements of science in the nineteenth century, their achievement reigns supreme.

10 Atoms and Molecules

Humphry Davy's work on gases; electrochemical research – John Dalton's atomic model; first talk of atomic weights – Jöns Berzelius and the study of elements – Avogadro's number – William Prout's hypothesis on atomic weights – Friedrich Wöhler: studies in organic and inorganic substances – Valency – Stanislao Cannizzaro: the distinction between atoms and molecules – The development of the periodic table, by Mendeleyev and others – The science of thermodynamics – James Joule on thermodynamics – William Thomson (Lord Kelvin) and the laws of thermodynamics – James Clerk Maxwell and Ludwig Boltzmann: kinetic theory and the mean free path of molecules – Albert Einstein: Avogadro's number, Brownian motion and why the sky is blue

Although the figure of Charles Darwin dominates any discussion of nineteenth-century science, he is something of an anomaly. It is during the nineteenth century – almost exactly during Darwin's lifetime – that science makes the shift from being a gentlemanly hobby, where the interests and abilities of a single individual can have a profound impact, to a well-populated profession, where progress depends on the work of many individuals who are, to some extent, interchangeable. Even in the case of the theory of natural selection, as we have seen, if Darwin hadn't come up with the idea, Wallace would have, and from now on we will increasingly find that discoveries are made more or less simultaneously by different people working independently and largely in ignorance of one another. The other side of this particular coin, unfortunately, is that the growing number of scientists brings with it a growing inertia and resulting resistance to change, which means that all too often when some brilliant individual does come up with a profound new insight into the way the world works, this is not accepted immediately on merit and may take a generation to work its way into the collective received wisdom of science.

We shall shortly see this inertia at work in the reaction (or lack of reaction) to the ideas of John Dalton about atoms; we can also conveniently see the way science grew in terms of Dalton's life. When

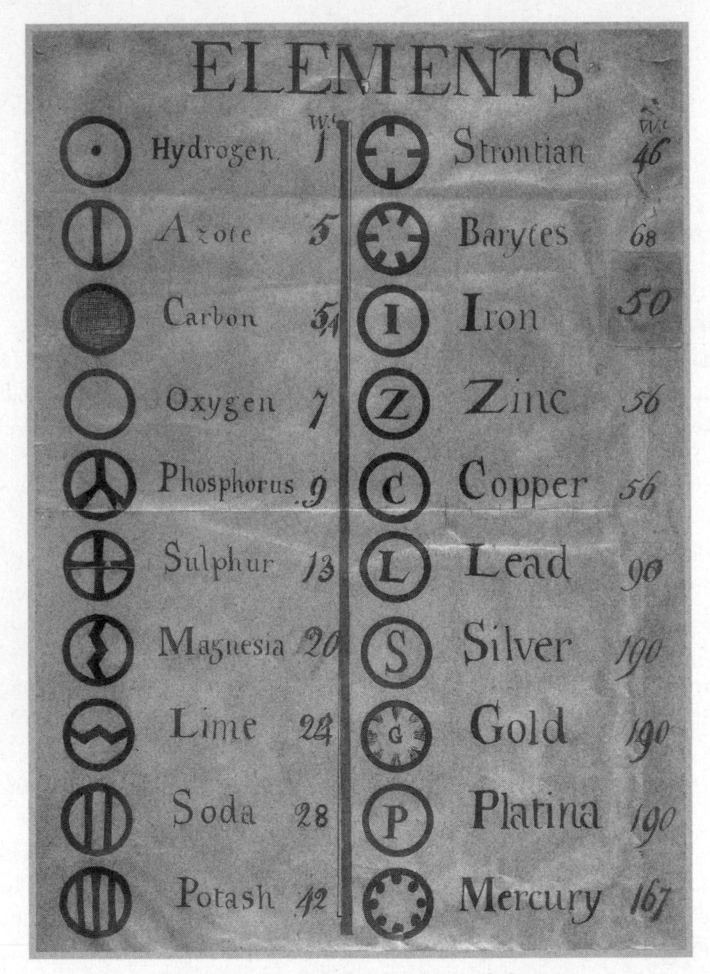

30. *Dalton's symbols for the chemical elements.*

Dalton was born, in 1766, there were probably no more than 300 people who we would now class as scientists in the entire world. By 1800, when Dalton was about to carry out the work for which he is now remembered, there were about a thousand. By the time he died, in 1844, there were about 10,000, and by 1900 somewhere around 100,000. Roughly speaking, the number of scientists doubled every fifteen years during the nineteenth century. But remember that the whole population of Europe doubled, from about 100 million to about 200 million, between 1750 and 1850, and the population of Britain alone doubled between 1800 and 1850, from roughly 9 million to roughly 18 million. The number of scientists did increase as a proportion of the population, but not as dramatically as the figures for scientists alone suggest at first sight.[1]

Humphry Davy's work on gases; electrochemical research

The transition from amateurism to professionalism is nicely illustrated by the career of Humphry Davy, who was actually younger than Dalton, although he had a shorter life. Davy was born in Penzance, Cornwall, on 17 December 1778. Cornwall was then still almost a separate country from England, and the Cornish language had not entirely died out; but from an early age Davy had ambitions that extended beyond the confines of his native county. Davy's father, Robert, owned a small farm and also worked as a woodcarver, but was never financially successful. His mother, Grace, later ran a milliner's shop with a French woman who had fled the Revolution, but in spite of her contribution the financial circumstances of the family were so difficult that at the age of 9, Humphry (the eldest of five children) went to live with his mother's adoptive father, John Tonkin, a surgeon. When Humphry's father died in 1794, leaving his family nothing but debts, it was Tonkin who advised Davy, who had been educated at Truro Grammar School without showing any conspicuous intellectual ability, to become the apprentice of a local apothecary, with the ultimate ambition of going to Edinburgh to study medicine. Around this time, Davy also learned French, from a refugee French priest, a skill that was soon to prove invaluable to him.

Davy seems to have been a promising apprentice and embarked on a programme of self-education reminiscent of Benjamin Thompson at a similar age. He might well have become a successful apothecary, or even a doctor. But the winter of 1797/8 marked a turning point in the young man's life. At the end of 1797, shortly before his nineteenth birthday, he read Lavoisier's *Traité Elémentaire* in the original French and became fascinated by chemistry. A few weeks earlier, Davy's

1. Figures from Greenaway.

widowed mother, still struggling to make ends meet, had taken in a lodger for the winter, a young man who suffered from consumption (TB) and had been sent to the relatively mild climate of Cornwall for the winter for his health. He happened to be Gregory Watt, the son of James Watt, and had studied chemistry at Glasgow University. Gregory Watt and Humphry Davy formed a friendship which lasted until Watt's death, in 1805, at the age of 27. Watt's presence in Penzance that winter gave Davy someone to share his developing interest in chemistry with; in 1798, through carrying out his own experiments, he developed his ideas about heat and light (still very much in the province of chemistry in those days) in a lengthy manuscript. Many of these ideas were naive and do not stand up to close scrutiny today (although it is noteworthy that Davy discarded the idea of caloric), but this was still an impressive achievement for a self-taught provincial 19-year-old. It was through Gregory Watt and his father James that Humphry Davy was initially introduced (by correspondence) to Dr Thomas Beddoes, of Bristol, and sent him his paper on heat and light.

Beddoes (1760–1808) had studied under Joseph Black in Edinburgh, before moving on to London and then Oxford, where he completed his medical studies and taught chemistry from 1789 to 1792. He then became intrigued by the discovery of different gases and decided to set up a clinic to investigate the potential of these gases in medicine. He had the (rather alarming to modern eyes) notion that breathing hydrogen might cure consumption, and moved to Bristol, where he practised medicine while obtaining funds for what became, in 1798, the Pneumatic Institute. Beddoes needed an assistant to help him with the chemical work and young Davy got the job. He left Penzance on 2 October 1798, still a couple of months short of his twentieth birthday.

It was in Bristol that Davy carried out the experiments with the gas now known as nitrous oxide, which made his name known more widely. Seeing no other way to find out how it affected the human body, he prepared 'four quarts' of nitrous oxide and breathed it from a silk bag, having first emptied his lungs as far as possible. He immediately discovered the intoxicating properties of the gas, which soon gave it the name 'laughing gas' as it became a sensation among the pleasure-seeking classes. A little later, while suffering considerable discomfort from a wisdom tooth, Davy also discovered, by accident, that the gas dulled the sensation of pain, and even wrote, in 1799, that 'it may probably be used with advantage during surgical operations'. At the time, unfortunately, this suggestion was not followed up, and it was left to the American dentist Horace Wells to pioneer the use of 'laughing gas' when extracting teeth, in 1844.

Davy continued to experiment on himself by inhaling various gases, with near-fatal results on one occasion. He was experimenting with the substance known as water gas (actually a mixture of carbon monoxide and hydrogen), produced by passing steam over hot charcoal. Carbon monoxide is extremely poisonous, rapidly but painlessly inducing a deep sleep which leads to death (which is why many suicides choose to kill themselves by breathing the exhaust fumes from a car engine). Davy just had time to drop the mouthpiece of the breathing bag he was using from his lips before collapsing, suffering nothing worse than a splitting headache when he awoke. But it was nitrous oxide that made his name.

After carrying out an intensive study of the chemical and physiological properties of the gas for about ten months, Davy wrote up his discoveries in a book more than 80,000 words long, completed in less than three months and published in 1800. The timing could not have been better for his career. In 1800, as the work on nitrous oxide drew to a close, Davy was becoming interested in electricity, as a result of the news of Volta's invention (or discovery) of the galvanic pile; starting out with the classic experiment in which water is decomposed into hydrogen and oxygen by the action of an electric current, Davy soon convinced himself there was a significant relationship between chemistry and electricity. While he was starting out on these studies, Count Rumford (as Benjamin Thompson now was) was trying to establish the Royal Institution (RI) in London. The RI had been founded in March 1799, but the first professor of chemistry appointed to the RI, Thomas Garnett, was not proving a success. His first lectures had gone down well, but a second series suffered from a lack of preparation and an unenthusiastic presentation. There were reasons for this – Garnett's wife had recently died and he seems to have lost his enthusiasm for everything, dying himself in 1802 at the age of only 36. Whatever the reasons for Garnett's failure, Rumford had to move quickly if the RI was to build on its promising beginning, and he invited Davy, the brightest rising star in the firmament of British chemistry, to come on board as assistant lecturer in chemistry and director of the laboratory at the RI, at an initial salary of 100 guineas per annum plus accommodation at the RI, with the possibility of succeeding Garnett in the top job. Davy accepted, and took up the post on 16 February 1801. He was a brilliant success as a lecturer, both in terms of the drama and excitement of his always thoroughly prepared and rehearsed talks, and in terms of his good looks and charisma, which had fashionable young ladies flocking to the lectures regardless of their content. Garnett (under pressure from Rumford) soon resigned, and Davy became top dog at the RI, appointed professor of chemistry in May 1802, shortly before Rumford left London and settled in Paris. Davy was still only 23 years old and had received no formal education beyond what Truro Grammar School could offer. In that sense, he was one of the last of the great amateur scientists (although not, strictly speaking, a gentleman); but as a salaried employee of the RI, he was also one of the first professional scientists.

Although usually remembered as a 'pure' scientist, Davy's greatest contemporary achievements were in promoting science, both in a general way at the RI and in terms of industrial and (in particular) agricultural applications. For example, he gave a famous series of lectures, by arrangement with the Board of Agriculture, on the relevance of chemistry to agriculture. It is a measure both of the importance of the subject and Davy's presentational skills that he was later (in 1810) asked to repeat the lectures (plus a series on electrochemistry) in Dublin, for a fee of 500 guineas; a year later he gave another series of lectures there for a fee of 750 guineas – more than seven times his starting salary at the RI in 1801. He was also awarded the honorary degree of Doctor of Laws by Trinity College, Dublin; this was the only degree he ever received.

We can gain an insight into Davy's method of preparing lectures from an account recorded by John Dalton when he gave a series of lectures at the RI in December 1803 (the year Davy became a Fellow of the Royal Society). Dalton tells us[1] how he wrote out his first lecture

1. See Hartley.

entirely, and how Davy took him into the lecture hall the evening before he was to talk and made him read it all out while Davy sat in the furthest corner and listened; then Davy read the lecture while Dalton listened. 'Next day I read it to an audience of about 150 to 200 people . . . they gave me a very generous plaudit at the conclusion.' But Davy, like so many of his contemporaries (as we shall see), was reluctant to accept the full implications of Dalton's atomic model.

Davy's electrochemical research led him to produce a masterly analysis and overview of the young science in the Royal Society's Bakerian Lecture in 1806 – so masterly and impressive, indeed, that he was awarded a medal and prize by the French Academy the following year, even though Britain and France were still at war. Shortly afterwards, using electrical currents passed through potash and soda, he isolated two previously unknown metals and named them potassium and sodium. In 1810, Davy isolated and named chlorine. Accurately defining an element as a substance that cannot be decomposed by any chemical process, he showed that chlorine is an element and established that the key component of all acids is hydrogen, not oxygen. This was the high point of Davy's career as a scientist, and in many ways he never fulfilled his real potential, partly because his lack of formal training led him into sometimes slapdash work lacking proper quantitative analysis, and partly because his head was turned by fame and fortune, so that he began to enjoy the social opportunities provided by his position more than his scientific work. He was knighted in 1812, three days before marrying a wealthy widow and a few months before appointing Michael Faraday (his eventual successor) as an assistant at the RI. The same year, he resigned his post as professor of chemistry at the RI, being succeeded by William Brande (1788–1866), but kept his role as director of the laboratory. An extensive tour of Europe with his bride (and Faraday) soon followed, thanks to a special passport provided by the French to the famous scientist (the war, of course, continued until 1815). It was after the party returned to England that Davy designed the famous miners' safety lamp which bears his name, although the work that led to this design was so meticulous and painstaking (and therefore so unlike Davy's usual approach) that some commentators suggest that Faraday must have had a great deal to do with it. In 1818, Davy became a baronet, and in 1820 he was elected President of the Royal Society, where he took great delight in all the ceremonial attached to the post and became such a snob that in 1824 he was the only Fellow to oppose the election of Faraday to the Royal. From 1825 onwards, however, Davy became chronically ill; he retired as director of the laboratory at the RI and played no further part in the affairs of British science. From 1827 onwards, he travelled in France and Italy, where the climate was more beneficial to his health, and died, probably of a heart attack, in Geneva on 29 May 1829, in his fifty-first year.

If Humphry Davy was one of the first people to benefit from the slow professionalization of science, the scientific career (such as it was) of John Dalton shows just how far this process still had to go in the first decades of the nineteenth century. Dalton was born in the village of Eaglesfield, in Cumberland, probably in the first week of September and definitely in 1766. He came from a Quaker family, but for some reason his exact birth date was not recorded in the Quaker register. We know that he had three siblings who died young and two, Jonathan and Mary, who survived; Jonathan was the eldest, but it is not recorded whether Mary was older or younger than John. Their father was a weaver, and the family occupied a two-roomed cottage, one room for working and the rest of the daily routine, the other for sleeping. Dalton attended a Quaker school where, instead of being force-fed with Latin, he was allowed to develop his interest in mathematics, and he came to the attention of a wealthy local Quaker, Elihu Robinson, who gave him access to his books and periodicals. At the age of 12, Dalton had to begin making a contribution to the family income and at first he tried teaching even younger children (some of them physically bigger than him) from his home and then from the Quaker Meeting House, for a modest fee; but this was not a success and he turned instead to farm work. In 1781, however, he was rescued from a life on the land when he went to join his brother Jonathan, who was helping one of their cousins, George Bewley, to run a school for Quakers in Kendal, a prosperous town with a large Quaker population. The two brothers took over the school when their cousin retired in 1785, with their sister keeping house for them, and John Dalton stayed there until 1793. He gradually developed his scientific interests, both answering and setting questions in popular magazines of the day, and beginning a long series of meteorological observations, which he recorded daily from 24 March 1787 until he died.

Frustrated by the limited opportunities and poor financial rewards offered by his dead-end job, Dalton developed an ambition to become either a lawyer or a doctor, and worked out how much it would cost to study medicine in Edinburgh (as a Quaker, of course, he could not then have attended either Oxford or Cambridge regardless of the cost). He was certainly up to the task intellectually; but friends advised him that there was no hope of obtaining the funds to fulfil such an ambition. Partly to bring in a little extra money, partly to satisfy his scientific leanings, Dalton began to give public lectures for which a small fee was charged, gradually extending his geographical range to include Manchester. Partly as a result of the reputation he gained through this work, in 1793 he became a teacher of mathematics and natural philosophy at a new college in Manchester, founded in 1786 and prosaically named New College. A book of his meteorological observations, written in Kendal, was published shortly after he moved to Manchester. In an appendix to that book, Dalton discusses the nature of water vapour and its relationship to air, describing the vapour in terms of particles which exist between the particles of air, so that the 'equal and opposite pressures' of the surrounding air particles on a particle of vapour 'cannot bring it nearer to another particle of vapour, without which no condensation can take place'. With hindsight, this can be seen as the precursor to his atomic theory.[1]

Manchester was a boom town in the 50 years Dalton lived there, with the cotton industry moving out of the cottages and into factories in the cities. Although he was not directly involved in industry, Dalton was part of this boom, because the very reason for the existence of educational establishments such as New College was to cater for the need of a growing population to develop the technical skills required by the new way of life. Dalton taught there until 1799; by then, he was well enough known to make a decent living as a private tutor, and he stayed in Manchester for the rest of his life. One reason why Dalton became well known, almost as soon as he arrived in Manchester, was

1. Really a model, at that time, but for historical reasons I will follow the convention of calling it a theory.

that he was colour blind. This condition had not previously been recognized, but Dalton came to realize that he could not see colours the same way most other people could, and found that his brother was affected in the same way. Blue and pink, in particular, were indistinguishable to both of them. On 31 October 1794 Dalton read a paper to the Manchester Literary and Philosophical Society describing his detailed analysis of the condition, which soon became known as Daltonism (a name still used in some parts of the world).

Over the next ten years or so, Dalton's keen interest in meteorology led him to think deeply about the nature of a mixture of gases, building from the ideas about water vapour just quoted. He never developed the idea of a gas as made up of huge numbers of particles in constant motion, colliding with one another and with the walls of their container, but thought instead in static terms, as if a gas were made of particles separated from one another by springs. But even with this handicap, he was led to consider the relationship between the volumes occupied by different gases under various conditions of temperature and pressure, the way gases dissolved in water and the influence of the weights of the individual particles in the gas on its overall properties. By 1801, he had come up with the law of partial pressures, which says that the total pressure exerted by a mixture of gases in a container is the sum of the pressures each gas would exert on its own under the same conditions (in the same container at the same temperature).

John Dalton's atomic model: first talk of atomic weights

It isn't possible to reconstruct Dalton's exact train of thought, since his notes are incomplete, but in the early 1800s he became convinced that each element was made up of a different kind of atom, unique to that element, and that the key distinguishing feature which made one element different from another was the weight of its atoms, with all the atoms of a particular element having the same weight, being indistinguishable from one another. Elementary atoms themselves, could be neither created nor destroyed. These elementary atoms could, however, combine with one another to form 'compound atoms' (what we would now call molecules) in accordance with specific rules. And Dalton even came up with a system of symbols to represent the different elements, although this idea was never widely accepted and soon became superceded by the now familiar alphabetical notation based on the names of the elements (in some cases, on their Latin names). Perhaps the biggest flaw in Dalton's model is that he did not realize that elements such as hydrogen are composed of molecules (as we would now call them), not individual atoms – H_2 not H. Partly for this reason, he got some molecular combinations wrong, and in modern notation he would have thought of water as HO, not H_2O.

Although parts of Dalton's model were described in various papers and lectures, the first full account of the idea was presented in the lectures at the RI in December 1803 and January 1804, where Davy helped Dalton to hone his presentation. The system was described (without being singled out as having any particular merit) by Thomas Thomson in the third edition of his book *System of Chemistry* in 1807, and Dalton's own book, *A New System of Chemical Philosophy*, which included a list of estimated atomic weights, appeared in 1808 (the very first table of atomic weights had been presented at the end of a paper by Dalton in 1803).

In spite of the seeming modernity and power of Dalton's model, though, it did not take the scientific world by storm at the end of the first decade of the nineteenth century. Many people found it hard, sometimes on philosophical grounds, to accept the idea of atoms (with the implication that there was nothing at all in the spaces between atoms) and even many of those who used the idea regarded it as no more than a heuristic device, a tool to use in working out how elements behave as *if* they were composed of tiny particles, without it necessarily being the case that they *are* composed of tiny particles. It took almost half a century for the Daltonian atom to become really fixed as a feature of chemistry, and it was only in the early years of the twentieth century (almost exactly a hundred years after Dalton's insight) that definitive proof of the existence of atoms was established. Dalton himself made no further contribution to the development of these ideas, but was heaped with honours during his long life (including becoming a Fellow of the Royal Society in 1822 and succeeding Davy as one of only eight Foreign Associates of the French Academy in 1830). When he died, in Manchester on 27 July 1844, he was given a completely over-the-top funeral, totally out of keeping with his Quaker lifestyle, involving a procession of a hundred carriages – but by then, the atomic theory was well on the way to becoming received wisdom.

Jöns Berzelius and the study of elements

The next key step in developing Dalton's idea had been made by the Swedish chemist Jöns Berzelius, who was born in Väversunda on 20 August 1779. His father, a teacher, died when Berzelius was 4, and his mother then married a pastor, Anders Ekmarck. Berzelius went to live with the family of an uncle when his mother also died, in 1788, and in 1796 began to study medicine at the University of Uppsala, interrupting his studies in order to work to pay his fees, but graduating with an MD in 1802. After he graduated, Berzelius moved to Stockholm, where he worked at first as an unpaid assistant to the chemist Wilhelm Hisinger (1766–1852) and then in a similar capacity as assistant to the professor of medicine and pharmacy at the College of Medicine in Stockholm; he did this so well that when the professor died in 1807, Berzelius was appointed to replace him. He soon gave up medicine and concentrated on chemistry.

His early work had been in electrochemistry, inspired, like Davy, by the work of Volta, but thanks to his formal training Berzelius was a much more meticulous experimenter than Davy. He was one of the first people to formulate the idea (true up to a point, but not universally so) that compounds are composed of electrically positive and electrically negative parts, and he was an early enthusiast for Dalton's atomic theory. From 1810 onwards, Berzelius carried out a series of experiments to measure the proportions in which different elements combine with one another (studying 2000 different compounds by 1816), which went a long way towards providing the experimental underpinning that Dalton's theory required and enabled Berzelius to produce a reasonably accurate table of the atomic weights of all forty elements known at the time (measured relative to oxygen, rather than to hydrogen). He was also the inventor of the modern alphabetical system of nomenclature for the elements, although it took a long time for this to come into widespread use. Along the way, Berzelius and his colleagues in Stockholm isolated and identified several 'new' elements, including selenium, thorium, lithium and vanadium.

It was around this time that chemists were beginning to appreciate that elements could be grouped in 'families' with similar chemical properties, and Berzelius gave the name 'halogens' (meaning salt-formers) to the group which includes chlorine, bromine and iodine; a dab hand at inventing names, he also coined the terms 'organic chemistry', 'catalysis' and 'protein'. His *Textbook of Chemistry*, first published in 1803, went through many editions and was highly influential. It's a measure of his importance to chemistry and the esteem in which he was held in Sweden that on his wedding day, in 1835, he was made a baron by the King of Sweden. He died in Stockholm in 1848.

Avogadro's number

But neither Berzelius nor Dalton (nor hardly anyone else, for that matter) immediately picked up on the two ideas which together carried the idea of atoms forward, and which were both formulated by 1811. First, the French chemist Joseph Louis Gay-Lussac (1778–1850) realized in 1808, and published in 1809, that gases combine in simple proportions by volume, and that the volume of the products of the reaction (if they are also gaseous) is related in a simple way to the volumes of the reacting gases. For example, two volumes of hydrogen combine with one volume of oxygen to produce two volumes of water vapour. This discovery, together with experiments which showed that all gases obey the same laws of expansion and compression, led the Italian Amadeo Avogadro (1776–1856) to announce, in 1811, his hypothesis that at a given temperature, the same volume of any gas contains the same number of particles. He actually used the word 'molecules', but where Dalton used 'atom' to mean both what we mean by atoms and what we mean by molecules, Avogadro used 'molecule' to mean both what we mean by molecules and what we mean by atoms. For simplicity, I shall stick to the modern terminology. Avogadro's hypothesis explained Gay-Lussac's discovery if, for example, each molecule of oxygen contains two atoms of the element, which can be divided to share among the molecules of hydrogen. This realization that oxygen (and other elements) could exist in polyatomic molecular form (in this case, O_2 rather than O) was a crucial step forward. So two volumes of hydrogen contain twice as many molecules as one volume of oxygen, and when they combine, each oxygen molecule provides one atom to each pair of hydrogen molecules, making the same number of molecules as there were in the original volume of hydrogen.

Using modern notation, $2H_2 + O_2 \rightarrow 2H_2O$.

Avogadro's ideas fell on stony ground at the time, and progress in developing the atomic hypothesis barely moved forward for decades,

William Prout's hypothesis on atomic weights

handicapped by the lack of experiments which could test the hypothesis. Ironically, the experiments were just good enough to cast considerable doubt on another good idea which emerged around this time. In 1815, the British chemist William Prout (1785–1850), building from Dalton's work, suggested that the atomic weights of all elements are exact multiples of the atomic weight of hydrogen, implying that in some way heavier elements might be built up from hydrogen. But the experimental techniques of the first half of the nineteenth century were good enough to show that this relationship did not hold exactly, and many atomic weights determined by chemical techniques could not be expressed as perfect integer multiples of the atomic weight of hydrogen. It was only in the twentieth century, with the discovery of isotopes (atoms of the same element with slightly different atomic weights, but each isotope having an atomic weight a precise multiple of the weight of one hydrogen atom) that the puzzle was resolved (since chemically determined atomic weights are an average of the atomic weights of all the isotopes of an element present) and Prout's hypothesis was seen as a key insight into the nature of atoms.

But while the subtleties of chemistry at the atomic level failed to yield much to investigation for half a century, a profound development was taking place in the understanding of what happens at a higher level of chemical organization. Experimenters had long been aware that everything in the material world falls into one of two varieties of chemical substances. Some, like water or common salt, can be heated and seem superficially to change their character (glowing red hot, melting, evaporating or whatever), but when cooled, revert back to the same chemical state they started from. Others, such as sugar or wood, are completely altered by the action of heat, so that it is very difficult, for example, to 'unburn' a piece if wood. It was in 1807 that Berzelius formalized the distinction between the two kinds of material. Since the first group of materials is associated with non-living systems, while the second group is closely related to living systems, he gave them the names 'inorganic' and 'organic'. As chemistry developed, it became clear that organic materials are, by and large, made up of much more complex compounds than inorganic materials; but it was also thought that the nature of organic substances was related to the presence of a 'life force' which made chemistry operate differently in living things than in non-living things.

Friedrich Wöhler: studies in organic and inorganic substances

The implication was that organic substances could *only* be made by living systems, and it came as a dramatic surprise when in 1828 the German chemist Friedrich Wöhler (1800–1882) accidentally discovered, during the course of experiments with a quite different objective, that urea (one of the constituents of urine) could be made by heating the simple substance ammonium cyanate. Ammonium cyanate was then regarded as an inorganic substance, but in the light of this and similar experiments which manufactured organic materials from substances which had never been associated with life, the definition of 'organic' changed. By the end of the nineteenth century it was clear that there was no mysterious life force at work in organic chemistry, and that there are two things which distinguish organic compounds from inorganic compounds. First, organic compounds are often complex, in the sense that each molecule contains many atoms of (usually) different elements. Second, organic compounds all contain carbon (which is, in fact, the reason for their complexity, because, as we shall see, carbon atoms are able to combine in many interesting ways with lots of other atoms and with other carbon atoms). This means that ammonium cyanate, which does contain carbon, is now regarded as an organic substance – but this doesn't in any way diminish the significance of Wöhler's discovery. It is now even possible to manufacture complete strands of DNA in the laboratory from simple inorganic materials.

The usual definition of an organic molecule today is any molecule that contains carbon, and organic chemistry is the chemistry of carbon and its compounds. Life is seen as a product of carbon chemistry, obeying the same chemical rules that operate throughout the world of atoms and molecules. Together with Darwin's and Wallace's ideas about evolution, this produced a major nineteenth-century shift in the view of the place of humankind in the Universe – natural selection tells us that we are part of the animal kingdom, with no evidence of a uniquely human 'soul'; chemistry tells us that animals and plants are part of the physical world, with no evidence of a special 'life force'.

Valency

By the time all this was becoming clear, however, chemistry had at last got to grips with atoms. Among the key concepts that emerged out of the decades of confusion, in 1852 the English chemist Edward Frankland (1825–1899) gave the first reasonably clear analysis of what became known as valency, a measure of the ability of one element to combine with another – or, as soon became clear, the ability of *atoms* of a particular element to combine with other atoms. Among many terms used in the early days to describe this property, one, 'equivalence', led through 'equivalency' to the expression used today. In terms of chemical combinations, in some sense two amounts of hydrogen are equivalent to one of oxygen, one of nitrogen to three of hydrogen, and so on. In 1858, the Scot Archibald Couper (1831–1892) wrote a paper in which he introduced the idea of bonds into chemistry, simplifying the representation of valency and the way atoms combine. Hydrogen is now said to have a valency of 1, meaning that it can form one bond with another atom. Oxygen has a valency of 2, meaning that it can form two bonds. So, logically enough, each of the two bonds 'belonging to' one oxygen atom can link up with one hydrogen atom, forming water – H_2O, or, if you prefer, H–O–H, where the dashes represent bonds. Similarly, nitrogen has a valency of three, making three bonds, so it can combine with three hydrogen atoms at a time, producing ammonia, NH_3. But the bonds can also form between two atoms of the same element, as in oxygen O_2, which can be represented as O=O. Best of all, carbon has a valency of four, so it can form four separate bonds with four separate atoms, including other atoms of carbon, at the same time.[1] This property lies at the heart of carbon chemistry, and Couper was quick to suggest that the complex carbon compounds which form the basis of organic chemistry might consist of a chain of carbon atoms 'holding hands' in this way with other atoms attached to the 'spare' bonds at the sides of the chain. The publication of Couper's paper was delayed, and the same idea was published first, independently, by the German chemist Friedrich August Kekulé (1829–1896); this overshadowed Couper's work at the time. Seven years later, Kekulé had the inspired insight that carbon atoms could also link up in rings (most commonly in a ring of six carbon atoms forming a hexagon) with bonds sticking out

1. Or it can form 'double bonds' (even triple bonds) with a smaller number of partners, as in carbon dioxide, CO_2, which we can represent as O=C=O.

from the ring to link up with other atoms (or even with other rings of atoms).

Stanislao Cannizzaro: the distinction between atoms and molecules

With ideas like those of Couper and Kekulé in the air at the end of the 1850s, the time was ripe for somebody to rediscover Avogadro's work and put it in its proper context. That somebody was Stanislao Cannizzaro, and although what he actually did can be described fairly simply, he had such an interesting life that it is impossible to resist the temptation to digress briefly to pick out some of the highlights from it. Cannizzaro, the son of a magistrate, was born in Palermo, Sicily, on 13 July 1826. He studied at Palermo, Naples, Pisa and Turin, before working as a laboratory assistant in Pisa from 1845 to 1847. He then returned to Sicily to fight in the unsuccessful rebellion against the ruling Bourbon regime of the King of Naples, part of the wave of uprisings which has led historians to dub 1848 'the year of revolutions' in Europe (Cannizzaro's father was at this time Chief of Police, which must have made life doubly interesting). With the failure of the uprising, Cannizzaro, sentenced to death in his absence, went into exile in Paris, where he worked with Michel Chevreul (1786–1889), professor of chemistry at the Natural History Museum. In 1851, Cannizzaro was able to return to Italy, where he taught chemistry at the Collegio Nazionale at Alessandria, in Piedmont,[1] before moving on to Genoa in 1855 as professor of chemistry. It was while in Genoa that he came across Avogadro's hypothesis and set it in the context of the progress made in chemistry since 1811. In 1858, just two years after Avogadro had died, Cannizzaro produced a pamphlet (what would now be called a preprint) in which he drew the essential distinction between atoms and molecules (clearing up the confusion which had existed since the time of the seminal work by Dalton and Avogadro), and explained how the observed behaviour of gases (the rules of combination of volumes, measurements of vapour density and so on) together with Avogadro's hypothesis could be used to calculate atomic and molecular weights, relative to the weight of one atom of hydrogen; and he drew

1. Italy was still a patchwork of small statelets at this time, so Cannizzaro's status as a failed revolutionary in Sicily did not automatically prevent him from finding refuge on the Italian mainland.

up a table of atomic and molecular weights himself. The pamphlet was widely circulated at an international conference held in Karlsruhe, Germany, in 1860, and was a key influence on the development of the ideas which led to an understanding of the periodic table of the elements.

Cannizzaro himself, however, was distracted from following up these ideas. Later in 1860, he joined Giuseppe Garibaldi's forces in the invasion of Sicily which not only ejected the Neapolitan regime from the island but quickly led to the unification of Italy under Victor Emmanuel II of Sardinia. In 1861, after the fighting, Cannizzaro became professor of chemistry in Palermo, where he stayed until 1871 before moving to Rome, where as well as being professor of chemistry at the university he founded the Italian Institute of Chemistry, became a senator in the Parliament and later Vice President of the Senate. He died in Rome on 10 May 1910, having lived to see the reality of atoms established beyond reasonable doubt.

The development of the periodic table, by Mendeleyev and others

The story of the discovery (or invention) of the periodic table is a curious mixture, highlighting the way in which, when the time is ripe, the same scientific discovery is likely to be made by several people independently, but also demonstrating the common reluctance of the old guard to accept new ideas. Hot on the heels of Cannizzaro's work, in the early 1860s both the English industrial chemist John Newlands (1837–1898) and the French mineralogist Alexandre Béguyer de Chancourtois (1820–1886) independently realized that if the elements are arranged in order of their atomic weight, there is a repeating pattern in which elements at regular intervals, with atomic weights separated by amounts that are multiples of eight times the atomic weight of hydrogen, have similar properties to one another.[1] Béguyer's work, published in 1862, was simply ignored (which may have been partly his own fault, for failing to explain his idea clearly and not even providing an explanatory diagram to illustrate it), but when Newlands, who knew nothing of Béguyer's work, published a series of papers

1. It is particularly appropriate that Newlands should have been one of the people to pick up on Cannizzaro's work. Newlands's mother was of Italian descent, and like Cannizzaro, Newlands fought with Garibaldi in Sicily in 1860.

touching on the subject in 1864 and 1865, he suffered the even worse fate of being savagely ridiculed by his peers, who said that the idea of arranging the chemical elements in order of their atomic weight was no more sensible than arranging them in alphabetical order of their names. The key paper setting out his idea in full was rejected by the Chemical Society and only published in 1884, long after Dmitri Mendeleyev had been hailed as the discoverer of the periodic table. In 1887, the Royal Society awarded Newlands their Davy Medal, although they never got around to electing him as a Fellow.

But Mendeleyev wasn't even the third person to come up with the idea of the periodic table. That honour, such as it is, belongs to the German chemist and physician Lothar Meyer (1830–1895), although, as he himself later acknowledged, to some extent he lacked the courage of his convictions, which is why the prize eventually went to Mendeleyev. Meyer made his name in chemistry through writing a textbook, *The Modern Theory of Chemistry*, which was published in 1864. He was an enthusiastic follower of Cannizzaro's ideas, which he expounded in the book. While he was preparing the book, he noticed the relationship between the properties of a chemical element and its atomic weight, but was reluctant to promote this novel and as yet untested idea in a textbook, so he only hinted at it. Over the next few years, Meyer developed a more complete version of the periodic table, intended to be included in a second edition of his book, which was ready in 1868, but did not get into print until 1870. By that time, Mendeleyev had propounded his version of the periodic table (in ignorance of all the work along similar lines that had been going on in the 1860s), and Meyer always acknowledged Mendeleyev's priority, not least because Mendeleyev had the courage (or chutzpah) to take a step that Meyer never took, predicting the need for 'new' elements to plug gaps in the periodic table. But Meyer's independent work was widely recognized, and Meyer and Mendeleyev shared the Davy medal in 1882.

It's a little surprising that Mendeleyev was out of touch with all the developments in chemistry in Western Europe in the 1860s.[1] He was

1. Mendeleyev may have known of Béguyer's work, though not of Newlands's, but this doesn't detract from his own achievement, which went far beyond the rather confused account of recurring patterns in Béguyer's 1862 paper.

born at Tobol'sk, in Siberia, on 7 February 1834 (27 January on the Old Style calendar still in use in Russia then), the youngest of fourteen children. Their father, Ivan Pavlovich, who was the head of the local school, went blind when Dmitri was still a child, and thereafter the family was largely supported by his mother, Marya Dmitrievna, an indomitable woman who set up a glass works to provide income. Mendeleyev's father died in 1847, and a year later the glass works was destroyed by fire. With the older children more or less independent, Marya Dmitrievna determined that her youngest child should have the best education possible and, in spite of their financial difficulties, took him to St Petersburg. Because of prejudice against poor students from the provinces, he was unable to obtain a university place, but enrolled as a student teacher in 1850, at the Pedagogical Institute, where his father had qualified. His mother died just ten weeks later, but Dmitri seems to have been as determined as she had been. Having established his credentials by completing his training and working as a teacher for a year in Odessa, he was allowed to take a master's degree in chemistry at the University of St Petersburg, graduating in 1856. After a couple of years working in a junior capacity at the university, Mendeleyev went on a government-sponsored study programme in Paris and Heidelberg, where he worked under Robert Bunsen and Gustav Kirchoff. He attended the 1860 meeting in Karlsruhe where Cannizzaro circulated his pamphlet about atomic and molecular weights, and met Cannizzaro. On his return to St Petersburg, Mendeleyev became professor of general chemistry in the city's Technical Institute and completed his PhD in 1865; in 1866 he became professor of chemistry at the University of St Petersburg, a post he held until he was forced to 'retire' in 1891, although only 57 years old, after taking the side of the students during a protest about the conditions in the Russian academic system. After three years he was deemed to have purged his guilt and became controller of the Bureau of Weights and Measures, a post he held until he died in St Petersburg on 2 February 1907 (20 January, Old Style). He just missed being an early recipient of the Nobel prize – he was nominated in 1906, but lost out by one vote to Henri Moissan (1852–1907), the first person to isolate fluorine. He died before the Nobel Committee met again (as, indeed, did Moissan).

Mendeleyev, like Meyer, made his name by writing a textbook,

Principles of Chemistry, which was published in two volumes in 1868 and 1870. Also like Meyer, he arrived at an understanding of the relationship between the chemical properties of the elements and their atomic weights while working on his book, and in 1869 published his classic paper 'On the Relation of the Properties to the Atomic Weights of Elements'.[1] The great thing about Mendeleyev's work, which singles him out from the pack of other people who had similar ideas at about the same time, is that he had the audacity to rearrange the order of the elements (slightly) in order to make them fit the pattern he had discovered, and to leave gaps in the periodic table, as it became known, for elements which had not yet been discovered. The rearrangements involved really were very small. Putting the elements exactly in order of their atomic weights, Mendeleyev had come up with an arrangement in a grid, rather like a chess board, with elements in rows of eight, one under another, so that those with similar chemical properties to one another lay underneath one another in the columns of the table. With this arrangement in strict order of increasing atomic weights (lightest at the top left of the 'chess board', heaviest at the bottom right), there were some apparent discrepancies. Tellurium, for example, came under bromine, which had completely different chemical properties. But the atomic weight of tellurium is only a tiny bit more than that of iodine (modern measurements give the atomic weight of tellurium as 127.60, while that of iodine is 126.90, a difference of only 0.55 per cent). Reversing the order of these two elements in the table put iodine, which has similar chemical properties to bromine, under bromine, where it clearly belongs in chemical terms.

Mendeleyev's bold leap of faith was entirely justified, as became clear in the twentieth century with the investigation of the structure of the nucleus, which lies at the heart of the atom. It turns out that the chemical properties of an element depend on the number of protons in the nucleus of each atom (the atomic number), while its atomic weight depends on the total number of protons plus neutrons in the nucleus. The modern version of the periodic table ranks the elements

1. Exactly ten years after Charles Darwin published *On the Origin of Species*; there was so much scientific work going on in parallel in the nineteenth century that it is sometimes hard to keep track of who was doing what, and when!

Reihen	Gruppe I. — R^2O	Gruppe II. — RO	Gruppe III. — R^2O^3	Gruppe IV. RH^4 RO^2	Gruppe V. RH^3 R^2O^5	Gruppe VI. RH^2 RO^3	Gruppe VII. RH R^2O^7	Gruppe VIII. — RO^4
1	H=1							
2	Li=7	Be=9,4	B=11	C=12	N=14	O=16	F=19	
3	Na=23	Mg=24	Al=27,3	Si=28	P=31	S=32	Cl=35,5	
4	K=39	Ca=40	—=44	Ti=48	V=51	Cr=52	Mn=55	Fe=56, Co=59, Ni=59, Cu=63.
5	(Cu=63)	Zn=65	—=68	—=72	As=75	Se=78	Br=80	
6	Rb=85	Sr=87	?Yt=88	Zr=90	Nb=94	Mo=96	—=100	Ru=104, Rh=104, Pd=106, Ag=108.
7	(Ag=108)	Cd=112	In=113	Sn=118	Sb=122	Te=125	J=127	
8	Cs=133	Ba=137	?Di=138	?Ce=140	—	—	—	— — — —
9	(—)	—	—	—	—	—	—	
10	—	—	?Er=178	?La=180	Ta=182	W=184	—	Os=195, Ir=197, Pt=198, Au=199.
11	(Au=199)	Hg=200	Tl=204	Pb=207	Bi=208	—	—	
12	—	—	—	Th=231	—	U=240	—	— — — —

31. *Mendeleyev's early version of the table of the elements, 1871.*

in order of increasing atomic number, not increasing atomic weight; but in by far the majority of cases, elements with higher atomic number also have greater atomic weight. In just a few rare cases the presence of an extra couple of neutrons makes the ranking of elements by atomic weight slightly different from the ranking in order of atomic number.

If that was all Mendeleyev had done, though, without the knowledge of protons and neutrons that only emerged decades later, his version of the periodic table would probably have got as short shrift as those of his predecessors. But in order to make the elements with similar chemical properties line up beneath one another in the columns of his table, Mendeleyev also had to leave gaps in the table. By 1871, he had refined his table into a version incorporating all of the 63 elements known at the time, with a few adjustments like the swapping of tellurium and iodine, and with three gaps, which he said must correspond to three as yet undiscovered elements. From the properties of the adjacent elements in the columns of the table where the gaps occurred, he was able to predict in some detail what the properties of these elements would be. Over the next fifteen years, the three elements needed to plug the gaps in the table, with just the properties predicted by Mendeleyev, were indeed discovered – gallium in 1875, scandium in 1879 and germanium in 1886. Although Mendeleyev's periodic table had not gained universal acclaim at first (and had, indeed, been criticized for his willingness to interfere with nature by changing the order of the elements), by the 1890s it was no longer possible to doubt that the periodicity, in which elements form families that have similar chemical properties to one another and within which the atomic weights of the individual elements differ from one another by multiples of eight times the atomic weight of hydrogen, represented a deep truth about the nature of the chemical world. It was also a classic example of the scientific method at work, pointing the way for twentieth-century scientists. From a mass of data, Mendeleyev found a pattern, which led him to make a prediction that could be tested by experiment; when the experiments confirmed his prediction, the hypothesis on which the prediction was based gained strength.

Surprising though it may seem to modern eyes, though, even this was not universally accepted as proof that atoms really do exist in the form of little hard entities that combine with one another in well-defined ways. But while chemists had been following one line in the investigation of the inner structure of matter and arriving at evidence which at the very least supported the atomic hypothesis, physicists had been following a different path which ultimately led to incontrovertible proof that atoms exist.

The science of thermodynamics

The unifying theme in this line of nineteenth-century physics was the study of heat and motion, which became known as thermodynamics. Thermodynamics both grew out of the industrial revolution, which provided physicists with examples (such as the steam engine) of heat at work, inspiring them to investigate just what was going on in those machines, and fed back into the industrial revolution, as an improved scientific understanding of what was going on made it possible to design and build more efficient machines. At the beginning of the nineteenth century, as we have seen, there was no consensus about the nature of heat, and both the caloric hypothesis and the idea that heat is a form of motion had their adherents. By the middle of the 1820s, thermodynamics was beginning to be recognized as a scientific subject (although the term itself was not coined until 1849, by William Thomson), and by the middle of the 1860s, the basic laws and principles had been worked out. Even then, it took a further forty years for the implications of one small part of this work to be used in a definitive proof of the reality of atoms.

The key conceptual developments that led to an understanding of thermodynamics involved the idea of energy, the realization that energy can be converted from one form to another but can neither be created nor destroyed, and the realization that work is a form of energy (as Rumford had more than hinted at with his investigation of the heat produced by boring cannon). It's convenient to date the beginning of the science of thermodynamics from the publication of a book by the Frenchman Sadi Carnot (1796–1832) in 1824. In the book (*Réflexions sur la puissance motive du feu*), Carnot analysed the efficiency of engines in converting heat into work (providing a scientific definition of work along the way), showed that work is done as heat passes from a higher temperature to a lower temperature (implying an early form of the second law of thermodynamics, the realization that heat always flows from a hotter object to a colder object, not the other way around) and even suggested the possibility of the internal combustion engine. Unfortunately, Carnot died of cholera at the age of 36, and although his notebooks contained further developments of these ideas, they had not been published at the time of his death. Most of the manuscripts were burned, along with his personal effects, because of the cause of his death; only a few surviving pages hint at what he had probably achieved. But it was Carnot who first clearly appreciated that heat and work are interchangeable, and who worked out for the first time how much work (in terms of raising a certain weight through a vertical distance) a given amount of heat (such as the heat lost when 1g of water cools by 1 °C) can do. Carnot's book was not very influential at the time, but it was mentioned in 1834 in a discussion of Carnot's work in a paper by Émile Clapeyron (1799–1864), and through that paper, Carnot's work became known to and influenced the generation of physicists who completed the thermodynamic revolution, notably William Thomson and Rudolf Clausius.

If Carnot's story sounds complicated, the way physicists became aware of the nature of energy is positively tortuous. The first person who actually formulated the principle of conservation of energy and published a correct determination of the mechanical equivalent of heat'[1] was actually a German physician, Julius Robert von Mayer (1814–1878), who arrived at his conclusions from his studies of human beings, not steam engines, and who, mainly as a result of coming at it from the 'wrong' direction as far as physicists were concerned, was largely ignored (or at least overlooked) at the time. In 1840, Mayer, newly qualified, was working as ship's doctor on a Dutch vessel which visited the East Indies. Bleeding was still popular at the time, not only to (allegedly) alleviate the symptoms of illness, but as a routine measure in the tropics, where, it was believed, draining away a little blood would help people to cope with the heat. Mayer was well aware of Lavoisier's work which showed that warm-blooded animals are kept warm by the slow combustion of food, which acts as fuel, with oxygen in the body; he knew that bright red blood, rich in oxygen, is carried around the body from the lungs in arteries, while dark purple blood, deficient in oxygen, is carried back to the lungs by veins. So when he opened the vein of a sailor in Java, he was astonished to find that the blood was as brightly coloured as normal arterial blood. The same proved true of the venous blood of all the rest of the crew, and of himself. Many other doctors must have seen the same thing before, but Mayer, only in his mid-twenties and recently qualified, was the one who had the wit to understand what was going on. Mayer realized that the reason why the venous blood was rich in oxygen was that in the heat of the tropics the body had to burn less fuel (and therefore consume less oxygen) to keep warm. He saw that this implied that all forms of heat and energy are interchangeable – heat from muscular exertion, the heat of the Sun, heat from burning coal, or whatever – and that heat, or energy, could never be created but only changed from one form into another.

Mayer returned to Germany in 1841 and practised medicine. But alongside his medical work he developed his interest in physics, reading widely, and from 1842 onwards published the first scientific papers drawing attention (or trying to draw attention) to these ideas. In 1848, he developed his ideas about heat and energy into a discussion of the age of the Earth and the Sun, which we shall discuss shortly. But all of his work went unnoticed by the physics community and Mayer became

1. A cruder version had been published in 1839 by Marc Séguin (1786–1875).

so depressed by his lack of recognition that he attempted suicide in 1850, and was confined in various mental institutions during the 1850s. From 1858 onwards, however, his work was rediscovered and given due credit by Hermann von Helmholtz (1821–1894), Clausius and John Tyndall (1820–1893). Mayer recovered his health, and was awarded the Copley Medal of the Royal Society in 1871, seven years before he died.

James Joule on thermodynamics

The first physicist who really got to grips with the concept of energy (apart from the unfortunate Carnot, whose work was nipped in the bud) was James Joule (1818–1889), who was born in Salford, near Manchester, the son of a wealthy brewery owner. Coming from a family of independent means, Joule didn't have to worry about working for a living, but as a teenager he spent some time in the brewery, which it was expected he would inherit a share in. His first-hand experience of the working machinery may have fired his interest in heat, just as the gases produced in brewing had helped to inspire Priestley's work. As it happened, Joule's father sold the brewery in 1854, when James was 35, so he never did inherit it. Joule was educated privately, and in 1834 his father sent James and his older brother to study chemistry with John Dalton. Dalton was by then 68 and in failing health, but still giving private lessons; however, the boys learned very little chemistry from him, because he insisted on first teaching them Euclid, which took two years of lessons at a rate of two separate hours per week, and then stopped teaching in 1837 because of illness. But Joule remained friendly with Dalton, who he visited from time to time for tea until Dalton's death in 1844. In 1838, Joule turned one of the rooms in the family house into a laboratory, where he worked independently. He was also an active member of the Manchester Literary and Philosophical Society, where he usually sat (even before he became a member) next to Dalton at lectures, so he was very much in touch with what was going on in the scientific world at large.

Joule's early work was with electromagnetism, where he hoped to invent an electric motor that would be more powerful and efficient than the steam engines of his day. The quest was unsuccessful, but it drew him into an investigation of the nature of work and energy. In 1841, he produced papers on the relationship between electricity and heat for the *Philosophical Magazine* (an earlier version of this paper was rejected by the Royal Society, although they did publish a short abstract summarizing the work) and the Manchester Literary and Philosophical Society. In 1842, he presented his ideas at the annual meeting of the British Association for the Advancement of Science (the BA), a peripatetic event which happened that year to be in Manchester. He was still only 23 years old. Joule's greatest work, in the course of which he developed the classic experiment which shows that work is converted into heat by stirring a container of water with a paddle wheel and measuring the rise in temperature, was produced over the next few years, but leaked out in a slightly bizarre way. In 1847, he gave two lectures in Manchester which, among other things, set out the law of conservation of energy and its importance to the physical world. As far as Joule was aware, of course, nobody had done anything like this before. Eager to see the ideas in print, with the help of his brother he arranged for the lectures to be published in full in a newspaper, the *Manchester Courier* – to the bafflement of its readers and without spreading the word to the scientific community. But later that year the BA met in Oxford, where Joule gave a résumé of his ideas, the importance of which was immediately picked up by one young man in the audience, William Thomson (then 22). The two became friends and scientific collaborators, working on the theory of gases and, in particular, the way in which gases cool when they expand (known as the Joule–Thomson effect, this is the principle on which a refrigerator operates). From the perspective of the atomic hypothesis, he published another important paper in 1848, in which he estimated the average speed with which the molecules of gas move. Treating hydrogen as being made up of tiny particles bouncing off one another and the walls of their container, he calculated (from the weight of each particle and the pressure exerted by the gas) that at a temperature of 60 °F and a pressure corresponding to 30 inches of mercury (more or less the conditions in a comfortable room), the particles of the gas must be moving at 6225.54 feet per second. Since oxygen molecules weigh sixteen times as much as hydrogen molecules, and the appropriate relationship depends on one over the square root of the mass, in ordinary air under the same conditions the oxygen molecules are moving at one quarter of this speed, 1556.39 feet per second. Joule's work on gas dynamics and, particularly, the law of conservation of energy, was widely recognized at the end of the 1840s (he even got to read a key paper on the subject to the Royal in 1849, no doubt ample recompense for the rejection of his earlier paper), and in 1850 he was elected a Fellow of the Royal Society. Now in his thirties, he achieved nothing else to rank with the importance of his early work, as is so often the case, and the torch passed to Thomson, James Clerk Maxwell and Ludwig Boltzmann.

William Thomson (Lord Kelvin) and the laws of thermodynamics

If Joule had been born with a silver spoon in his mouth and, as a result, never worked in a university environment, Thomson was born with a different kind of silver spoon in his mouth and, as a result, spent almost his entire life in a university environment. His father, James Thomson, was professor of mathematics at the Royal Academical Institution in Belfast (the forerunner of the University of Belfast) when William was born on 26 June 1824. He had several siblings, but their mother died when William was 6: William and his brother James (1822–1892), who also became a physicist, were educated at home by their father, and after the elder James Thomson became professor of mathematics at the University of Glasgow in 1832, both boys were allowed to attend lectures there, officially enrolling at the university (matriculating) in 1834, when William was 10 – although not with the aim of completing a degree but rather to regularize the fact that they were attending lectures. William moved on to the University of Cambridge in 1841, and graduated in 1845, by which time he had won several prizes for his scientific essays and had already published a series of papers in the *Cambridge Mathematical Journal*. After graduating, William worked briefly in Paris (where he became familiar with the work of Carnot), but his father's dearest wish was that his brilliant son would join him at the University of Glasgow, and by the time the professor of natural philosophy in Glasgow died (not unexpectedly; he was very old) in 1846, had already begun an ultimately successful campaign to see William elected to the post. James Thomson didn't live long to enjoy the situation, though; he died of cholera in 1849. William Thomson remained professor of natural philosophy in Glasgow from 1846 (when he was 22) until he retired, aged 75, in 1899; he then enrolled as a research student at the university to keep his hand in, making him possibly both the youngest student and the oldest student ever to attend the

University of Glasgow. He died at Largs, in Ayrshire, on 17 December 1907, and was buried next to Isaac Newton in Westminster Abbey.

Thomson's fame and the honour of his resting place were by no means solely due to his scientific achievements. He made his biggest impact on Victorian Britain through his association with applied technology, including being responsible for the success of the first working transatlantic telegraph cable (after two previous attempts without the benefit of his expertise had failed), and made a fortune from his patents of various kinds. It was largely for his part in the success of the cable (as important in its day as the Internet is at the beginning of the twenty-first century) that Thomson was knighted in 1866, and it was as a leading light of industrial progress that he was made Baron Kelvin of Largs in 1892, taking his name from the little river that runs through the site of the University of Glasgow. Although the peerage came long after his important scientific work, Thomson is often referred to even in scientific circles as Lord Kelvin (or simply Kelvin), partly in order to distinguish him from the physicist J. J. Thomson, to whom he was not related. The absolute, or thermodynamic, scale of temperature is called the Kelvin scale in his honour.

Although he also worked in other areas (including electricity and magnetism, which feature in the next chapter), Thomson's most important work was, indeed, in establishing thermodynamics as a scientific discipline at the beginning of the second half of the nineteenth century. Largely jumping off from Carnot's work, it was as early as 1848 that Thomson established the absolute scale of temperature, which is based on the idea that heat is equivalent to work, and that a certain change in temperature corresponds to a certain amount of work. This both defines the absolute scale itself and carries with it the implication that there is a minimum possible temperature (–273 °C, now written as 0 K) at which no more work can be done because no heat can be extracted from a system. Around this time, Carnot's ideas were being refined and developed by Rudolf Clausius (1822–1888) in Germany (Carnot's work certainly needed an overhaul; among other things, he had used the idea of caloric). Thomson learned of Clausius's work in the early 1850s, by which time he was already working along similar lines. They each arrived, more or less independently, at the key principles of thermodynamics.

The first law of thermodynamics, as it is grandly known, simply says that heat is work, and it provides an intriguing insight into the way nineteenth-century science developed that it was necessary, in the 1850s, to spell this out as a law of nature. The second law of thermodynamics is actually much more important, arguably the most important and fundamental idea in the whole of science. In one form, it says that heat cannot, of its own volition, move from a colder object to a hotter object. In that form, it sounds obvious and innocuous. Put an ice cube in a jug of warm water, and heat flows from the warm water to the cold ice, melting it; it doesn't flow from the ice into the water, making the ice colder and the water hotter. Put more graphically, though, the universal importance of the second law becomes more apparent. It says that things wear out – *everything* wears out, including the Universe itself. From another perspective, the amount of disorder in the Universe (which can be measured mathematically by a quantity Clausius dubbed 'entropy') always increases overall. It is only possible for order to be preserved or to increase in local regions, such as the Earth, where there is a flow of energy from outside (in our case, from the Sun) to feed off. But it is a law of nature that the decrease in entropy produced by life on Earth feeding off the Sun is smaller than the increase in entropy associated with the processes that keep the Sun shining, whatever those processes might be. This cannot go on for ever – the supply of energy from the Sun is not inexhaustible. It was this realization that led Thomson to write, in a paper published in 1852, that:

> Within a finite period of past time the earth must have been, and within a finite period of time to come the earth must again be unfit for the habitation of man as at present constituted, unless operations have been or are to be performed which are impossible under the laws to which the known operations going on at present in the material world are subject.

This was the first real scientific recognition that the Earth (and by implication, the Universe) had a definite beginning, which might be dated by the application of scientific principles. When Thomson himself applied scientific principles to the problem, he worked out the age of the Sun by calculating how long it could generate heat at its present rate by the most efficient process known at the time, slowly contracting under its own weight, gradually converting gravitational energy into heat. The answer comes out as a few tens of millions of years – much less than the timescale already required by the geologists in the 1850s, and which would soon be required by the evolutionists. The resolution of the puzzle, of course, came with the discovery of radioactivity, and then with Albert Einstein's work showing that matter is a form of energy, and including his famous equation $E = mc^2$. All this will be discussed in later chapters, but the conflict between the timescales of geology and evolution and the timescales offered by the physics of the time rumbled throughout the second half of the nineteenth century.

This work also brought Thomson into conflict with Hermann von Helmholtz (1821–1894), who arrived at similar conclusions independently of Thomson. There was an unedifying wrangle over priority between supporters of the two men, which was particularly pointless since not only the unfortunate Mayer but also the even more unfortunate John Waterston had both got there first. Waterston (1811–1883?) was a Scot, born in Edinburgh, who worked as a civil engineer on the railways in England before moving to India in 1839 to teach the cadets of the East India Company. He saved enough to retire early, in 1857, and came back to Edinburgh to devote his time to research into what soon became known as thermodynamics, and other areas of physics. But he had long been working in science in his spare time, and in 1845 he had written a paper describing the way in which energy is distributed among the atoms and molecules of a gas in accordance with statistical rules – not with every molecule having the same speed, but with a range of speeds distributed in accordance with statistical rules around the mean speed. In 1845, he sent the paper describing this work from India to the Royal Society, which not only rejected the paper (their referees didn't understand it and therefore dismissed it as nonsense) but promptly lost it. The paper included calculations of the properties of gases (such as their specific heats) based on these ideas, and was essentially correct; but Waterston had neglected to keep a copy and never reproduced it, although he did publish related (and largely ignored) papers on his return to England. He also, ahead of both Thomson and von Helmholtz but roughly at the same time as Mayer, had the same insight about the way heat to keep the Sun hot might be generated by gravitational means. With none of his work gaining much recognition, like Mayer, Waterston became ill and depressed. On 18 June 1883, he walked out of his house and never came back. But there is a (sort of) happy ending to the story – in 1891, Waterston's missing manuscript was found in the vaults of the Royal Society and it was published in 1892.

James Clerk Maxwell and Ludwig Boltzmann: kinetic theory and the mean free path of molecules

By then, the kinetic theory of gases (the theory treating gases in terms of the motion of their constituent atoms and molecules) and the ideas of statistical mechanics (applying statistical rules to describe the behaviour of collections of atoms and molecules) had long been established. The two key players who established these ideas were James Clerk Maxwell (who features in another context in the next chapter) and Ludwig Boltzmann (1844–1906). After Joule had calculated the speeds with which molecules move in a gas, Clausius introduced the idea of a mean free path.[1] Obviously, the molecules do not travel without deviation at the high speeds Joule had calculated; they are repeatedly colliding with one another and bouncing off in different directions. The mean free path is the average distance that a molecule travels between collisions, and it is tiny. At the annual meeting of the BA in 1859 (held that year in Aberdeen), Maxwell presented a paper which echoed (without knowing it) much of the material in Waterston's lost paper. This time, the scientific world was ready to sit up and take notice. He showed how the speeds of the particles in a gas were distributed around the mean speed, calculated the mean velocity of molecules in air at 60 °F as 1505 feet per second and determined the mean free path of those molecules to be 1/447,000 of an inch. In other words, every second each molecule experiences 8,077,200,000 collisions – more than eight billion collisions per second. It is the shortness of the mean free path and the frequency of these collisions that gives the illusion that a gas is a smooth, continuous fluid, when it is really made up of a vast number of tiny particles in constant motion,

1. We should mention that an early version of the kinetic theory was developed by Bristol-born John Herapath (1790–1868) and published in 1821; although too far ahead of its time to be quantitatively accurate, Herapath's work was known to Joule and helped point him in the right direction.

with nothing at all in the gaps between the particles. Even more significantly, it was this work that led to a full understanding of the relationship between heat and motion – the temperature of an object is a measure of the mean speed with which the atoms and molecules that make up the object are moving – and the final abandonment of the concept of caloric.

Maxwell developed these ideas further in the 1860s, applying them to explain many of the observed properties of gases, such as their viscosity and, as we have seen, the way they cool when they expand (which turns out to be because the atoms and molecules in a gas attract one another slightly, so work has to be done to overcome this attraction when the gas expands, slowing the particles down and therefore making the gas cooler). Maxwell's ideas were taken up by the Austrian Ludwig Boltzmann, who refined and improved them; in turn, in a constructive feedback, Maxwell took on board some of Boltzmann's ideas in making further improvements to the kinetic theory. One result of this feedback is that the statistical rule describing the distribution of the velocities (or kinetic energies) of the molecules in a gas around their mean is now known as the Maxwell–Boltzmann distribution.

Boltzmann made many other important contributions to science, but his greatest work was in the field of statistical mechanics, where the overall properties of matter (including the second law of thermodynamics) are derived in terms of the combined properties of the constituent atoms and molecules, obeying simple laws of physics (essentially, Newton's laws) and the blind working of chance. This is now seen as fundamentally dependent on the idea of atoms and molecules, and it was always seen that way in the English-speaking world, where statistical mechanics was brought to its full flowering through the work of the American Willard Gibbs (1839–1903) – just about the first American (given that Rumford regarded himself as British) to make a really significant contribution to science. Even at the end of the nineteenth century, however, these ideas were being heavily criticized in the German-speaking world by anti-atomist philosophers, and even scientists such as Wilhelm Ostwald (1853–1932), who insisted (even into the twentieth century) that atoms were a hypothetical concept, no more than a heuristic device to help us describe the observed properties of chemical elements. Boltzmann, who suffered from depression anyway, became convinced that his work would never receive the recognition it deserved. In 1898, he published a paper detailing his calculations, in the express hope 'that, when the theory of gases is again revived, not too much will have to be rediscovered'. Soon afterwards, in 1900, he made an unsuccessful attempt on his own life (probably not his only unsuccessful suicide attempt), in the light of which this paper can be seen as a kind of scientific suicide note. He seemed to recover his spirits for a time, and travelled to the United States in 1904, where he lectured at the World's Fair in St Louis and visited the Berkeley and Stanford campuses of the University of California, where his odd behaviour was commented on as 'a mixture of manic ecstasy and the rather pretentious affectation of a famous German professor'.[1] The mental improvement (if this behaviour can be regarded as an improvement) did not last, however, and Boltzmann hanged himself while on a family holiday at Duino, near Trieste, on 5 September 1906. Ironically, unknown to Boltzmann, the work which would eventually convince even such doubters as Ostwald of the reality of atoms had been published the year before.

Albert Einstein: Avogadro's number, Brownian motion and why the sky is blue

The author of that work was the most famous patent clerk in history, Albert Einstein. We shall explain shortly how he came to be a patent clerk; what is relevant here is that in the early 1900s, Einstein was a brilliant young scientist (26 in 1905) working independently of the usual academic community, who was obsessed with the idea of proving that atoms are real. As he later wrote in his *Autobiographical Notes*,[2] at that time his interest was focused on the search for evidence 'which would guarantee as much as possible the existence of atoms of definite finite size'. This search was being carried out in the context of Einstein trying to obtain a PhD, which by the beginning of the twentieth century was already being seen as the scientist's meal ticket, an essential requirement for anyone hoping to pursue a career in university research. Einstein had graduated from the Eidgenössische Technische Hochschule (ETH – the Swiss Federal Institute of Technology) in

1. See Cercignani.
2. Available in an edition edited and translated by P. A. Schilpp and published by Open Court, La Salle, Illinois, 1979.

Zurich in 1900, but although he had done well in his final examinations, his attitude had not endeared him to the professors at the ETH (one of his tutors, Hermann Minkowski (1864–1909) described young Albert as a 'lazy dog' who 'never bothered about mathematics at all'), and he was unable to get a job as one of their assistants, and equally unable to get a decent reference from them for a junior academic post. So he had a variety of short-term and part-time jobs before becoming a patent officer in Bern in 1902. He spent a lot of his time working on scientific problems (not just his spare time, but also time at his desk when he should have been working on patent applications) and published several papers between 1900 and 1905. But his most important project was to obtain that PhD and reopen the doors to academia. The ETH did not award doctorates itself, but there was an arrangement whereby graduates from the ETH could submit a doctoral thesis to the University of Zurich for approval, and this is the path Einstein took. After an abortive attempt on a piece of work which he decided in the end not to submit, he was ready in 1905 with a paper that would prove entirely satisfactory to the examiners in Zurich,[1] and was the first of two papers in which he established the reality of atoms and molecules beyond reasonable doubt.

Scientists who accepted the idea of atoms had already found several rough and ready ways to estimate the sizes of these little particles, going right back to the work of Thomas Young (of whom more in Chapter 11) in 1816. Young worked out how to estimate the sizes of water molecules from studying the surface tension in a liquid – the elasticity on the surface of a glass of water that makes it possible (with great care!) to 'float' a steel needle on the surface. In terms of molecules, the surface tension is explained because the molecules of the liquid attract one another – they are sticky, if you like. In the bulk of the liquid, the attraction is felt evenly all around, but at the surface, molecules do not have any neighbours above them to pull them up, and the attraction only pulls them sideways and downwards, locking them into an elastic skin on the surface of the liquid. Young reasoned

1. Later, Einstein used to delight in telling listeners that the examiners had only one objection to the thesis, that it was too short. He claimed that he then added one sentence and it was accepted. This story should perhaps be taken with a pinch if salt.

that the strength of the resulting tension must be related to the range of the attractive force, which, as a first guess, could be regarded as the same as the size of the molecules. He calculated, from measurements of surface tension, that the size of what he called 'particles of water' must be 'between the two thousand and the ten thousand millionth of an inch', which corresponds to between 5000 and 25,000 millionths of a centimetre, only about ten times bigger than modern estimates. An impressive achievement coming just a year after the Battle of Waterloo, but not accurate enough or convincing enough to sway doubters.

Several more accurate estimates were made in the second half of the nineteenth century, but one will suffice as an example. In the mid-1860s, the Austrian chemist Johann Loschmidt[1] used a technique which is strikingly simple in principle. He argued that in a liquid, all the molecules are touching their neighbours, with no empty space between them, and the volume of the liquid is equal to the volume of all the molecules added together. When the same amount of liquid is vaporized to make a gas, the volume of all the molecules stays the same, but there is now empty space between them. From calculations of the mean free path, related to the measurable pressure of the gas and to Avogadro's number, he had an independent way to work out how much of the gas is actually empty space. The difficulty with this otherwise elegant idea in the 1860s, was liquefying gases such as nitrogen, and the densities of these liquids (the volume of a given mass) had to be estimated in various ways. Even so, combining the two sets of calculations, Loschmidt came up with an estimate of the size of molecules in air (a few millionths of a millimetre) and a value for Avogadro's number, about 0.5×10^{23} (or a five followed by twenty-two zeroes). He also defined another number important in the study of gases and related to Avogadro's number – the number of molecules in one cubic metre of gas under standard conditions of temperature and pressure. Now known as the Loschmidt number, modern measurements give this as 2.686763×10^{25}.

1. Loschmidt (1821–1895) just missed being even more famous than he is. He independently came up with many important ideas about the structure of organic molecules, but he had only published this work in a privately circulated pamphlet in 1861, which was overlooked when Kekulé went public with his similar ideas in the mid-1860s.

The way Einstein tackled the problem of determining the sizes of molecules in his thesis didn't use gases, however, but solutions – specifically, solutions of sugar in water. It also used the understanding of thermodynamics that had developed in the second half of the nineteenth century. It is a rather surprising fact when first encountered, but nonetheless true, that the molecules of a solution behave in some ways very much like the molecules of a gas. The way Einstein made use of this involves a phenomenon known as osmosis. Imagine a container which is half full of water, divided by a barrier which has holes in it just large enough to allow molecules of water to pass through. On average, the same number of molecules will pass through the barrier in each direction every second, and the level of liquid in the two halves of the container will be the same. Now put sugar into one side of the container, making a solution. Because sugar molecules are much bigger than water molecules, they cannot pass through the semi-permeable membrane, as the barrier is called. What happens to the level of the liquid on either side of the barrier? The first time they encounter this problem, most people guess that the presence of the sugar increases the pressure on that side of the barrier, pushing more water molecules through the membrane and making the level rise on the side without sugar. In fact, the opposite happens, in accordance with the second law of thermodynamics.

The simple version of the second law, that heat flows from a hotter object to a cooler object, is a specific example of the tendency of all differences in the Universe to average out (which is why things wear out). Heat flows out of hot stars, for example, into the cold of space in an attempt to even out the temperature of the entire Universe. A system in which there is a clear pattern (or even a vague pattern) has more order, and therefore less entropy, than a system in which there is no pattern (a black-and-white chessboard has lower entropy than a similar board painted a uniform shade of grey). You could express the second law by saying that 'nature abhors differences'. So, in the example just described, water moves through the membrane *into* the solution, diluting the strength of the sugar solution and thereby making it less different from the pure water remaining on the other side of the barrier. The level of liquid actually *rises* on the side of the barrier where the sugar is, and *falls* on the side where there is pure water. This continues until the extra pressure caused by the difference in height between the solution on one side of the barrier and the water on the other side is enough to counterbalance the pressure of water trying to pass through the membrane (the osmotic pressure). So the osmotic pressure can be determined simply by measuring the height difference once the system has settled down. The osmotic pressure itself depends on the number of molecules of the solute (in this case, sugar) in the solution; the more concentrated the solution the greater the pressure. The size of the molecules comes into the calculation in terms of the fraction of the volume of solution actually occupied by those molecules. And, once again, the mean free path of the molecules comes into the story, through its relation to the speed with which molecules diffuse through the membrane. Putting everything together, Einstein calculated in his thesis (which was published in a slightly revised form in 1906) that Avogadro's number was 2.1×10^{23} and that water molecules must be a few hundred millionths of a centimetre across. In the version published in 1906, with new data from more accurate experiments, he refined the value of Avogadro's number upwards, to 4.15×10^{23} and by 1911 to 6.6×10^{23} – but by then other experiments jumping off from another important paper by Einstein had already pinned Avogadro's number down to considerable accuracy.[1]

This second piece of work aimed at providing evidence 'which would guarantee as much as possible the existence of atoms of definite finite size' was completed and published in 1905, and provided a much simpler physical image of what was going on, which is one reason why it became the decisive evidence that eventually persuaded the last doubters of the reality of atoms. But it also introduced statistical techniques which were to have a profound relevance in many areas of physics over the following decades.

This classic paper by Einstein concerned the phenomenon known as Brownian motion – although Einstein did not set out with the intention of explaining Brownian motion, but rather worked out from

1. Perhaps it is worth reiterating that knowing Avogadro's number and, for example, the relative volumes of the same mass of a substance in its liquid and gaseous forms automatically gives you a measure of the size of the molecules, so when we refer to measurements of Avogadro's number, it is taken as read that we are describing measurements of the sizes of molecules.

first principles (his usual approach to any problem) how the existence of atoms and molecules might show up on a scale large enough to see, and *then* suggested that what he had described might correspond to the known phenomenon. He made his position clear in the opening paragraph of the paper:

> In this paper it will be shown that, according to the molecular-kinetic theory of heat, bodies of a microscopically visible size suspended in liquids must, as a result of thermal molecular motions, perform motions of such magnitude that they can be easily observed with a microscope. It is possible that the motions to be discussed here are identical with so-called Brownian molecular motion; however, the data available to me on the latter are so imprecise that I could not form a judgement on the question.[1]

Brownian motion gets its name from the Scottish botanist Robert Brown (1773–1858), who noticed the phenomenon while studying pollen grains through a microscope in 1827. He observed that these grains (which typically have a diameter less than half a hundredth of a millimetre) move around in a jerky, zig-zag fashion when floating in water. At first, it seemed that this was because the grains were alive, and swimming in the water; but it soon became clear that tiny grains of anything suspended in liquid (or in air) moved in the same way, even when the particles (such as smoke particles floating in air) clearly had no connection with living things. In the 1860s, as the atomic hypothesis gained strength, several people suggested that the motion might be caused by the impact of molecules with the grains – but for an individual molecule to produce a measurable 'kick' on a pollen grain it would have to be a good fraction of the size of the grain, which was clearly ridiculous. Later in the nineteenth century, the French physicist Louis-Georges Gouy (1854–1926) and William Ramsay (1852–1916) in England independently suggested that a better way to explain the Brownian motion might be in statistical terms. If a particle suspended in water or air is constantly being bombarded by large numbers of molecules from all sides, on average, the force it feels will

1. Translation from *Einstein's Miraculous Year*, ed. John Stachel, which reprints in English, with commentary, all of Einstein's classic papers from 1905.

be the same from all directions. But from time to time, just by chance, there will be more molecules hitting it on one side than another, making the grain jerk away from the side where there is excess pressure. But they did not follow this idea through in any detail, and when Einstein developed his similar ideas in a proper statistical fashion he was almost certainly unaware of these earlier suggestions (Einstein was notorious for working ideas out for himself from first principles without reading up the background to the subject thoroughly in the published literature).

The reason why Einstein's paper was so influential was precisely because it was precise – it gave an accurate mathematical and statistical account of the problem. You might think that because, on average, the pressure on a pollen grain is the same from all sides, it ought to stay in more or less the same place, jiggling on the spot. But each jerky movement is random, so after the grain has jerked a little bit in one direction, there is an equal probability of it being jerked again in the same direction, or back to where it came from, or in any other direction. The result is that it follows a zig-zag path in which the distance it travels from its starting point (measured in a straight line across all the zigs and zags) is always proportional to the square root of the time that has elapsed since the first kick. This is true wherever you start measuring from (whichever jerk you count as the first kick). The process is now known as a 'random walk', and the statistics behind it (worked out by Einstein) turn out to be important, for example, in describing the decay of radioactive elements.

Einstein had put numbers into the calculation and made a prediction, based on 'the molecular-kinetic theory', that could be tested by observations if any microscopist were up to the job of observing Brownian motion in sufficient detail. Einstein came up with an equation that linked Avogadro's number with the speed with which molecules move and the measurable rate at which particles drift away from their starting point through Brownian motion. Using a value for Avogadro's number of 6×10^{23} (not pulled out of a hat, but based on yet another piece of work he had been carrying out in 1905, of which more late[1]) for a particle one

1. The modern value for Avogadro's number (now also called the Avogadro constant) is 6.022×10^{23}.

thousandth of a millimetre across suspended in water at 17 °C, he predicted a drift of six thousandths of a millimetre in one minute (the particle will move twice as far in four minutes, four times as far in sixteen minutes and so on). The challenge of measuring such slow drift to the required accuracy was taken up by the Frenchman Jean Perrin (1870–1942), who published his results in the later years of the first decade of the twentieth century, prompting Einstein to write to him, 'I would have considered it impossible to investigate Brownian motion so precisely; it is a stroke of luck for this subject that you have taken it up.' And it's a sign of just how important this proof of the reality of atoms and molecules was at the time that in 1926 Perrin received the Nobel prize for this work.

Einstein himself still wasn't quite finished with the search for proof of the existence of atoms and the quest for Avogadro's number. In October 1910, he wrote a paper explaining how the blue colour of the sky is caused by the way light is scattered by the molecules of gas in the air itself. Blue light is more easily scattered in this way than red or yellow light, which is why the blue light from the Sun comes to us from all directions on the sky (having been bounced from molecule to molecule across the heavens), while the direct light from the Sun is orange. Back in 1869, John Tyndall had discussed this kind of light scattering, but in terms of the effect of particles of dust in the air on the light – this scattering by dust, taking even more of the blue out of sunlight, is the reason why the Sun seems even redder at sunrise and sunset. Other scientists correctly suggested that it is the molecules of the air, not dust suspended in air, that causes the sky to be blue; but it was Einstein who put the numbers into the calculation, using the blueness of the sky to calculate Avogadro's constant in yet another way, and at the same time provide supporting evidence, if any were still needed in 1910, for the reality of atoms and molecules.

Charming though this piece of work is, however, it pales alongside the work for which Einstein is best remembered, which also deals with light but in a much more fundamental way. In order to set the special theory of relativity in context, we need to go back to look at the way an understanding of the nature of light developed in the nineteenth century, and how this led Einstein to appreciate the need for a modification to the most hallowed maxims of science, Newton's laws of motion.

Let There be Light

The wave model of light revived – Thomas Young: his double-slit experiment – Fraunhofer lines – The study of spectroscopy and the spectra of stars – Michael Faraday: his studies in electromagnetism – The invention of the electric motor and the dynamo – Faraday on the lines of force – Measuring the speed of light – James Clerk Maxwell's complete theory of electromagnetism – Light is a form of electromagnetic disturbance – Albert Michelson and Edward Morley: the Michelson–Morley experiment on light – Albert Einstein: special theory of relativity – Minkowski: the geometrical union of space and time in accordance with this theory

Until the end of the eighteenth century, Newton's conception of light as a stream of particles dominated over the rival wave model of light, as much through the influence of Newton's stature as the scientific oracle as through any evidence that the particle model really was better than the wave model. But over the next hundred years or so a new understanding of light developed which first showed that even Newton was by no means infallible in his pronouncements and then, in the early twentieth century, that even his laws of motion were not the last word in mechanics. Newton's influence really did hold back progress in this regard since quite apart from the work of Huygens which we mentioned earlier, by the end of the eighteenth century there was plenty of observational evidence which, had it been followed up with more enthusiasm, might have led to the establishment of the wave model a couple of decades sooner than actually happened. Indeed, there was already some evidence that light travels as a wave even before Newton came on the scene, although the significance of the work was not widely appreciated at the time. It came from the Italian physicist Francesco Grimaldi (1618–1663), professor of mathematics at the Jesuit college in Bologna, who, as Newton was to do later, studied light by letting a beam of sunlight into a darkened room through a small hole. He found that when the beam of light was passed through a second small hole and on to a screen, the image on the screen formed by the spot of light had coloured fringes and was slightly larger than it should be if the light had travelled in straight lines through the hole. He concluded (correctly) that the light had been bent outwards slightly as it passed through the hole, a phenomenon to which he gave the name 'diffraction'. He also found that when a small object (such as a knife edge) was placed in the beam of light, the shadow cast by the object had coloured edges where light had been diffracted around the edge of the object and leaked into the shadow.[1] This is direct evidence that light travels as a wave, and the same sort of effect can be seen when waves on the sea, or on a lake, move past obstructions or through gaps in obstructions. But when light is involved, because the wavelengths are so small, the effects are tiny and can only be detected by very careful measurement. Grimaldi's work was not published until two years after he died, in a book called *Physico-mathesis de lumine, coloribus, et iride*; he was not around to promote or defend his ideas, and the few people who noticed the book at the time may have been unable (or unwilling) to carry out the delicate experiments needed to confirm the results. One reader of the book who should perhaps have realized its significance, was Newton himself, who was 21 when Grimaldi died – but he seems not to have appreciated the power of Grimaldi's proofs that neither reflection nor refraction could explain the observed phenomena. It is tantalizing, but ultimately fruitless, to speculate how science might have developed had Newton been converted to the wave model by Grimaldi's book.

The wave model of light revived

After Newton died in 1727, although the particle model of light dominated thinking for the rest of the eighteenth century, there were people who considered the alternative, most notably the Swiss mathematician Leonhard Euler, who we mentioned earlier. Euler is usually remembered for his work in pure mathematics, where he developed the idea of the principle of least action (which says, in effect, that nature is lazy; one manifestation of this is that light travels in straight lines, by the shortest route). This pointed the way for the work of Joseph Lagrange (1736–1813), which in turn provided the basis for a mathematical description of the quantum world in the

1. The fringes are coloured, like a rainbow, because different wavelengths of light are bent by different amounts, and different wavelengths correspond to different colours.

twentieth century. Euler, as we mentioned, introduced mathematical notations such as π, e and i, and he is also the archetypal example of the dangers of looking directly at the Sun. In 1733, when he was professor of mathematics in St Petersburg, this piece of foolishness cost him the sight of his right eye. This was doubly unfortunate since in the late 1760s he went blind in his left eye because of a cataract, but none of this even slowed down his prodigious mathematical output.

Euler published his model of light in 1746, while he was working at Frederick the Great's Academy of Sciences in Berlin (he was later called back to St Petersburg by Catherine the Great, and finished his life there). Much of the power of Euler's argument lies in the way he carefully marshalled all of the difficulties with the corpuscular model, including the difficulty of explaining diffraction in this way, as well as spelling out the evidence in support of the wave model. He specifically made the analogy between light waves and sound waves, and in a letter written in the 1760s he said that sunlight is 'with respect to the ether, what sound is with respect to air' and described the Sun as 'a bell ringing out light'.[1] The analogy, although graphic, is at best imperfect, and shows us now how far the development of the wave model still had to go in the middle of the eighteenth century; it is no real surprise that the world of physics was not persuaded to shift its view on the nature of light until the improved experimental techniques of the nineteenth century put the matter beyond doubt. But the first person to give a significant nudge to that shift of opinion was already ten years old when Euler died in 1773.

Thomas Young: his double-slit experiment

Thomas Young was born at Milverton, Somerset, on 13 June 1773. He was an infant prodigy who read English at the age of 2, Latin when he was 6, and rapidly moved on to Greek, French, Italian, Hebrew, Chaldean, Syriac, Samarian, Arabic, Persian, Turkish and Ethiopic – all by the time he was 16. Coming from a wealthy family (he was the son of a banker), Young had the freedom to do more or less as he liked, and he received little formal education as a child and teenager. He clearly did not need it, studying widely on his own, with an early interest (as the list of languages suggests) in ancient history and the archeology of the Middle

1. Quoted by Zajonc.

East, but also learning physics, chemistry and much more besides. At the age of 19, influenced by his great-uncle Richard Brocklesby (1722–1797), a distinguished physician, Young started to train for the medical profession, intending to join (and in due course take over) his great-uncle's practice in London. He studied in London, Edinburgh and Göttingen, where he was awarded his MD in 1796, and travelled in Germany for several months before settling for a time in Cambridge (his great-uncle having just died). By then, he was already well known in scientific circles, having explained the focusing mechanism of the eye (the way muscles change the shape of the lens of the eye) when in his first year as a medical student, and being elected a Fellow of the Royal Society at the age of 21 as a result. During two years in Cambridge, where he resided at Emmanuel College, he gained the nickname 'Phenomenon' Young for his ability and versatility. But Richard Brocklesby had left Young his London house and a fortune, and in 1800 the young man, now 27, returned there to set himself up in medical practice. Although he remained active in medicine for the rest of his life, becoming a physician at St George's Hospital from 1811 until his death (on 10 May 1829), this did not prevent him from continuing to make wide-ranging and important contributions to science. There was, though, just one hint of fallibility – from 1801 to 1803 Young gave lectures at the RI, but these were not a success, going over the heads of most of his audience.

Among his many interests, Young correctly explained astigmatism as caused by uneven curvature in the cornea of the eye, was the first person to appreciate that colour vision is produced by a combination of three primary colours (red, green and blue) which affect different receptors in the eye (and thereby explained colour blindness as due to a failure of one or more sets of these receptors), estimated the sizes of molecules (as we saw in the previous chapter), served as Foreign Secretary to the Royal and played a leading part in deciphering the Rosetta Stone, although he did not immediately receive full credit for this because the work was published anonymously, in 1819. But what matters to us here is the work for which Young is best remembered, his experiments involving light which proved that light travels as a wave.

Young began to experiment with the phenomenon of interference

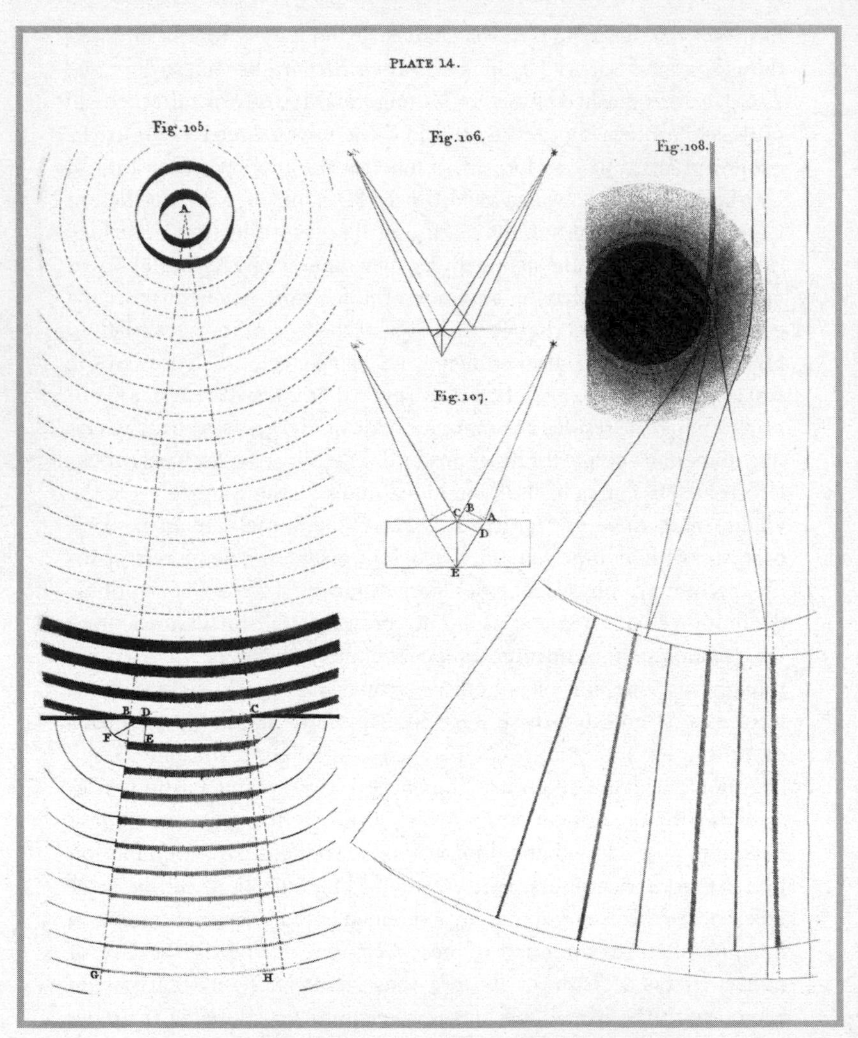

32. *Young's drawings showing how light waves propagate. From Young's* A Course of Lectures on Natural Philosophy and Mechanical Arts, *1807.*

in light while he was at Cambridge at the end of the 1790s. In 1800, in his *Outlines of Experiments and Enquiries Respecting Sound and Light*, he compared and contrasted the rival models of Newton and Huygens and 'came out' in support of the wave model of Huygens, proposing that different colours of light correspond to different wavelengths. In 1801, he announced his key contribution to the debate, the idea of interference in light waves. This is exactly like the way waves on a pond interfere with one another (for example, if two pebbles are tossed into a still pond at the same time but in different places) to produce a complicated pattern of ripples. Young first explained how phenomena Newton had observed himself, such as Newton's rings, could be explained by interference, and used Newton's own experimental data to calculate the wavelength of red light as 6.5×10^{-7} metres (in modern units) and violet light as 4.4×10^{-7} metres. These numbers agree well with modern measurements, which shows how good an experimenter Newton was and how good a theorist Young was. Young then went on to devise and carry out the experiment which bears his name, Young's double-slit experiment.

In the double-slit experiment, light (ideally of a pure colour – that is, a single wavelength – although this is not absolutely necessary) is passed through a narrow slit in a piece of card ('narrow' means the slit has to be about as wide as the wavelength of the light, roughly a millionth of a metre, so a slit made by a razor is appropriate). Light from the slit spreads out and falls on a second piece of card in which there are two similar parallel slits. Light from these two slits spreads out in its turn and falls on a screen where it makes a pattern of light and shade, called an 'interference pattern'. Young explained that there is light where the waves arriving from each of the two slits march in step, so that the peaks in both waves add together; there is dark where the waves from the two slits are out of step with each other (out of phase) so the peak in one wave is cancelled by the trough in the other wave. The exact spacing of the pattern seen on the screen depends on the wavelength of the light, which can be calculated by measuring the spacing of the stripes in the pattern. There is absolutely no way to explain this phenomenon by treating light as like a stream of tiny cannon balls whizzing through space. Young essentially completed this work by about 1804, and in 1807 he wrote:

The middle (of the pattern) is always light, and the bright stripes on each side are at such distances, that the light coming to them from one of the apertures must have passed through a longer space than that which comes from the other by an interval which is equal to the breadth of one, two, three, or more of the supposed undulations, while the intervening dark spaces correspond to a difference of half a supposed undulation, of one and a half, of two and a half, or more.[1]

Ten years later, Young refined his model further by suggesting that light waves are produced by a transverse 'undulation', moving from side to side, rather than being longitudinal (push–pull) waves like those of sound. Far from convincing his peers, though, Young's work on light brought him abuse from his physicist colleagues in Britain, angered by the suggestion that anything Newton said could be wrong, who ridiculed the idea that you could make darkness by *adding* two beams of light together. Young, who had many other irons in various fires, suffered little as a result, and the progress of science was not held up because similar evidence in support of the wave model came almost immediately from (perhaps appropriately) Britain's bitterest foe at the time, France.

Augustin Fresnel was born on 10 May 1788 in Broglie, Normandy. He was the son of an architect, who retreated to his country estate near Caen to escape the turmoil of the French Revolution (shades of the d'Héricy family and Georges Cuvier), and was educated at home until he was 12. He then studied at the École Centrale in Caen, before moving to Paris in 1804 to study engineering. In 1809, he qualified as a civil engineer, working for the government on road projects in various parts of France, while developing an interest in optics alongside his day job. But as Fresnel was outside the circle of academic scientists in Paris, he seems not to have learned of Young's work; more surprisingly, he seems also to have been unaware of the work of Huygens and Euler, eventually developing his own wave model of light completely from scratch. Fresnel got the opportunity to develop that model partly through politics. Although he had worked as a government employee

1. From Young's work *A Course of Lectures on Natural Philosophy and the Mechanical Arts*, quoted by Baierlein.

under the Napoleonic regime without making waves, when Napoleon was defeated by the allies and exiled to Elba, Fresnel, like many of his contemporaries, revealed himself to a be a Royalist. When Napoleon returned briefly from exile for the Hundred Days in 1815, Fresnel was either sacked from his post or quit in protest (there are conflicting accounts), and was sent home to Normandy, where he was placed under house arrest. It was there that he had just enough time to develop his ideas before Napoleon was finally overthrown and Fresnel was able to go back to his work as an engineer, leaving the optics to be relegated, once again, to what amounted to a hobby.

Fresnel's approach to a wave model of light was also based on diffraction, but using a single narrow slit to cast light on to a screen. If the slit is narrow enough, this produces its own characteristic stripy pattern of light and shade on the screen. Without going into details, the simplest way to think of how this happens is to imagine light being bent around either side of the slit slightly, spreading out from each edge and travelling by two slightly different paths, each corresponding to a different number of wavelengths, to the screen. But you can also turn this experiment inside out by placing a small obstruction (such as a needle) in the path of a beam of light. The light then bends around the obstruction (just as waves in the sea bend around a rock sticking up out of the water) to produce a diffraction pattern in the shadow of the obstruction.

It shows how little Young's work had been appreciated that in 1817, the French Academy, although aware of that work, offered a prize for whoever could provide the best experimental study of diffraction and back it up with a theoretical model to explain what was going on. The competition produced only two entries. One was so clearly nonsense that the Academy has not even kept a record of the name of the person who entered it, let alone the details of the entry itself. The other came from Fresnel, in the form of a paper 135 pages long. It had one rather large obstacle to overcome – it was, of course, a wave model, and the three judges for the competition, the mathematician Siméon-Denis Poisson (1781–1840), the physicist Jean Baptiste Biot (1774–1862) and the astronomer-mathematician Pierre Simon Laplace, were all confirmed Newtonians and therefore favoured the corpuscular model. Their efforts focused on finding a flaw in Fresnel's model, and Poisson, a considerable mathematician, thought he had one. He calculated that according to Fresnel's wave model of light, if a small round object (like a piece of lead shot) was placed in the path of a beam of light, the light bending around the object should produce a bright spot exactly behind the centre of the object, where common sense said there ought to be the darkest shadow. This seemed as crazy to him as the idea that adding two beams of light together could make darkness had to the British opponents of Young's work. But the calculations were unambiguous. As Poisson himself wrote:

Let parallel light impinge on an opaque disc, the surroundings being perfectly transparent. The disc casts a shadow – of course – but the very centre of the shadow will be bright. Succinctly, there is no darkness anywhere along the central perpendicular behind an opaque disc (except immediately behind the disc). Indeed, the intensity grows continuously from zero right behind the thin disc. At a distance behind the disc equal to the disc's diameter, the intensity is already 80 per cent of what the intensity would be if the disc were absent. Thereafter, the intensity grows more slowly, approaching 100 per cent of what it would be if the disc were not present.[1]

To the judges, that seemed absurd, but it was what Fresnel's model predicted. As good scientists, in the best Newtonian tradition, they and the chairman of the panel set up to oversee the competition, the physicist François Arago (1786–1853), arranged for an experiment to be carried out to test the prediction. The predicted bright spot was exactly where Poisson, using Fresnel's model, had predicted it would be. In March 1819, Arago reported to the Council of the Academy of Sciences that:

One of your commissioners, M. Poisson, had deduced from the integrals reported by the author [Fresnel] the singular result that the centre of the shadow of an opaque circular screen must . . . be just as illuminated as if the screen did not exist. The consequence has been submitted to the test of direct experiment, and observation has perfectly confirmed the calculation.

1. See Baierlein. The next quote is from the same source.

The very scientific method which Newton had made the foundation of the investigation of the world, the 'test of direct experiment', had proved that Newton was wrong, and light travels as a wave. From that moment on, the wave model of light had to be upgraded to a theory from a hypothesis. Fresnel's reputation was assured, and although only a part-time scientist, he carried out important work with Arago on developments of the wave theory of light, was elected to the French Academy in 1823 and was made a Fellow of the Royal Society in 1825. In 1827, he received the Rumford Medal, a hundred years after the death of Newton and a few days before his own death, from TB, on 14 July that year. It took several decades for the wave theory of light to be properly worked out, and in particular for physicists to appreciate just what it was that was waving. But this did not inhibit practical progress in using light. Fresnel himself developed, originally for use in lighthouses, an efficient lens made of concentric annular rings of glass, each with a slightly different curvature (the Fresnel lens); and light itself was becoming possibly the most valuable tool in science, through the young discipline of spectroscopy.

Spectroscopy is such a valuable and important tool of science, a comfortable part of the furniture, that it is almost a surprise to learn that it has not always been available and only began to be understood at the beginning of the nineteenth century. It's almost like being told that before 1800, nobody knew that the Pope was a Catholic. But like so much scientific progress, spectroscopy had to await the development of the appropriate technology for the job – in this case, the combination of a prism or other system of spreading light out into its rainbow spectrum of colours and a microscope that could be used to scan that spectrum in detail.

Fraunhofer lines

When light is studied in this way, it can be seen that there are many distinct, sharp lines in the spectrum – some bright and some dark. The first person to notice this was the English physicist and chemist William Wollaston (1766–1828), who passed light from the Sun through a prism and studied the magnified spectrum in which, in 1802, he saw some dark lines. Wollaston was a good all-round scientist of the second rank, who discovered the elements rhodium and palladium, and was an early supporter of Dalton's atomic theory, but never quite made a major contribution to science. Somehow, he never followed up his discovery of dark lines in the spectrum of light from the Sun, and it was left for the German industrial physicist Josef von Fraunhofer (1787–1826) to make the same discovery independently in 1814 and (crucially) to follow the discovery up with a proper investigation of the phenomenon – which is why the lines in the solar spectrum are now called Fraunhofer lines, not Wollaston lines. Fraunhofer also, in 1821, invented another technique for spreading light out into a spectrum, the diffraction grating (which, as its name suggests, depends for its action entirely on the wave nature of light). All of this, though, came about because Fraunhofer worked in the optical laboratory of the Munich Philosophical Instrument Company, where he was trying to improve the quality of glass used in lenses and prisms for scientific work and in the high-tech industries of the day. His skill made the fortunes of the company and laid the foundations for Germany to become pre-eminent in the manufacture of optical systems for the best part of a century.

The study of spectroscopy and the spectra of stars

One of the first spectroscopic discoveries he made was that there are two bright yellow lines, each (it soon became clear) at a particular well-defined wavelength, in the spectrum of light from a flame. In 1814, Fraunhofer was using the two bright yellow lines (now known to be produced by sodium, and responsible for the colour of yellow street lights) as a source of pure monochromatic light with which to test the optical properties of different kinds of glass. It was when he compared the effect of the glass on this light with its effect on sunlight that he noticed the dark lines in the solar spectrum, and thanks to the superior quality of his instruments he saw many more than Wollaston had, counting a total of 576 between the red and violet ends of the spectrum, and mapping the wavelengths of each of them. He also noticed similar lines in the spectra of Venus and the stars. And by showing that the same lines are present, at the same wavelengths, in spectra obtained using diffraction gratings, he proved that they were a property of the light itself, not a phenomenon produced by the glass in the prisms as the light passed through. He never found out what made the lines, but it was Fraunhofer who perfected the use of spectroscopy in science.

Although many people investigated the newly discovered phenomenon, the key developments were also made in Germany, by Robert

Bunsen (1811–1899) and Gustav Kirchoff (1824–1887), working together at Heidelberg in the 1850s and 1860s. It is no coincidence that this is the same Robert Bunsen who lent his name to perhaps the most familiar piece of laboratory equipment, since the eponymous burner was a key tool in the development of spectroscopy.[1] When a substance is heated in the clear flame of a bunsen burner, it provides a characteristic colour to the flame, depending on what is being heated (you should not be surprised to learn that substances containing sodium, such as common salt, colour the flame yellow). Even without spectroscopy, this provides a simple way to test whether particular elements are present in a compound. But with the aid of spectroscopy, you can go further than saying that one element colours the flame yellow, another green and a third pink; you can see that each element, when hot, produces a characteristic pattern of bright lines in the spectrum, like the pair of yellow lines associated with sodium. Then, anywhere that you see those lines in a spectrum, you know that the element associated with those lines is present – even if, as was still the case in the nineteenth century, you do not know how the atoms involved make those lines. Each pattern is as distinctive as a fingerprint or a barcode. Where a substance is hot, it produces bright lines as it radiates light; when the same substance is present but cold, it produces dark lines in the spectrum[2] as it absorbs background light at precisely the same wavelengths where it would radiate light if it were hot. By running a succession of 'flame tests' in the laboratory with different elements, it was straightforward to build up a library of the characteristic spectral patterns associated with each of the known elements. In 1859, Kirchoff identified the characteristic sodium lines in light from the Sun – proof that there is sodium present in the atmosphere of our neighbourhood star. Other lines in the solar spectrum, and then in the spectra of stars, were soon identified with other elements. In the most striking example of the power of spectroscopy,

1. Although, in fact, the basic form of the burner was invented by Michael Faraday, and was improved by Bunsen's assistant Peter Desdega, who had the nous to market it under his boss's name.

2. 'Cold' is a relative term. The Fraunhofer lines are dark because although the gas in the atmosphere of the Sun is hot, it is not as hot as the surface of the Sun itself, where the light is coming from.

it enabled astronomers to find out what the stars are made of. In one striking reversal of this process, during a solar eclipse in 1868, the French astronomer Pierre Jansen (1824–1907) and the English astronomer Norman Lockyer (1836–1920) found a pattern of lines in the solar spectrum that did not match the 'fingerprint' of any known element on Earth; Lockyer inferred that they must belong to a previously unknown element, which he dubbed 'helium', from Helios, the Greek word for the Sun. Helium was only identified on Earth in 1895. By then, though, the puzzle of the nature of light seemed to have been completely resolved, thanks to the understanding of electricity and magnetism which had grown out of the work of Humphry Davy's former assistant, Michael Faraday, and been completed by James Clerk Maxwell in what was regarded as the most profound piece of new physics since the time of Newton.

Michael Faraday: his studies in electromagnetism

Faraday is almost unique among the major names in science in that he did nothing of any significance before he was 30, then made one of the most important contributions of anyone in his generation (or, indeed, any other generation), with his best work being carried out after he was 40. There have been rare examples of scientists who remained active at the highest level in later years (Albert Einstein is the obvious example), but even they had shown signs of unusual ability when they were in their twenties. Almost certainly, given his later achievements, Faraday would have done so as well; but his circumstances prevented him from even getting a start in scientific research until he was 25 – an age when Einstein had already produced not only the work on atoms mentioned in the previous chapter but also both his special theory of relativity and the work for which he would later receive the Nobel prize.

The Faraday family came from what was then Westmorland, in the north of England. Michael's father, James, was a blacksmith, who moved south with his wife Mary and two small children (Robert, born in 1788, and Elizabeth, born in 1787) in search of work in 1791. The family settled briefly in Newington, which was then a village in Surrey but has now been swallowed up by London, where Michael was born on 22 September 1791. Soon, though, the family moved into London itself, settling in rooms over a coach house in Jacob's Well Mews, near Manchester Square, where another child, Margaret, was born in 1802.

Although James Faraday was a good blacksmith, he suffered from ill health and was often unable to work (he died in 1810), so the children were raised in poverty, with no money to spare for luxuries such as education beyond the level of the basics of reading, writing and arithmetic (even that, though, distinguished them from the poorest people of the time). But the family was a close and loving one, sustained in no small measure by their religious faith as members of a sect, the Sandemanians, which had originated in the 1730s as a breakaway from the Scottish Presbyterians. Their firm belief in salvation helped to make it easier to endure the hardships of life on Earth, and the sect's teaching of modesty, abhorrence of ostentation or showing off, and commitment to unobtrusive charitable work all helped to colour Faraday's life.

When he was 13, Michael started to run errands for George Riebeau, a bookseller, bookbinder and newsagent who had a shop in Blandford Street, just off Baker Street, not far from where the Faraday family lived. A year later, he was apprenticed to Ribeau to learn bookbinding, and soon moved in to live above the shop. Although little is known about Faraday's life over the next four years, some idea of the happy family atmosphere in Ribeau's shop (and his benevolence as an employer) can be gleaned from the fact that of his three apprentices at the time, one went on to become a professional singer, a second later made a living as a comedian in the music halls, while Faraday read voraciously from the piles of books available to him and went on to become a great scientist. His fascination with electricity, the area in which he later made his greatest contributions to science, was, for example, first stirred by reading an article on the subject in a copy of the third edition of the *Encyclopaedia Britannica*, which had been brought to the shop for binding.

In 1810, the year his father died,[1] Faraday became a member of the City Philosophical Society, which (in spite of its grand name) was a group of young men eager for self-improvement, who met to discuss the issues of the day, including the exciting new discoveries in science, taking turns to lecture on a particular topic (Faraday's subscription of

1. Faraday's mother lived until 1838, amply long enough to see her son become one of the greatest scientists of his generation.

a shilling was paid by his brother Robert, about to become head of the family and working as a blacksmith). Through his discussions and correspondence with the friends that he made there, Faraday began to develop both his scientific knowledge and his personal skills, working assiduously to improve his grammar, spelling and punctuation. He carried out experiments in both chemistry and electricity, discussing them with his fellow 'City Philosophers', and also took detailed notes of the topics discussed at the meetings, which he carefully bound up. By 1812, when he was approaching the age of 21 and facing up to the end of his apprenticeship, he had four volumes of this work, which the indulgent Ribeau, delighted by the presence of the young philosopher in his household, used to show off to his friends and customers. One of those customers was a Mr Dance, who was so impressed that he asked to borrow the books to show to his father, who had an interest in science; the elder Mr Dance was so impressed that he gave Faraday tickets to attend a series of four lectures on chemistry being given by Humphry Davy at the RI (as it turned out, the last lecture course Davy gave there), in the spring of 1812. These in turn received the Faraday treatment, and were written up carefully, with accompanying diagrams, and bound in a book, which was also shown to the elder Mr Dance, who was delighted at the response to his generosity.

But although the lectures had confirmed Faraday's burning desire to become a scientist, there seemed no way in which the dream could become a reality. His apprenticeship ended on 7 October 1812, and he began to work as a bookbinder for a Mr De La Roche, who has come down to us as a difficult employer, but was probably just an ordinary businessman who expected his employees' minds to be on their work. Faraday's certainly was not – he wrote to everyone he could think of (including the President of the Royal Society, Sir Joseph Banks, who didn't bother to reply), asking how he could get even the most menial job in science, but to no avail. Within a few weeks, however, he had the stroke of luck that would change his life. Davy was temporarily blinded by an explosion in his lab and needed somebody with a little knowledge of chemistry who could act as his secretary for a few days. Faraday got the job (very probably on the recommendation of the elder Mr Dance). There is no record of how he got time off from work to carry out these duties, but the fact that he did, suggests that Mr De La Roche was not as black as he is sometimes painted. When Faraday had to go back to his trade after Davy recovered, he sent Davy the bound notes from the lectures he had attended in the spring, with a letter asking (virtually begging) to be considered for even the most modest job at the RI.

There were no openings – but then came the second instalment of the stroke of luck. In February 1813, William Payne, the laboratory assistant at the RI and a man with a fondness for drink, had to be dismissed after assaulting the instrument maker (we do not know the reason for the brawl). Davy offered Faraday the post, with the warning that 'science was a harsh mistress, and in a pecuniary point of view but poorly rewarding those who devoted themselves to her service'.[1] Faraday didn't care. He accepted the job for a guinea a week, plus accommodation in two rooms at the top of the RI building in Albermarle Street, with candles and fuel for his fire included (the pay was actually slightly less than he had been earning as a bookbinder). He took up the post on 1 March 1813 and became (among other things) literally Humphry Davy's bottle washer. But he was from the outset always much more than just a bottle washer and worked with Davy on just about all the experiments Davy carried out in his remaining time at the RI.

Faraday's value as an assistant can be seen from the fact that barely six months later Davy asked Faraday to accompany him and his wife on their tour of Europe, as Davy's scientific assistant. The reason why the French were willing to provide passports for Davy's party was that this was presented as a scientific expedition to investigate, among other things, the chemistry of volcanic regions. It certainly was a scientific expedition, but the presence of Lady Davy made it something of a honeymoon as well, and posed some problems for Faraday. Because Davy's valet refused, at the last moment, to venture into Napoleonic France, Faraday was asked to double up his work and carry out those duties, as well as assisting with the chemistry. This would probably have worked out reasonably well if Davy had not been accompanied by his wife, but Lady Davy seems to have taken the mistress/servant relationship seriously, making life so hard for Faraday that at times he

1. See Hartley.

was seriously tempted to abandon the expedition and return home. But he stuck it out in the end, and the experience changed his life for the better.

Before the party set off, on 13 October 1813, Faraday was a naive young man who had never travelled more than 20 kilometres from the centre of London. By the time they returned, a year and a half later, he had met many of the leading scientists in France, Switzerland and Italy, seen mountains and the Mediterranean (as well as the telescope used by Galileo to discover the moons of Jupiter), and become Davy's scientific collaborator, not just his assistant. He had learned to read French and Italian, and to speak French adequately. His improved abilities were immediately recognized back at the RI. In order to go on the trip, Faraday had been obliged to resign his post at the RI, only six months after taking it up, but with a guarantee that he would be re-employed on no less favourable terms when he returned. In fact, he was appointed superintendent of the apparatus and assistant in the laboratory and Mineralogical Collection, with his pay increased to 30 shillings a week and better rooms at the RI. As Davy, as we have seen, withdrew from the day-to-day work at the RI, Faraday grew in stature, establishing a reputation as a solid, reliable chemist, but not yet showing any signs of brilliance. On 12 June 1821, in his thirtieth year, he married Sarah Barnard, another Sandemanian (the Sandemanians don't seem to have got out much; five years later Michael's sister Margaret married Sarah's brother John) and the couple lived together 'above the shop' in Albermarle Street until 1862 (they never had children). It was about this time that Faraday first investigated the electrical phenomena that would make him famous, although even then the work would not be followed up for a decade.

In 1820, the Dane Hans Christian Oersted (1777–1851) had discovered that there is a magnetic effect associated with an electric current. He noticed that when a magnetic compass needle is held over a wire carrying an electric current, the needle is deflected to point across the wire, at right angles. This was utterly unexpected, because it suggested that there was a magnetic force operating in a circle (or series of circles) around the wire, quite different from the familiar push–pull forces with which bar magnets attract and repel one another, and the way static electricity and gravity also operate as straightforward forces of attraction (in the case of static electricity, repulsion as well). As the sensational news spread around Europe, many people repeated the experiment and tried to come up with an explanation for what was going on. One of them was William Wollaston, who came up with the idea that electric current travels in a helix down a wire, like a child sliding down a helter-skelter, and that it was this twisting current that gave rise to the circular magnetic force. According to his argument, a wire that was carrying an electric current ought to spin on its axis (like a very thin spinning top) when brought near to a magnet. In April 1821, Wollaston visited the RI and carried out some experiments with Davy to search for this effect, but failed to find it. Faraday, who had not been present for the experiments, joined in their discussion afterwards.

Later in 1821, Faraday was asked by the journal *Annals of Philosophy* to write a historical account of Oersted's discovery and its aftermath. Being a thorough man, in order to do the job properly he repeated all of the experiments that he intended to describe in the article. In the course of this work, he realized that a wire carrying an electric current should be forced to move in a circle around a fixed magnet, and devised both an experiment to demonstrate this and an experiment in which a magnet moved around a fixed wire carrying electric current. 'The effort of the wire,' he wrote, 'is always to pass off at a right angle from the pole [of the magnet], indeed to go in a circle round it.' This was quite different from the (non-existent) effect discussed by Wollaston, but when Faraday's paper was published in October 1821 some people who had only a vague idea what Wollaston had been on about (and even Davy, who ought to have known better) thought that Faraday had either merely proved Wollaston right, or that he was trying to steal the credit for Wollaston's work. This unpleasantness may have been a factor in Davy attempting to prevent Faraday becoming a Fellow of the Royal Society in 1824; the fact that Faraday was elected by such an overwhelming majority shows, though, that more perspicacious scientists fully appreciated the significance and originality of his work. Indeed, the discovery, which forms the basis of the electric motor, made Faraday's name across Europe. It's a sign of just how important the discovery was, and of the pace of technological change at the time, that just sixty years after Faraday's toy demonstration of a single wire circling around a stationary magnet, electric trains were running in Germany, Britain and the USA.

Faraday did very little else in electricity and magnetism in the 1820s (at least he made no real progress when he did, from time to time, briefly try to tackle the subject) but carried out sound work in chemistry, being the first person to liquefy chlorine (in 1823) and discovering the compound now called benzene (in 1825), which is important because it turned out to have the archetypal ring structure later explained by Kekulé and was found in the twentieth century to be of key importance in the molecules of life. He also became director of the laboratory at the RI (which effectively meant he was running the place) in 1825, as Davy's successor, and in the later 1820s enhanced the fortunes of the RI by introducing new series of popular lectures (many of which he gave himself) and the Christmas lectures for children. The wonder is not that he did not return to studying electricity and magnetism thoroughly for so long, but that he found time to do any research at all. It's a significant sign of how science was changing, that as early as 1826 Faraday would write:

> It is certainly impossible for any person who wishes to devote a portion of his time to chemical experiment, to read all the books and papers that are published in connection with his pursuit; their number is immense, and the labour of winnowing out the few experimental and theoretical truths which in many of them are embarrassed by a very large proportion of uninteresting matter, of imagination, and of error, is such, that most persons who try the experiment are quickly induced to make a selection in their reading, and thus inadvertently, at times, pass by what is really good.[1]

It was a problem that could only get worse, and the response of many of the best scientists (as we have already seen in the case of Einstein) has often been to make no attempt at all to 'keep up with the literature'. In 1833, a new endowment made Faraday, in addition to his post as director, Fullerian professor of chemistry at the Royal Institution – but by then, although now in his forties, he had returned successfully

1. Quoted by Crowther, *British Scientists of the Nineteenth Century*.

33. *Faraday lecturing at the Royal Institution.*
From The Illustrated London News, *1846.*

to the work on electricity and magnetism which was to be his greatest achievement.

The invention of the electric motor and the dynamo

The question that had been nagging at many minds, including Faraday's, in the 1820s was this: if an electric current can induce a magnetic force in its vicinity, can a magnet induce an electric current? There had been one key discovery, in 1824, but nobody had interpreted it correctly by the time Faraday returned to the problem in the 1830s. François Arago had found that when a magnetic compass needle was suspended by a thread over a copper disc and the disc was rotated (like a CD spinning in its player), the needle was deflected. A similar effect had been noticed by the English physicists Peter Barlow (1776–1862) and Samuel Christie (1784–1865), but they had used iron discs. Since iron is magnetic, while copper is not, Arago's discovery was more surprising, and in the end would prove more insightful. We now explain the phenomenon as a result of the conducting disc moving relative to the magnetic needle. This induces an electric current in the disc, and this current induces a further magnetic influence, which affects the needle. This explanation is entirely due to Faraday's work in the 1830s.

By the time Faraday tackled the problem in 1831, it was clear that an electric current passing through a wire wound in a helix (usually referred to as a coil, although that is not strictly accurate) would make it act like a bar magnet, with a north pole at one end of the coil and a south pole at the other. If the coil of wire were wound around an iron rod, the rod would become a magnet when the current was switched on. To see if the effect would work the other way round, with a magnetized iron rod making current flow in the wire, Faraday carried out an experiment with an iron ring about 15 centimetres in diameter, with the iron itself about 2 centimetres thick. He wound two coils of wire on opposite sides of the ring and connected one to a battery (magnetizing the iron as current flowed through the coil) and the other to a sensitive meter (a galvanometer, itself based on the electric motor effect Faraday had described in 1821) to detect any current induced when the iron was magnetized. The key experiment took place on 29 August 1831. To his surprise, Faraday noticed that the galvanometer needle flickered just as the first coil was connected to the battery, then fell back to zero. When the battery was being disconnected, it flickered again. When a *steady* electric current was flowing, producing a *steady* magnetic influence in the ring, there was no induced electric current. But during the brief moment when the electric current was *changing* (either up or down) and the magnetic influence was also changing (either increasing or decreasing) there was an induced current. In further experiments, Faraday soon found that moving a bar magnet in and out of a coil of wire was sufficient to make a current flow in the wire. He had discovered that just as moving electricity (a current flowing in a wire) induces magnetism in its vicinity, so a moving magnet induces an electric influence in its vicinity, a neatly symmetrical picture which explains Arago's experiment, and also why nobody had ever been able to induce an electric current using static magnets. Along the way, having already in effect invented the electric motor, Faraday had now invented the electric generator, or dynamo, which uses the relative motion of coils of wire and magnets to generate electric currents. This group of discoveries, announced in a paper read to the Royal Society on 24 November 1831, placed Faraday in the top rank of contemporary scientists.[1]

Faraday on the lines of force

Faraday continued to carry out work involving electricity and chemistry (electrochemistry), much of it with important industrial applications, and introduced by now familiar terms including 'electrolyte', 'electrode', 'anode', 'cathode' and 'ion'. But he also made a key contribution to the scientific understanding of the forces of nature, which is more relevant to our present story, although for a long time he kept his deep thoughts on these important matters to himself. He first used the term 'lines of force' in a scientific paper in 1831, developing the concept from the familiar schoolday experiment in which iron filings are sprinkled on a piece of paper above a bar magnet and form curved lines linking the two poles. The idea of these lines, reaching outwards from magnetic poles or electrically charged particles, is particularly illuminating in visualizing magnetic and electric induction. If a conductor is stationary relative to a magnet, it is stationary relative to the lines of force, and no current flows. But if it moves past the magnet (or the magnet moves past the conductor, it is all the same), the conductor cuts through the lines of force as it moves and it is this that creates the current in the conductor. When a magnetic field builds up from zero, as in the iron ring experiment, the way Faraday envisaged the process involved lines of force pushing out from the magnet to take up their positions and cutting through the other coil in the ring, producing a brief flicker of current before the pattern of force lines stabilized.

Faraday hesitated to publish these ideas, but wanted to lay claim to them (rather in the way that Darwin later hesitated to publish his theory of natural selection, but wanted to establish his priority). On 12 March 1832 he wrote a note which was placed in a sealed, dated

1. The American Joseph Henry (1797–1878), then teaching at the Albany Academy in New York, discovered electromagnetic induction a little ahead of Faraday, but had not published his results, which were unknown in Europe in 1831.

and witnessed envelope, and put in a safe at the Royal Society, to be opened after his death. In part, the note read:

> When a magnet acts upon a distant magnet or a piece of iron, the influencing cause (which I may for the moment call magnetism) proceeds gradually from the magnetic bodies, and requires time for its transmission . . . I am inclined to compare the diffusion of magnetic forces from a magnetic pole, to the vibrations upon the surface of disturbed water, or those of air in the phenomena of sound: i.e. I am inclined to think the vibratory theory will apply to these phenomena, as it does to sound, and most probably to light.

As early as 1832, Faraday was suggesting that magnetic forces take time to travel across space (rejecting the Newtonian concept of instantaneous action at a distance), proposing that a wave motion was involved, and even making an (albeit tenuous) connection with light. But because of his background, he lacked the mathematical skills necessary to carry these ideas forward, which is one reason why he hesitated about publishing them. On the other hand, because he lacked those mathematical skills, he was forced to develop physical analogies to get his ideas across and did eventually present them to the public in that form. But that was only after he suffered a severe nervous breakdown, brought on by overwork, at the end of the 1830s. It was after he recovered from this breakdown[1] that, perhaps acknowledging that he wouldn't live for ever and needed to leave posterity with something more than the sealed note in the vaults of the Royal Society, Faraday first aired his ideas, at a Friday Evening Discourse (part of the programme of lectures he had introduced in the late 1820s) at the Royal Institution.

The date was 19 January 1844, and Faraday was 52 years old. The theme of his talk was the nature of atoms, which he was not then alone in regarding as heuristic devices, but which Faraday had clearly thought more deeply about than many of his contemporary opponents of the atomic hypothesis. Instead of treating an atom as a physical entity that lay at the centre of a web of forces and was the reason for the existence of those forces, Faraday proposed to his audience that it

1. 'Recovered' is also a relative term; he was never fully his old self again.

made more sense to regard the web of forces as having the underlying reality, with the atoms only existing as concentrations in the lines of force making up the web – in modern terminology, the field of the force. Faraday made it clear that he wasn't just thinking about electricity and magnetism. In a classic 'thought experiment', he asked his audience to imagine the Sun sitting alone in space. What would happen if the Earth were suddenly dropped into place, at its proper distance from the Sun? How would the Sun 'know' that the Earth was there? How would the Earth respond to the presence of the Sun? According to the argument Faraday put forward, even before the Earth were put in place, the web of forces associated with the Sun – the field – would spread throughout space, including the place where the Earth was about to appear. So as soon as the Earth appeared, it would 'know' that the Sun was there and would react to the field it encountered. As far as the Earth is concerned, the field *is* the reality it experiences. But the Sun would not 'know' that the Earth had arrived until there had been time (Faraday had no way of guessing how much time) for the Earth's gravitational influence to travel across space (like the magnetic lines of force spreading out from the coil as the connection to the battery was made) and reach the Sun. Magnetic, electric and gravitational lines of force filled the Universe, according to Faraday, and were the reality with which the seemingly material entities that make up the world are interconnected. The material world, from atoms to the Sun and Earth (and beyond), was simply a result of knots in the various fields.

These ideas were so far ahead of their time that they made no impact in 1844, although they neatly describe (without the mathematics) the way modern theoretical physicists view the world. But in 1846, Faraday returned to his theme of lines of force, in another Friday Evening Discourse. This time, the ideas he presented would bear fruit within a couple of decades. The occasion owed something to chance, although clearly Faraday had spent a long time working things out. The speaker booked to appear at the RI on 10 April 1846, one James Napier, had to cancel the engagement a week before the event, leaving Faraday with no time to find any substitute except himself. Happy to fill the breach, he spent most of his time that evening summarizing some work by Charles Wheatstone (1802–1875), who was professor of experimental physics at King's College in London, and had, among other things, carried out interesting and important work on sound. Since Wheatstone was notoriously diffident about lecturing, Faraday knew that he would be doing his friend a favour by describing his work. But this didn't take up the whole of the time available, and at the end of the lecture Faraday presented some more of his own ideas about lines of force. Now, he suggested that light could be explained in terms of the vibrations of electric lines of force, removing the need for the old idea of a fluid medium (the aether) to carry waves of light:

> The view which I am so bold as to put forth considers, therefore, radiation as a high species of vibration in the lines of force which are known to connect particles, and also masses of matter, together. It endeavours to dismiss the aether, but not the vibrations.

Faraday went on to point out that the kind of vibrations he was referring to were latitudinal, side-to-side ripples running along the lines of force, not push–pull waves like sound waves, and he emphasized that this propagation would take time, speculating that gravity must operate in a similar way and also take time to travel from one object to another.

Faraday remained active in his late fifties and beyond as a consultant, adviser to the government on science education and in other areas. True to his Sandemanian principles, he turned down the offer of a knighthood, and twice declined the invitation to become President of the Royal Society – although these offers must have brought a warm glow to the heart of the former bookbinder's apprentice. With his mental abilities failing, he offered to resign from the RI in 1861, when he was 70, but was asked to stay on in the (largely nominal) post of superintendent. He maintained a link with the RI until 1865, giving his last Friday Evening Discourse on 20 June 1862, the year that he and Sarah moved out of Albermarle Street and into a grace and favour house at Hampton Court provided by Queen Victoria at the suggestion of Prince Albert. He died there on 25 August 1867. Just three years earlier, James Clerk Maxwell had published his complete theory of electromagnetism, directly jumping off from Faraday's ideas about lines of force, and definitively explaining light itself as an electromagnetic phenomenon.

Measuring the speed of light

By the time Maxwell developed his theory of electromagnetism and light, one further crucial piece of experimental evidence (or rather, two related pieces of evidence) had come in. At the end of the 1840s, the French physicist Armand Fizeau (1819–1896), who was, incidentally, the first person to study the Doppler effect for light, had made the first really accurate ground-based measurement of the speed of light. He sent a beam of light through a gap (like the gaps in the battlements of a castle) in a rotating toothed wheel, along a path 8 kilometres long between the hilltop of Suresnes and Montmartre, off a mirror and back through another gap in the toothed wheel. This only worked if the wheel was rotating at the right speed. Knowing how fast the wheel was rotating, Fizeau was able to measure how long it took for light to make the journey, getting an estimate of its speed within 5 per cent of the modern determination. In 1850, Fizeau also showed that light travels more slowly through water than through air, a key prediction of all wave models of light, and, it seemed, the final nail in the coffin of the corpuscular model, which predicted that light would travel faster in water than in air. Léon Foucault (1819–1868), who had worked with Fizeau on scientific photography in the 1840s (they obtained the first detailed photographs of the surface of the Sun together), was also interested in measuring the speed of light and developed an experiment devised by Arago (and based on an idea by Wheatstone), initially with apparatus that he took over from Arago after Arago's sight failed in 1850. This experiment involved light being bounced from a rotating mirror on to a stationary mirror and back again, to bounce off the rotating mirror a second time. The amount by which the light beam is deflected shows how much the rotating mirror has moved while the light has been bouncing off the stationary mirror, and again, knowing how fast the mirror is rotating gives you the speed of light. Reversing the approach followed by Fizeau, in 1850 Foucault first used this method to show (slightly before Fizeau did) that light travels more slowly in water than in air. Then he measured the speed of light. By 1862 he had refined the experiment so much that he came up with a speed of 298,005 km/s, within 1 per cent of the modern value (which is 299,792.5 km/s). This accurate measurement of the speed of light turned out to be invaluable in the context of Maxwell's theory.

Maxwell was descended from not one but two prominent and moderately affluent Scottish families, the Maxwells of Middlebie and the Clerks of Penicuik, linked through two intermarriages in the eighteenth century. It was the Middlebie property (about 1500 acres of farmland near Dalbeattie in Galloway, in the southwestern corner of Scotland) that descended to Maxwell's father, John Clerk, who took the name Maxwell as a result; the Penicuik property was inherited by John Clerk's elder brother George (the inheritance was legally arranged so that both properties could not descend to the same person), who, as Sir George Clerk, was a Member of Parliament for Midlothian and served in government under Robert Peel. The Middlebie inheritance was not much to write home about, consisting of poor land and lacking even a proper house for its owner, and John Clerk Maxwell lived mostly in Edinburgh, practising law in a desultory fashion but more interested in keeping up to date with what was going on in science and technology (which, as we have seen, was quite a lot in Edinburgh in the first decades of the nineteenth century). But in 1824 he married Frances Cay, had a house built at Middlebie, took up residence there and set about improving the land, having boulders removed from the fields to prepare them for cultivation.

James Clerk Maxwell was born on 13 June 1831 – not in Galloway, but in Edinburgh, where his parents had gone to ensure proper medical attention at the birth. This was particularly important since Mrs Maxwell was now in her fortieth year, and a previous baby, Elizabeth, born a couple of years earlier, had died after a few months. James, who remained an only child, was brought up at the new house, Glenlair, where he played with the locals and grew up with a thick Galloway accent, in spite of his mildly aristocratic antecedents. Dalbeattie really was out in the sticks when he was a child – although Glasgow was only 110 kilometres away, this was a full day's journey, and Edinburgh was two days away until the Glasgow–Edinburgh railway line opened in 1837. As the need to clear large stones from the fields before cultivation could begin suggests, the situation of the Maxwell family was in some ways more like a pioneering family in the western United States of the time than that of a comparable English family living a few score kilometres from Birmingham.

Maxwell's mother died of cancer when she was 48 and he was just

8, which removed a possible restraining influence on the development of his uncouth ways. He had a happy and close relationship with his father, who encouraged the boy's curiosity about the world and intellectual development, but had some bizarre notions, including designing their clothes and shoes in what may well have been practical but were certainly not fashionable styles. The only black cloud at Glenlair was the presence of a young tutor (scarcely more than a boy himself) who was employed to teach James around the time of his mother's final illness. This tutor seems to have literally tried to beat knowledge into the boy, a situation which lasted for about two years, since James stubbornly refused to complain to his father about the treatment. But at the age of 10 James was sent to the Edinburgh Academy to receive a proper education, and lived with one of his aunts during term time.

The boy's appearance at the Academy (where he arrived after term had started), wearing weird clothes and speaking with a broad country accent, elicited the response you might expect from the other boys, and even after some of the initial confrontations were resolved by fisticuffs, Maxwell was stuck with the nickname 'Dafty', an allusion to his strange appearance and manners, not to imply presumed lack of intellect. He made a few close friends and learned to tolerate the rest, and enjoyed his father's visits to Edinburgh, when young James was often taken to see scientific demonstrations – at the age of 12 he saw a demonstration of electromagnetic phenomena and the same year attended a meeting of the Edinburgh Royal Society with his father. Within a few years, Maxwell was showing unusual mathematical ability, and at the age of 14 invented a way of drawing a genuine oval (not an ellipse) using a looped piece of string. Although original, this was not actually an Earth-shattering achievement, but through John Clerk Maxwell's connections, James had the work published in the *Proceedings of the Royal Society of Edinburgh* – his first scientific paper. In 1847, when he was 16 (the usual age then for entry to a Scottish university), Maxwell moved on to the University of Edinburgh, where he studied for three years but left without graduating and went up to Cambridge (at first to Peterhouse, but he shifted to Trinity, Newton's old college, at the end of his first term), where he graduated (second in his year) in 1854. As an outstanding student,

Maxwell became a Fellow of Trinity, but only stayed there until 1856, when he became professor of natural philosophy at Marischal College, in Aberdeen.

This brief period as a Cambridge Fellow was just long enough for Maxwell to carry out two important pieces of work – one, developing from Young's work on the theory of colour vision, in which he showed how a few basic colours could 'mix' together to fool the eye into seeing many different colours (the classic experiment to demonstrate this is with different segments of a spinning top painted in different colours, so that when the top is spun the colours merge), and an important paper *On Faraday's Lines of Force* which comprehensively spelled out how much was known about electromagnetism and how much remained to be discovered, and laid the groundwork for his later studies. Maxwell's work on colour vision (developed further by him in later years) was the foundation for the method of making colour photographs by combining monochrome pictures taken through three different filters (red, green and blue); his work is also the basis of the system used in colour TV and computer monitors today, and in colour inkjet printers.

Maxwell's father died on 2 April 1856, shortly before James was appointed to the Aberdeen post; but he was not alone for long, since in 1858 he married Katherine Mary Dewar, the daughter of the principal of the college, who was seven years older than him. They never had children, but Katherine acted as Maxwell's assistant in much of his work. The family connection proved to no avail, however, in 1860, when Marischal College merged with King's College in Aberdeen (forming the nucleus of what would become the University of Aberdeen). The merged institution needed only one professor of natural philosophy, and as Maxwell was junior to his colleague at King's (who happened to be a nephew of Faraday, though that had nothing to do with the choice), he had to go. The most valuable work Maxwell carried out in Aberdeen was a theoretical study of the nature of the rings of Saturn, which proved that the rings must be made of myriad small particles, or moonlets, each in their own orbit round the planet, and could not be solid. It seems likely that the mathematical treatment of the many particles required to prove this helped set Maxwell on the track of his contributions to the kinetic theory, mentioned in the previous chapter, after his interest was roused by reading the work of Clausius. When colour images of the rings of Saturn were sent back to Earth from space probes in the late twentieth century, they used Maxwell's three-colour photographic technique to provide pictures of the moonlets predicted by Maxwell – and they sent the images back by radio waves, another prediction (as we shall see) made by Maxwell.

James Clerk Maxwell's complete theory of electromagnetism

From Aberdeen, Maxwell and his wife headed back to Glenlair, where he suffered an attack of smallpox but recovered in time to apply for and obtain the post of professor of natural philosophy and astronomy at King's College in London. It was there that he completed his great work on electromagnetic theory, but he had to resign the post in 1866 on the grounds of ill health – while out riding he had grazed his head on a tree branch and the wound had led to a severe attack of erysipelas, an inflammatory disease (caused, we now know, by streptococcal infection) characterized by severe headaches, vomiting and raised purple lesions on the face. There is some inference that the severity of this attack may have been related to the earlier bout of smallpox.

Maxwell's great work had been gestating for some ten years, since his earlier interest in Faraday's lines of force. In the 1840s, William Thomson had found a mathematical analogy between the way heat flows through a solid and the patterns made by electric forces. Maxwell picked up on these studies and looked for similar analogies, communicating with Thomson in a series of letters that helped him to clarify his ideas. Along the way, Maxwell came up with an intermediate model based on a now very strange-looking idea in which the forces of electricity and magnetism were conveyed by the interactions of vortices, whirlpools that spun in a fluid filling all of space. But the strangeness of this physical model didn't hold back the development of his ideas, because as Maxwell correctly commented, all such physical images are less important than the mathematical equations that describe what is going on. In 1864, he would write:

For the sake of persons of different types of mind, scientific truth should be presented in different forms and should be regarded as equally scientific whether it appears in the robust form and vivid colouring of a physical illustration or in the tenuity and paleness of a symbolic expression.[1]

This is almost the most important thing Maxwell ever wrote. As science (in particular, quantum theory) developed in the twentieth century it became increasingly clear that the images and physical models that we use to try to picture what is going on on scales far beyond the reach of our senses are no more than crutches to our imagination, and that we can only say that in certain circumstances a particular phenomenon behaves 'as if' it were, say, a vibrating string, not that it *is* a vibrating string (or whatever). As we shall see later, there are circumstances where it is quite possible for different people to use different models to image the same phenomenon, but for them each to come up with the same predictions, based on the mathematics, of how the phenomenon will respond to certain stimuli. Getting ahead of our story only slightly, we will find that although it is quite right to say that light behaves like a wave in many circumstances (particularly when travelling from A to B), under other circumstances it behaves like a stream of tiny particles, just as Newton thought. We cannot say that light *is* a wave or *is* corpuscular; only that under certain circumstances it is *like* a wave or *like* a particle. Another analogy, also drawing on twentieth-century science, may help to make the point. I am sometimes asked if I believe that there 'really was' a Big Bang. The best answer is that the evidence we have is consistent with the idea that the Universe as we see it today has evolved from a hot, dense state (the Big Bang) about 13 billion years ago. In that sense, I believe there was a Big Bang. But this is not the same kind of belief as, for example, my belief that there is a large monument to Horatio Nelson in Trafalgar Square. I have seen that monument and touched it; I believe it is there. I have not seen or touched the Big Bang, but the Big Bang model is the best way I know of picturing what the Universe was like long ago, and that picture matches the available observations and mathematical calculations.[2]

1. The great 1864 paper and most of Maxwell's other scientific papers can be found in *The Scientific Papers of J. Clerk Maxwell* (see Bibliography).

2. And there is a third kind of belief, in some religions, where the whole point is that the religious story is believed without any evidence at all, on faith.

These are all important points to absorb as we move on from the classical science of Newton (dealing, broadly speaking, with things you can see and touch) to the ideas of the twentieth century (dealing, in some sense, with things that cannot be seen or touched). Models are important, and helpful; but they are not the truth; in so far as there is scientific truth, it resides in the equations. And it was equations that Maxwell came up with.

In 1861 and 1862 he published a set of four papers *On Physical Lines of Force*, still using the image of vortices but looking, among other things, at how waves would propagate in such circumstances. The speed with which those waves move depends on the properties of the medium, and by putting in the appropriate properties to match what was already known about electricity and magnetism Maxwell found that the medium should transmit waves at the speed of light. His excitement at the discovery shines through in his own words in one of the 1862 papers, where he emphasizes the significance of the discovery with italics: 'We can scarcely avoid the inference that *light consists in the transverse undulations of the same medium which is the cause of electric and magnetic phenomena*.'[1]

Light is a form of electromagnetic disturbance

Refining the mathematics of his theory, Maxwell soon found that he could abandon the ideas of vortices and an intervening medium altogether. The physical image had helped him to construct the equations, but once they had been constructed they stood alone – the obvious analogy is with a great medieval cathedral, built with the aid of a ramshackle wooden scaffolding, but which stands alone in all its glory, without external support, once the scaffolding has been removed. In 1864 Maxwell published his *tour de force* paper, 'A Dynamical Theory of the Electromagnetic Field', which summed up everything that it is possible to say about classical electricity and magnetism in a set of four equations, now known as Maxwell's equations. Every problem involving electricity and magnetism can be solved using these equations, except for certain quantum phenomena. Since one set of equations solves all electric and magnetic problems, Maxwell had also completed the possibility first hinted at by Faraday's work, of unifying two forces into one package; where

1. *op. cit.*

there had been electricity and magnetism, now there was just one field, the electromagnetic field. All of this is why Maxwell is placed alongside Newton in the pantheon of great scientists. Between them, Newton's laws and his theory of gravity, and Maxwell's equations, explained everything known to physics at the end of the 1860s. Without doubt, Maxwell's achievement was the greatest piece of physics since the *Principia*. And there was some icing on the cake. The equations contain a constant, c, which represents the speed with which electromagnetic waves move, and this constant is related to measurable electrical and magnetic properties of matter. In experiments to measure those properties, as Maxwell put it, 'the only use made of light . . . was to see the instruments'. But the number that came out of the experiments (the value of c) was, to within experimental error, exactly the same as the (by then well-determined) speed of light.

> This velocity is so nearly that of light that it seems we have strong reason to conclude that light itself (including radiant heat and other radiations, if any) is an electromagnetic disturbance in the form of waves propagated through the electromagnetic field according to electromagnetic laws.[1]

The reference to 'other radiations' is significant; Maxwell predicted that there could be forms of electromagnetic waves with wavelengths much longer than those of visible light – what we now call radio waves. In the late 1880s, the German physicist Heinrich Hertz (1857–1894) carried out experiments which confirmed the existence of these waves, showing that they travel at the speed of light, and that, like light, they can be reflected, refracted and diffracted. It was further proof that Maxwell's theory of light was right.

Compelling though the equations and the experimental proof are, it is, as Maxwell appreciated, helpful to have a colourful model to think about – as long as we remember that the model is not the reality, it is simply a construct to help us picture what is going on. In this case, one way to imagine the propagation of light (or other electromagnetic radiation) is to think of a stretched rope that can be wiggled at one end. Remember that, as Faraday discovered, a moving magnetic field

1. Maxwell, *op. cit.*

creates an electric field, and a moving electric field creates a magnetic field. If energy is put in by jiggling the stretched rope (equivalent to putting energy into an electromagnetic field by making a current flow first in one direction and then in the other in a long wire or an antenna system), you can send ripples down the rope. Jiggle the rope up and down to make vertical ripples, and jiggle it sideways to make horizontal ripples. One of the things Maxwell's equations tell us is that the equivalent electric and magnetic ripples in an electromagnetic wave are at right angles to one another – if, say, the electric ripples are vertical, then the magnetic ripples are horizontal. At any point along the path of the wave (along the rope), the electric field is constantly changing as the ripples pass by. But this means that there must be a constantly changing magnetic field, produced by the electric field. So at every point along the path of the wave there is a constantly changing magnetic field, which produces a constantly changing electric field. The two sets of ripples run along in step, as a beam of light (or radio waves), drawing on energy being fed in at the source of the radiation.

With his great work completed – the last great piece of classical science, in the Newtonian tradition – Maxwell (still only 35 in 1866) settled comfortably into life in Galloway, keeping in touch with his many scientific friends by correspondence and writing a great book, his *Treatise on Electricity and Magnetism*, published in two volumes in 1873. He turned down the offer of several prestigious academic posts, but was tempted back to Cambridge in 1871 when asked to become the first Cavendish professor of experimental physics and (much more important) set up and head the Cavendish Laboratory.[1] It opened in 1874. Maxwell lived long enough to set his stamp on the laboratory, which became the most important centre for research into the new discoveries in physics in the scientifically revolutionary decades that followed. But he became seriously ill in 1879, and on 5 November that year he died of the same illness that had killed his mother (cancer) and at the same age (48). That same year, on 14 March, the person who was to be the first to see the full implications of Maxwell's

1. He also edited *The Unpublished Electrical Writings of the Honorable Henry Cavendish*, which was published in 1879.

equations was born at Ulm, in Germany. His name, of course, was Albert Einstein.

Einstein's connection with the world of electromagnetism began, in a way, the year after his birth, when the family moved to Munich. There his father, Hermann, joined forces with his uncle, Jakob (with the aid of funds from the family of Albert's mother, Pauline), to set up an electrical engineering business – a fine example of how Faraday's discoveries had by then been turned to practical use. Technically, the company was a success, at one time employing 200 people and installing electric light in small towns; but they were always underfunded and eventually lost out to the firms that became the giants of the German electrical industry, including Siemens and the German Edison Company, and they went under in 1894. Seeking a more congenial business environment, the brothers moved to northern Italy, where their firm had previously carried out contract work, but where they achieved only modest success; this move involved leaving 15-year-old Albert behind to complete his education in the German school system.

This was not a good idea. Albert was an intelligent and independently minded youth, who did not fit in to the rigidly disciplined school system of his native country, newly unified and ruled by a militaristic Prussian tradition which included compulsory army service for all young men. Just how Einstein engineered his removal from the gymnasium (high school) is not clear; according to some accounts he was expelled after a period of rebelliousness, while according to others he arranged things entirely off his own bat. Either way, he persuaded the family doctor to certify that he was suffering from a nervous disorder which required a complete rest, and armed with this certificate he set off to join the family (his parents and younger sister Maja, his only sibling), arriving in Italy early in 1895. Renouncing his German citizenship (the only sure way to avoid that military service), he spent some time working in the family business and a lot more time taking in the delights of Italy before taking the entrance examinations for the Swiss Federal Institute of Technology (the Eidgenössische Technische Hochschule, or ETH) in Zürich, where he could obtain a degree – not as good as a degree from one of the great German universities, but at least a qualification. In the autumn of 1895, Albert was a full 18 months younger than the age at which students usually entered the

ETH (18), and had left the gymnasium without any kind of diploma except a letter from a teacher testifying to his mathematical ability. To us, it is hardly surprising that he failed the entrance examination, although it seems to have come as a shock to the cocky young man. It was only after a year in a Swiss secondary school at Aaru, south of Zürich, that Albert finally gained entrance to the ETH in 1896. It turned out to be one of the happiest years of his life, lodging with the principal of the school he was attending, Jost Winteler, and making lifelong friends among the Winteler family (Albert's sister Maja later married Jost Winteler's son Paul).

In Zürich, ostensibly studying maths and physics, Einstein enjoyed life to the full (which eventually included getting his girlfriend Mileva Maric pregnant; the illegitimate baby was adopted) and did the minimum amount of work required to satisfy his teachers, while reading and studying widely outside the official curriculum. As confident as ever in his own abilities, he expected to do brilliantly in his final examinations and to get a job in a junior capacity at the ETH itself or some university. In fact, he did well in the exams, graduating in July 1900, but not brilliantly – certainly not brilliantly enough to outweigh the reluctance of the professors to employ someone they saw as temperamentally unsuited for serious hard work. Which is how Albert Einstein came to be working as a patent officer in Bern in 1905, having married Mileva in 1903 and with a baby son, Hans Albert, born on 14 May 1904 (a second legitimate child, Eduard, was born on 28 July 1910).

The foundation stone for Einstein's special theory of relativity, published in 1905, was the constancy of the speed of light. By the time he developed his theory, there was experimental evidence that the measured speed of light is always the same, irrespective of how the person doing the measuring is moving. But it is important to appreciate that Einstein, although he knew of this work, was not influenced by it. The thing which marked out his approach to the problem was that it started from Maxwell's equations. The equations contain a constant, c, identified with the speed of light. There is no provision in the equations, as far as the determination of c is concerned, to take account of how an observer is moving relative to the light. According to Maxwell's equations, all observers will measure the same light speed, *c*, whether they are stationary, moving towards the source of the light or moving away from the source of the light (or, indeed, moving at any angle across the beam of light). This flies in the face of both common sense and the way velocities add up in Newtonian mechanics. If a car is moving towards me along a straight road at 100 km an hour and I am driving in the opposite direction at 50 km an hour, the car is approaching me at a relative speed of 150 km an hour; if I am driving at 50 km an hour and the other car, just ahead of me, is going in the same direction as me at 100 km an hour, then the other car is travelling at 50 km an hour relative to me. But according to Maxwell's equations, in either situation, the speed of light coming from the headlights or taillights of the other car is always *c*, both relative to me *and* relative to the driver of the other car (and, indeed, relative to any bystanders at the roadside). As soon as you think about it, it is clear that Newton's laws of motion and Maxwell's equations cannot both be right. Most people who did think about it before 1905 assumed that there must be something not quite right about Maxwell's theory, the new kid on the block. Einstein, always iconoclastic, had the temerity to consider the alternative – that Maxwell was right and Newton, at least in this limited case, was wrong. That was the basis of his great insight. But it does no harm to look at the experimental evidence as well, which confirms in full measure just how right Maxwell was.

Albert Michelson and Edward Morley: the Michelson–Morley experiment on light

Even though Faraday had attempted to 'dismiss the aether' back in 1846, the concept had refused to die, and in an article published in the *Encyclopaedia Britannica* in 1878 (just a year before he died), Maxwell himself had suggested an experiment to measure the velocity of the Earth relative to the ether (to use the more modern spelling), with beams of light being used to do the experiment. The experiment would involve splitting a beam of light in two and sending each of the two resulting beams on a journey between two mirrors, one set of mirrors lined up in the direction of the Earth's motion through space (and, presumably, through the ether) and the other at right angles to it. After bouncing between their respective mirrors, the beams of light could be brought back together and allowed to interfere. If the experiment was set up so that each beam of light covered the same distance, then because of the Earth's motion through the ether they should take different times to complete their journeys, and get out of step with one another, producing an interference pattern like the one seen in the double-slit experiment. The challenge of carrying out such an experiment to the precision required to test this prediction was taken up by the American physicist Albert Michelson (1852–1931), at first on his own (while working at the laboratory of Hermann Helmholtz in Berlin in 1881) and later in collaboration with Edward Morley (1838–1923), in Ohio in 1887. They found, to very high precision, that there was no evidence that the Earth moves relative to the ether – or, to put it another way, the measured speed of light is the same in the direction of the Earth's motion as it is at right angles to the direction of the Earth's motion. Indeed, it is the same in *all* directions. They could rotate the apparatus without getting any result; they could carry out the experiment at different times of day (at different stages in the Earth's rotation); the could carry it out at different times of year (different stages in the Earth's orbit around the Sun). Always the answer was the same – no interference between the two beams.

Michelson, who had something of an obsession with light, repeatedly devised and carried out better and better experiments to measure the speed of light itself (the Michelson–Morley experiment did not have to measure the actual speed, of course, since it was only looking for *differences* between the two light beams); he received the Nobel prize in 1907 for the superb precision of all this work, but even then he was far from finished with light. In the ultimate manifestation of his work, carried out when Michelson was 73, in 1926, the light went on a two-way journey between two mountain peaks in California. He determined its speed to be 299,796 ± 4 km/s, matching the best modern value of 299,792.458 km/s to within the limits of experimental error. In fact, the modern figure is now *defined* as the speed of light, which means that the length of the standard metre is specified by these measurements.[1]

Very soon after Michelson and Morley reported their definitive experimental results, the Irish mathematician and physicist George

1. In a sane world, with the measurement now so precise, we would make a tiny adjustment in the length of the metre, defining it so that the speed of light was exactly 300,000 km/s.

Fitzgerald (1851–1901), who worked at Trinity College, Dublin, suggested an explanation. Fitzgerald had been one of the first people to take Maxwell's equations seriously, and he elaborated on the theme of what we now call radio waves before Hertz carried out his experiments. In 1889, Fitzgerald suggested that the failure of the Michelson–Morley experiment to measure any change in the speed of light, regardless of which way the apparatus was oriented with respect to the Earth's motion through space, could be explained if the whole apparatus (indeed, the whole Earth) shrank by a tiny amount in the direction of motion – by an amount which depended on its speed and could be calculated precisely from the fact that the experiment gave a null result. The same idea was put forward independently in the 1890s by the Dutch physicist Hendrik Lorentz (1853–1928), who worked in Leiden and developed the idea more fully (not least because he had the good fortune to outlive Fitzgerald, who died young from the effects of a stomach ulcer brought on by overwork), producing the definitive form of what are known as the Lorentz transformation equations in 1904. Rather neglecting the historical priority, the shrinking effect is now known as Lorentz–Fitzgerald contraction.

Albert Einstein: special theory of relativity

This work is sometimes represented as somehow pre-empting Einstein's special theory of relativity, implying that all he did was dot the *i*s and cross the *t*s. But this is far from being the case. The kind of shrinking Fitzgerald and Lorentz envisaged involved the individual electrically charged particles (atoms) in a substance getting closer together as the force of attraction between them increased because of their motion – not a completely off-the-wall idea, given Faraday's discoveries about how electricity and magnetism were affected by motion, but now known to be wrong. On the other hand, starting out from first principles, based on the fact that Maxwell's equations specify a unique speed for light, Einstein came up with equations that were mathematically identical to the Lorentz transformation equations, but which envisaged the space occupied by an object itself shrinking in line with the motion of the object relative to an observer. The equations also describe time dilation (moving clocks run slow, relative to time measured by the stationary observer) and an increase in mass of moving objects. The special theory reveals that no object which starts out moving at less than the speed of light can be accelerated to above the speed of light (one way to think of this is that its mass would become infinite at the speed of light, so infinite energy would be needed to make it go any faster). And, related to the way mass depends on velocity, the theory reveals the equivalence of mass and energy, in the most famous equation in science, $E=mc^2$.

But who are all these measurements made relative to? The other key feature of the special theory, alongside the constancy of the speed of light, is that there is no preferred state of rest in space. Einstein saw that there is no preferred frame of reference in the Universe – no 'absolute space' against which motion can be measured. All motion is relative (hence the name of the theory), and any observer who is not being accelerated is entitled to regard himself (or herself) as at rest and to measure all other motion relative to his or her frame of reference. The theory is 'special' in the sense that it is restricted – a special case, in which accelerations are not considered. All observers moving at constant velocities relative to one another (inertial observers) are equally entitled to say they are at rest and to measure all motion relative to themselves.

There is a pleasing, and essential, symmetry in the equations, which means that observers in different frames of reference (moving relative to one another) get the same answers to experiments when they compare notes, even if they disagree on how they got those answers. For example, if I watch a spaceship travel at a large fraction of the speed of light to a star 10 light years away, it will seem to me that the time taken to complete the journey is less than 10 years *according to clocks on the spaceship*, without the spaceship going faster than light, because the moving clocks run slow. To the crew of the spaceship, it also seems that the journey has taken the same time that I calculate; but they say their clocks are working as they have always worked, and their journey has literally been shortened because the space between here and the distant star has been shrunk, thanks to the relative motion of all the stars in the Universe 'past' the spaceship, which they are entitled to say is at rest. If any observer A sees that observer B's clock is running slow and his measuring rods have shrunk, then observer B sees observer A's clocks and measuring rods affected in exactly the same way and to exactly the same degree, with neither of them noticing anything odd about their own measuring apparatus. One curious result of all this is that for anything travelling at the speed of light time stands still. From the point of view of a photon (a quantum of light, to be discussed in Chapter 13), it takes no time at all to cross the 150 million kilometres from the Sun to the Earth. From our point of view, this is because any clock riding with the photon would stand still; from the photon's point of view, it is because at such high speeds (remember, the photon is entitled to regard itself as at rest, with the Earth hurtling towards it) the space between the Sun and the Earth shrinks away to nothing, so obviously it takes no time at all to cross it. It is, of course, crucially important that, bizarre though such inferences may seem, the predictions of the special theory of relativity have been confirmed many times in experiments (for example, using beams of particles accelerated to close to the speed of light) to an accuracy of many decimal places. That is why it is a theory, not merely a hypothesis. It is because these effects only show up if things are indeed moving at a sizeable fraction of the speed of light that we are not aware of them in everyday life, so they are not common sense. But they are none the less proven to be real.

Minkowski: the geometrical union of space and time in accordance with this theory

It would be wrong to imply that the special theory was not understood by Einstein's contemporaries in 1905; the fact that Michelson received his Nobel prize a couple of years later is a significant reflection of the fact that many physicists understood the importance of both the Lorentz transformation equations and Einstein's work. But it is true that Einstein's ideas only really began to have a big impact, and the significant differences between his work and that of Lorentz and Fitzgerald began to be fully appreciated, after 1908, when Einstein's old teacher Minkowski (the one who described Einstein as a 'lazy dog') presented the idea not just in terms of mathematical equations but in terms of four-dimensional geometry, the geometry of space and time (now fused as spacetime). In a lecture he gave in Cologne that year, Minkowski (who was born in 1864 and died from complications following an attack of appendicitis only a year after giving that lecture) said that:

> Henceforth space by itself, and time by itself, are doomed to fade into mere shadows, and only a kind of union of the two will preserve an independent reality.

Although Einstein was not, at first, happy with this geometrization of his ideas, as we shall see, it was precisely this geometrical union of space and time that would lead to what is widely regarded as his greatest achievement, the general theory of relativity.

After 1905, physics would never be the same again (and we have still to discuss what I consider to be Einstein's most important achievement of his *annus mirabilis*, the work for which he received the Nobel prize, and which laid the foundations of quantum theory). Fundamental physics in the twentieth century would develop in ways that could not have been imagined by the classical pioneers such as Newton, or even Maxwell. But classical science (and classical physics in particular) still had one great triumph to come, which emerged from the application of essentially pre-1905 ideas to the grandest human-scale puzzle of them all, the nature of the origin and evolution of the Earth itself.

12

The Last Hurrah! of Classical Science

Contractionism: our wrinkling planet? – Early hypotheses on continental drift – Alfred Wegener: the father of the theory of continental drift – The evidence for Pangea – The radioactive technique for measuring the age of rocks – Holmes's account of continental drift – Geomagnetic reversals and the molten core of the Earth – The model of 'sea-floor spreading' – Further developments on continental drift – The 'Bullard fit' of the continents – Plate tectonics – The story of Ice Ages: Jean de Charpentier – Louis Agassiz and the glacial model – The astronomical theory of Ice Ages – The elliptical orbit model – James Croll – The Milankovitch model – Modern ideas about Ice Ages – The impact on evolution

The last great triumph of classical science did depend on one discovery which, with hindsight, belongs in the twentieth century post-classical world ('post-classical' in the scientific, not the literary or art-historical, sense, meaning that it is based on relativity theory and quantum mechanics). That was the discovery of radioactivity (itself made in the nineteenth century), which provided a source of heat that could prevent the interior of the Earth from cooling into a solid, inert lump on the sort of timescales required by the uniformitarian ideas developed by Lyell and his predecessors. It would take the theories of relativity and quantum physics to progress from the discovery of radioactivity to an explanation of the phenomenon and an understanding of how the conversion of mass into energy keeps the stars shining. But, just as Galileo could study the way pendulums swing and balls roll down inclined planes without knowing how gravity worked, all that geophysicists needed to know about radioactivity was that it did provide a way to keep the Earth warm inside – that there was a source of energy to drive the physical processes which have shaped the surface of the planet over an immense span of time and continue to do so today. Armed with that knowledge they could develop geology into geophysics, explaining the origin of the continents and ocean basins, the occurrence of earthquakes, volcanoes and mountain building, the wearing away of land by erosion and much more besides; all in terms of the kind of science that would have been well understood by Isaac Newton or Galileo Galilei, let alone William Thomson or James Clerk Maxwell.

Contractionism: our wrinkling planet?

In spite of the importance of Lyell's influence (especially in the English-speaking world, and most especially on Charles Darwin), you shouldn't run away with the idea either that uniformitarianism swept the board after the publication of his *Principles of Geology* or that most geologists of the nineteenth century actually cared all that much about the debate concerning the physical causes which had shaped the globe. Indeed, you couldn't really say that there was a debate; different people put forward different models, each of which had their adherents, but the rivals didn't meet to discuss the merits of their rival models or engage much in any kind of confrontation in print. The first task, still very much at the forefront throughout the nineteenth century, was to carry out the field work which put the strata in order and gave geologists a *relative* timescale to work with, so that they knew which rocks were older and which were younger. As far as investigating ideas about the origin of those strata went, there were even shades of uniformitarianism, and it was widely thought that although the same *kinds* of forces had been at work in the past as today (earthquakes and volcanoes, for example), they may have been more powerful in the past, when the Earth was younger and presumed to have been hotter. Lyell's uniformitarianism said that continents could be converted into sea floor and ocean floor raised up to make continents; but another (still uniformitarian) school of thought, known as permanentism, held that continents had always been continents and oceans had always been oceans. The permanentists were particularly strong in North America, where James Dana (1850–1892), professor of natural history and geology at Yale University, was its leading advocate. He linked the hypothesis with the (not unreasonable, given the state of knowledge at the time) idea that the Earth was gradually shrinking, contracting as it cooled, and that mountain ranges such as the Appalachians were produced, in effect, by the wrinkling of the Earth's crust as it shrank.

In Europe, the idea of contractionism was developed along different lines, as a variation on catastrophism. This idea culminated in the final decades of the nineteenth century in a synthesis of older ideas developed by Eduard Suess (1831–1914), who was born in London (the son of a German wool merchant) but moved with his family as a child first to Prague and then to Vienna, where he eventually became professor of geology at the university. Suess's model saw contraction as the driving force for rapid bursts of dramatic change, separated by long intervals of relative calm, on a cooling and contracting Earth. He suggested that the present-day land masses of Australia, India and Africa were fragments of a much greater land mass (which he dubbed Gondwanaland, after a region of India) that had once existed in the southern hemisphere, much of which had sunk into the cooling interior. The wrinkling crust of the Earth had, in this picture, formed folds (mountain ranges and rifts) and large chunks (such as the Atlantic, as well as in the southern hemisphere) had subsided into space made available in the interior as it cooled and contracted, forming new ocean basins between formerly connected land masses; but this happened in sudden bursts, not as a gradual, continuing process. The model failed to stand up to proper investigation. For example, the amount of wrinkling and folding required to produce the Alps alone, squeezing (according to the Suess synthesis) 1200 kilometres of crust into 150 kilometres of mountains, corresponded to a cooling of 1200 °C. Even greater cooling would be required to produce the shrinking alleged to have given rise to the Himalayas, the Rockies and the Andes, which formed at essentially the same time as the mountains of the Alps. But the key blow to all such models was the discovery of radioactivity, made at almost the same time Suess was developing his synthesis, which showed that the Earth's interior was not, in fact, cooling dramatically at all. The story of Suess's synthesis is, though, significant for two reasons. First, it highlights the lack of any 'standard model' of Earth history at the beginning of the twentieth century; second, it gave us a name, Gondwana, which would become familiar as the idea of continental drift became established. But although that idea itself had also been aired in the nineteenth century, it would not become established until well into the second half of the twentieth century – less than fifty years ago.

Early hypotheses on continental drift

Among the variations on the theme of continental drift put forward in the nineteenth century there was the idea that the continents might be sitting on magnetized crystalline foundations and were being swept northwards by a magnetic flow, and the suggestion that the Earth had originally been not only smaller than it is today but tetrahedral in shape, with the continents originally nestling close together but being ripped apart from one another in a catastrophic expansion event which also flung the Moon out of the Mediterranean basin and into its orbit. In 1858 (the year before the publication of the *Origin of Species*), Antonio Snider-Pellegrini, an American working in Paris, published a book, *La Création et ces mystères devoilés*, which put forward a bizarre model based on his interpretation of the Bible. This involved a series of catastrophes taking place on a rapidly shrinking Earth at the beginning of its history. It is only worth mentioning because the book marked the first publication of a map bringing together the continents on both sides of the Atlantic Ocean, which was used to explain the similarities between fossils found in coal deposits on opposite sides of the ocean. The map has been widely reprinted, giving the misleading impression that Snider-Pellegrini actually had a sensible model of continental drift. A somewhat more scientific (but still catastrophic) version of continents in motion was raised by Osmond Fisher in a paper published in the science journal *Nature* on 12 January 1882. He took up an idea proposed by the astronomer George Darwin (1845–1912; one of Charles Darwin's sons) that the Moon had formed when the young Earth split into two unequal parts. Fisher suggested that the Pacific basin marked the wound where the Moon had been torn out of the Earth, and that continental material on the other side of the world would have cracked and the fragments been pulled apart as the remaining surface of the Earth slowly moved in the direction of the hole as it began to fill in.

Alfred Wegener: the father of the theory of continental drift

In the first decades of the twentieth century, other versions of continental drift were also proposed. But the one which (eventually) made its mark and influenced the development of the Earth sciences was the one put forward by the German meteorologist Alfred Wegener, initially in 1912. Coming from a different scientific discipline (he originally trained as an astronomer), Wegener seems to have known little of the plethora of older ideas about continental drift (probably just as well, seeing how harebrained some of them were). His ideas became influential not just because he developed a more complete model than those predecessors, but because he campaigned for it over a period of decades, seeking out more evidence in support of his idea, defending the model in the light of criticism and publishing a book which ran into four editions before his untimely death in 1930. Wegener kicked up a fuss about continental drift, rather than just publishing his ideas and leaving them to sink or swim on their own. Although many of his detailed ideas were incorrect, his overall concept has stood the test of time, and Wegener is now rightly regarded as the father of the theory (as it now is) of continental drift.

Wegener was born in Berlin on 1 November 1880, and studied at the universities of Heidelberg, Innsbruck and Berlin, obtaining his doctorate in astronomy from Berlin in 1905. He then joined the Prussian Aeronautical Observatory at Tegel, where he worked for a time alongside his brother Kurt (literally alongside on one occasion, when the brothers undertook a balloon flight lasting 52½ hours, a record at the time, to test instruments). From 1906 to 1908 Wegener worked as meteorologist to a Danish expedition to the interior of Greenland, and on his return he joined the University of Marburg as a lecturer in meteorology and astronomy. He published a meteorological textbook in 1911, but by then was already developing his ideas on continental drift, which first appeared in print in 1912 in a pair of papers based on talks he had given in Frankfurt am Main and Marburg in January that year. As Wegener later recalled, in 1910 one of his colleagues at Marburg had been given a new world atlas, and while looking at it Wegener was struck (like others before him) by the way the east coast of South America and the west coast of Africa looked as if they ought to fit together, like pieces in a jigsaw puzzle, as if they had once been joined. Although intrigued, he regarded the idea as improbable and didn't take it forward until the spring of 1911, when he came across a report discussing the paleontological similarities between the strata of Brazil and Africa. The evidence was presented in that report in support of the idea of a former land bridge linking the two continents; but Wegener saw things differently. As he wrote in the first edition of what became his masterwork, *Die Entstehung der Kontinente und Ozeane*, published in 1915:[1]

1. By Friedrich Viewege, Brunswick; for the definitive English translation of the fourth edition, see Bibliography.

This induced me to undertake a cursory examination of relevant research in the fields of geology and palaeontology, and this provided immediately such a weighty corroboration that a conviction of the fundamental soundness of the idea [of continental drift] took root in my mind.

One other piece of evidence helped to persuade Wegener that he was on to something – the jigsaw-puzzle-like fit of the continents is even better if they are matched up not along the present-day shorelines (which depend on the height of the ocean today), but along the edges of the continental shelf, the true edge of the continents, where there is a steep plunge down to the ocean floor. But although the idea had taken root, several distractions delayed its full flowering. Shortly after the presentation of his first drift ideas in those talks in January 1912, Wegener set out on another Greenland expedition, returning in 1913 and marrying Else Köppen.[1] Any plans they had for a quiet academic life were shattered by the First World War, in which Wegener was called up as a reserve lieutenant and served on the Western Front, where he was wounded twice in the early months; unfit for further active service, he worked in the meteorological service of the army after his recovery. It was while on convalescent leave that he wrote the first version of his famous book (whose title translates as *The Origin of Continents and Oceans*). This made very little impact at the time. It was published in 1915, at the height of the war, and was scarcely more than a pamphlet, only running to 94 pages. After the war, Wegener worked for the German Marine Laboratory in Hamburg (again alongside his brother), and was also a lecturer in meteorology at the then-new University of Hamburg. He established a reputation as a distinguished meteorologist, but also continued to work on his model of continental drift, producing new editions of his book (each larger than its predecessor) in 1920 and 1922. Friends worried that this might damage his reputation, but whatever people thought about the drift idea, Wegener was such a good meteorologist that in 1924 he was appointed professor of meteorology at the University of Graz, in Austria. The same year, he published (with Wladimir Köppen) the first

1. The daughter of Russian-born meteorologist Wladimir Köppen (1846–1940), a friend and colleague of Wegener.

attempt at an explanation of past climates based on continental drift, and both French and English translations of the third (1922) edition of *Die Entstehung* appeared. But just when Wegener seemed to be gaining an audience for his ideas, the opportunity to promote them further was snatched from him, although he did prepare a fourth edition of his book, responding to criticisms of the third edition from the English-speaking world that had now been introduced to his ideas, which was published in 1929. In 1930, Wegener set off on yet another Greenland expedition, this time (at the age of 49) as its head; the aim of the expedition was to gather evidence in support of the drift hypothesis. The expedition ran into trouble on the desolate Greenland ice cap, and with supplies running low at an inland camp, on 1 November 1930 (his fiftieth birthday) Wegener set off for the main base on the coast with an Inuit companion. He never made it. The following spring, his body was found on the ice cap on the route between the two camps, neatly wrapped in his sleeping bag and marked with his upright skies; his companion was never seen again. Now, the continental drift idea would have to sink or swim without the aid of its chief proponent.

The evidence for Pangea

Wegener's model envisaged the Earth as made up of a series of layers, increasing in density from the crust to the core. He saw that the continents and the ocean floors are fundamentally different, with the continents explained as blocks of light granitic rock (known as sial, from the name silica–alumina which describes their composition) essentially floating on denser basaltic rock (sima, from silica–magnesium), which (underneath a layer of sediment) forms the rock of the ocean floor. He said that the present-day continental blocks still have essentially the same outlines as they have had since the breakup of a single supercontinent, Pangea, which contained all the land surface of the planet at the end of the Mesozoic era (about 150 million years ago by modern dating). One big weakness of Wegener's model is that he had no reason for the breakup of Pangea and could only invoke rather vague ideas such as a 'retreat from the pole' caused by centrifugal forces, or possible tidal effects, to produce continental drift. But he went further than his predecessors in pointing to the sites of rift valleys (such as the East African rift valley) as locations of incipient continental breakup, indicating that whatever the process is that drives continental drift, it is still continuing today, and thereby making his version of the drift idea a uniformitarian one. Crucially, he also based his ideas on an Earth of constant size, with no catastrophic (or even gradual) contraction or expansion. One of the weakest features of the model is that Wegener envisaged the continents ploughing through the sima of the sea floor, which geologists (rightly) found hard to swallow. But he linked his ideas with the way mountains had formed along the eastern edges of the North and South American continents as they had drifted away from Europe and Africa, with the continents being crumpled up as they ploughed through the sima. Mountain ranges such as the Himalayas, in the hearts of land masses, could be explained by the collision of continents.

Like the curate's egg, the details of Wegener's hypothesis were good in parts. Where it was particularly good was in the evidence he marshalled from paleoclimatology, showing how glaciation had occurred in the distant past simultaneously on continents that are now far apart from one another, and far from the polar regions; where it was particularly bad (apart from the fact that he often ignored evidence which did not support his case, which made geologists suspicious of the whole package) was in his belief that continental drift was happening so rapidly that Greenland had broken away from Scandinavia only 50,000 to 100,000 years ago, and was moving westward at a rate of 11 metres per year. This suggestion came from geodetic surveys carried out in 1823 and 1907, and the measurements were simply inaccurate; today, using laser range finding with satellites, we know that the Atlantic is actually widening at a rate of a couple of centimetres per year (it was, incidentally, in pursuit of improved geodetic data that Wegener made his last, fatal, trip to the Greenland ice cap). But the most valuable contribution he made to the development of the idea of continental drift was his synthesis, gathering evidence to support the former existence of the supercontinent of Pangea by linking mountain ranges, sedimentary rocks, evidence from the scars of ancient glaciations, and the distribution of both fossil and living plants and animals. In a telling analogy, Wegener made a comparison with a sheet of printed paper torn into fragments. If the fragments could be reassembled so that the printed words joined up to make coherent sentences, it would be compelling evidence that the fit of the pieces was correct; in the same way, the kind of evidence he gathered formed a coherent geological 'text' when the pieces of Pangea were reassembled. It is this broad sweep of evidence which made the case for continental drift, even before the mechanisms were fully understood.

The radioactive technique for measuring the age of rocks

In fact, even the key component of the mechanism of continental drift was already in place by the end of the 1920s, as one geologist in particular appreciated. That geologist was Arthur Holmes (1890–1965), who by the 1920s had become a leading expert on radioactive decay and was at the forefront of efforts to measure the age of the Earth using radioactive techniques. More than any other individual, he was, indeed, 'the man who measured the age of the Earth'. Holmes came from an unremarkable family in Gateshead, in the northeast of England (his father was a cabinet maker and his mother worked as a shop assistant). He went to the Royal College of Science in London in 1907, after passing examinations for a National Scholarship which provided him with thirty shillings a week (£1.50) during the academic year. This was not enough to live on even in 1907, and there was no prospect of financial support from his parents; Holmes just had to make do as best he could.

Around this time, both radioactivity and the age of the Earth were hot scientific topics, and the American Bertram Boltwood (1870–1927) had recently developed the technique for dating samples of rock from the proportions of lead and uranium isotopes they contain. Since the radioactive decay of uranium eventually produces lead, with (as we shall see in Chapter 13) a characteristic timescale, measuring these ratios can reveal the ages of rocks. As his undergraduate project in his final year, Holmes used the technique to date samples of Devonian rock from Norway, coming up with an age of 370 million years. Scarcely ten years into the twentieth century, even an undergraduate could get a date for a piece of rock, which was clearly by no means the oldest rock in the Earth's crust, that was far in excess of the timescale for the Solar System allowed by the idea of heat being released from the Sun only as a result of its gravitational collapse. Graduating in 1910 with a glowing reputation but a burden of debt from his undergraduate days, Holmes was delighted to obtain a well-paid post for six months as a prospecting geologist in Mozambique, at £35 per month. A serious bout of blackwater fever delayed his return home, and he also contracted malaria (a blessing in disguise, since it prevented him from joining the army in the First World War). With his finances now in order (he made a profit on the trip of £89 7s 3d), Holmes was able to join the staff of Imperial College (as the Royal College of Science had metamorphosed into in 1910), where he stayed until 1920, receiving a doctorate in 1917. He then worked in Burma for an oil company, returning to Britain in 1924 to become professor of geology at Durham University. He moved on to the University of Edinburgh in 1943, and retired in 1956. By then, he had firmly established the radioactive technique for measuring the ages of rocks, coming up with an age for the Earth itself of 4,500 ± 100 million years.[1] Along the way, he produced an influential textbook, *Principles of Physical Geology* (the title deliberately chosen as a nod to Lyell), which was first published in 1944 and, in revised editions, has been a standard text ever since. Part of its success may be explained by the way Holmes tackled the task of making geology intelligible. As he later wrote to a friend, 'to be widely read in English-speaking countries think of the most stupid student you have ever had then think how you would explain the subject to him'.[2]

Holmes's account of continental drift

Holmes's interest in continental drift was almost certainly aroused before 1920 by one of his colleagues at Imperial, John Evans, who read German fluently and became an early enthusiast for Wegener's ideas (he later wrote the Foreword for the first English edition of Wegener's book). The third edition of the book had just been published in England when Holmes returned from Burma, and this seems to have been the stimulus for him to take up the idea, during a break from the uranium–lead work, as soon as he had established himself in Durham. Although he started out favouring the contraction hypothesis, his understanding of radioactivity and the potential this provided to generate heat inside the Earth soon led him to change his views. The idea that convection might be associated with

1. It took so long to pin the age down accurately because although the principles of the technique were known by 1910, the technology required to make measurements of the required precision took several decades to develop. As ever, science needs technology to progress, as much as technology needs science.
2. Quoted by Lewis.

mountain building and continental drift was planted in his mind by the discussion of them in the Presidential Address given by A. J. Bull to the Geological Society in London in 1927 (just a hundred years after Charles Darwin went up to Cambridge University intending to become a clergyman). In December that year Holmes presented a paper to the Edinburgh Geological Society which took up these ideas. He suggested that although the continents indeed floated on denser material, more or less as Wegener proposed, they did not move through the sima. Rather, this denser material itself moved around very slowly, stirred by convection currents produced by the heat within the Earth, cracking apart in some places (such as the ocean ridge down the centre of the Atlantic Ocean) and pushing the continents on either side of the crack apart, while they collided in other parts of the globe. Apart from the radioactive heating, the key component of Holmes's model was time – 'solid' rock, warmed from beneath, could indeed stretch and flow, like very thick treacle (or like the 'magic putty' you can find in some toy shops), but only *very* slowly. It is no surprise that one of the first geologists to espouse continental drift was one of the first people to appreciate quantitatively the immense age of the Earth, and to be actively involved in measuring it. In 1930, Holmes produced his most detailed account of continental drift, describing how convection currents operating inside the Earth as a result of heat generated by radioactive decay could have caused the breakup of Pangea, first into two large land masses (Gondwanaland in the southern hemisphere; Laurasia in the north), which in turn fragmented and drifted to form the pattern of land that we see on the surface of the Earth today. All this was published in the *Transactions of the Geological Society of Glasgow*, including an estimate very much in line with present-day measurements that the convection currents would move continents about at a rate of some 5 centimetres a year – enough to produce the Atlantic basin from a crack in the crust over an interval of about 100 million years.

Very many pieces of the modern version of continental drift were already in place in 1930, and Holmes presented the evidence for drift in the final chapter of his *Principles* in 1944, clearly arguing the case, but honestly pointing out the flaws in Wegener's own presentation:

Wegener marshalled an imposing collection of facts and opinions. Some of his evidence was undeniably cogent, but so much of his advocacy was based on speculation and special pleading that it raised a storm of adverse criticism. Most geologists, moreover, were reluctant to admit the possibility of continental drift, because no recognized natural process seemed to have the remotest chance of bringing it about . . . Nevertheless, the really important point is not so much to disprove Wegener's particular views as to decide from the relevant evidence whether or not continental drift is a genuine variety of earth movement. Explanations may safely be left until we know with greater confidence what it is that needs to be explained.

And right at the end of his chapter on continental drift, after making the case for convection as the driving force for the process, Holmes wrote:

It must be clearly recognized, however, that purely speculative ideas of this kind, specially invented to match the requirements, can have no scientific value until they acquire support from independent evidence.

I wonder if Holmes knew how closely he was echoing the words put into the mouth of his fictional namesake by Arthur Conan Doyle in *A Scandal in Bohemia*:

It is a capital mistake to theorise before one has data. Insensibly, one begins to twist facts to suit theories, instead of theories to suit facts.

Virtually nothing had happened to strengthen the case for continental drift between 1930 and 1944 precisely because there were no new facts to go on. Of course, there was some resistance among the old guard to the new ideas, just because they were new – there are always people reluctant to throw out everything they have been taught in order to espouse a new understanding of the world, no matter how compelling the evidence. But in the context of the 1930s and 1940s, the evidence in support of continental drift was persuasive (maybe very persuasive, if you took on board Holmes's work) rather than compelling. There were still other well-regarded rival ideas, notably permanentism, and with Wegener dead and Holmes concentrating on his dating techniques, nobody went in to bat for continental drift, which gradually fell out of what favour it had had (to the point where just about the only criticism of Holmes's great book came from people who said that he should not have included a chapter supporting such cranky ideas). What made continental drift first respectable, and then an established paradigm, the standard model of how the Earth works, was indeed new evidence – new evidence that emerged in the 1950s and 1960s thanks to new technology, itself developed partly as a result of the dramatic boost to all the technological sciences provided by the Second World War. This is also the first example we shall encounter in the present book of the way science became a discipline where real progress could only be achieved by large numbers of almost interchangeable people working on big projects. Even a Newton could not have obtained all the information needed to make the breakthrough that converted the continental drift hypothesis into the theory of plate tectonics, although he undoubtedly would have been able to put the evidence together to form a coherent model.

Although the technological advances stemming from the Second World War eventually helped to provide the key evidence in support of continental drift, during the 1940s many geologists were working on war-related projects, serving in the armed forces or living in occupied countries where there was little opportunity for global scientific research. In the immediate aftermath of the war, rebuilding in Europe and the dramatically changed relationship between science and government in the United States helped to delay the development and application of the new techniques. Meanwhile, although papers discussing drift (both for and against) were published, it largely remained a backwater of the geological sciences. The idea was, however, ready and waiting in the wings when that new evidence, that might otherwise have proved extremely puzzling and difficult to explain, began to come in.

Geomagnetic reversals and the molten core of the Earth

The first new evidence came from the study of fossil magnetism – the magnetism found in samples of rock from old strata. The impetus for this work itself came originally from the investigation of the Earth's magnetic field, whose origin was still a puzzle in the 1940s. Walter Elsasser (1904–1991), one of many German-born scientists who left Germany when Adolf Hitler came to power and ended up in the United States, began, in the late 1930s, to develop the idea that the Earth's magnetism is generated by a natural internal dynamo, and he published his detailed ideas almost as soon as the war ended, in 1946. The idea was taken up by the British geophysicist Edward Bullard (1907–1980), who had worked during the war on techniques for demagnetizing ships ('degaussing') to protect them from magnetic mines. In the late 1940s, Bullard was working at the University of Toronto, where he developed further the model of the Earth's magnetic field as the product of circulating conducting fluids in the hot fluid core of the planet (crudely speaking, convection and rotation in molten iron). In the first half of the 1950s, as Director of the UK National Physical Laboratory in London, he used their early electronic computer for the first numerical simulations of this dynamo process.

By that time, measurements of fossil magnetism had shown that the Earth's magnetic field had had the same orientation relative to the rocks for the past 100,000 years. The rocks are magnetized when they are laid down, as molten material flowing from volcanoes or cracks in the Earth's crust, and once set they preserve the pattern of the magnetic field in which they formed, becoming like bar magnets. But British researchers in particular (notably small groups based at the universities of London, Cambridge and Newcastle upon Tyne) had found that in older rocks the direction of the fossil magnetism could be quite different from the orientation of the present-day geomagnetic field, as if either the field or the rocks had shifted position after the strata solidified. Even more strange, they found that there seemed to be occasions in the geological past when the geomagnetic field had the opposite sense to that of today, with north and south magnetic poles swapped. It was this paleomagnetic evidence that made the debate about continental drift hot up at the beginning of the 1960s, with some people using the magnetic orientations of the rocks from particular times in the geological past as the 'lines of print' to be matched up across the joins of continental reconstructions, and finding that those reconstructions broadly matched the ones made by Wegener.

Alongside all this, there had been a huge development in knowledge of the sea bed, which makes up two-thirds of the crust of the Earth's surface. Before the First World War, this was still largely a mysterious and unexplored world. The need for ways to counter the submarine menace encouraged development of the technology for identifying what lay beneath the surface of the ocean (particularly echo-location, or sonar) and the incentive to use the technology not just for detecting submarines directly but, after the war, to map the sea bed, partly out of scientific curiosity but also (as far as governments holding the purse strings were concerned) to locate hiding places for submarines. It was this technology which had begun, by the end of the 1930s, to fill in the outline features of sea floors, most notably indicating the presence of a raised system, a mid-ocean ridge, not just running down the Atlantic Ocean, but also forming a kind of spine down the centre of the Red Sea. The Second World War saw a huge improvement in the technology used for this kind of work, and the Cold War encouraged a continued high level of funding, as nuclear armed submarines became the primary weapons systems. In the United States, for example, the Scripps Institution of Oceanography had a budget of just under $100,000 in 1941, employed 26 staff and owned one small ship. In 1948, it had a budget of just under $1,000,000, a staff of 250 and four ships.[1] What the resources of Scripps and other ocean researchers found was quite unexpected. Before the 1940s, geologists had assumed that the sea floor represented the most ancient crust of the Earth – even supporters of continental drift thought this way. Because they were assumed to be ancient, the sea floors were also assumed to be covered with huge amounts of ancient sediment worn off the land over the eons, forming an essentially featureless layer perhaps 5 or 10 kilometres thick. And the crust itself, beneath the sediment, was assumed to be tens of kilometres thick, like the crust of the continents. When samples were obtained from the ocean bed and surveys were carried out, they showed that all of these ideas were wrong. There is only a thin layer of sediment, and hardly any at all away from the edges of the continents. All of the rocks of the sea floor are young, with the youngest rocks found next to the ocean ridges, which are geologically active features where underwater volcanic activity marks the line of a crack in the Earth's crust (so some of the rocks there have literally been born yesterday, in the sense that that is when they solidified from molten magma). And seismic surveys showed that the thickness of the Earth's crust is only

1. Figures from Le Grand.

about 5–7 kilometres under the oceans, compare with an average of 34 kilometres for the continental crust (in places, the continental crust is 80 or 90 kilometres thick).

The model of 'sea-floor spreading'

The pieces of the puzzle were put together in a coherent fashion by the American geologist Harry Hess (1906–1969), of Princeton University, in 1960. According to this model, which goes by the name of 'sea-floor spreading',[1] the ocean ridges are produced by convection currents in the fluid material of the mantle (the layer of treacly rock just below the solid crust) welling up from deeper below the surface. This warm material is not liquid in the sense that the water of the oceans themselves is liquid, but is hot enough to flow slowly as a result of convection, a little like warm glass.[2] The volcanic activity associated with the ocean ridges marks the place where this hot material breaks through to the surface. It then spreads out on either side of the ridge, pushing the continents on either side of the ocean basin apart, with the youngest rocks solidifying next to the ridges today and with the older rocks, laid down tens or hundreds of millions of years earlier, further away from the ridges, where they have been pushed to make room for the new material. And there is no need for the continents to be ploughing through the ocean crust – which is just as well, since surveys of the sea floor show no evidence of this. New oceanic crust created in this way is widening the Atlantic at a rate of about 2 centimetres a year, roughly half the speed that Holmes suggested. There are some echoes of Holmes's ideas in Hess's model, but the crucial difference is that where Holmes could only talk in general terms based on the fundamental laws of physics, Hess had direct evidence of what was going on and could put numbers derived from measurements of the ocean crust into his calculations. Holmes largely ignored the ocean basins in his model, for the very good reason that very little was known about them at the time; after Hess's work

1. The term 'spreading sea-floor theory' appeared in a paper published in 1961, and this was soon adapted to the more snappy 'sea-floor spreading'.

2. The broad features of the internal structure of the Earth are now quite well known, because the Earth's interior has been probed by studying seismic waves produced in earthquakes and by nuclear bombs during the Cold War era of underground testing; the details, alas, are among the many detailed aspects of modern science that we do not have room to discuss.

had been fully assimilated, which took most of the 1960s, the ocean basins were seen as *the* sites of action in continental drift, with the continents themselves being literally carried along for the ride as a result of activity associated with the crust of the ocean floors.

Although the Atlantic is getting wider, this does not mean that the Earth is expanding at the rate required to explain the formation of the entire Atlantic basin in a couple of hundred million years, roughly 5 per cent of the age of the Earth, the rate required by those measurements. Convection currents go up in some places, but down in others. The second key component of Hess's model of sea-floor spreading was the suggestion that in some parts of the world (notably along the western edge of the Pacific Ocean), thin oceanic crust is being forced down under the edges of thicker continental crust, diving back down into the mantle below. This explains both the presence of very deep ocean trenches in those parts of the world and the occurrence of earthquakes and volcanoes in places such as Japan – islands like those of Japan, indeed, are explained as being entirely produced by the tectonic activity[1] associated with this aspect of sea-floor spreading. The Atlantic Ocean is getting wider, but the Pacific Ocean is narrowing. Eventually, if the process continues, America and Asia will collide to form a new supercontinent; meanwhile, the Red Sea, complete with its own spreading ridge, marks the site of a new region of upwelling activity, cracking the Earth's crust and beginning to splinter Africa away from Arabia to the east.

As the model developed, it was also able to explain features such as the San Andreas Fault in California, where the widening of the Atlantic has pushed America westwards to overrun a less active spreading zone that formerly existed in what was, hundreds of millions of years ago, an even wider Pacific basin. Faults like the San Andreas also provided circumstantial evidence in support of the new ideas, as some geologists were quick to point out. There, blocks of the Earth's crust are moving past one another at a rate of a few centimetres per year, roughly the same speeds required for the new version of continental drift and proof that the 'solid' Earth was by no means fixed in one permanent geographical pattern. The traditional analogy,

1. Literally, building activity, from the Greek word for builder.

which has never been bettered, is that sea-floor spreading is like a slow conveyor belt, endlessly looping round and round. Over the entire surface of the globe, everything cancels out and the planet stays the same size.[1]

Hess's model, and the evidence on which it was based, inspired a new generation of geophysicists to take up the challenge of building a complete theory of how the Earth works, from this beginning. One of the leading players in what was very much a team game was Dan McKenzie (born 1942), of the University of Cambridge, who recalls[2] that it was a talk given by Hess in Cambridge in 1962, when McKenzie was still an undergraduate, that fired his imagination and set him thinking about the problems remaining to be solved by the model, and seeking other evidence to support it. Slightly more senior geophysicists in Cambridge were similarly inspired by that talk, and two of them, graduate student Frederick Vine (1939–1988) and his thesis supervisor Drummond Matthews (born 1931), combined the following year in a key piece of work linking the evidence for geomagnetic reversals with the sea-floor spreading model of continental drift.

By the early 1960s, as well as the growing mass of data about the magnetic history of the Earth obtained from the continents, the pattern of magnetism over parts of the sea bed had begun to be mapped, using magnetometers towed by survey ships. One of the first of these detailed surveys was carried out in the northeast Pacific, off Vancouver Island, around a geological feature known as the Juan de Fuca Ridge. Such surveys had shown a stripy pattern of magnetism in the rocks of the sea bed, with the stripes running more or less north–south; in one stripe the rocks would be magnetized in line with the present-day geomagnetic field, but in adjacent stripes the rocks would have the opposite magnetism. When plotted on a map, with one orientation shaded black and the other in white, the pattern resembles a slightly distorted bar code. Vine and Matthews suggested that these patterns were produced as a result of sea-floor spreading. Molten rock flowing

1. There is some evidence that geographical reconstructions of ancient supercontinents fit better if the Earth has actually expanded by a very small amount since the breakup of Pangea. This is intriguing, but even if the evidence stands up, the effect is only one of detail, not a major factor in driving continental drift.
2. Conversation with JG, circa 1967.

from an oceanic ridge and setting would be magnetized with the magnetism corresponding to the Earth's field at the time. But the continental evidence showed that the Earth's magnetic field reversed direction from time to time.[1] If Vine and Matthews were correct, it meant two things. First, the pattern of magnetic stripes on the ocean floor should be correlated with the pattern of geomagnetic reversals revealed by continental rocks, providing a way to check the two patterns against one another and refine the magnetic dating of rocks. Second, since, according to Hess, crust spread out evenly on both sides of an oceanic ridge, the pattern of magnetism seen on one side of such a ridge should be the mirror image of the pattern seen on the other side of the ridge. If so, this would be striking confirmation that the sea-floor spreading model was a good description of how the Earth worked.

Further developments on continental drift

With the limited data available in 1963, the arguments put forward by Vine and Matthews could only be suggestive, not conclusive evidence in support of sea-floor spreading and continental drift. But Vine, in collaboration with Hess and with the Canadian geophysicist Tuzo Wilson (1908–1993), developed the idea further, taking on board new magnetic data that were coming in from around the world, and soon made the case compelling. Among the key contributions Wilson made was the realization that a spreading ridge like the one running down the Atlantic Ocean need not be a continuous feature, but could be made up of narrow sections which got displaced sideways from one another (along so-called transform faults), as if they were not one wide conveyor belt but a series of narrow conveyor belts lying side by side; he also played a major part in packaging many of the ideas of the new version of continental drift into a coherent whole. He was a leading advocate for these ideas, and coined the term 'plate' for one of the rigid portions of the Earth's crust (oceanic, continental or a combination of both)

1. It is still not known exactly why this occurs, but it is thought to be a result of the dynamo effect operating in the Earth's fluid core dying away to nothing, then building up again in the opposite sense. Intriguingly, the Sun, which is thought to have a similar internal dynamo, undergoes a similar pattern of magnetic reversals, but much more rapidly and much more regularly, associated with the roughly 11-year long sunspot cycle.

that are being moved around by the forces associated with sea-floor spreading and continental drift.

The clinching evidence in support of the sea-floor spreading model came in 1965, when a team on board the research vessel *Eltanin* carried out three transverse magnetic surveys across a ridge known as the East Pacific Rise. The surveys showed a striking similarity between the magnetic stripes associated with the East Pacific Rise and those of the Juan de Fuca Ridge, further to the north – but they also showed such a pronounced lateral symmetry, from the pattern on one side of the Rise to a mirror image of that pattern on the other side of the Rise, that when the charts were folded along the line denoting the ridge the two plots lay one on top of the other. The results were announced in April 1966 at a meeting of the American Geophysical Union, held in Washington, DC; they were then published in a landmark paper in the journal *Science*.[1]

The 'Bullard fit' of the continents

Meanwhile, the traditional approach to gathering evidence in favour of continental drift had been getting a boost. In the early 1960s, Bullard (by now head of the Department of Geodesy and Geophysics in Cambridge) championed the case that geological evidence in support of the idea of drift had by then overcome the difficulties the model had encountered in the 1920s and 1930s, and he presented the case for drift to a meeting of the Geological Society in London in 1963. The next year, he helped to organize a two-day symposium on continental drift at the Royal Society, where all the latest work was discussed, but where, ironically, the greatest impact was made by a new version of a very old idea – a jigsaw-puzzle-like reconstruction of Pangea. This reconstruction used what was presented as an objective method, based on a mathematical rule for moving things about on the surface of a sphere (Euler's theorem), and with the actual reconstruction to provide the 'best fit', defined mathematically, carried out by an electronic computer to provide an unbiased, objective matching. The result is strikingly similar to Wegener's fit of the continents and, in truth, said little that was new. But in 1964 people were still impressed by electronic computers and, surely even more significantly, were also, as they had not been

1. W. C. Pitman and J. P. Heirtzler, *Science*, volume 154, pp. 1164–71, 1966.

forty years previously, disposed by the gathering weight of other evidence to take continental drift seriously. Whatever the psychological reasons, the 'Bullard fit' of the continents, published in 1965,[1] has gone down as a defining moment in the story of the development of the theory of continental drift.

Plate tectonics

By the end of 1966, the evidence in support of continental drift and sea-floor spreading was compelling, but had not yet been assembled into a complete package. Most of the Young Turks of geophysics tackled the problem, racing to be first to publish. The race was won by Dan McKenzie (fresh from the award of his PhD in 1966) and his colleague Robert Parker, who published a paper in *Nature* in 1967[2] introducing the term plate tectonics for the overall package of ideas and using it to describe in detail the geophysical activity of the Pacific region – the Pacific plate, as it is now known – in terms of the way plates move on the surface of a sphere (Euler's theorem again). Jason Morgan, of Princeton University, came up with a similar idea, published a few months later, and although many details remained to be filled in (and are still being tackled today), what is sometimes referred to as 'the revolution in the Earth sciences'[3] had been completed by the end of the year. The essence of plate tectonics is that seismically quiet regions of the globe are quiet because they form rigid plates (six large plates and about a dozen small ones, between them covering the entire surface of the globe). An individual plate may be made up of just oceanic crust or just continental crust, or both; but most of the interesting geological activity going on at the surface of the Earth happens at the boundaries between plates – plate margins. Constructive margins are regions where, as we have seen, new oceanic crust is being made at ocean ridges and spreading out on either side. Destructive margins are regions where one plate is being pushed under the edge of another, diving down at an angle of about 45 degrees and melting back into the magma below. And conservative

1. Blackett, Bullard and Runcorn, *A Symposium on Continental Drift*.

2. Volume 216, pp. 1276–80.

3. Of course, it was *not* a revolution; we hope we have made clear the way in which the ideas evolved, with new models patiently building on new data in the usual way that science progresses. The idea of scientific revolutions is essentially a myth beloved by sociologists who have never worked at the scientific coal face.

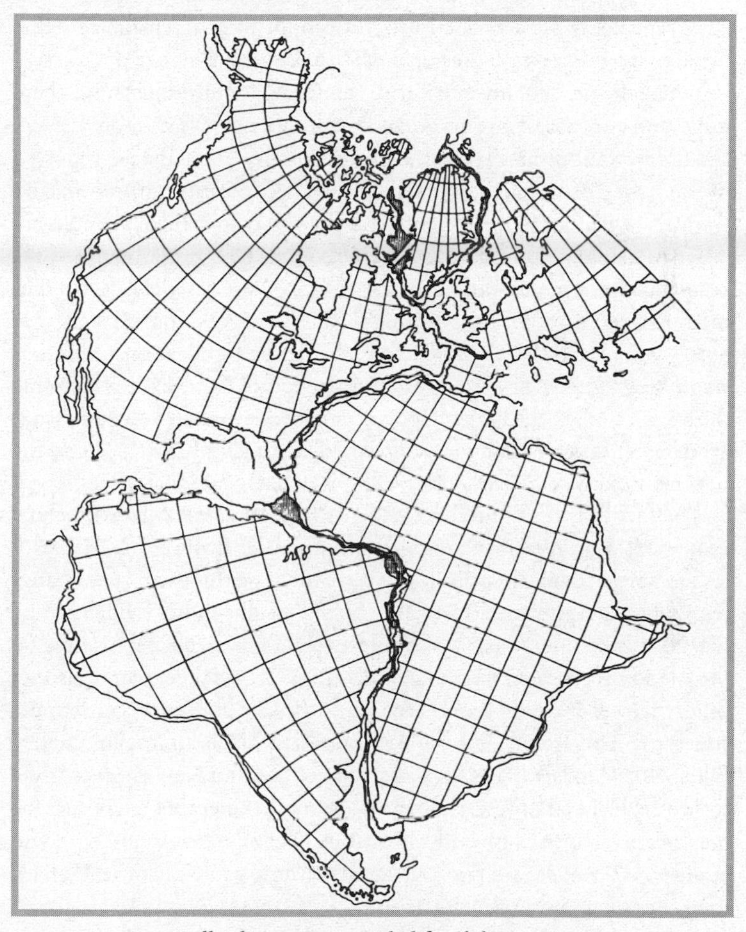

34. *Bullard's computer-aided fit of the continents prior to the opening of the Atlantic.*

margins are regions where crust is neither being created nor destroyed, but the plates are rubbing sideways past one another as they rotate, as is happening along the San Andreas Fault today. Evidence such as the existence of ancient mountain ranges and former sea beds in the hearts of continents today shows that all this tectonic activity has been going on since long before the breakup of Pangea, and that supercontinents have repeatedly been broken up and rebuilt in different patterns by the activity going on on the surface of the restless Earth.

When the Open University was founded in Britain in 1969, these ideas and the rest of the package that made up plate tectonics were already becoming familiar to the professionals, and had been reported in the pages of popular magazines such as *Scientific American* and *New Scientist*, but had not yet found their way into the textbooks. In order to be bang up to date, in keeping with the vibrant image of the young institution, the staff at the Open University rapidly put together their own text, the first built around the global theory of plate tectonics. Since we have to draw a line somewhere, the publication of *Understanding the Earth*[1] in 1970, neatly at the end of the decade which saw the 'revolution' in the Earth sciences, can be conveniently (if somewhat arbitrarily) taken as the moment when continental drift became the new orthodoxy – the last great triumph of classical science.

The fact that the continents have drifted, once established, helped to provide a new basis for understanding other features of the Earth, and in particular the relationship between living things and the changing global environment. The value of the insight this provides can be highlighted by one example. Alfred Russel Wallace, during his time on the islands of the Malay Archipelago, noticed that there was a distinct difference between the species to the northwest and the species to the southeast. This region between Asia and Australia is almost completely filled with islands, ranging in size from Borneo and New Guinea down to tiny atolls, and at first sight provides no insuperable barrier to the movement of species in both directions. Yet Wallace found that you could mark a narrow band on the map (now known as the Wallace line), running more or less from southwest to northeast between Borneo and New Guinea, with a distinctive Asian fauna to the northwest of this transition zone and a distinctive Australian fauna to the southeast, with little blurring between the two. This was a great puzzle at the time, but can be explained within the context of plate tectonics, where modern studies reveal that during the breakup of the southern supercontinent of Gondwanaland, Indo-Asia broke away first and moved to the northwest, where natural selection applied evolutionary

1. Edited by Gass, Smith and Wilson.

pressures different from those at work in Australia-Antarctica that had been left behind. In a later phase of tectonic activity, Australia-New Guinea broke away from Antarctica, and moved northwards rapidly (by the standards of continental drift), eventually catching up with Asia. The two continents have only recently come into close proximity once again, and there has not yet been time for species from either side to mingle significantly across the Wallace line. Wegener himself commented on this possibility (writing in the third edition of his book, in 1924, only sixty-five years after Darwin and Wallace published their theory of natural selection, and just eleven years after Wallace had died); but it took plate tectonic theory to prove the point.

Continental drift is relevant to many aspects of the evolution of life on Earth, and is especially relevant to our theme of how science has refined our understanding of the relationship between human beings and the Universe at large, and our continual displacement from centre stage by new discoveries. Like Wallace, Charles Darwin explained how evolution works, but before he did that he was a geologist, and he would surely have been intrigued and delighted to learn of the modern understanding of the way in which continental drift and climate have come together to shape our species. It begins with the story of Ice Ages.

The story of Ice Ages: Jean de Charpentier

Even before the beginning of the nineteenth century, there were people who thought that glaciation in Europe had been much more extensive in the past than it is today. The most obvious evidence of this is the presence of huge boulders dumped far from the strata where they belong, having been carried there by glaciers which have since melted back and retreated – so it is hardly surprising that one of the first people to draw attention to these so-called 'erratics' was a Swiss, Bernard Kuhn, in 1787. But it is rather more surprising that he had this idea even though he was a clergyman; the received wisdom of the time was that all such phenomena could be explained by the effects of the Biblical Flood, whatever mountain folk might think from the evidence of their everyday contact with the effects of glaciers. Almost everyone subscribed to the received wisdom, and supporters of glaciation as the explanation of erratics were very much in the minority for decades to come. They included James Hutton, who was convinced by the evidence he saw on a visit to the

Jura mountains; the Norwegian Jens Esmark, writing in the 1820s; and the German Reinhard Bernhardi, who knew of Esmark's work and published an article in 1832 suggesting that the polar ice cap had once extended as far south as central Germany. This was just a year before Charles Lyell came up with the idea that erratics had indeed been transported by ice, but not by glaciers – he suggested, in the third volume of *Principles of Geology*, that great boulders could have been carried embedded in icebergs or resting on ice rafts, which floated on the surface of the Great Flood. But the chain which eventually led to a proper model of Ice Ages began not with any of the great names of nineteenth-century science, but with a Swiss mountaineer, Jean-Pierre Perraudin.[1]

Perraudin observed, up in the now ice-free mountain valleys, the way in which hard rock surfaces that did not weather easily had been scarred by something pressing down strongly on them, and realized that the most likely explanation was that they had been gouged by rocks scraped over them by ancient glaciers. In 1815, he wrote about these ideas to Jean de Charpentier, as he then was, a mining engineer who was also a well-known naturalist with geological interests that extended beyond the strict requirements of his profession. He was born Johann von Charpentier, in 1786, in the German town of Freiberg, but moved to Switzerland in 1813 and adopted a French version of his name; he stayed there, at Bex in the valley of the Aar, for the rest of his life, dying in 1855. De Charpentier found the idea of erratics being transported by glaciers too extravagant to accept at the time, although he was equally unimpressed by the idea that they had been carried to their present locations by water. Undaunted, Perraudin continued to present the evidence to anyone who would listen and found a sympathetic audience in the form of Ignace Venetz, a highway engineer whose profession, like de Charpentier's, encouraged a broad knowledge of geology. Venetz gradually became persuaded by the evidence, including piles of debris found several kilometres beyond the end of the Flesch glacier, which seemed to be terminal moraines (heaps of geological rubbish left at the ends of glaciers) from a time when the glacier extended

1. Not a mountaineer in the modern sporting sense, but somebody who made his living on the mountains, in this case by hunting chamois.

further down the valley. In 1829, he presented the case for former glaciation to the annual meeting of the Swiss Society of Natural Sciences, where just about the only person he convinced was de Charpentier, an old acquaintance with whom he had already discussed some of these ideas. It was de Charpentier who then picked up the baton, gathering more evidence over the next five years and presenting an even more carefully argued case to the Society of Natural Sciences in 1834. This time, nobody at all seems to have been persuaded (perhaps partly because Lyell's ice-rafting model seemed to solve some of the puzzles involved in explaining erratics in terms of the Great Flood). Indeed, one member of the audience, Louis Agassiz, was so irritated by the notion that, in the best scientific tradition, he set out to disprove it and stop people discussing this nonsense once and for all.

Louis Agassiz and the glacial model

Agassiz (who was christened Jean Louis Rodolphe, but was always known as Louis) was a young man in a hurry. He was born at Motier-en-Vuly, in Switzerland, on 28 May 1807, and studied medicine at Zurich, Heidelberg and Munich before moving on to Paris, in 1831, where he was influenced by Georges Couvier (then nearing the end of his life). He had already turned his attention to paleontology and soon became the world's leading expert on fossil fishes. In 1832, Agassiz returned to Switzerland, where he was appointed professor of natural history at a new college and natural history museum being established in Neuchâtel, the capital of the region where he had been brought up. This part of Switzerland had a curious double status at the time. From 1707 onwards, although French-speaking, it had been part of the domain of the King of Prussia (except for a brief Napoleonic interregnum). In 1815, Neuchâtel joined the Swiss Confederation, but the Prussian link was neither formally acknowledged nor formally revoked (one reason why Agassiz studied in Germany) and the new college was supported by Prussian funds. When he took up the post, Agassiz already knew de Charpentier (they had met when Agassiz was still a schoolboy in Lausanne), who he liked and respected, and had visited the older man for busmen's holidays during which they had probed the geology of the region around Bex. De Charpentier tried to persuade Agassiz that there had been a great glaciation; Agassiz tried to find evidence that there had not.

After another summer geologizing with de Charpentier around Bex in 1836, Agassiz was completely convinced, and took up the cause with all the evangelical enthusiasm of a convert. On 24 June 1837 he stunned the learned members of the Swiss Society of Natural Sciences (meeting, as it happened, at Neuchâtel) by addressing them, in his role as President, not with an anticipated lecture on fossil fishes, but with a passionate presentation in support of the glacial model, in which he used the term Ice Age (*Eizeit*; Agassiz lifted the term from the botanist Karl Schimper, one of his friends and colleagues). This time the idea really made waves. Not that people were convinced, but because Agassiz's enthusiasm, and his position as President, meant that it could not be ignored. He even dragged the reluctant members of the Society up into the mountains to see the evidence for themselves, pointing out the scars on the hard rocks left by glaciation (which some of those present still tried to explain away as produced by the wheels of passing carriages). His colleagues were unimpressed, but Agassiz went ahead anyway, determined to find compelling evidence in support of the Ice Age model. To this end, Agassiz set up a small observing station (basically a little hut) on the Aar glacier to measure the movement of the ice by driving stakes into it and noting how quickly they moved. To his surprise, on summer visits over the next three years he discovered that the ice moved even faster than he had anticipated and that it could indeed carry very large boulders along with it. Fired by these discoveries, in 1840 Agassiz published (privately, at Neuchâtel) his book (*Études sur les Glaciers*), *Studies on Glaciers* which thrust the Ice Age model firmly into the arena of public debate.

In fact, Agassiz went completely over the top. It's very difficult not to sit up and take notice of (for or against) a scientist who argues that the entire planet had once been sheathed in ice, and who makes his case in this kind of language:

> The development of these huge ice sheets must have led to the destruction of all organic life at the Earth's surface. The ground of Europe, previously covered with tropical vegetation and inhabited by herds of giant elephants, enormous hippopotami, and gigantic carnivora became suddenly buried under a vast expanse of ice covering plains, lakes, seas and plateaus alike. The silence of death followed . . . springs dried up, streams ceased to flow, and sunrays

rising over that frozen shore . . . were met only by the whistling of northern winds and the rumbling of the crevasses as they opened across the surface of that huge ocean of ice.

These wild exaggerations managed to annoy even de Charpentier, who published his own, more sober (and less entertaining) account of the Ice Age model in 1841. It also planted Agassiz's version of Ice Ages firmly in the catastrophist camp, as that 'suddenly' highlights, reducing its chances of receiving a welcome from Lyell and his followers. But evidence continued to accumulate, and as it did so, the fact that there had been at least one great Ice Age could no longer be ignored; before too long even Lyell was convinced that the model could be shorn of its catastrophist trappings and made acceptable to uniformitarians.

Several years earlier, Agassiz had visited Britain to study collections of fossil fish, staying for part of the time in Oxford with William Buckland, Lyell's old mentor (but still a confirmed catastrophist), with whom he became friends. A year after Agassiz had startled his colleagues with his talk at Neuchâtel, Buckland attended a scientific meeting in Freiberg, where he heard Agassiz expound his ideas, then travelled on with his wife to Neuchâtel to see the evidence for himself. He was intrigued, but not immediately convinced. In 1840, however, Agassiz made another trip to Britain to study fossil fish and took the opportunity to attend the annual meeting of the British Association for the Advancement of Science (that year held in Glasgow) to present his Ice Age model. Following the meeting, Agassiz joined Buckland and another geologist, Roderick Murchison (1792–1871), on a field trip through Scotland, where the evidence in support of the model finally persuaded Buckland that Agassiz was right. Agassiz then went on to Ireland, while Buckland paid a visit to Kinnordy, where Charles and Mary Lyell had gone to stay after the Glasgow meeting. Within days, taking advantage of the evidence of former glaciation in the immediate vicinity, he had convinced Lyell, and on 15 October 1840 wrote to Agassiz:

Lyell has adopted your theory *in toto!!* On my showing him a beautiful cluster of moraines within two miles of his father's house, he instantly accepted it, as solving a host of difficulties that have all his life embarrassed him. And not these only, but similar moraines and detritus of moraines that cover half of the adjoining counties are explicable on your theory, and he has consented to my proposal that he should immediately lay them all down on a map of the county and describe them in a paper to be read the day after yours at the Geological Society.'[1]

The conversion of Lyell was not quite as dramatic as it seems, since (as this passage shows) he had already been worrying about the origin of these geological features; he had also visited Sweden in 1834, where he cannot have failed to notice the evidence of glaciation, even if he did not interpret it that way immediately. Compared with a Great Flood (still the preferred alternative), glaciation *was* uniformitarian – after all, there are glaciers on Earth today.

Buckland was referring in that letter to a forthcoming meeting of the Geological Society in London, where Agassiz was already listed as a speaker. In the end, papers by Agassiz himself, Buckland and Lyell, all in support of the Ice Age model, were spread over two meetings of the Society, and were read on 18 November and 2 December. It would be another twenty years or more before the model became fully accepted, but for our purposes we can cite those meetings where such established geological luminaries as Buckland and Lyell started preaching the gospel as the moment when the Ice Age model came in from the cold. The next big question to be answered would be, what made the Earth colder during an Ice Age? But before we look at how that question was answered, we should take a brief look at what happened to Agassiz after 1840.

In 1833, Agassiz had married Cécile Braun, a girl he had met as a student in Heidelberg. The couple were initially very happy, and had a son (Alexander, born in 1835) and two younger daughters (Pauline and Ida). But by the middle of the 1840s their relationship had deteriorated, and in the spring of 1845 Cécile had left Switzerland to stay with her brother in Germany, taking her two young daughters but leaving the older son to complete the current phase of his education in Switzerland. About this time (and a contributory factor to the breakup of the marriage), Agassiz was in severe financial difficulties because of

1. Quoted in Elizabeth Carey Agassiz.

an unwise involvement in an unsuccessful publishing venture. It was against this background that, in 1846, he left Europe for what was supposed to be a year-long trip to the United States, seeing geological features of the New World with his own eyes and giving a series of lectures in Boston. He was delighted both by the abundant evidence of glaciation that he saw – some within walking distance of the docks at Halifax, Nova Scotia, where the ship stopped before moving on to Boston – and by the discovery that his ideas about Ice Ages had not only preceded him across the Atlantic but had been widely accepted by American geologists. American geologists were equally delighted with Agassiz, and decided they wanted to keep him. In 1847, a new chair was established at Harvard for his benefit, solving his financial problems as well as giving him a secure academic base. He became professor of zoology and geology and stayed there for the rest of his life, establishing the Museum of Comparative Zoology in 1859 (the year Darwin's *Origin* was published). Agassiz was a major influence on the development of the way his subjects were taught in the United States, emphasizing the need for hands-on investigation of natural phenomena; he was also a popular lecturer who helped to spread interest in science outside the campuses. Nobody is perfect, though, and to the end of his life Agassiz did not accept the theory of natural selection.

The American move proved timely for political, as well as for personal, reasons. In 1848, the wave of revolutionary activity in Europe engulfed Neuchâtel and the links with Prussia were finally severed. The college (which had actually been elevated to the status of an Academy in 1838, with Agassiz as its first rector) lost its funding and closed. The turmoil across Europe encouraged many naturalists to cross the Atlantic, some coming to work with Agassiz and boosting the status of the work going on at Harvard. Also in 1848, news came from Europe that Cécile had died, of tuberculosis. Pauline and Ida went to stay with their Swiss grandmother, Rose Agassiz, while their brother (who had joined the family in Freiburg just a year before) stayed with his uncle in Freiburg to finish his schooling. In 1849, Alexander joined Louis in Cambridge, Massachusetts; he eventually became a distinguished naturalist and founded the American branch of the Agassiz family. In 1850, Louis got married for the second time, to Elizabeth Cary, and brought his two daughters, then aged 13 and 9, over from Europe to join the family. He enjoyed nearly a quarter of a century of both domestic happiness and academic success in his new country, and died in Cambridge on 14 December 1873.

The astronomical theory of Ice Ages

The roots of what is sometimes called the astronomical theory of Ice Ages go back to Johannes Kepler's discovery, early in the seventeenth century, that the orbits of the planets (including the Earth) around the Sun are elliptical, not circular. But the story really begins with the publication in 1842, soon after Agassiz had published his own book on Ice Ages, of a book called *Révolutions de la mer* (*Revolutions of the Sea*) by the French mathematician Joseph Adhémar (1797–1862). Because the Earth moves in an ellipse around the Sun, for part of its orbit (part of the year) it is closer to the Sun than it is at the other end of its orbit (the other half of the year). In addition, the axis of the spinning Earth is tilted relative to a line joining the spinning Earth with the Sun, at an angle of 23½ degrees from the vertical. Because of the gyroscopic effect of the Earth's rotation, on a timescale of years or centuries, this tilt always points in the same direction relative to the stars, which means that as we go around the Sun, first one hemisphere and then the other is leaning towards the Sun and gets the full benefit of the Sun's warmth. That is why we have seasons.[1] On 4 July each year the Earth is at its furthest distance from the Sun and on 3 January it is at its closest – but the difference amounts to less than 3 per cent if its 150 million kilometres average distance from the Sun. The Earth is furthest from the Sun in northern hemisphere summer, and is therefore moving at its slowest in its orbit then (remember Kepler's laws). Adhémar reasoned (correctly) that because the Earth is moving more slowly during southern winter, the number of hours of total darkness experienced at the South Pole in winter is longer than the number of hours of continual daylight that the same region receives in southern summer, when the Earth is at the opposite end of its orbit and moving fastest. He believed that this meant that the south polar region was getting colder as the centuries passed, and

1. On timescales of thousands and tens of thousands of years, the tilt is affected by various wobbles, which we are coming on to shortly.

that the Antarctic ice cap (which he believed to be still growing) was proof of this.

The elliptical orbit model

But the same thing can happen in reverse. Like a spinning top, the Earth wobbles as it rotates, but, being much bigger than a child's top, the wobble (known as the precession of the equinoxes) is slow and stately. It causes the direction in which the axis of rotation of the Earth points relative to the stars to describe a circle on the sky once every 22,000 years.[1] So 11,000 years ago, the pattern of the seasons relative to the elliptical orbit was reversed – *northern* winters occurred when the Earth was furthest from the Sun and moving most slowly. Adhémar envisaged an alternating cycle of Ice Ages, with first the southern hemisphere and then the northern hemisphere, 11,000 years later, being covered by ice. At the end of an Ice Age, as the frozen hemisphere warmed up, he imagined the seas gnawing away at the base of a huge ice cap, eating it into an unstable mushroom shape, until the whole remaining mass collapsed into the ocean and sent a huge wave rushing into the opposite hemisphere – which is where the title of his book came from. In fact, the entire basis of Adhémar's model was as shaky as he imagined those collapsing ice sheets to be. The idea that one hemisphere of the Earth is getting warmer while the other is getting colder is just plain wrong. As the German scientist Alexander von Humboldt (1769–1859) pointed out in 1852, astronomical calculations dating back more than a hundred years to the work of the French mathematician Jean d'Alembert (1717–1783) showed that the cooling effect Adhémar relied on is exactly balanced (it has to be exact, because both effects depend on the inverse square law) by the extra heat that the same hemisphere receives in summer, when the Earth is closest to the Sun. The total amount of heat received by each hemisphere during the course of a year is always the same as the total amount of heat received over the year by the opposite hemisphere. In the twentieth century, of course, as the understanding of the geological record improved and radioactive dating techniques became available, it became clear that there

1. Modern calculations reveal that the cycle actually varies from 23,000 to 26,000 years, on still longer timescales, as a result of gravitational interactions with other bodies in the Solar System.

is no pattern of alternating southern and northern glaciations 11,000 years apart. But although Adhémar's model was wrong, his book was the trigger which set the next person in the story thinking about orbital influences on climate.

James Croll

James Croll was born in Cargill, Scotland, on 2 January 1821. The family owned a tiny piece of land, but their main source of income was from the work of Croll's father as a stonemason. This meant that he travelled for much of the time, leaving his family to cope with the farming. The boy received only a basic education, but read avidly and learned the basics of science from books. He tried working at various trades, starting out as a millwright, but found that 'the strong natural tendency of my mind towards abstract thinking somehow unsuited me for the practical details of daily work'.[1] The situation was complicated further when his left elbow, injured in a boyhood accident, stiffened almost to the point of uselessness. This restricted Croll's opportunities for work, but gave him even more time to think and read. He wrote a book, *The Philosophy of Theism*, which was published in London in 1857 and, amazingly, made a small profit. Two years later, he found his niche with a job as janitor at the Andersonian College and Museum in Glasgow. 'Taking it all in all,' he wrote, 'I have never been in a place so congenial to me ... My salary was small, it is true, little more than sufficient to enable me to subsist; but this was compensated by advantages for me of another kind.' He meant access to the excellent scientific library of the Andersonian, peace and quiet, and lots of time to think. One of the things Croll read there was Adhémar's book; one of the things he thought about was the way changes in the shape of the Earth's orbit might affect the climate.

This idea built from the detailed analysis of the way the Earth's orbit changes with time, which had been carried out by the French mathematician Urbain Leverrier (1811–1877). Leverrier is best remembered for his work that led to the discovery of the planet Neptune in 1846 (the same calculations were made independently in England by John Couch Adams (1819–1892)). This was a profound

1. For Croll's autobiographical sketch, see Irons. Other quotes from Croll are from the same source.

piece of work which predicted the presence of Neptune on the basis of Newton's laws and the way in which the orbits of other planets were being perturbed by an unseen gravitational influence, after allowance had been made for the gravitational influence of all the known planets on each other. It was far more profound than the discovery of Uranus by William Herschel (1738–1822) in 1781, even though that caused much popular excitement as the first planet discovered since the time of the Ancients. Herschel's discovery was lucky (in so far as building the best telescope in the world and being a superb observer was lucky). Neptune's existence was (like the return of Halley's comet in 1758) predicted mathematically, a great vindication of Newton's laws and the scientific method. But the prediction involved horrendously laborious calculations with paper and pencil in those pre-computer days, and one of the fruits of those labours was the most accurate analysis yet of how the shape of the Earth's orbit changes, on a timescale roughly 100,000 years long. Sometimes, the orbit is more elliptical and sometimes it is more circular. Although the *total* amount of heat received by the whole planet over an entire year is always the same, when the orbit is circular, the amount of heat received by the planet from the Sun each week is the same throughout the year; when the orbit is elliptical, more heat is received in a week when it is close to the Sun than in a week when it is at the other end of its orbit. Could this, Croll wondered, explain Ice Ages?

The model he developed assumed that an Ice Age would occur in whichever hemisphere suffered very severe winters, and he combined both the changes in ellipticity calculated by Leverrier and the effect of the precession of the equinoxes to produce a model in which alternating Ice Ages in each hemisphere are embedded in an Ice Epoch hundreds of thousands of years long. According to this model, the Earth had been in an Ice Epoch from about 250,000 years ago until about 80,000 years ago, since when it had been in a warm period between Ice Epochs, dubbed an Interglacial. Croll went into much more detail, including a sound discussion of the role of ocean currents in climate, in a series of papers which began with his first publication on Ice Ages in the *Philosophical Magazine* in 1864, at the age of 43. His work immediately attracted considerable attention, and Croll soon realized his lifelong ambition of becoming a full-time scientist. In 1867, he accepted a post with the Geological Survey of Scotland, and in 1876, the year after the publication of his book *Climate and Time*, he was elected as a Fellow of the Royal Society (possibly the only former janitor to receive this honour). Another book, *Climate and Cosmology*, followed in 1885, when Croll was 64. He died in Perth on 15 December 1890, having seen his Ice Age model become widely accepted and influential, even though, in fact, there was very little hard geological evidence to back it up.

In *Climate and Time*, Croll had pointed the way for further improvements to the astronomical model of Ice Ages by suggesting that changes in the tilt of the Earth might also play a part. This is the tilt, now 23½ degrees, that causes the seasons. In Croll's day, it was known that the tilt changes (with the Earth nodding up and down, between extremes of tilt of about 22 degrees and 25 degrees from the vertical), but nobody, not even Leverrier, had calculated precisely how much it nods and over what timescale (it actually takes about 40,000 years to nod from its most vertical all the way down and back up to where it started). Croll speculated that when the Earth was more upright an Ice Age would be more likely, since both polar regions would be getting less heat from the Sun – but this was no more than a guess. By the end of the nineteenth century, however, the whole package of ideas had begun to fall into disfavour, as geological evidence began to pile up indicating that the latest Ice Age had ended not 80,000 years ago, but between about 10,000 and 15,000 years ago, completely out of step with Croll's hypothesis. Instead of warming 80,000 years ago, the northern hemisphere was then plunging into the coldest period of the latest Ice Age, just the opposite of what Croll's model required (and an important clue, which nobody picked up on at the time). At the same time, meteorologists calculated that the changes in the amount of solar heating produced by these astronomical effects, although real, were too small to explain the great temperature differences between Interglacials and Ice Ages. But the geological evidence did by then show that there had been a succession of Ice Ages, and if nothing else, the astronomical model did predict a repeating rhythm of cycles of ice. The person who took up the daunting task of improving the astronomical calculations and seeing if the cycles did fit the geological patterns was a Serbian engineer, Milutin Milankovitch, who was born in Dalj on 28 May 1879 (making him just a couple of months younger than Albert Einstein).

The Milankovitch model

In those days, Serbia had only recently (1882) become an independent kingdom after centuries of outside domination (mainly by the Turks), although it had been an autonomous principality under Turkish suzerainty since 1829, part of a ferment of Balkan states gaining increasing independence in the gap between the crumbling Turkish Empire to the south and the scarcely healthier Austro-Hungarian Empire to the north. Milankovitch, unlike Croll, received a conventional education and graduated from the Institute of Technology in Vienna with a PhD in 1904. He stayed in Vienna and worked as an engineer (on the design of large concrete structures) for five years before becoming professor of applied mathematics at the University of Belgrade, back in Serbia, in 1909. This was very much a provincial backwater compared with the bright lights of Vienna, where Milankovitch could have carved out a good career; but he wanted to help his native country, which needed more trained engineers, and thought he could do this best as a teacher. He taught mechanics, of course, but also theoretical physics and astronomy. Somewhere along the line, he also picked up his grand obsession with climate. Much later,[1] he romantically dated the moment when he decided to develop a mathematical model to describe the changing climates of Earth, Venus and Mars to a drunken conversation over dinner in 1911, when Milankovitch was 32, although this can perhaps be taken with a pinch of salt. What matters is that about that time, Milankovitch did indeed set out on a project to calculate not only what the present-day temperatures ought to be at different latitudes on each of these three planets today (providing a way to test the astronomical model by comparison with observations, at least on Earth), but also how these temperatures had changed as a result of the changing astronomical rhythms – actual temperatures, not just the more vague claim that at certain times of these cycles one hemisphere of the Earth was cooler than at other times in the cycles. And all of this, it cannot be overemphasized, with no mechanical aids – just brain power, pencil (or pen) and paper – and

1. See *Durch ferne Welten und Zeiten*; this is the principal (if probably biased) source for biographical information about Milankovitch.

not just for one planet, but three! This went far beyond anything that Croll had even contemplated, and even though Milankovitch started out with an enormous bonus when he discovered that the German mathematician Ludwig Pilgrim had already (in 1904) calculated the way the three basic patterns of eccentricity, precession and tilt had changed over the past million years, it still took him three decades to complete the task.

Climate is determined by the distance of a planet from the Sun, the latitude of interest and the angle at which the Sun's rays strike the surface at that latitude.[1] The calculations are straightforward in principle but incredibly tedious in practice, and became a major part of Milankovitch's life, occupying him at home for part of every evening. The relevant books and papers even travelled with him on holidays with his wife and son. In 1912, the first of a series of Balkan wars broke out. Bulgaria, Serbia, Greece and Montenegro attacked the Turkish Empire, swiftly achieving victory and gaining territory. In 1913, in a squabble over the spoils, Bulgaria attacked her former allies and was defeated. And all of this turmoil in the Balkans contributed, of course, to the outbreak of the First World War in 1914, following the assassination of Franz Ferdinand by a Bosnian Serb at Sarajevo on 18 June that year. As an engineer, Milankovitch served in the Serbian army in the First Balkan War, but not in the front line, which gave him plenty of time to think about his calculations. He began to publish papers on his work, showing in particular that the tilt effect is even more important than Croll had suggested, but since these were published in Serbian at a time of political upheaval, little notice was taken of them. When the First World War broke out, Milankovitch was visiting his home town of Dalj, when it was overrun by the Austro-Hungarian army. He became a prisoner of war, but by the end of the year his status as a distinguished academic had earned him release from prison, and he was allowed to live in Budapest and work on his calculations for the next four years. The fruits of these labours, a mathematical description of the present-day climates of Earth, Venus and Mars,

1. And by the composition of the atmosphere, which is where the greenhouse effect comes in; but for these calculations we assume that the atmosphere has had the same composition over the past few million years.

were published in a book in 1920 to widespread acclaim. The book also included mathematical evidence that the astronomical influences could alter the amount of heat arriving at different latitudes sufficiently to cause Ice Ages, although Milankovitch had not yet worked out the details. This aspect of the work was, however, immediately picked up by Wladimir Köppen, and led to a fruitful correspondence between Köppen and Milankovitch, and then to the incorporation of these ideas in Köppen's book on climate with Alfred Wegener.

Köppen brought one key new idea to the understanding of how the astronomical rhythms affect climate on Earth. He realized that what matters is not the temperature in winter, but the temperature in summer. At high latitudes (he was thinking particularly in terms of the northern hemisphere), it is always cold enough for snow to fall in winter. What matters is how much of the snow stays unmelted in summer. So the key to Ice Ages is cool summers, not extra-cold winters, even though those cool summers go hand in hand with relatively mild winters. This is exactly the opposite of what Croll had thought, immediately explaining why the latest Ice Age was intense about 80,000 years ago and ended about 10,000–15,000 years ago. When Milankovitch took detailed account of this effect, calculating temperature variations on Earth at three latitudes (55 degrees, 60 degrees and 65 degrees North) he got what seemed to be a very good match between the astronomical rhythms and the pattern of past Ice Ages as indicated by the geological evidence available in the 1920s.

With the publicity given to these ideas by Köppen and Wegener in their book *Climates of the Geological Past*, for a time it seemed that the astronomical model of Ice Ages had graduated into a full-blown theory. In 1930, Milankovitch published the results of even more calculations, this time for eight different latitudes, before moving on to calculate, over the next eight years, just how the ice sheets would respond to these changes in temperature. A book summing up his life's work, *Canon of Insolation and the Ice Age Problem*, was in the press when German forces invaded Yugoslavia (which had been established following the First World War and incorporated Serbia) in 1941. At the age of 63, Milankovitch decided that he would fill in time during the occupation by writing his memoirs, which were eventually published by the Serbian Academy of Sciences in 1952. After a quiet retirement, Milankovitch died on 12 December 1958. But by then, his model had fallen out of favour because new and more detailed (though still very incomplete) geological evidence no longer seemed to match it as well as the older, even less accurate evidence had.

Modern ideas about Ice Ages

In truth, the geological data simply were not good enough for any conclusive comparison with the now highly detailed astronomical model, and whether or not a particular set of data matched the model did not reveal any deep truth about the way the world works. As with the idea of continental drift, the true test would only come from much better measurements of the geological record, involving new techniques and technologies. These culminated in the 1970s, by which time the astronomical model itself (now often called the Milankovitch model) had been improved to an accuracy he can never have hoped for, using electronic computers. The key geological evidence came from cores of sediment extracted from the sea bed, where layers of sediment have been laid down year by year, one on top of the other. These sediments can be dated, using now-standard techniques including radioactive and geomagnetic dating, and are found to contain traces of tiny creatures that lived and died in the oceans long ago. Those traces come in the form of the chalky shells left behind when these creatures die. At one level, the shells reveal which species of these creatures flourished at different times, and that in itself is a guide to climate; at another level, analysis of isotopes of oxygen from these shells can give a direct indication of the temperature at the time those creatures were active, because different isotopes of oxygen are taken up by living creatures in different proportions according to the temperature and how much water is locked away in ice sheets. All three astronomical rhythms clearly show up in these records as the pulsebeat of the changing climate over the past million years or more. The moment when the model finally became established is generally taken as the publication of a key paper summarizing the evidence in the journal *Science* in 1976,[1] exactly a hundred years after the publication of *Climate and Time*. But that left one intriguing question, which turns out to be crucially important for our own

1. J. D. Hays, J. Imbrie and N. J. Shackleton, 'Variations in the earth's orbit: pacemaker of the ice ages', *Science*, volume 194, pp. 112–1132, 1976.

existence. *Why* is the Earth so sensitive to these admittedly small changes in the amount of sunshine reaching different latitudes?

The answer brings us back to continental drift. Taking a step back from the close-up on climatic change provided by the Milankovitch model, the now well-understood (and well-dated) geological record tells us that the natural state of the Earth throughout most of its long history has been completely ice-free (except, perhaps, for the tops of high mountains). As long as warm ocean currents can get to the polar regions, it doesn't matter how little sunlight they receive, since the warm water prevents sea ice from forming. But occasionally, as the scars of ancient glaciations reveal, at intervals separated by hundreds of millions of years, one or other hemisphere is plunged into a period of cold lasting for several million years; we can call this an Ice Epoch, lifting the term used by Croll and giving it a similar meaning but longer timescale. For example, there was an Ice Epoch which lasted for about 20 million years back in the Permian Era; this Ice Epoch ended around 250 million years ago. The explanation for such an event is that from time to time continental drift carries a large land surface over or near one of the poles. This does two things. First, it cuts off (or at least impedes) the supply of warm water from lower latitudes, so in winter the affected region gets very cold indeed. Second, the continent provides a surface on which snow can fall, settle and accumulate to build up a great ice sheet. Antarctica today provides a classic example of this process at work, which produces an Ice Epoch that is only slightly affected by the astronomical rhythms.

After the Permian Ice Epoch ended (which happened because continental drift opened the way once again for warm water to reach the polar regions), the world enjoyed about 200 million years of warmth, the time during which the dinosaurs flourished. But a gradual cooling began to set in about 55 million years ago, and by 10 million years ago glaciers returned, first on the mountains of Alaska but soon afterwards on Antarctica, where the ice sheet grew so much that by five million years ago it was bigger than it is today. The fact that glaciers spread in both hemispheres at the same time is an important insight. While Antarctica covered the South Pole and glaciers built up there in the way we have just described, the north polar region also cooled and eventually froze, even though the Arctic Ocean, not land, covered the pole. The reason for this is that continental drift gradually closed off an almost complete ring of land around the Arctic Ocean, shutting out much of the warm water that would otherwise have kept it ice-free. Notably, the presence of Greenland today deflects the Gulf Stream eastwards, where it warms the British Isles and northwestern mainland Europe instead. A thin sheet of ice formed over the polar ocean, and from about three million years ago much more ice lay over the surrounding land. This situation, with a polar ocean surrounded by land on which snow can fall and settle, but where it will melt away in hot summers, turns out to be particularly sensitive to the astronomical rhythms. For the past five million years or so, the Earth has been in what may be a unique state for its entire history, with ice caps over both poles, produced by two distinctly different geographical arrangements of land and sea. This, and in particular the geography of the northern hemisphere, makes the planet sensitive to the astronomical rhythms, which show up strongly in the geological record from recent geological times.

The impact on evolution

Within the current Ice Epoch, the effect of this pulsebeat of climate is to produce a succession of full Ice Ages, each very roughly 100,000 years long, separated by warmer conditions like those of the present day, Interglacials some 10,000 years long. By this reckoning, the present Interglacial would be coming to an end naturally within a couple of thousand years – less time than the span of recorded history. But the future is beyond the scope of this book. There are also lesser ripples of climate change superimposed on this principal pattern by the combination of rhythms investigated by Milankovitch. The sequence of these Ice Ages, dated by the radioactive technique using isotopes of potassium and argon, set in a little more than 3.6 million years ago. At just that time, our ancestors were living in the Great Rift Valley of East Africa (itself a product of plate tectonic activity), where an ancestral form of hominid was giving rise to three modern forms, the chimpanzee, the gorilla and ourselves.[1] It is just at

1. The dating of the split of the human line from the other African apes comes from direct measurements of their DNA, which provides a kind of 'molecular clock'. This was finally established in the 1990s, as told by John Gribbin and Jeremy Cherfas in *The First Chimpanzee* (Penguin, London, 2001).

this time that the fossil record provides direct evidence of a hominid that walked upright, in the form both of footprints made in soft ground which then set hard (like those handprints of the stars on the sidewalk in Hollywood) and of fossil bones. Although nobody can be sure, without the aid of a time machine, exactly what it was that turned one East African hominid from three to four million years ago into *Homo sapiens*, it is easy to make a case that the pulsebeat of climate played a key role, and hard to escape the conclusion that it was at least partially responsible. In East Africa, what mattered was not so much the temperature fluctuations, which were so important at high latitudes, as the fact that during a full Ice Age the oceans are so chilled that there is less evaporation and therefore less rainfall, the Earth is drier and the forests retreat. This would have increased competition among woodland hominids (including our ancestors), with some being forced out of the woods and on to the plains. There, there was intense selection pressure on these individuals, and only the ones who adapted to the new way of life would survive. Had the situation continued unchanged indefinitely, they might well have died out in competition with better-adapted plains dwellers. But after a hundred thousand years or so, conditions eased and the descendants of the survivors of that winnowing process of natural selection had a chance to take advantage of the expanding forest, breed in safety from the plains predators and build up their numbers. Repeat that process ten or twenty times, and it is easy to see how a ratcheting effect would have selected for intelligence and adaptability as the key requirements for survival on the fringes of the forest – while back in the centre of the forest, the most successful hominid lines adapted ever more closely to life in the trees, and became chimps and gorillas.

The story is perhaps as plausible as the idea of continental drift was in Arthur Holmes's day. But even if the details are incorrect, it is hard to see how the close match between the climatic pattern which set in between three and four million years ago and the development of human beings from woodland apes, which also set in between three and four million years ago, can be a coincidence. We owe our existence to the combination of continental drift, establishing rare conditions ideal for the astronomical cycles to affect the Earth's climate, and those astronomical rhythms themselves. The package involves basic physics (as basic as an understanding of convection, which drives continental drift), Newtonian dynamics and gravity (which explains the astronomical cycles and makes them predictable), chemistry (in analysing the samples from the sea bed), electromagnetism (for geomagnetic dating), an understanding of species and the living world built from the work of people like Ray and Linnaeus, and the Darwin–Wallace theory of evolution by natural selection. It is an insight which both puts us in perspective as just one form of life on Earth, created by the same process of natural selection that has created all other species, and triumphantly crowns three centuries of 'classical' science that began with the work of Galileo Galilei and Isaac Newton. Follow that, you might think. But by the end of the twentieth century, much of science had not so much followed that, as gone beyond classical science, changed in ways that are alien to the very foundations of the Newtonian world view. It all began right at the end of the nineteenth century, with the quantum revolution[1] that completely altered the way physicists thought about the world on very small scales.

1. Perhaps the only scientific 'revolution' that actually justifies the use of the term!

Book Five

MODERN TIMES

13 Inner Space

Invention of the vacuum tube – 'Cathode rays' and 'canal rays' – William Crookes: the Crookes tube and the corpuscular interpretation of cathode rays – Cathode rays are shown to move far slower than light – The discovery of the electron – Wilhelm Röntgen and the discovery of X-rays – Radioactivity; Becquerel and the Curies – Discovery of alpha, beta and gamma radiation – Rutherford's model of the atom – Radioactive decay – The existence of isotopes – Discovery of the neutron – Max Planck and Planck's constant, black-body radiation and the existence of energy quanta – Albert Einstein and light quanta – Niels Bohr – The first quantum model of the atom – Louis de Broglie – Erwin Schrödinger's wave equation for electrons – The particle-based approach to the quantum world of electrons – Heisenberg's uncertainty principle: wave–particle duality – Dirac's equation of the electron – The existence of antimatter – The strong nuclear force – The weak nuclear force; neutrinos – Quantum electrodynamics – The future? Quarks and string

Invention of the vacuum tube

The biggest revolution in the history of science began with the invention of a better kind of vacuum pump, in the middle of the nineteenth century. To put the significance of this development, which may seem trivial compared with modern technology, in perspective, consider the kind of equipment that Michael Faraday worked with when he wanted to investigate the behaviour of electricity in the absence of air. At the end of the 1830s, Faraday was investigating such discharges using a glass jar in which there was a single fixed electrode. The opening of the jar was 'sealed' (if that is quite the right word) by a cork, through which a metal pin, the other electrode, was pushed and could be moved in and out to different positions. The whole apparatus was far from airtight, and the pressure inside the vessel could only be kept low (and even then nowhere near a vacuum) by constant pumping using equipment which differed little in principle from the pumps used by Otto von Guericke two centuries earlier (and which is essentially the same as a modern bicycle pump). It was the German Heinrich Geissler (1814–1879) who made the breakthrough in Bonn in the late 1850s. His improved vacuum pump used mercury to make airtight contact, sealing all the connections and taps involved in the process of evacuating air from a glass vessel. The vessel being evacuated was connected by a tube to one branch of a two-way tap linking it to a bulb, which was itself connected by a flexible tube to a container filled with mercury. The other branch of the two-way tap connected the bulb to the open air. With the bulb connected to the open air, the mercury reservoir was raised, with the pressure of the mercury forcing air out of the bulb. Then, the tap was switched to the other branch and the mercury reservoir lowered, encouraging air to flow out of the glass vessel into the bulb. Repeat this process often enough and you end up with a very 'hard' vacuum in the glass jar. Even better, Geissler, who trained as a glassblower, devised the technique of sealing two electrodes into the evacuated glass vessel, creating a tube in which there was a permanent vacuum. He had invented the vacuum tube. The technology was improved over the following years and decades, by Geissler himself and others, until by the middle of the 1880s it was possible to make vacuum tubes in which the pressure was only a few ten-thousandths of the pressure of air at sea level on Earth. It was this technology that led to the discovery of electrons ('cathode rays') and X-rays, and thereby encouraged the work which led to the discovery of radioactivity.

'Cathode rays' and 'canal rays'

In the 1860s, Julius Plücker (1801–1868), an otherwise undistinguished professor of physics at the University of Bonn, had the advantage of being one of the first people to have access to Geissler's new vacuum-tube technology. He carried out a series of experiments investigating the nature of the glow seen in such tubes when an electric current flows between the electrodes (essentially, this is the technology of the neon tube), and it was one of Plücker's students, a Johann Hittorf (1824–1914), who first noticed that glowing rays emitted from the cathode (negative electrode) in such tubes seem to follow straight lines. In 1876, Eugen Goldstein (1850–1930), then working with Hermann Helmholtz in Berlin, gave these glowing lines the name 'cathode rays'. He showed that these rays could cast shadows and (like several of his contemporaries) that they are deflected by magnetic fields, but he thought that they were electromagnetic waves similar to light. In 1886, Goldstein discovered another form of 'ray' being emitted from holes in the anodes (positive electrodes) of the discharge tubes he was using at the time; he dubbed them 'canal rays', from the German term for these holes. We now know that these 'rays' are streams of positively charged ions, atoms from which one or more electrons have been stripped.

As early as 1871, in a paper published by the Royal Society, the electrical engineer Cromwell Fleetwood Varley (1828–1883) suggested that cathode rays might be 'attenuated particles of matter, projected from the negative pole by electricity',[1] and this idea of a corpuscular explanation for the rays was taken up by William Crookes (1832–1919).

Wilhelm Crookes: the Crookes tube and the corpuscular interpretation of cathode rays

Crookes, who was born on 17 June 1832 in London, was the eldest of 16 children of a tailor and businessman. He had an unusual scientific career, and little is known of his early education, but by the end of the 1840s he was assistant to August von Hoffmann at the Royal College of Chemistry. He worked in the meteorological department of the Radcliffe Observatory in Oxford in 1854 and 1855, then as a lecturer in chemistry at Chester Training College for the academic year 1855–6. But he then inherited enough money from his father to be financially independent and returned to London, where he set up a private chemistry laboratory and founded the weekly *Chemical News*, which he edited until 1906. Crookes had wide interests (including spiritualism), but we shall only touch on his part in the story of the discovery of electrons. This hinged upon his development of an improved vacuum tube (known as the Crookes tube) with an even better (harder) vacuum than those of his contemporaries in mainland Europe. With a better vacuum, Crookes was able to carry out experiments which seemed to him to provide definitive proof of the corpuscular nature of cathode rays. These included placing a metal Maltese Cross in the tube and obtaining a sharply defined shadow of the cross in the glow made where the rays struck the wall of the glass tube behind it. He also placed a tiny paddle wheel in the beam of the rays, showing, since the impact of the rays made the wheel turn, that they carried momentum. By 1879 he was championing the corpuscular interpretation of cathode rays, and this quickly became the established interpretation of most

1. *Proceedings of the Royal Society*, volume 19, p. 236, 1871.

British physicists. Things were rather different on the Continent, though, and especially in Germany, where in the early 1880s Heinrich Hertz carried out experiments which seemed to show that an electric field had no effect on the rays (we now know this was because his vacuum tubes contained too much residual gas, which became ionized and interfered with the electrons), and the idea that the rays were a form of electromagnetic wave became firmly established. Partly because physicists were diverted by the discovery of X-rays (of which much more shortly), it wasn't until the end of the 1890s that the situation was finally resolved.

Cathode rays are shown to move far slower than light

Evidence that cathode rays could not simply be a form of electromagnetic radiation came in 1894, when J. J. Thomson, in England, showed that they move much more slowly than light (remember Maxwell's equations tell us that all electromagnetic radiation moves at the speed of light). By 1897 there was an increasing weight of evidence that cathode rays carried electric charge. Experiments carried out in 1895 by Jean Perrin (who we met in Chapter 10) were among those which showed that the rays are deflected sideways by a magnetic field just as a beam of electrically charged particles should be, and he had also shown that when cathode rays hit a metal plate the plate becomes negatively charged. In 1897, he was working on experiments to probe the properties of the particles in these 'rays' when he was pre-empted by other workers – Walter Kaufmann in Germany and, crucially, J. J. Thomson in England. Kaufmann, working in Berlin, was studying the way cathode rays were deflected by electric and magnetic fields in vacuum tubes which contained different residual pressures of gas, and different kinds of gas. In these experiments, he was able to infer the ratio of the charge of the particles to their mass – e/m. He expected to find different values for this ratio for different gases, because he thought that he was measuring the properties of what we now call ions, atoms which had become charged from contact with the cathode. He was surprised to find that he always got the same value for e/m. Thomson (no relation, by the way to William Thomson, who became Lord Kelvin) also measured e/m, using a neat technique in which a beam of cathode rays was deflected one way by a magnetic field and then the opposite way by an electric field, so that the two effects exactly cancelled out. But he was *not* surprised to discover that he always got the same value for this number, because from the start he thought that he was dealing with streams of identical particles emitted by the cathode. Expressing his result the other way up, as *m/e*, he pointed out that the smallness of the number he obtained, compared with the equivalent result for hydrogen (what we now know to be hydrogen ions, equivalent to single protons), meant either that the mass of the particle involved was very small or that the charge was very large, or some combination of these effects. In a lecture to the Royal Institution on 30 April 1897, Thomson commented that 'the assumption of a state of matter more finely divided than the atom is a somewhat startling one'.[1] And he later wrote that 'I was told long afterwards by a distinguished colleague who had been present at my lecture that he thought I had been "pulling their legs".'[2]

The discovery of the electron

In spite of all this, 1897 is often regarded as the year of the 'discovery' of the electron. But the real discovery came two years later, in 1899, when Thomson succeeded in measuring the electric charge itself, using a technique in which droplets of water are electrically charged and monitored using electric fields. It was this measurement of *e* that enabled him to provide an actual value for *m*, showing that the particles (he called them corpuscles) which make up cathode rays each have only about one two-thousandth of the mass of a hydrogen atom and are 'a part of the mass of the atom getting free and becoming detached from the original atom'.[3] In other words, startling though the discovery might be, the atom definitely was *not* indivisible. But who was the man who dropped this bombshell?

Thomson was born in Cheetham Hill, near Manchester, on 18 December 1856. He was christened Joseph John, but in adult life was always known simply by his initials, as 'J. J.'. When he was 14, he started to study engineering at Owens College (the forerunner of Manchester University), but his father, an antiquarian bookseller, died two years later and the resulting constraints on the family finances

1. The lecture is reprinted in Bragg and Porter.

2. J. J. Thomson, *Recollections and Reflections*.

3. J. J. Thomson, *Philosophical Magazine*, volume 48, p. 547, 1899. The name 'electron' in its modern context was soon applied to Thomson's 'corpuscles' by the Dutch physicist Hendrik Lorentz, of Lorentz–Fitzgerald fame.

meant that he had to switch to a course in physics, chemistry and mathematics, for which he was awarded a scholarship. He moved on to Trinity College, Cambridge (again on a scholarship), in 1876, graduating in mathematics in 1880, and stayed there (apart from short visits to Princeton) for the rest of his working life. From 1880, Thomson worked at the Cavendish Laboratory; he succeeded Lord Rayleigh as the head of the Cavendish in 1884 (the University had wanted William Thomson, who preferred to stay in Glasgow) and held the post until 1919, when he resigned after becoming the first scientist to be appointed Master of Trinity, a post he held until his death on 30 August 1940. He received the Nobel prize (for his work on electrons) in 1906 and was knighted in 1908.

The choice of Thomson, a mathematician, to be professor of experimental physics and head of the Cavendish was either an inspired one or a lucky fluke. He had an uncanny ability to devise experiments which would reveal fundamental truths about the physical world (like the experiment to measure *e/m*), and he could also use this ability to point out why experiments devised by other people were not working as expected – even when the person who built the experiment could not see what had gone wrong. But he was notoriously clumsy when it came to handling delicate equipment, so much so that it was said that his colleagues tried to avoid having him enter the labs where they were working (unless they needed his insight into the foibles of a recalcitrant experiment). You might almost say that J. J. was the ultimate theoretical experimentalist. It is a measure of his ability, and of the way the Cavendish lured many of the best physicists to work in Cambridge at the end of the nineteenth and beginning of the twentieth centuries, that seven of the physicists who worked as his assistants received Nobel prizes. Thomson played a significant part, as a teacher, guide and much-liked head of department, in all this success.

Although the Cavendish by no means cornered the market in this kind of success, as the discoveries of X-rays and radioactivity show, even when the breakthroughs were achieved elsewhere Thomson's team were usually quick to exploit the implications. Great discoveries in science are usually made by young men in a hurry, full of bright ideas. But in the late nineteenth century, thanks to the improving technology of vacuum tubes, the science of what became known as atomic physics was itself young, with the technology opening up new avenues of science for investigation. Under those conditions, with discoveries almost literally lying around waiting to be made, experience and access to the new technology counted for as much as youth and enthusiasm. Wilhelm Röntgen, for example, was hardly in the first flush of youth when he made the discovery of X-rays.

Wilhelm Röntgen and the discovery of X-rays

Röntgen was born at Lennep, in Germany, on 27 March 1845 and followed a conventional path through the academic system to become professor of physics at the University of Würzburg in 1888. He was a good, solid physicist who worked in several areas of his subject without making any particularly distinguished mark. In November 1895, however, when he was 50 years old, Röntgen was studying the behaviour of cathode rays using an improved vacuum-tube design (the various designs were known by the names of the pioneers of the technology, as Hittorf tubes or Crookes tubes, but the principles were the same for all of them). In 1894, Philipp Lenard had shown, building from work by Hertz, that cathode rays could pass through sheets of thin metal foil without leaving any holes. At the time, this was interpreted as evidence that the 'rays' must be waves, since particles, it was assumed, ought to leave evidence of their passage (they were thinking, of course, in terms of particles at least the size of atoms). In following up this discovery, Röntgen was working with a vacuum tube completely covered in thin black card. His idea was to block out the glow of the light inside the tube in order to be able to detect any faint trace of cathode rays penetrating the glass of the tube itself. One of the standard ways to detect cathode rays was to use a paper screen painted with barium platinocyanide, which would fluoresce when struck by the rays. On 8 November 1895, Röntgen had such a screen, which was nothing to do with his current experiment, lying to one side of his apparatus, out of the line of fire of the cathode rays. To his astonishment, he noticed the screen fluoresce brightly when the vacuum tube was in operation in the darkened laboratory. After carrying out careful investigations to make sure that he had indeed discovered a new phenomenon, Röntgen submitted his discovery paper to the *Würzburg Physikalisch-Medizinische Gesellschaft* on 28 December, and it was published in January 1896. The discovery of what Röntgen himself termed X-rays, but which were

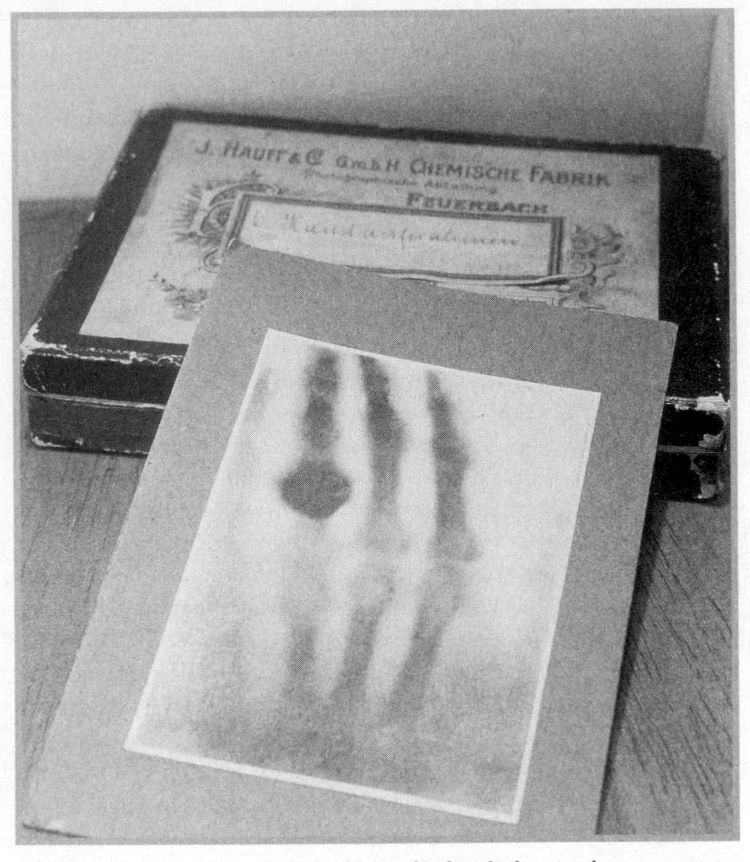

35. *Röntgen's X-ray of his wife's hand, showing her wedding ring, 1895.*

often known in the German-speaking world as Röntgen rays, caused a sensation, not least because of the ability of the rays to penetrate human flesh and provide photographic images of the underlying skeleton. Preprints of the discovery paper, including X-ray photographs of (among other things) his wife's hand, were sent out on 1 January 1896, and reports began to appear in newspapers within a week. On 13 January, Röntgen demonstrated the phenomenon to Emperor Wilhelm II in Berlin, while English translations of his paper appeared in *Nature* on 23 January (the same day that Röntgen gave his only public lecture on the subject, in Würzburg) and in *Science* on 14 February. In March 1896, Röntgen published two more papers on X-rays, but these were his last contributions to the subject, although he remained scientifically active, becoming professor of physics at Munich in 1900 and living until 10 February 1923. His one great contribution to science earned him the first-ever Nobel prize for physics, in 1901.

Almost from the beginning, physicists knew a great deal about the behaviour of X-rays, even though they did not know what these rays were. The rays were produced where the cathode rays struck the glass wall of a vacuum tube (so there was no mystery about where the energy they carried came from) and spread out from this source in all directions. Like light, they travelled in straight lines and affected photographic material, and they were not deflected by electric or magnetic fields. Unlike light, though, they did not seem to be reflected or refracted, and for many years it was not clear whether they were waves or particles. This did not prevent these X-rays becoming widely used in the decade or so following their discovery, both in the obvious medical applications (although with sometimes unfortunate side effects because of the unappreciated dangers of over-exposure) and in physics, where, for example, they proved ideal for ionizing gases. It was only after about 1910 that it became clear that X-rays are indeed a form of electromagnetic wave with wavelengths much shorter than visible light (or even ultraviolet light), and that they do indeed reflect and refract, given appropriate targets. In terms of the development of atomic physics, however, the most important thing about the discovery of X-rays is that it led, almost immediately, to the discovery of yet another, far more puzzling kind of radiation.

Radioactivity; Becquerel and the Curies

If ever a scientist was in the right place at the right time, it was Henri Becquerel. Henri's grandfather, Antoine Becquerel (1788–1878), was a pioneer in the study of electric and luminescent phenomena, so successful that in 1838 a chair of physics was inaugurated for him at the French Museum of Natural History. Antoine's third son, Alexandre-Edmond Becquerel (1820–1891), worked alongside his father in Paris and became interested in the behaviour of phosphorescent solids – crystals that glow in the dark. When Antoine died in 1878, Edmond (as he was usually known) succeeded him as professor. By then, his own son, Henri Becquerel (1852–1908), was already following in the family tradition of physics, and he gained his PhD from the Faculty of Sciences in Paris in 1888. When Edmond died in 1891, Henri became professor of physics at the Museum, although he also served as chief engineer to the Department of Bridges and Highways in Paris alongside this post. When Henri died, he was in turn succeeded as professor by his own son, Jean (1878–1953); and it was only in 1948, when Jean retired without having produced an heir, that the chair at the Museum was, for the first time since its inception 110 years earlier, awarded to anyone other than a Becquerel. In the middle of this dynasty, Henri Becquerel was at a meeting of the French Academy of Sciences on 20 January 1896 when he heard details of the hot news about X-rays, including the discovery that they originated from a bright spot on the glass wall of a cathode ray tube, where the cathode rays struck the glass and made it fluoresce. This suggested to him that phosphorescent objects, which also glow in the dark, might produce X-rays, and he immediately set out to test this hypothesis using the variety of phosphorescent materials that had accumulated at the museum since his grandfather's time.

The key feature of these phosphorescent materials was that they had to be exposed to sunlight to make them glow. This exposure, in some unknown way, charged them up with energy, so that afterwards they would glow in the dark for a while, fading away as the energy from the sunlight was used up. In his search for X-rays, Becquerel carefully wrapped photographic plates in two sheets of thick, black paper, so that no light could penetrate, and placed the wrapped plate under a dish of phosphorescent salts that had been 'charged up' by exposure to sunlight. Sure enough, he found that for some of these salts, when the photographic plates were unwrapped and developed, they showed an outline of the phosphorescent material – and if a metal object such as a coin was placed between the dish of phosphorescent salts and the wrapped plates, then when the plates were developed they showed an outline of the metal object. X-rays, it seemed, could be produced by the action of sunlight on phosphorescent salts as well as by the action of cathode rays on glass, and these results were duly reported to the scientific community.

But at the end of February 1896 Becquerel prepared another experiment. He place a piece of copper in the shape of a cross between the wrapped photographic plates and the dish of phosphorescent salts (a compound of uranium) and waited for the Sun to come out. Paris remained overcast for several days, and on 1 March, tired of waiting, Becquerel developed the plate anyway (it isn't clear whether this was a whim or a deliberate decision to carry out a control experiment). To his astonishment, he found the outline of the copper cross. Even when the phosphorescent salts were not glowing, and even though they had not been charged up by sunlight, at least in the case of uranium compounds they produced what seemed to be X-rays.[1] The most dramatic aspect of the discovery was that the salts were producing energy, as it seemed, out of nothing at all, apparently contradicting one of the most cherished tenets of physics, the law of conservation of energy.

The discovery didn't have the popular impact of the discovery of X-rays, because outside the circle of scientific experts (and even to many scientists) it looked like just another version of X-rays. Becquerel himself soon moved on to other work, although he did also make some studies of the properties of the radiation he had discovered and showed in 1899 that it could be deflected by a magnetic field, so that it could not be X-radiation and must be composed of charged particles. But the detailed investigation of the phenomenon was taken up by Marie and Pierre Curie, in Paris (with whom Becquerel shared the Nobel prize in 1903), and by Ernest Rutherford (of whom more later), initially at the Cavendish.

It is Marie Curie's name that is most strongly linked in the popular mind with the early investigation of radioactivity (a term she coined). This is partly because her role really was important, partly because she was a woman, and by providing one of the few role models for girls in science was assured of a good press, and partly because of the difficult conditions under which she worked, adding an element of romance to the story. This even seems to have affected the Nobel committee, which managed to give her the prize twice for essentially

1. Exactly the same discovery was made at almost exactly the same time by Silvanus Thompson, in England, but Becquerel published first.

the same work – in physics in 1903, and in chemistry in 1911. Born in Warsaw on 7 November 1867, and originally named Marya Sklodowska, she had no hope of attending university in what was then the Russian part of a divided Poland, and had great difficulty scraping together funds to move to Paris to study at the Sorbonne in 1891. As an undergraduate, she almost literally starved in a garret. It was at the Sorbonne that she met (and in 1895 married) Pierre Curie, who was born the son of a doctor on 15 May 1859 and was already established as a highly regarded expert on the properties of magnetic materials. The marriage soon led to pregnancy, and as a result it was only in September 1897 that Marie settled down to her PhD work on 'uranium rays'. At that time, no woman had yet completed a PhD at any European university, although Elsa Neumann, in Germany, would soon do so. As a pioneering female scientist, Marie was rather grudgingly allowed the use of a leaky shed for her work – she was banned from the main laboratories for fear that the sexual excitement of her presence there might prevent any research getting done.

Marie made her first great discovery in February 1898 – that pitchblende (the ore from which uranium is extracted) is *more* radioactive than uranium, and must contain another, highly radioactive, element. The discovery was so dramatic that Pierre abandoned his own current research project and joined Marie in the effort to isolate this previously unknown element; after intense labours they actually found two, one which they called 'polonium' (an overtly political gesture to her homeland, which officially, remember, did not exist) and the other which they dubbed 'radium'. It took until March 1902 to isolate one-tenth of a gram of radium from tonnes of pitchblende, enough for it to be analysed chemically and located in its place in the periodic table. Marie's PhD was awarded a year later – the same year she received her first Nobel prize. It was Pierre who measured the astonishing output of energy from radium – enough from each gram of radium to heat a gram and a third of water from freezing point to boiling point in an hour. There seemed no end to this activity, with a single gram of radium capable of repeatedly heating gram after gram of water to boiling point in this way – something for nothing, in violation of the law of conservation of energy. It was as important a discovery as the discovery of radium itself, and brought the team more acclaim. But just as the Curies were beginning to enjoy a more comfortable life as a result of their success, on 19 April 1906 Pierre was killed when he slipped crossing a road in Paris and his skull was crushed under the wheels of a horse-drawn wagon. It seems extremely likely that the slip was a result of dizzy spells he was experiencing, caused by what is now thought to have been radiation sickness. Marie lived until 4 July 1934, when she died of leukemia at a clinic in the Haute-Savoie, also (eventually) a victim of radiation sickness. Her laboratory notebooks are still so radioactive that they are kept in a lead-lined safe, and only removed occasionally and with careful safety precautions.

The discoveries of X-rays and 'atomic' radiation, even the identification of the electron, represented only the first discovery stage in the development of an understanding of the subatomic world – the discovery that there was a subatomic world to explore. The person who, more than anyone else, gave shape to the subatomic world, putting these discoveries into some sort of order and gaining the first understanding of the structure of the atom, was Ernest Rutherford. Rutherford was born in a rural community on the South Island of New Zealand on 30 August 1871. At the time there was little except rural communities in New Zealand, which had only been claimed by Britain in May 1840, largely in order to pre-empt the establishment of a French colony there. Both Rutherford's parents had come to New Zealand as children with their parents (Scottish on his father's side; English on his mother's), in the first wave of settlers. As is often the case in pioneering communities, families tended to be large; Ernest (whose name was actually registered as Earnest, through a clerical error, although this spelling was never used by him) was one of twelve siblings, and had four uncles on his mother's side of the family and three uncles and three aunts on his father's side. He was born in the parish of Spring Grove, near the town of Nelson, but because of boundary changes the site of his birth is now in the parish of Brightwater. The family moved a few kilometres, to Foxhill, when Ernest was five and a half years old.

Rutherford was an able but not outstanding scholar as a child, who always seemed to do just well enough, by dint of hard work, to be able to scrape together scholarships to proceed to the next stage of his education. In this way, he obtained a BA degree from Canterbury

College, Christchurch, in 1892 (the curriculum involved both arts and science subjects), and an MA, partly based on original research in electricity and magnetism, in 1893 (that year, he was one of only fourteen postgraduate students in New Zealand). By this time, he was shining academically, but even so found it impossible to get a teaching job (his first choice) and had just about exhausted the educational opportunities in New Zealand. He developed a plan to continue his studies in Europe, on a scholarship; but in order to apply for such funding he had to be a registered student at the university, so in 1894 he enrolled for a rather superfluous BSc course and carried out more research while supporting himself initially with tutoring work (and probably a little financial help from his family). Fortunately for Rutherford, a teacher at the Christchurch Boys' High School fell ill in November 1894 and Rutherford was able to take over some of his duties.

The scholarship Rutherford sought was part of a British scheme set up in 1851 to celebrate the Great Exhibition. The scholarship provided funding for two years (at a modest £150 per year) for research students from Britain, Ireland, Canada, Australia and New Zealand to study anywhere in the world, but they were strictly limited in number and not available to every country every year. There was only a single scholarship offered to New Zealand in 1895, and only two candidates, with the decision between them made in London on the basis of dissertations describing their research projects. The award went to James Maclaurin, a chemist from Auckland. But Maclaurin had a job in Auckland and had recently got married. When it came to the crunch, he decided that he would not be able to take up the offer after all. So Rutherford received the award and in the autumn of 1895 joined the Cavendish Laboratory as the first person to enter the University of Cambridge as a research student – previously, the only way to become a member of that exclusive community was to start as an undergraduate and work your way up. This was just a couple of months before Röntgen discovered X-rays, and a couple of years before Thomson measured *e/m* for the electron. Rutherford was both the right man in the right place at the right time *and* a young man in a hurry; the combination would make him spectacularly successful in science.

Rutherford's early research in New Zealand had involved the magnetic properties of iron, which he probed using high-frequency radio

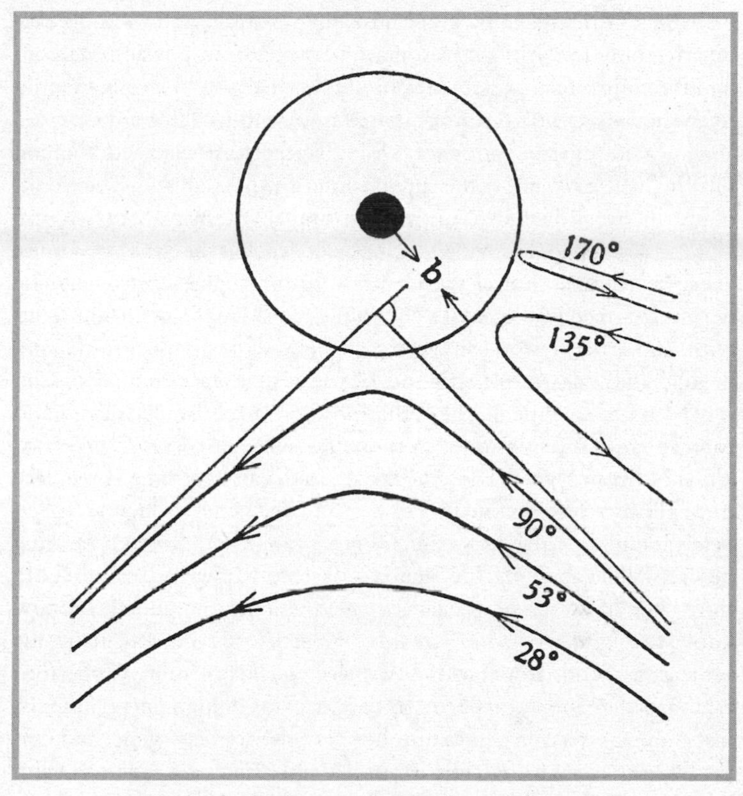

36. *Rutherford's diagram showing how alpha particles are deflected when they pass close to a heavy nucleus. From Rutherford's* A Newer Alchemy, *1937.*

waves (just six years after Hertz had discovered radio waves). As part of this work, he had built a sensitive (by the standards of the day) detector for these waves – one of the first radio receivers. His research initially continued along similar lines in Cambridge, where he carried out experiments in long-range radio transmission (eventually over distances of a couple of miles) at about the same time that Guglielmo Marconi was carrying out similar experiments in Italy – although various claims have been made, it is impossible now to tell who was actually first to achieve this sort of range. While Rutherford was interested in the scientific aspects of this research, and soon became diverted into the exciting investigation of subatomic physics, Marconi had the commercial possibilities of wireless telegraphy firmly in mind from the outset, with results that are familiar to us all.

Discovery of alpha, beta and gamma radiation

By the spring of 1896, Rutherford was working on X-rays, under the supervision of J. J. Thomson. Their joint work included an investigation of the way X-rays ionize gases and provided strong evidence that the 'rays' are a more energetic (that is, shorter wavelength) form of light, electromagnetic waves described by Maxwell's equations (thus tying in, with hindsight, with Rutherford's work on electromagnetic waves in the radio band, beyond the long-wavelength end of the visible spectrum). He soon moved on to investigate the radiation discovered by Becquerel, and found that it is made up of two components, one (which he called alpha radiation) which has a short range and can be stopped by a piece of paper or a few centimetres of air, and the other (which he called beta radiation) which has a much longer range and more penetrating power. While working in Canada in 1900, Rutherford identified a third type of radiation, which he called gamma radiation.[1] We now know that alpha rays are streams of particles, each essentially the same as a helium atom lacking two electrons (something Rutherford established in 1908); that beta rays are streams of high-energy (that is, fast-moving) electrons, like cathode rays but more energetic; and that gamma rays are an energetic form of electromagnetic radiation with wavelengths even shorter than X-rays.

Rutherford's move to Canada was largely a result of a quirk in the rules for the new research students. His 1851 scholarship provided support for only two years, but the Cambridge rules said that, regardless of merit, it was only possible to apply for a Fellowship after four years at the university, a hangover from the time when everybody in the Cambridge system came up through the undergraduate route. Although Rutherford did obtain funding from another scholarship for one more year, he was more or less forced to move on in 1898 (the Fellowship rules were changed the following year). Fortunately, a chair

1. The identification of alpha and beta rays was carried out in Cambridge, but reported in a paper published in 1899, after Rutherford had left for Canada.

became vacant at McGill University in Montreal, and Rutherford got the job. He was 27 when he took up the post, and although he had carried out first-rate research in Cambridge he did not have a PhD, which was not, in those days, essential for an academic career in science.[1] It was in Montreal, working with English-born Frederick Soddy (1877–1956) that Rutherford found that during the process which releases the radiation discovered by Becquerel (now known as radioactive decay), an atom is converted into an atom of a different element. When alpha or beta particles are emitted by an atom (strictly speaking, and getting only slightly ahead of the chronology, by the nucleus of an atom), what is left behind is a different kind of atom. The Rutherford–Soddy collaboration also solved the puzzle of the seemingly inexhaustible supply of energy from radioactive material such as radium. They found that this transformation of atoms follows a clear rule, in which a certain proportion of the atoms originally present in a sample will decay in a certain time. This is most commonly expressed in terms of the half-life. In radium, for example, careful measurements of the rate at which the radiation decreases in the laboratory show that in 1602 years, half of the atoms will have decayed into atoms of the gas radon as alpha particles are emitted. In the next 1602 years, half of the remaining atoms (a quarter of the original sample) will decay, and so on. This means two things. First, radium found on Earth today cannot have been there since the planet formed, but must have been produced more or less *in situ* (we now know, from the decay of the much longer-lived radioactive uranium); second, the supply of energy from radium and other radioactive elements is not, after all, inexhaustible. Even a radium-powered water heater will eventually run out of usable energy, and radium represents a finite storehouse of energy (in the same sort of way that an oilfield represents a finite storehouse of energy) rather than a violation of the law of conservation of energy. It was Rutherford who pointed out that this storehouse of energy gave the Earth a possible lifetime of at least hundreds of millions of years, directly inspiring the work of Bertram

1. The security offered by the job enabled Rutherford to marry his fiancée, May Newton, in 1900. She had been waiting patiently in New Zealand ever since 1895, meeting up with Ernest only on holidays.

Boltwood (who heard Rutherford give a talk on radioactivity at Yale) and paving the way for the work of Arthur Holmes, mentioned in Chapter 12.

Although happy and successful in Canada, Rutherford was concerned at being cut off from the mainstream development of research in physics in Europe. Turning down a lucrative offer from Yale, he returned to England in 1907 as professor of physics at the University of Manchester, where there were outstanding research facilities. It's a sign of how rapidly physics was progressing at the time that within a year Rutherford's team had established that alpha particles are the same as helium atoms which have lost two units of negative electric charge (because, we now know, they have lost two electrons). And a year after that, in 1909, alpha particles themselves, produced by natural radioactivity, were being used to probe the structure of the atom.[1] This is probably the work for which Rutherford is best remembered, although the experiments were actually carried out (under Rutherford's direction) by Hans Geiger (1882–1945) and a student called Ernest Marsden (1889–1970). It's no coincidence that this was the same Hans Geiger who developed the eponymous radiation detector, since, of course, these experiments depended on being able to detect alpha particles at different locations after they had interacted with atoms – in the classic experiments carried out by Geiger and Marsden, the atoms in a thin sheet of gold foil at which the alpha particles were directed.

Rutherford's model of the atom

Before these experiments were carried out, the most popular model of the atom was probably one developed by J. J. Thomson, in which the atom was seen as something like a watermelon – a sphere of positively charged material in which negatively charged electrons were embedded, like pips in the melon. But when positively charged alpha particles were fired at gold foil, most went straight through, some were deflected slightly to one side and a few bounced straight back, like a tennis ball hitting a brick wall.

1. Along the way, Rutherford received the Nobel prize, in 1908, but for chemistry, not physics. Chemists regarded radioactivity as falling in their province in those days; but the award caused some hilarity among his colleagues, because Rutherford was known to regard chemistry as an inferior branch of science.

Because the alpha particles carry two units of positive charge, this could only mean that occasionally they were being repelled by approaching head on other concentrations of mass carrying positive charge. It was Rutherford who interpreted these results as implying that most of the mass and charge of an atom is concentrated in a tiny central nucleus, surrounded by a cloud of electrons. Most alpha particles never come into contact with the central nucleus (a name coined, in this context, by Rutherford in 1912, a year after the Geiger–Marsden results were officially announced) but brush straight through the electron cloud. An individual alpha particle weighs 8000 times as much as an individual electron, which has no hope of deflecting the alpha particle. If an alpha particle comes near the nucleus of an atom (which, in the case of gold, weighs 49 times as much as an alpha particle), it is nudged sideways by the positive charge. And only on rare occasions does it head straight for a nucleus, being repelled back from where it came.

Later experiments showed that the nucleus occupies only about one hundred-thousandth of the diameter of an atom; typically, a nucleus 10^{-13} centimetres across is embedded in an electron cloud 10^{-8} centimetres across. In very rough terms, the proportions are those of a grain of sand relative to Carnegie Hall. Atoms are mostly empty space, filled with a web of electromagnetic forces linking the positive and negative charges. Which means (which surely would have delighted Faraday) that everything we think of as solid matter, including the book you are reading and the chair you are sitting on, is mostly empty space, filled with a web of electromagnetic forces linking the positive and negative charges.

Radioactive decay

Rutherford still had a distinguished career ahead of him, but nothing else he achieved would be as important as his model of the atom, for which he surely should have received a second Nobel prize, this time in physics. During the First World War, he worked on techniques for detecting submarines using sound (including the forerunner of what became Asdic and Sonar), and in 1919 he succeeded Thomson as Cavendish professor and head of the Cavendish Laboratory. The same year, following up earlier experiments by Marsden, he found that nitrogen atoms bombarded with alpha particles were converted into a form of oxygen, with the ejection of a hydrogen nucleus (a proton; Rutherford also coined this term, which first appeared in print in 1920). This was the first artificial transmutation of an element. It was clear that this process involved a change in the nucleus of the atom, and it marked the beginning of nuclear physics. In experiments with James Chadwick (1891–1974) between 1920 and 1924, Rutherford showed that most of the lighter elements ejected protons when bombarded with alpha particles. From then until his early death (on 19 October 1937, from complications resulting from a long-standing hernia problem) his role was chiefly as the guiding influence to a new generation of physicists at the Cavendish; he had been knighted in 1914, and became Baron Rutherford of Nelson in 1931, just a year before the nuclear model of the atom was completed by Chadwick's discovery (or identification) of the neutron.

The existence of isotopes

Between 1912, when Rutherford gave the nucleus its name, and 1932, when Chadwick identified the neutron, apart from the discovery that nuclei of one element could be transformed into nuclei of another element, the most important development in the understanding of the atom was the discovery that individual elements come in different varieties. This was made by Francis Aston (1877–1945), working with Thomson at the Cavendish at the end of the second decade of the twentieth century. Frederick Soddy, then working at the University of Glasgow, had suggested in 1911 that some puzzling features of chemical behaviour might be explained if elements came in different varieties, which had identical chemical properties but different atomic weights. In 1913, he gave these varieties the name 'isotopes' (among other things, as we have mentioned, the existence of isotopes explains some of the rearrangements Mendeleyev had to make to the periodic table). The proof of the existence of isotopes came from Aston's work, which involved monitoring the way the positively charged 'rays' produced in discharge tubes (actually ions, atoms stripped of some of their electrons) are deflected by electric and magnetic fields. This was a development of the technique used by Thomson for measuring e/m for the electron; Aston was measuring e/m for ions, and with e known, that meant he was measuring their mass. For the same electric charge, particles moving at the same speed in the same electric field will be deflected sideways less for each metre they move forward if they weigh more, and more if they weigh less. This is the basis of what is known as the mass spectrograph, which Aston used to show that elements such as oxygen do indeed come in different varieties, with atoms of different mass. An atom of the most common form of oxygen, for example, has a mass 16 times the mass of a hydrogen atom; but the oxygen atoms made when Rutherford bombarded nitrogen with alpha particles each have a mass 17 times that of a hydrogen atom. Just why this should be so only became clear after Chadwick's work in the 1930s, although it was important enough in its own right to earn Aston the Nobel prize for chemistry in 1922 (Soddy had received the same prize in 1921). In 1900, as we have seen, there was still considerable opposition to the idea of atoms as real, physical entities; in the first decade of the twentieth century, even Einstein's compelling evidence in favour of the reality of atoms was based on statistical effects, involving large numbers of these particles. But by 1920, experiments involving just a few atoms (very nearly at the level of individual atoms) were becoming routine.

Discovery of the neutron

The work for which Chadwick received his own Nobel prize (for physics, in 1935) was carried out in 1932, following up discoveries made by Walter Bothe (1891–1957) in Germany and the Joliot-Curies, Frédéric (1900–1958) and Irène (1897–1956), in France.[1] Bothe had found, in 1930, that beryllium exposed to alpha particles produced a new form of radiation, which he tried to explain in terms of gamma rays. The Joliot-Curies took things a stage further. They reported late in January 1932 that they had found that when beryllium was bombarded with alpha particles, some form of uncharged radiation, which was difficult to detect, was emitted from the target atoms (actually, as they realized, from the nuclei). This radiation in turn caused protons, which are easy to detect, to be ejected from paraffin (from the nuclei of the atoms in the paraffin). They also thought that this artificial radioactivity induced in the beryllium was an intense form of gamma radiation, but Chadwick realized that what was happening was that the alpha radiation was knocking neutral particles out of the beryllium nuclei, and that these neutral particles

1. Irène was the daughter of Pierre and Marie Curie; when she married fellow physicist Frédéric Joliot in 1926, they both took the surname Joliot-Curie.

were in turn knocking protons (hydrogen nuclei) out of the paraffin, which contains a lot of hydrogen atoms. In further experiments using boron as the target, Chadwick confirmed the existence of this neutral particle, and measured its mass, which is slightly greater than that of the proton.

It's slightly ironic that Chadwick's greatest work was carried out in the course of a few hectic days in February 1932, stimulated by the announcement from Paris. Throughout the 1920s, the Cavendish team, and Chadwick in particular, had been searching, off and on, for a neutral entity made up of a proton and an electron bound tightly together, which seemed to be necessary to explain how alpha particles (which at that time were thought to be composed of four protons bound up with two electrons) and nuclei in general could exist. Rutherford had even used the term 'neutron' to refer to such a bound state of a proton and an electron, probably as early as 1920, although it first appeared in print in this context only in 1921.[1] This explains why Chadwick was able to move so swiftly to what we now know to be the correct conclusion when the news came from Paris. With the discovery of the neutron, all the ingredients of the atom that we learned about in school had been identified, just three score years and ten before the publication of the present book. But in order to understand how the pieces of the atom are put together, and in particular why the negatively charged electron cloud doesn't collapse down on to the positively charged nucleus, we have to go back once more to the end of the nineteenth century and yet another puzzle about the nature of light.

That puzzle concerned the nature of electromagnetic radiation from a perfect radiator, a black body. A perfect black body is one that absorbs all the radiation that falls on it, and when such an object is hot it emits radiation in a manner that is entirely independent of what the object is made of, but depends only on temperature. If you have a sealed container with a small hole in it, the hole acts like a black body; when the container is heated, radiation bounces around inside it and gets thoroughly mixed up before escaping from the hole as black-body

1. The name itself had been proposed several times previously for other hypothetical neutral particles, but this was the first reference to 'our' neutron.

radiation. This gave physicists both a tool with which to study such radiation and another name for it, 'cavity radiation'. But many objects, such as a lump of iron, will behave roughly like a black body when they are heated and radiate energy. Black-body radiation was described in this way and studied by Robert Kirchoff (1824–1887) in the late 1850s, but over the following decades it proved difficult, although many researchers tried, to come up with a mathematical model that would accurately describe the spectrum of radiation from a black body revealed by experiments. Although it would be inappropriate to go into details here, the key features of this black-body spectrum are that it has a peak in a certain band of wavelengths, with less energy radiated both at longer and at shorter wavelengths, and that the position of this peak in the electromagnetic spectrum shifts to shorter wavelengths as the temperature of the black body increases. So, for example, the fact that a red-hot lump of iron is cooler than a lump of iron which glows yellow is related to the fact that iron radiates more or less like a black body. This relationship between colour and temperature is vital in astronomy, where it is used to measure the temperatures of the stars.

Max Planck and Planck's constant, black-body radiation and the existence of energy quanta

One of the physicists who struggled to find a mathematical model for black-body radiation was Max Planck (1858–1947), who became professor of theoretical physics at the University of Berlin in 1892. His background was in thermodynamics, and from 1895 onwards he tried to find a way to derive the black-body radiation law in terms of the entropy of an array of electromagnetic oscillators (remember that at this time the electron had not been identified, and Planck and his contemporaries were largely in the dark as to what exactly these oscillators might be). Planck's model went through successive refinements in his attempt to obtain a perfect match between theory and experiment. He eventually succeeded, but only at the cost of including in his model the idea of what he called 'energy elements', by analogy with the chemical elements. In this model, the total energy of all these oscillators in a black body is divided into a finite (but very large) number of equal (but tiny!) parts, determined by a constant of nature which he labelled *h*. This became known as Planck's constant. Planck announced this version of his model at a meeting of the Berlin Academy of Sciences on 14 December 1900. Partly because of the neat calendrical coincidence that this 'revolutionary' idea was presented at the beginning of the twentieth century, that date is often given as the beginning of the quantum revolution in physics. But neither Planck himself nor the colleagues who listened to his presentation thought that way. They did not regard these energy quanta as real, but as a temporary feature of the mathematics that would disappear when a better model was developed. After all, Planck's model had already gone through many changes; why shouldn't it continue to be refined? There was no suggestion at that time, from Planck or anyone else, that there was any physical reality to the notion of energy quanta. The real quantum revolution began five years later, when Albert Einstein made his first, dramatic, contribution to the debate.

Albert Einstein and light quanta

Of all the papers Einstein published in 1905, the one that he personally singled out as 'very revolutionary'[1] was the one on light quanta (he was not alone in this judgement, since this is the work for which he eventually received the Nobel prize). Einstein used a different thermodynamic approach from Planck, building from Boltzmann's probabilistic derivation of entropy, and found that electromagnetic radiation behaves 'as if it consisted of mutually independent energy quanta'.[2] He calculated that when an 'oscillator' (that is, an atom) emitted or absorbed electromagnetic radiation it would do so in discrete units which were multiples of $h\nu$, where ν is the frequency of the radiation being emitted or absorbed (frequency is essentially the inverse of wavelength). In the same short paper, Einstein discussed how electromagnetic radiation could knock electrons out of the surface of a piece of metal – the photoelectric effect. In 1902, Philipp Lenard, following up earlier studies of the photoelectric effect, had found that when light of a particular wavelength (colour) is shone on such a surface, the electrons emitted all have the same energy, but that the energy is different for different wavelengths of light. It didn't matter if the light source was bright or faint – you got more electrons emitted when the light was brighter,

1. In a letter to his friend Conrad Habicht. See John Stachel, *Einstein's Miraculous Year*.

2. See John Stachel, *Einstein's Miraculous Year*.

but they still each had the same energy. This could be explained, on Einstein's model, if light of a particular wavelength (frequency) consisted of a stream of individual light quanta, each with the same energy *hv*. They could each give the same amount of energy to an electron in the metal, which is why the ejected electrons all have the same energy. The phenomenon discovered by Lenard could not, though, be explained at all on the wave model of light. Although Einstein emphasized the provisional nature of his ideas (even the title of the paper was 'On a Heuristic Point of View Concerning the Production and Transformation of Light'), unlike Planck he seems to have been convinced in his own mind that light quanta (they were only given the name 'photons' in 1926, by the American chemist Gilbert Lewis) were real. He fully accepted that this was a revolutionary idea, and spelled out that:

> According to the assumption considered here, in the propagation of a light ray emitted from a point source, the energy is not distributed continuously over ever-increasing volumes of space, but consists of a finite number of energy quanta localized at points of space that move without dividing, and can be absorbed or generated only as complete units.

That sentence marks the true beginning of the quantum revolution. Depending on which experiments you used, light could be seen behaving either as a wave (the double-slit experiment) or as a stream of particles (the photoelectric effect). How could this be?

Einstein's contemporaries were well aware of the revolutionary implications of his suggestion, but not by any means convinced. One person in particular was infuriated by what he regarded as a piece of nonsense, and was in a position to do something about it. Robert Millikan (1868–1953) was an American experimental physicist working at the University of Chicago. He could not accept the idea of real light quanta, and set out to prove that Einstein's interpretation of the photoelectric effect was wrong. In a long series of difficult experiments he succeeded only in proving that Einstein was right, and along the way derived a very accurate measurement of Planck's constant, as 6.57×10^{-27}. In the best traditions of science, it was this experimental confirmation of Einstein's hypothesis (all the more impressive since it was obtained by a sceptic trying to prove the idea wrong) that established clearly, by about 1915, that there was something in the idea of light quanta. As Millikan commented ruefully towards the end of his life, 'I spent ten years of my life testing that 1905 equation of Einstein's and contrary to all my expectations, I was compelled in 1915 to assert its unambiguous verification in spite of its unreasonableness.'[1] Millikan's consolation was the Nobel prize for physics in 1923, for this work and for his superbly accurate measurement of the charge on the electron; it is no coincidence that Einstein's Nobel prize had been awarded in 1922 (although it was actually the 1921 prize, which had been held over for a year). But by then, the idea of quanta had already proved its worth in explaining the behaviour of electrons in atoms, even though there was still no complete understanding of the quantum phenomenon.

The problem with Rutherford's model of the atom as a tiny central nucleus surrounded by a cloud of electrons moving in empty space was that there was nothing to stop the electrons falling into the nucleus. After all, the nucleus has positive charge and the electrons have negative charge, so they must attract each other. In seeking how to stabilize such a system, it is possible to make an analogy with the planets in orbit around the Sun – but unfortunately, the analogy doesn't hold up. To be sure, the planets are attracted to the Sun by gravity, and would 'like' to fall towards it, but are kept in their orbits by their motion, in a sense balancing centrifugal force against the pull of gravity. But electrons cannot orbit the nucleus of the atom in the same way because they have electric charge, and because they have to change direction to move in an orbit around the nucleus they are accelerated – as for the Moon orbiting the Earth, acceleration means a change in either the speed or direction of motion, or both. An electric charge which accelerates will radiate energy away in the form of electromagnetic waves, and as it loses energy in this way, an electron 'in orbit' around a nucleus would spiral into the nucleus and the atom would collapse, on a timescale of about one

1. See *Reviews of Modern Physics*, volume XXI, p. 343, 1949.

ten-billionth of a second.[1] There is no way to avoid this dilemma within the framework of the classical physics of Newton and Maxwell. The reason why atoms are stable is entirely thanks to quantum physics, and the first person to appreciate how this might happen was the Dane Niels Bohr.

Niels Bohr

Bohr was born in Copenhagen on 7 October 1885. He came from an academic family (his father was professor of physiology at the University of Copenhagen and Niels's brother Harald became professor of mathematics at the same university) and received a good scientific education, culminating in a PhD in physics from the University of Copenhagen in 1911 (his father had died of a heart attack a few months earlier). In September that year he went to work for a year under J. J. Thomson at the Cavendish Laboratory, but found it difficult to fit in there, partly because of his own imperfect English and diffident nature, partly because his research interests did not entirely mesh with those of the Cavendish at the time and partly because J. J., now in his mid-fifties, was no longer quite as receptive to new ideas as he had once been. In October, however, Rutherford gave a talk in Cambridge, describing his latest work and making a strong impression on the young Bohr. A month later, Bohr visited one of his father's former colleagues in Manchester; this family friend (at Bohr's instigation) invited Rutherford to join them at dinner. In spite of the language barrier, Rutherford and Bohr hit it off immediately (apart from their common scientific interests, Rutherford of all people understood what it was like to be an outsider in Cambridge at the start of your career), and the upshot was that in March 1912 Bohr moved to Manchester for the final six months of his visit to England. It was there that he worked out the first quantum model of the atom, directly based on Rutherford's model, although it took more than six months to complete his work.

The first quantum model of the atom

1. Strictly speaking, a planet in orbit around a star, moving through the star's gravitational field, will generate gravitational radiation and slowly lose energy in a similar way; but gravity is such a feeble force (after all, it takes the gravity of the entire Earth to overcome the electric forces between a few atoms in the stalk of an apple and make the apple fall from a tree) that this produces no measurable effect on the orbit of a planet like the Earth even after billions of years.

Bohr returned to Denmark in the summer of 1912, married his fiancée Margrethe Nørlund on 1 August and took up a junior teaching post at the University of Copenhagen in the autumn. It was there that he completed a trilogy of papers on the structure of the atom, which were all published before the end of 1913, and which formed the basis of the work for which he would receive the Nobel prize in 1922. Bohr's great genius, or knack, throughout his career, lay in his ability to stitch together whatever bits of physics were needed to make a working model of some phenomenon. He didn't worry too much about the internal consistency of the model, as long as it was useful in giving a picture in your head of what was going on, and (crucially) as long as it provided predictions which matched up with the results of experiments. The Rutherford–Bohr model of the atom, for example, contains pieces of classical theory (the idea of orbiting electrons) and pieces of quantum theory (the idea that energy is only emitted or absorbed in discrete quanta, $h\nu$); but nevertheless it contained just enough physical insight to tide physicists over until they could come up with something better. Indeed, the physical insight it provides is so good that it is still essentially the model of the atom that we learn about in school, and scarcely needs rehearsing in detail here. Bohr said that electrons had to stay in their orbits around the nucleus because they are not physically capable of continuously emitting radiation, as they would be if classical laws applied. An electron can only emit quanta of energy, one at a time, and this would correspond to it jumping down from one orbit to another, rather as if the planet Mars suddenly emitted a burst of energy and appeared in the orbit of the Earth. Stable orbits correspond to certain fixed amounts of energy, but there are no in-between orbits, so inward spiralling is impossible. Why, then, don't all the electrons jump right down into the nucleus? Bohr argued (completely *ad hoc*) that each allowed orbit had 'room' for only a certain number of electrons, and that electrons further out from the nucleus cannot jump inwards if the inner orbits are already full (so Mars could not, on this analogy, jump into the Earth's orbit because the Earth is already there). The electrons closest to the nucleus were simply forbidden to jump right into the centre of the atom, but the explanation for this would have to wait (it came a little more than ten years later, as we shall see, when Werner Heisenberg discovered the uncertainty principle).

All of this, of course, is mere hand-waving, a pretty model without any structural foundations. But Bohr did much better than that. Each 'jump' of an electron from one orbit to another corresponds to the release of a precise quantum of energy, which corresponds to a precise wavelength of light. If a large number of individual atoms (for example, a sample of hydrogen gas) are all radiating in this way, the quanta (photons) will add together to make a bright line in the spectrum at that wavelength. When Bohr put the mathematics into the model and calculated the way in which energy would be emitted when electrons jumped downwards (or, conversely, how energy would be absorbed when electrons jumped upwards from one permitted orbit to another), he found that the positions of the spectral lines predicted by the model precisely matched the positions of the lines in the observed spectra.[1] Quantum physics had explained why, and how, each element produces its own unique spectral fingerprint. The model might be a crazy patchwork of ideas old and new, but it worked.

The Rutherford–Bohr model raised about as many questions as it answered, but it showed that the way forward must involve quantum physics and, together with Einstein's theoretical work and Millikan's experiments, it pointed the way for progress towards a complete quantum theory, which would be developed in the 1920s. Bohr himself became a hot property as soon as news of this work leaked out, even before the three papers were published. At the beginning of 1914, the University of Copenhagen offered to create the post of professor of theoretical physics for Bohr, should he care to take it. Then, Rutherford wrote to offer him a two-year appointment in Manchester as a reader (a post which, as the name implies, leaves the holder free to do research with no teaching or administrative duties). Persuading Copenhagen to wait, Bohr (still only 29) seized the opportunity to work alongside Rutherford for a time. In spite of the outbreak of war (Denmark remained neutral in the First World War), the Bohrs made the journey to England safely by ship, and in due course made the return journey in 1916. In spite of several offers, including a permanent post in

1. At least, they did for hydrogen, the simplest atom; it proved very difficult to carry the calculations through for complex atoms, but this was enough to show that the model worked.

Manchester, Bohr preferred to stay in Denmark, where his prestige enabled him to obtain funding for a research Institute for Theoretical Physics in Copenhagen – now known as the Niels Bohr Institute. The Institute attracted most of the great physicists of the day for shorter or longer visits over the following years, providing a forum where ideas about the new quantum physics could be thrashed out. In the 1930s, Bohr himself became interested in nuclear physics and the possibility of obtaining energy by nuclear fission, and when Denmark was occupied by German forces in the Second World War he became concerned about the possibility of the Nazis obtaining atomic weapons and escaped via Sweden to Britain. With his son Aage Bohr (who also received a Nobel prize, in 1975) he acted as an adviser on the Manhattan Project. After the war, Niels Bohr promoted peaceful uses of atomic energy and was a leading figure in the foundation of CERN, the European particle physics research centre in Switzerland. He died on 18 November 1962, and was succeeded as Director of the Copenhagen Institute by Aage Bohr.

One of the best things about the Bohr model of the atom and the refinements made to it in the 1920s was that it provided the basis for an understanding of chemistry – how and why some elements react with one another to form compounds, while others do not. But we shall save that story for the next chapter, where we look at the chemistry of life. Here, we want to continue our journey into the atom, to see how the new quantum physics led to an understanding of the nucleus and opened up a new world of particle physics.

Louis de Broglie

Leaving aside a lot of false trails and paths that turned out to be blind alleys, and the detail of some important but technical work on the statistics of light quanta, the next big step in quantum physics came in 1924, when the French physicist Louis de Broglie (1892–1987) put forward in his doctoral thesis at the Sorbonne (published in 1925) the idea that, just as electromagnetic waves could be described in terms of particles, all material particles, such as electrons, could be described in terms of waves. De Broglie was a late starter in physics (in his thirties when he submitted that thesis) both because his aristocratic family had intended him for a career as a diplomat, so that he started studying history at the Sorbonne in 1909 before switching to physics very much against the wishes of his father, and because of the First World War, during which he served as a radio specialist, based at the Eiffel Tower. But he certainly made up for lost time, producing a key insight into the subatomic world which earned him the Nobel prize in 1929. The idea is blindingly simple when put into words, but flies completely in the face of common sense.

De Broglie started out from two equations that apply to light quanta (from now on, we will call them photons, even though the term was only applied a couple of years after this work). One of these we have already met – $E = h\nu$. The other, Einstein had derived from relativity theory, relating the momentum of a photon (p, since m is already used for mass) to the speed with which it moves (c, the speed of light) and the energy it carries – $E = pc$. Putting this pair of equations together, de Broglie had $h\nu = pc$, or $p = h\nu/c$. Since the wavelength of electromagnetic radiation (usually denoted by the Greek letter lambda, λ) is related to frequency by $\lambda = c/\nu$. This meant that $p\lambda = h$. Or in plain English, the momentum of a 'particle' multiplied by its wavelength is equal to Planck's constant. By 1924, this wasn't such a startling idea as far as light was concerned, but de Broglie suggested that it also applied to more traditional particles, and in particular to electrons. On this basis, he devised a model of the atom in which the electrons were represented by waves running around the 'orbits', like a wriggling snake biting its own tail. He said that the different energy levels of the electrons in the atom correspond to different harmonics of these waves, like notes played on a plucked guitar string, and that only orbits in which these harmonics fitted exactly, with the peaks and troughs of the wave reinforcing one another rather than cancelling out, were allowed. His thesis supervisor, Paul Langevin (1872–1946) was nonplussed by all this, and showed the thesis to Einstein, who said that it was sound work and represented more than a mere mathematical trick.

De Broglie got his PhD, and when asked at the oral examination how his idea might be tested, pointed out that according to his equation, electrons ought to have just the right wavelengths to be diffracted from crystal lattices. In two independent experiments carried out in 1927 (one by Clinton Davisson (1881–1958) and Lester Germer (1896–1971) in the US, one by George Thomson (1892–1975) in Aberdeen) de Broglie's prediction was confirmed. Davisson and Thomson shared the Nobel prize in 1937; Germer missed out, presumably because he was 'only' a student when he carried out the work with Davisson. As has often been pointed out, though, the share of the Nobel prize awarded to George Thomson, the son of J. J., deliciously highlights the non-commonsensical nature of the quantum world. J. J. got the prize for proving that electrons are particles. George got the prize for proving that electrons are waves. And both were right.

By that time, what is regarded as the definitive proof that photons exist had also been provided, by the work of Arthur Compton (1892–1962), first at Washington University, St Louis, and then in Chicago. In a series of experiments involving X-rays scattered from electrons in atoms, Compton established by the end of 1923 that this scattering could only be explained in terms of an exchange of momentum between particles, and he received the Nobel prize for this in 1927. In another example of the bizarre logic of the quantum world, it was this work, which treated electrons as particles to establish that electromagnetic radiation is both wave and particle, that helped to inspire de Broglie to show that electrons can also behave like waves! What de Broglie's equation tells us is that *everything* has dual wave–particle character. Because momentum is related to mass (except for light, which is so often a special case and where photons have no mass in the everyday sense of the term), and because Planck's constant is so small, the 'waviness' of an everyday object, such as you or me, a house, or a football, is so utterly tiny that it can never be detected. The waviness only becomes important when the mass of an object is (in the appropriate units) about the same size as, or less than, Planck's constant. Which means that the wave aspect of wave–particle duality scarcely matters at all above the molecular level, cannot entirely be ignored for whole atoms, is an important factor in describing the behaviour of protons and neutrons, and is absolutely crucial in trying to describe the behaviour of electrons, inside or outside atoms. This also tells us that we have no hope of understanding what an electron 'really is' in terms of our everyday, common-sense experience. It is literally like nothing we have ever seen. All we can hope to do is to find equations – mathematical models – which tell us how electrons behave in different circumstances, sometimes more like a wave, sometimes more like a particle. That is exactly what happened in quantum mechanics, almost before the ink was dry on de Broglie's thesis.

Erwin Schrödinger's wave equation for electrons

The particle-based approach to the quantum world of electrons

Just such a complete mathematical model describing the behaviour of electrons in atoms was developed not once, but twice in the months following the publication of de Broglie's ideas. The direct line from de Broglie ran to the Austrian physicist Erwin Schrödinger (1887–1961), then professor of physics in Zürich, who developed a model entirely based on waves, and was delighted to think that he had restored some sanity to the strange world of subatomic physics by explaining it in terms of something as comfortable and familiar as a wave equation. But by the time his work was published, in 1926, he had just been beaten into print by another complete mathematical description of the behaviour of electrons in atoms, which essentially emphasized the particle approach and quantum-jumping from one energy level to the next. This approach was initiated by the German Werner Heisenberg (1901–1976), promptly followed up by his colleagues at the University of Göttingen, Max Born (1882–1970) and Pascual Jordan (1902–1980), and seized upon by the young British physicist Paul Dirac (1902–1984). Dirac first developed a more abstract mathematical formalism to describe the behaviour of electrons in atoms (a third complete quantum theory!), and then showed that both of the other two approaches were contained within that formalism, and were mathematically equivalent to one another, just as whether you choose to measure a distance in miles or in kilometres doesn't change the distance you are measuring. All of these people except Jordan (a mysterious aberration by the Nobel committee) eventually received Nobel prizes for their various contributions to quantum theory.

The upshot of this flurry of activity was that by 1927 physicists had a choice of mathematical models which they could use in calculating the behaviour of quantum entities such as electrons. Most, like Schrödinger, preferred the cosy familiarity of working with a wave equation; but this should never be seen as implying that the wave version of quantum reality contains any deeper truth than the particle version (if anything, by its very familiarity the wave mechanics approach tends to conceal the true nature of the quantum world). They are simply different facets of a whole which is something unlike anything in our everyday world, but which can sometimes behave like a particle and sometimes like a wave. People still argue about what all this 'really means', but for our purposes it is sufficient to take the pragmatic approach and say that quantum mechanics works, in the sense of making predictions that are confirmed by experiments, so it doesn't matter what it means.

Heisenberg's uncertainty principle: wave–particle duality

Heisenberg made another contribution to quantum physics, though, which is worth discussing here – his famous uncertainty principle. This is related to the idea of wave–particle duality, and says that certain pairs of quantum properties, such as position and momentum, can never both be precisely defined at the same time; there is always a residue of uncertainty (related to the size of Planck's constant, so again these effects only show up on very small scales) in the value of at least one of these parameters. The more accurately one member of the pair is constrained, the less accurately the other one is constrained. This is *not* simply a matter of our imperfect measuring apparatus disturbing the quantum world when we try to measure it, so that, for example, if we try to measure the position of an electron we give it a nudge and change its momentum. It is a fundamental feature of the quantum world, so that an electron itself does not 'know' both precisely where it is and precisely where it is going at the same time. As Heisenberg himself put it, in a paper published in 1927, 'we *cannot* know, as a matter of principle, the present in all its details'.

It turned out that this is such a fundamental aspect of the way the world works that it is possible to construct the entire edifice of quantum mechanics starting out from the uncertainty principle, although we won't go into that here. The power of the uncertainty principle can, though, be gleaned by returning to the puzzle of why the electrons in an atom don't all fall into the nucleus, even if they have to do it in a series of jumps, rather than by spiralling inwards. When an electron is in orbit around a nucleus, its momentum is very well determined by the properties of the orbit, so any uncertainty in the momentum-position pair is forced to be in its position. If an electron is somewhere in an orbit, there is indeed uncertainty in its position – it could be on one side of the orbit or the other (or it could, if you prefer that picture, be a wave spread around the orbit). But if it fell right into the nucleus, its position would be very well determined – within the volume of the nucleus. Its momentum would also be very well determined, since it would not be going anywhere. This would violate the uncertainty principle (if you like to think of it that way, you could say that a nucleus is too small for the wave associated with an electron to fit inside it). Put the appropriate numbers in, with the appropriate momentum for an electron in an atom, and it turns out that the size of the smallest electron orbit in an atom is as small as it can be without violating the uncertainty principle. The very sizes of atoms (and the fact that atoms exist at all!) are determined by the uncertainty principle of quantum mechanics.

Dirac's equation of the electron

It took a couple of decades after the breakthroughs of the mid-1920s for all the loose ends to be tied up, not least because of the disruption to scientific research caused by the Second World War. But before that disruption, two more key developments took place. In 1927, Dirac published a paper in which he presented a wave equation for the electron which fully incorporated the requirements of the special theory of relativity; it was the definitive last word on the subject, *the* equation of the electron. Curiously, though, the equation had two solutions, rather like the way in which the simple equation $x^2 = 4$ has two solutions. In this trivial case, either $x = 2$ or $x = -2$. But what did the much more complicated 'negative solution' to Dirac's equation mean? It seemed to describe a particle which had the opposite properties to an electron, including, most notably, positive charge instead of negative charge. At first, Dirac tried to make this solution fit the proton, which did indeed have positive charge but, of course, has far too much mass to be a 'negative electron'.[1] By 1931, he realized (along with other people) that the equation was actually predicting the existence of a previously unknown particle, with the same mass as an electron but positive charge. Further investigation of the equation suggested that if enough energy was available (for example, from an energetic gamma ray) then it could be converted, in line with Einstein's equation $E = mc^2$, into a *pair* of particles, an ordinary electron and a negative electron. The energy could not be converted into a single particle, or even two electrons, since that would

1. A negative electron must have positive charge, because an electron has negative charge and two negatives make a positive.

violate the conservation of electric charge; but by creating a positive–negative pair all the properties except mass (itself provided by the input of energy) would be cancelled out.

The existence of antimatter

In experiments carried out in 1932 and 1933, Carl Anderson (1905–1991), working at the California Institute of Technology, found the trace of just such a positively charged particle in his studies of cosmic rays. Although he did not realize that the positron, as he called it, had indeed been manufactured in the cloud chamber used to study the cosmic rays by the pair-production process predicted by Dirac, the connection was soon made by other people. Antimatter, as it came to be known, was a real feature of the physical world, and every type of particle is now known to have an antimatter equivalent with opposite quantum properties.

The strong nuclear force

To put the last key discovery of the 1930s in perspective, we have to step back about ten years, to the early 1920s. At that time, the neutron had yet to be discovered, and various models of the alpha particle existed which tried to explain it as a combination of four protons with two electrons. Clearly, such an entity ought to blow itself apart because of electrostatic repulsion. In a paper published in 1921, Chadwick and his colleague Etienne Bieler wrote that if this kind of model of the alpha particle were correct, it must be held together by 'forces of very great intensity', and concluded 'it is our task to find some field of force which will reproduce these effects'.[1] This conclusion applies with equal force to models of the alpha particle as being made up of two protons and two neutrons, and, indeed, to all nuclei, which are essentially balls of neutrons and protons with an overall positive charge. Some strong force, stronger than the electric force over the very short distances, represented by the diameter of an atomic nucleus, must overwhelm the electric repulsion and hold everything together. Rather prosaically, this became known as the strong nuclear force or just 'the strong force'. Experiments later showed that this force is about a hundred times stronger than the electric force, which is why there are about a hundred protons in the largest stable nuclei; any more, and electric repulsion overcomes the strong force and blows the nucleus apart. But the strong force, unlike

1. *Philosophical Magazine*, volume 42, p. 923, 1921.

electric, magnetic and gravitational forces, does not obey an inverse square law. It is very strong indeed over a limited range of about 10^{-13} centimetres, and essentially cannot be felt at all beyond that range. This is why nuclei have the size they do – if the strong force had longer range, nuclei would be correspondingly larger.

The final piece of the atomic jigsaw puzzle was slotted into place to solve a puzzle that had grown in importance during the 1920s. It concerned the process of beta decay, in which an atom (actually the nucleus of an atom) ejects an electron and is transformed in the process into an atom of the element next door in the periodic table. After the neutron was discovered, it became clear that this process actually involves a neutron being transformed (or rather, transforming itself) into a proton and an electron. Neutrons decay spontaneously in this way if left to their own devices outside an atomic nucleus. It is important to appreciate, though, that there is no sense in which the electron is 'inside' the neutron and escapes – as quantum uncertainty (among other things) makes clear, this is not possible. What happens is that the mass-energy of the neutron is converted into the mass-energy of an electron and a proton, with some left over to provide kinetic energy for the electron to speed away from the site of the decay.

The puzzle was that the electron speeding away from a nucleus in this way seemed able to carry any amount of energy, up to a well-defined maximum value. This was quite different from the behaviour of alpha particles ejected during alpha decay. In alpha decay, all the particles ejected from a particular kind of nucleus may emerge with the same kinetic energy, or they may emerge with a certain smaller energy, but accompanied by an energetic gamma ray. The energy carried by the alpha particle and the gamma ray always adds up to the same maximum energy for that particular kind of nucleus, and the energy liberated in this way is equal to the difference in mass-energy between the original nucleus and the one left after the decay – so energy is conserved. But the escaping alpha particles can only have certain discrete energies, because the gamma-ray photons are quantized, and can only carry certain discrete amounts of energy as their contribution to the total. Similarly, momentum and angular momentum are conserved in alpha decay. But in beta decay, although there was still a well-defined maximum energy for the electrons emitted from a particular kind of nucleus, they seemed able to emerge with any lesser amount of energy they liked, right down to almost zero, and there was no accompanying photon to carry off the excess. It seemed as if the process violated the law of conservation of energy. At first, it seemed the experiments must be in error – but by the end of the 1920s it was clear that there really was a continuous 'spectrum' of electron energies associated with beta decay. Other properties also seemed not to be conserved in the process, but there is no need for us to go into the details.

At the end of 1930, Wolfgang Pauli (1900–1958) came up with a speculative suggestion to explain what was going on. To appreciate how shocking this suggestion seemed to many of his colleagues, remember that at this time the only two traditional particles known to physics were the electron and the proton (the photon was not regarded as a particle in the same way, even then, and the neutron had yet to be discovered), so that any suggestion of another 'new' particle (let alone an essentially invisible one) was almost sacrilegious. In a letter dated 4 December 1930, Pauli wrote:

> I have come upon a desperate way out . . . To wit, the possibility that there could exist in the nucleus electrically neutral particles, which I shall call neutrons . . . The continuous beta-spectrum would then become understandable from the assumption that in beta-decay a neutron is emitted along with the electron, in such a way that the sum of the energies of the neutron and the electron is constant.[1]

In other words, Pauli's 'neutron' played the role of the gamma ray in alpha decay, but with the difference that it could carry any amount of kinetic energy, up to the maximum available, and was not quantized in the way that gamma-ray photons are quantized.

The weak nuclear force; neutrinos

It's a sign of how little impact Pauli's desperate remedy had that within two years the name 'neutron' had been applied to the nuclear particle identified by Chadwick, which was very definitely not the particle Pauli had in mind. But the problem of 'the continuous beta-spectrum' refused to go away, and in

1. Quoted by Pais, *Inward Bound*.

1933 Enrico Fermi (1901–1954), with the benefit of knowing that the neutron existed, took up Pauli's idea and developed it into a complete model, in which the decay process is triggered by the action of a new field of force, which soon became known (by contrast with the strong nuclear force) as the weak nuclear force. His model described how, in addition to the strong force holding protons and neutrons together in the nucleus, there was a weak, short-range force which could cause a neutron to decay into a proton and an electron *plus* another, uncharged particle, which he dubbed the 'neutrino' (from the Italian for 'little neutron'). Unlike Pauli's speculation, Fermi provided a mathematical model which indicated clearly the way the energy of the electrons emitted during beta decay was distributed, and agreed with the experiments. Even so, when Fermi sent his paper describing this work to the journal *Nature*, in London, it was rejected as 'too speculative' and he published it in an Italian journal. Although the paper proved sound and circumstantial evidence in support of the idea built up over the following years, the neutrino proved so elusive that it was only detected directly in the mid-1950s. To give you some idea of what an experimental *tour de force* this was, if a beam of neutrinos travelled through a wall of lead 3000 light years thick, only half of them would be captured by the nuclei of the lead atoms along the way.

The identification of the neutrino completes the set of particles and forces that are responsible for the way things behave in the everyday world. We are made of atoms. Atoms are made of protons, neutrons and electrons. The nucleus contains protons and neutrons, held together by the strong force, in which beta decay can take place as an effect of the weak force (and from which, in some cases, alpha particles may be ejected as a result of an internal readjustment of the nucleus). The electrons are in a cloud outside the nucleus, held in place by electromagnetic forces but only allowed to occupy certain energy states by the rules of quantum physics. On large scales, gravity is important in holding bigger lumps of matter together. This gives us four particles (proton, neutron, electron and neutrino) to worry about (plus their associated antiparticles) and four forces (electromagnetism, the strong and weak nuclear forces, and gravity) to consider. That is sufficient to explain everything that is detectable to our senses, from why the stars shine to how your body digests food, from the explosion of a hydrogen bomb to the way ice crystals form snowflakes.

Quantum electrodynamics

Indeed, apart from gravity, and the limited ways in which the weak nuclear force affects us through radioactivity, almost everything in the human world is affected almost entirely by the interactions of electrons with one another, with the positively charged nuclei of atoms and with electromagnetic radiation. Those interactions are governed by the laws of quantum mechanics, which were pieced together into a complete theory of light (electromagnetic radiation) and matter in the 1940s. That theory is known as quantum electrodynamics, or QED, and is probably the most successful scientific theory yet developed. In fact, QED was developed independently by three different scientists. The first to come up with a complete theory was Sin-itiro Tomonaga (1906–1979), who was working under first difficult and then appalling conditions in Tokyo during and immediately after the Second World War; because of these difficulties, his work only appeared in print at about the same time as the papers describing the work of the other two pioneers, the Americans Julian Schwinger (1918–1994) and Richard Feynman (1918–1988). All three shared the Nobel prize in 1965. Tomonaga and Schwinger both worked within what might be called the traditional mathematical framework of quantum mechanics at the time (a tradition going back all of two decades), building directly on the work which had been going on since the breakthroughs of the 1920s, and Dirac's work in particular. Feynman used a different approach, essentially reinventing quantum mechanics from scratch. All these approaches are mathematically equivalent, though, in the way that the Heisenberg–Born–Jordan, Schrödinger and Dirac versions of quantum mechanics are all mathematically equivalent. But we don't need to go into the details here since there is a neat physical picture which gives you a feel for what is going on.

When two charged particles, such as two electrons, or an electron and a proton, interact, they can be thought of as doing so by the exchange of photons. Two electrons, say, may move towards one another, exchange photons and be deflected on to new paths. It is this exchange of photons which produces the repulsion which shows up as an inverse square law, a law which emerges naturally from QED. The strong and weak nuclear forces can be described in terms of the exchange of photon-like particles in a similar way (the weak force with such success that it has now been incorporated into electromagnetism to form a single model, called the electroweak interaction; the strong force with slightly less success), and it is thought that gravity should also be described by the exchange of particles, called gravitons, although no complete model of quantum gravity has yet been developed. The accuracy of QED itself, however, can be gleaned from looking at just one property of the electron, called its magnetic moment.[1] In the early version of QED developed by Dirac at the end of the 1920s, with a suitable choice of units this property has the predicted value 1. In the same units, experiments measure the value of the electron magnetic moment as 1.00115965221, with an uncertainty of ± 4 in the last digit. This was already an impressive achievement, which convinced physicists in the 1930s that QED was on the right track. The final version of QED, though, predicts a value of 1.00115965246, with an uncertainty of ± 20 in the last digits. The agreement between theory and experiment is 0.00000001 per cent, which, Feynman used to delight in pointing out, is equivalent to measuring the distance from New York to Los Angeles to the thickness of a human hair. This is by far the most precise agreement between theory and experiment for any experiment carried out on Earth,[2] and is a genuine example of how well science can explain the behaviour of the physical world in which we live our everyday lives, and of how far we have come since people like Galileo and Newton began to compare theory with observation and experiment in a properly scientific way.

The future? Quarks and string

In the second half of the twentieth century, as physicists probed within the nucleus and investigated high-energy events using giant particle accelerators, they uncovered a world of subatomic particles, and found (at just the first level of this new world)

1. This is a typical example, not one chosen because it is the only good match between theory and experiment.

2. The general theory of relativity has been tested to similar precision, at the end of the twentieth century, by measuring changes in the observed properties of astronomical objects known as binary pulsars, many light years away from Earth; but although this is an impressive achievement it isn't quite the same as doing the experiments under controlled conditions down here in a laboratory on the surface of the Earth.

that protons and neutrons can be thought of as composed of entities called quarks, held together by the exchange of entities analogous to photons, and that the strong nuclear force is just an outward manifestation of this deeper force at work. At the beginning of the twenty-first century, many physicists are persuaded by the available evidence that all of these 'particles' may be better understood as the manifestations of even deeper layers of activity involving tiny loops of vibrating 'string'. But it is far too early yet to be able to write the history of all this work, and it seems appropriate to end this particular account at the level of nuclei and atoms – still, as yet, the deepest level which has any impact at all on our daily lives. In particular, as we shall explain in the next chapter, it is all we need to explain the way in which life itself operates.

14 The Realm of Life

The most complex things in the Universe – Charles Darwin and nineteenth-century theories of evolution – The role of cells in life – The division of cells – The discovery of chromosomes and their role in heredity – Intracellular pangenesis – Gregor Mendel: father of genetics – The Mendelian laws of inheritance – The study of chromosomes – Nucleic acid – Working towards DNA and RNA – The tetranucleotide hypothesis – The Chargaff rules – The chemistry of life – Covalent bond model and carbon chemistry – The ionic bond – Bragg's law – Chemistry as a branch of physics – Linus Pauling – The nature of the hydrogen bond – Studies of fibrous proteins – The alpha-helix structure – Francis Crick and James Watson: the model of the DNA double helix – The genetic code – The genetic age of humankind – Humankind is nothing special

The most complex things in the Universe

We are the most complicated things that we know about in the entire Universe. This is because, on the cosmic scale of things, we are middle-sized. As we have seen, small objects, like atoms, are composed of a few simple entities obeying a few simple laws. As we shall see in the next chapter, the entire Universe is so big that the subtleties of even objects as large as stars can be ignored, and the whole cosmos can be treated as a single object made up of a reasonably smooth distribution of mass-energy, again obeying a few very simple laws. But on scales where atoms are able to join together to make molecules, although the laws are still very simple, the number of compounds possible – the number of different ways in which atoms can join together to make molecules – is so great that a huge variety of different things with complicated structures can exist and interact with one another in subtle ways. Life as we know it is a manifestation of this ability for atoms to form a complex variety of large molecules. This complexity starts on the next scale up from atoms, with simple molecules such as water and carbon dioxide; it ends where molecules begin to be crushed out of existence by gravity, once we are dealing with the interiors of objects the size of large planets, and even atoms are entirely stripped of their electrons by the time we are dealing with objects the size of stars.

The exact size of a lump of matter needed to destroy the complexity on which life as we know it depends is determined by the different strengths of the electromagnetic and gravitational forces. The electrical forces that hold molecules together are 10^{36} times stronger than the gravitational forces that try to crush molecules out of existence in a lump of matter. When atoms are together in a lump of matter there is no overall electric charge, because each atom is electrically neutral. So each atom is essentially on its own when it comes to withstanding gravity through the strength of QED. But the strength of the inward gravitational force on each atom in the lump of matter increases with the addition of every extra atom that is contributed to the lump. The amount of mass in a sphere with a certain density is proportional to the cube of the radius (for constant density), but the strength of the gravitational force falls off in accordance with an inverse square law, so in terms of the radius of a lump of matter, gravity at the surface 'gains' on electric forces in accordance with a two-thirds power. This means, since 36 is two-thirds of 54, that when 10^{54} atoms are together in a single lump, gravity dominates and complicated molecules are broken apart.

Imagine starting out with a set of objects made up of 10 atoms, 100 atoms, 1000 atoms and so on, with each lump containing ten times more atoms than the one before. The twenty-fourth object would be as big as a sugar cube, the twenty-seventh would be about the size of a large mammal, the fifty-fourth would be the size of the planet Jupiter and the fifty-seventh would be about as big as the Sun, where even atoms are destroyed by gravity, leaving a mixture of nuclei and free electrons called a plasma. On this logarithmic scale, people are almost exactly halfway in size between atoms and stars. The thirty-ninth object in our collection would be equivalent to a rock about a kilometre in diameter, and the realm of life forms like ourselves can reasonably be said to be between the sizes of sugar lumps and large rocks. This is more or less the realm investigated by Charles Darwin and his successors in establishing the theory of evolution by natural selection. But the basis for the complexity of life that we see around us on these scales depends on chemical processes going on at a slightly deeper level, where, we now know, DNA is the key component of life. The story of how DNA was identified as the key to life is the second great story of twentieth-century science, and, like the story of quantum physics, it began almost exactly with the dawn of the new century, although in this case there had been a neglected precursor to the new discoveries.

Charles Darwin and nineteenth-century theories of evolution

From the time of the great debate stirred by the publication of the *Origin of Species* in 1859, understanding of the process of evolution by natural selection had at best marked time, and arguably went backwards, during the rest of the nineteenth century. One reason was the problem of the timescale required for evolution, which we have already mentioned, and which was only resolved in the twentieth century by an understanding of radioactivity. But although Darwin (and others) fought the case for the long timescale required by evolution, the strength of the case put forward by the physicists (in particular William Thomson/Lord Kelvin) put even Darwin on the defensive. The other, and even more important, reason was that Darwin and his contemporaries did not understand the mechanism by which characteristics are passed on from one generation to the next – the mechanism of heredity. That, too, would not become clear until well into the twentieth century.

Darwin's own ideas about heredity were first presented to the world in 1868, in a chapter at the end of his book *Variation of Animals and Plants under Domestication*; they indicate the way many biologists thought at the time, although Darwin offered the most complete model. He gave it the name 'pangenesis', from the Greek 'pan', to indicate that every cell in the body contributed, combined with 'genesis', to convey the idea of reproduction. His idea was that every cell in the body contributes tiny particles (which he called 'gemmules') which are carried through the body and are stored in the reproductive cells, egg or sperm, to be passed on to the next generation. The model also incorporated the idea of blending inheritance, which says that when two individuals combine to produce offspring, the offspring represent a blend of the characteristics of the parents. To modern eyes, it is startling to see Charles Darwin himself promoting this idea, which implies that, for example, the children of a tall woman and a short man should grow up to some intermediate height. This runs completely against the basic tenet of evolution by natural selection, the requirement of variation among individuals to select from, since in a few generations, blending inheritance would produce a uniform population. The fact that Darwin even considered such an idea shows how far biologists were from a true understanding of inheritance at the time. It is against this background that we see Darwin's many revisions of the *Origin* leaning more and more towards the Lamarckian position, while his opponents argued that evolution could not proceed by the series of tiny steps envisaged in the original version of natural selection, because intermediate forms (such as a proto-giraffe with a neck longer than that of a deer but too short for it to browse on treetops) would not be viable.[1] Critics of Darwin, such as the splendidly named Englishman St George Jackson Mivart (1827–1900), suggested that evolution required sudden changes in body plan from one generation to the next, with a deer, in effect, giving birth to a giraffe. But they had no mechanism for this process either (except for the hand of God), and Darwin was at least on the right lines when he highlighted the importance of individual cells in reproduction, and even with his idea that the reproductive cells contain tiny 'particles' which carry information from one generation to the next.

The role of cells in life

The division of cells

The role of cells as the fundamental component of living things had only become clear at the end of the 1850s, the same time that Darwin was presenting his theory of evolution by natural selection to a wide audience. The realization was driven largely by improving microscopic instruments and techniques. Matthias Schleiden (1804–1881) proposed in 1838 that all plant tissues are made of cells, and a year later Theodor Schwann (1810–1882) extended this to animals, suggesting that all living things are made up of cells. This led to the idea (suggested by, among others, John Goodsir (1814–1867)) that cells arise only from other cells, by division, and it was this idea that was taken up and developed by Rudolf Virchow (1821–1902) in a book, *Die Cellular-*

1. This is not the place to go into the details of why those critics were wrong, but if you want to know how evolution does work to, among other things, turn a deer into a giraffe, the best place to start is with Richard Dawkins' book *The Blind Watchmaker*.

pathologie, published in 1858. Virchow, then professor of pathology in Berlin, explicitly stated that 'every cell is derived from a preexisting cell', and applied this doctrine to his field of medicine, suggesting that disease is no more than the response of a cell (or cells) to abnormal conditions. In particular, he showed that tumours are derived from pre-existing cells in the body. This proved immensely fruitful in many ways, and produced an explosion of interest in the study of the cell; but Virchow put all of his theoretical eggs in one basket and was strongly opposed to the 'germ' theory of infection (he also rejected the theory of evolution by natural selection). This means that although he made many important contributions to medicine, served in the Reichstag (where he was an opponent of Otto von Bismarck) and worked on the archeological dig to discover the site of Homer's Troy in 1879, he made no further direct contribution to the story we have to tell here.

The discovery of chromosomes and their role in heredity

The microscopic techniques available at the time were more than adequate to show the structure of the cell as a bag of watery jelly with a central concentration of material, known as the nucleus. They were indeed so good that in the late 1870s both Hermann Fol (1845–1892) and Oskar Hertwig (1849–1922) independently observed the penetration of the sperm into the egg (they worked with sea urchins, which have the invaluable property of being transparent), with two nuclei fusing to form a single new nucleus, combining material provided by (inherited from) both parents. In 1879, yet another German, Walther Flemming (1843–1915), discovered that the nucleus contains thread-like structures which readily absorb coloured dyes used by microscopists to stain cells and highlight their structure; the threads became known as chromosomes. Flemming and the Belgian Edouard van Beneden (1846–1910) independently observed, in the 1880s, the way in which chromosomes were duplicated and shared between the two daughter cells when a cell divided. August Weismann (1834–1914), working at the University of Freiburg, took up this line of study in the 1880s. It was Weismann who pointed to chromosomes as the carriers of hereditary information, stating that 'heredity is brought about by the transmission from one generation to another of a substance with a definite chemical and, above all, molecular constitution'.[1] He gave this substance the name 'chromatin', and spelled out the two kinds of cell division that occur in species like our own. During the kind of cell division associated with growth and development, all the chromosomes in a cell are duplicated before the cell divides, so each daughter cell obtains a copy of the original set of chromosomes; during the kind of cell division that produces egg or sperm cells, the amount of chromatin is halved, so that a full set of chromosomes is only restored when two such cells fuse to create the potential for the development of a new individual.[2] It was Weismann who showed, by the early years of the twentieth century, that the cells responsible for reproduction are not involved with other processes going on in the body, and the cells that make up the rest of the body are not involved with the manufacture of reproductive cells, so that Darwin's idea of pangenesis is definitely wrong, and the Lamarckian idea of outside influences from the environment directly causing variations from one generation to the next could be ruled out (not that this stopped the Lamarckians from arguing their case well into the twentieth century). The later discovery that radiation can cause what are now known as mutations, by directly damaging the DNA in the reproductive cells, in no way diminishes the power of Weismann's argument, since these random changes are almost invariably deleterious, and certainly do not adapt the descendants of the affected organism more closely to their environment.

Intracellular pangenesis

At about the same time that Weismann was probing inside the cell to identify the chemical units that are the carriers of heredity, the Dutch botanist Hugo de Vries (1848–1935) was working with whole plants to gain an insight into the way characteristics are passed on from one generation to the next. In 1889, just seven years after Darwin had died, de Vries published a book, *Intracellular Pangenesis*, in which he tried to adapt Darwin's ideas to the picture of how cells worked that was then beginning to emerge. Combining this with observations of how heredity works in

1. Quoted by David Young in *The Discovery of Evolution*.

2. All this applies, of course, to sexual reproduction. Asexual reproduction is, by and large, much simpler, with daughter cells being exact replicas of the parent cell (but see Gribbin and Cherfas, *The Mating Game*; as a sexually reproducing species ourselves, however, it is sexual reproduction that is central to our own story.

plants, he suggested that the characteristics of a species must be made up from a large number of distinct units, each of them due to a single hereditary factor which was passed on from one generation to the next more or less independently of the others. He gave the hereditary factors the name 'pangens' (sometimes translated into English as 'pangenes'), from Darwin's term pangenesis; after the studies by Weismann (and others) which showed that the whole body is not involved in producing these hereditary factors, the 'pan' was quietly dropped, giving us the familiar modern term 'gene', first used by the Dane Wilhelm Johannsen, in 1909.

Gregor Mendel: father of genetics

In the 1890s, de Vries carried out a series of plant-breeding experiments in which he carefully recorded the way in which particular characteristics (such as the height of a plant or the colour of its flowers) could be traced down the generations. Similar studies were being carried out at the same time in England by William Bateson (1861–1926), who later coined the term 'genetics' to refer to the study of how heredity works. By 1899, de Vries was ready to prepare his work for publication, and while doing so carried out a survey of the scientific literature in order to place his conclusions in their proper context. It was only at this point that he discovered that almost all of the conclusions he had reached about heredity had been published already, in a seldom read, and even less frequently cited, pair of papers by a Moravian monk, Gregor Mendel. The work had actually been described by Mendel in two papers he read to the Natural Science Society in Brünn (as it then was; now Brno, in the Czech Republic) in 1865, and published a year later in its *Proceedings*. It is easy to imagine de Vries's feelings when he made this discovery. Perhaps a little disingenuously, he published his own findings in two papers which appeared early in 1900. The first, in French, made no mention of Mendel. But the second, in German, gives almost fulsome credit to his predecessor, commenting that 'this important monograph is so rarely quoted that I myself did not become acquainted with it until I had concluded most of my experiments, and had independently deduced the above propositions',[1] and summing up:

1. Translation from Iltis.

From this and numerous other experiments I drew the conclusion that the law of segregation of hybrids as discovered by Mendel for peas finds very general application in the plant kingdom and that it has a basic significance for the study of the units of which the species character is composed.

This was clearly an idea whose time had come. In Germany, Karl Correns (1864–1933), working along similar lines, had also recently come across Mendel's papers, and was preparing his own work for publication when he received a copy of de Vries' French paper. And in Austria, Erich Tschermak von Seysenegg suffered a similar fate.[1] The overall result was that the genetic basis of heredity soon became firmly established, and each of the three rediscoverers of the basic principles involved gave due credit to Mendel as the real discoverer of the laws of heredity. This was certainly true, but the ready acknowledgement of Mendel's priority should not be seen entirely as an act of selfless generosity – after all, with three people having a claim to the 'discovery' in 1900, it suited each of them to acknowledge a now-dead predecessor rather than get into an argument among themselves about who had done the work first. There is, though, an important historical lesson to be drawn from the story. Several people made similar discoveries independently at the end of the 1890s because the time was ripe and the groundwork had been laid by the identification of the nucleus and the discovery of chromosomes. Remember that the nucleus itself was only identified in the same year that the joint paper by Darwin and Wallace was read to the Linnean Society, 1858, while Mendel's results were published in 1866. It was an inspired piece of work, but it was ahead of its time and made little sense on its own, until people had actually seen the 'factors of heredity' inside the cell, and the way in which they were separated and recombined to make new packages of genetic information. But although Mendel's work, as it happens, had no influence at all on the development of biological science in the second half of the nineteenth century, it's worth taking a brief look at what he did, both to counter some of the misconceptions about the

1. Tschermak was a 26-year-old graduate student at the time, who certainly discovered Mendel's papers independently, but made only a minor contribution in his own right, compared with those of de Vries and Correns.

37. *Gregor Mendel*

man and to emphasize the really important feature of his work, which is often overlooked.

Mendel was not some rural gardener in monk's habit who got lucky. He was a trained scientist, who knew exactly what he was doing, and was one of the first people to apply the rigorous methods of the physical sciences to biology. Born on 22 July 1822 in Heinzendorf in Moravia (then part of the Austrian Empire) and christened Johann (he took the name Gregor when he joined the priesthood), Mendel was clearly an unusually intelligent child, but came from a poor farming family, which exhausted all its financial resources in sending the bright young man through high school (gymnasium) and a two-year course at the Philosophical Institute in Olmütz, intended to prepare him for university. With university itself beyond his financial means, in 1843 Mendel joined the priesthood as the only means of furthering his education, having been headhunted by the abbot of the monastery of St Thomas, in Brünn. The abbot, Cyrill Franz Napp, was in the process of turning the monastery into a leading intellectual centre where the priests included a botanist, an astronomer, a philosopher and a composer, all with high reputations outside the monastery wall. Abbot Napp was eager to add to his community of thinkers by recruiting bright young men with ability but no other prospects, and was introduced to Mendel by Mendel's physics professor at Olmütz, who had previously worked in Brünn. Mendel completed his theological studies in 1848, and worked as a supply teacher at the nearby gymnasium and later at the technical college, although because of severe examination nerves he repeatedly failed the examinations that would have regularized his position.

Mendel showed such ability that in 1851, at the age of 29, he was sent to study at the University of Vienna, where Christian Doppler was professor of physics (to put the date in another context relevant to that city, Johann Strauss the younger was 26 in 1851). He was allowed only two years away from the monastery for this privileged opportunity, but crammed into that time studies of experimental physics, statistics and probability, the atomic theory of chemistry and plant physiology, among others. He did not take a degree – that was never the abbot's intention – but returned to Brünn better equipped than ever for his role as a teacher. But this was not enough to satisfy his thirst for scientific knowledge. In 1856, Mendel began an intensive study of the way heredity works in peas,[1] carrying out painstaking and accurate experiments over the next seven years which led him to

1. He chose peas because he knew that they had distinctive characteristics that bred true and would be susceptible to statistical analysis.

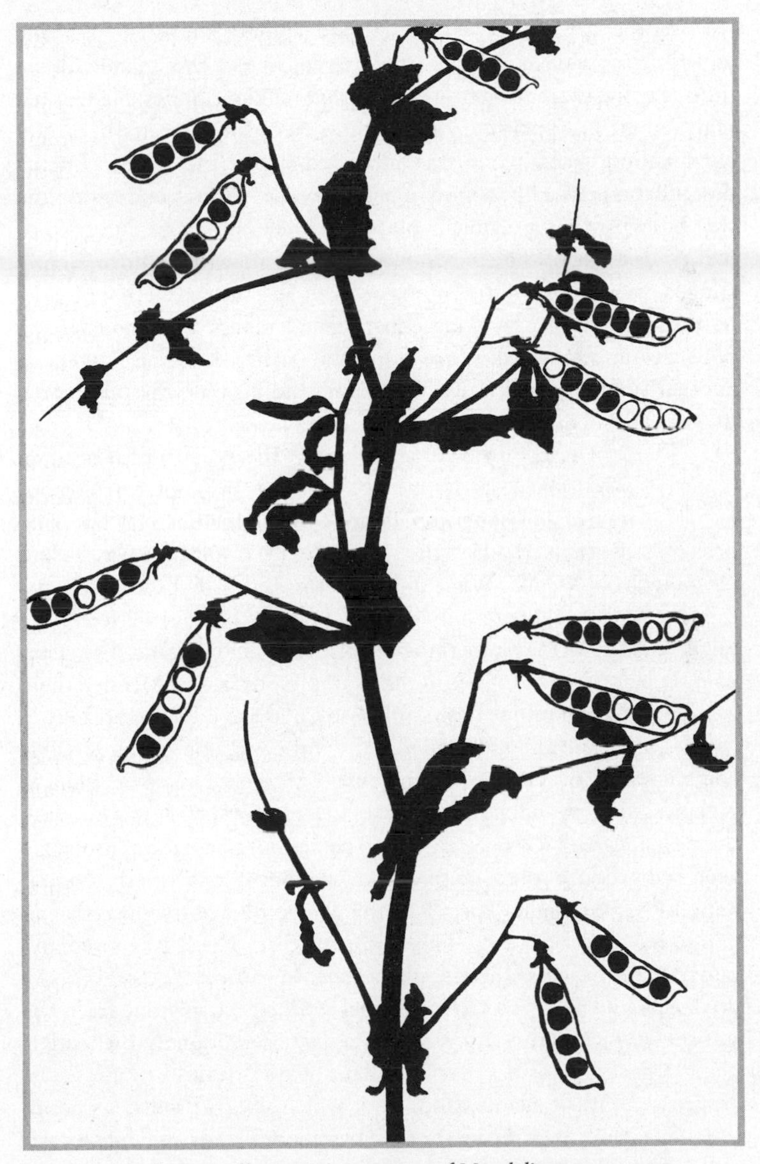

38. *A diagram illustrating an aspect of Mendel's unsung paper on heredity, showing pea plants.*

discover the way heredity works. He had a plot of land 35 metres long and 7 metres wide in the monastery garden, a greenhouse, and all the time he could spare from his teaching and religious duties. He worked with about 28,000 plants, from which 12,835 were subjected to careful examination. Each plant was identified as an individual, and its descendants traced like a human family tree, in marked contrast to the way biologists had previously planted varieties *en masse* and tried to make sense out of the confusion of hybrids that resulted (or simply studied plants in the wild). Among other things, this meant that Mendel had to pollinate each of his experimental plants by hand, dusting pollen from a single, known individual plant on to the flowers of another single, known individual plant, and keeping careful records of what he had done.

The Mendelian laws of inheritance

The key to Mendel's work – the point which is often overlooked – is that he worked like a physicist, carrying out repeatable experiments and, most significantly, applying proper statistical methods to the analysis of his results, the way he had been taught in Vienna. What the work showed is that there is something in a plant that determines the properties of its overall form. We might as well call that something by its modern name, gene. The genes come in pairs, so that (in one of the examples studied by Mendel) there is a gene S, which results in smooth seeds, and a gene R, that results in rough seeds, but any individual plant will carry one of the possible combinations SS, RR or SR. Only one of the genes in a pair, though, is expressed in the individual plant (in the 'phenotype', as it is known). If the plant carries RR or SS, it has no choice but to use the appropriate gene and produce rough or smooth seeds. But if it carries the combination RS, you might expect half the plants to have rough seeds and half to have smooth seeds. This is *not* the case. The R is ignored and only the S is expressed in the phenotype. In this case, S is said to be dominant and R is recessive. Mendel worked all this out from the statistics, which in this case start from the observation that when RR plants (that is, plants from a line that always produces rough seeds) are crossed with SS plants (from a line which always produces smooth seeds), 75 per cent of the offspring have smooth seeds and only 25 per cent rough. The reason, of course, is that there are two ways to make RS offspring (RS and SR), which are equivalent. So in the next generation the individuals are evenly distributed among four genotypes, RR, RS, SR and SS, of which only RR will have rough seeds. This is just the simplest example (and only looking at the first generation, whereas Mendel actually carried the statistics forward to the 'grandchildren' and beyond) of the kind of analysis Mendel (and later de Vries, Bateson, Correns, von Seysenegg and many other people) used in their studies. Mendel had shown conclusively that inheritance works not by blending characteristics from the two parents, but by taking individual characteristics from each of them. By the early 1900s, it was clear (from the work of people such as William Sutton at Columbia University) that the genes are carried on the chromosomes, and that chromosomes come in pairs, one inherited from each parent. In the kind of cell division that makes sex cells, these pairs are separated, but only (we now know) after chunks of material have been cut out of the paired chromosomes and swapped between them, making new combinations of genes to pass on to the next generation.

Mendel's discoveries were presented to a largely uncomprehending Natural Science Society in Brünn (few biologists had any understanding of statistics in those days) in 1865, when he was 42 years old. The papers were sent out to other biologists, with whom Mendel corresponded, but their importance was not appreciated at the time. Perhaps Mendel might have promoted his work more vigorously and ensured that it received more attention; but in 1868, Cyrill Franz Napp died, and Gregor Johann Mendel was elected Abbot in his place. His new duties gave him little time for science, and his experimental plant breeding programme was essentially abandoned by his forty-sixth year, although he lived until 6 January 1884.

The rediscovery of the Mendelian laws of inheritance at the beginning of the twentieth century, combined with the identification of chromosomes, provided the keys to understanding how evolution works at the molecular level. The next big step was taken by the American Thomas Hunt Morgan, who was born in Lexington, Kentucky, on 25 September 1866, and who became professor of zoology at Columbia University in 1904. Morgan came from a prominent family – one of his great-grandfathers was Francis Scott Key, who wrote the US National anthem; his father served for a time as US Consul at Messina, in Sicily; and one uncle had been a Colonel in the

Confederate army. In an echo of the way Robert Millikan was sceptical about Einstein's ideas concerning the photoelectric effect (and another shining example of the scientific method at work), Morgan had doubts about the whole business of Mendelian inheritance, which rested on the idea of hypothetical 'factors' being passed from one generation to the next. The possibility that these factors might be carried by chromosomes existed, but Morgan was not convinced, and began the series of experiments that would lead to him receiving the Nobel prize (in 1933) in the expectation of proving that the simple laws discovered by Mendel were at best a special case that applied only to a few simple properties of particular plants, and did not have general application to the living world.

The study of chromosomes

The organism Morgan chose to work with was the tiny fruit fly *Drosophila*. The name means 'lover of the dew', but it is actually the fermenting yeast, not dew, that attracts them to rotting fruit. In spite of the obvious difficulties of working with insects rather than plants, *Drosophila* have one great advantage for students of heredity. Whereas Mendel had to wait for a year to inspect the next generation of peas at each stage in his breeding programme, the little flies (each only an eighth of an inch long) produce a new generation every two weeks, with each female laying hundreds of eggs at a time. It was pure luck, though, that it turned out that *Drosophila* have only four pairs of chromosomes, which made Morgan's investigation of how characteristics are passed from one generation to the next much easier than it might have been.[1]

One pair of those chromosomes has a particular significance, as in all sexually reproducing species. Although the individual chromosomes in most pairs are similar in appearance to one another, in the pair that determines sex there is a distinct difference in the shapes of these chromosomes, and from these shapes they are known as X and Y. You might think that there are three possible combinations that may occur in a particular individual – XX, XY and YY. But in females the cells always carry the XX pair, while in males the combination is

1. People have 23 pairs of chromosomes; but there is no simple relationship between the complexity of the phenotype and number of chromosomes, and some ferns have more than 300 chromosome pairs in every cell.

XY.[1] So a new individual must inherit an X chromosome from its mother, and could inherit either X or Y from its father; if it inherits an X from its father it will be female; if it inherits a Y it will be male. The point of all this is that Morgan found a variety of flies among his *Drosophila* that had white eyes instead of the usual red eyes. A careful breeding programme and statistical analysis of the results showed that the gene (a term which Morgan soon took up and promoted) affecting the colour of the eyes of the insect must be carried on the X chromosome, and that it was recessive. In males, if the variant gene (the different varieties of a particular gene are known as alleles) was present on the single X chromosome they had white eyes. But in a female, the relevant allele had to be present on both X chromosomes for the white-eye characteristic to show up in the phenotype.

This first result encouraged Morgan to continue his work in the second decade of the twentieth century, in collaboration with a team of research students. Their work established that chromosomes carry a collection of genes like beads strung out along a wire, and that during the process that makes sperm or egg cells, paired chromosomes are cut apart and rejoined to make new combinations of alleles. Genes that are far apart in the chromosome are more likely to be separated when this process of crossing over and recombination occurs, while genes that are close together on the chromosome only rarely get separated; this (and a lot of painstaking work) provided the basis for mapping the order of genes along the chromosome. Although a great deal more work of this kind remained to be done, using the improving technology of the later twentieth century, the time when the entire package of Mendelian heredity and genetics finally came of age can be conveniently dated to 1915, when Morgan and his colleagues A. H. Sturtevant, C. B. Bridges and H. J. Muller published their classic book, *The Mechanism of Mendelian Heredity*. Morgan himself went on to write *The Theory of the Gene* (1926), moved to Caltech in 1928, received that Nobel prize in 1933 and died at Corona del Mar in California on 4 December 1945.

Evolution by natural selection only works if there is a variety of

1. The pattern is reversed in a few species, and there are other oddities, but these are not important here.

individuals to select from. So the understanding developed by Morgan and his colleagues, of how the constant reshuffling of the genetic possibilities provided by the process of reproduction encourages diversity, also explains why it is so easy for sexually reproducing species to adapt to changing environmental conditions. Asexual species do evolve, but only much more slowly. In human beings, for example, there are about 30,000 genes that determine the phenotype. Just over 93 per cent of these genes are homozygous, which means that they are the same on each chromosome of the relevant pair, in all human beings. Just under 7 per cent are heterozygous, which means that there is a chance that there are different alleles for that particular gene on the paired chromosomes of an individual person chosen at random. These different alleles have arisen by the process of mutation, of which more later, and sit in the gene pool having little effect unless they confer some advantage on the phenotype (mutations that cause a disadvantage soon disappear; that is what natural selection is all about). With some 2000 pairs of genes which come in at least two varieties (some have more than two alleles), that means that there are 2 to the power of 2000 ways (2^{2000} ways) in which two individual people could be different from one another. This is such a spectacularly large number that even astronomical numbers (like those we will encounter in the next chapter) pale by comparison, and it means not only that no two people on Earth are genetically identical (except for twins who share the same genotype because they come from the same fertilized egg), but that no two people who have ever lived have been exactly the same as each other. That is some indication of the variety on which natural selection operates. After 1915, as the nature of chromosomes, sex, recombination and heredity became increasingly clear, the big question was what went on at a deeper level, within the nucleus and within the chromosomes themselves. The way to answer that question would involve the latest developments in quantum physics and chemistry as scientists probed the secrets of life at the molecular level; but the first steps on the path to the double helix of DNA had been taken in a distinctly old-fashioned way almost half a century before.

Nucleic acid

The person who took those first steps was a Swiss biochemist, Friedrich Miescher (1844–1895). His father (also called Friedrich) was professor of anatomy and physiology in Basle from 1837 to 1844, before moving on to Bern, and young Friedrich's maternal uncle, Wilhelm His (1831–1904), held the same chair from 1857 to 1872. His was a particularly strong influence on his nephew, only 13 years his junior, who studied medicine at Basle before going to the University of Tübingen, where he studied organic chemistry under Felix Hoppe-Seyler (1825–1895) from 1868 to 1869, then he spent a spell in Leipzig before returning to Basle. When His moved the other way in 1872, leaving Basle for Leipzig, his chair was divided into two, one for anatomy and one for physiology; young Miescher got the physiology chair, clearly partly as a result of literal nepotism. He stayed in the post until he died, of tuberculosis, on 16 August 1895, just three days after his fifty-first birthday.

Miescher went to work in Tübingen because he was interested in the structure of the cell (an interest encouraged by his uncle, and very much in the mainstream of biological research at the time); Hoppe-Seyler had not only established the first laboratory devoted to what is now called biochemistry, but was a former assistant of Rudolf Virchow, with a keen interest in how cells worked – remember that Virchow had laid down the doctrine that living cells are created only by other living cells scarcely ten years before Miescher went to Tübingen. After discussing the possibilities for his own first research project with Hoppe-Seyler, Miescher settled on an investigation of the human white blood cells known as leucocytes. These had the great advantage, from a practical if not an aesthetic point of view, of being available in large quantities from the pus-soaked bandages provided by a nearby surgical clinic. Proteins were already known to be the most important structural substances in the body, and the expectation was that the investigation carried out by Miescher would identify the proteins that were involved in the chemistry of the cell, and which were, therefore, the key to life. Overcoming the difficulties of washing the intact cells free from the bandages without damaging them, and then subjecting them to chemical analysis, Miescher soon found that the watery cytoplasm that fills the volume of the cell outside the nucleus is indeed rich in proteins; but further studies showed that there was something else present in the cell as well. After removing all the outer material and collecting large numbers of undamaged nuclei free of cytoplasm (something nobody else had achieved before), Miescher was able to analyse the composition of the nucleus and found that it was significantly different from that of protein. This substance, which he called 'nuclein', contains a lot of carbon, hydrogen, oxygen and nitrogen, like other organic molecules; but he also found that it contained a significant amount of phosphorus, unlike any protein. By the summer of 1869 Miescher had confirmed that the new substance came from the nuclei of cells and had identified it not only in leucocytes from pus but in cells of yeast, kidney, red blood cells and other tissues.

News of Miescher's discovery did not create the sensation you might expect – indeed, it was a long time before anybody outside Hoppe-Seyler's lab learned of it. In the autumn of 1869 Miescher moved on to Leipzig, where he wrote up his discoveries and sent them back to Tübingen to be published in a journal that Hoppe-Seyler edited. Hoppe-Seyler found it hard to believe the results and stalled for time while two of his students carried out experiments to confirm the discovery. Then, in July 1870 the Franco-Prussian war broke out and the general turmoil of the war delayed publication of the journal. The paper eventually appeared in print in the spring of 1871, alongside the work confirming Miescher's results, and with an accompanying note from Hoppe-Seyler explaining that publication had been delayed due to unforeseen circumstances. Miescher continued his studies of nuclein after he became a professor at Basle, concentrating on the analysis of sperm cells from salmon. The sperm cell is almost all nucleus, with only a trace of cytoplasm, since its sole purpose is to fuse with the nucleus of a more richly endowed egg cell and contribute hereditary material for the next generation. Salmon produce enormous quantities of sperm, growing thin on their journey to the spawning grounds as body tissue is converted into this reproductive material. Indeed, Miescher pointed out that structural proteins from the body must be broken down and converted into sperm in this way, itself an important realization that different parts of the body can be deconstructed and rebuilt in another form. In the course of this work, he found that nuclein was a large molecule which included several acidic groups; the term 'nucleic acid' was introduced to refer to the molecules in 1889, by Richard Altmann, one of Miescher's students. But Miescher died without ever knowing the full importance of what he had discovered.

Like virtually all of his biochemical colleagues, Miescher failed to appreciate that nuclein could be the carrier of hereditary information. They were too close to the molecules to see the overall picture of the cell at work, and thought of these seemingly relatively simple molecules as some kind of structural material, perhaps a scaffolding for more complicated protein structures. But cell biologists, armed with the new staining techniques that revealed chromosomes, could actually see how genetic material was shared out when cells divided, and were much quicker to realize the importance of nuclein. In 1885, Oskar Hertwig wrote that 'nuclein is the substance responsible not only for fertilization but also for the transmission of hereditary characteristics',[1] while in a book published in 1896[2] the American biologist Edmund Wilson (1856–1939) wrote more fulsomely:

> Chromatin is to be regarded as the physical basis of inheritance. Now chromatin is known to be closely similar to, if not identical with, a substance known as nuclein ... a tolerably definite chemical compound of nucleic acid (a complex organic acid rich in phosphorus) and albumin. And thus we reach the remarkable conclusion that inheritance may, perhaps, be effected by the physical transmission of a particular compound from parent to offspring.

But it was to be a tortuous road before Wilson's 'remarkable conclusion' was confirmed.

Working towards DNA and RNA

Progress down that road depended on identifying the structure of nuclein, and the basic building blocks of the relevant molecules (though not, as yet, the details of how the building blocks were joined together) were all identified within a few years of Miescher's death – some even before he died. The building block which gives its name to DNA is ribose, a sugar whose central structure consists of four carbon atoms linked with an oxygen atom in a pentagonal ring, with other atoms (notably hydrogen–oxygen pairs, OH) attached at the corners. These attachments can be replaced by other

1. *Jenaische Zeitschrift für Medizin und Naturwizzenschaft*, volume 18, p. 276; translation from Lagerkvist.

2. *The Cell in Development and Inheritance*. Wilson was the professor of zoology at Columbia University, head of the department where Morgan would carry out his fruit fly experiments,

molecules, linking the ribose units to them. The second building block, which attaches in just this way, is a molecular group containing phosphorus, and is known as a phosphate group – we now know that these phosphate groups act as links between ribose pentagons in an alternating chain. The third and final building block comes in five varieties, called 'bases', known as guanine, adenine, cytosine, thymine and uracil, and usually referred to simply by their initials, as G, A, C, T and U. One base, it was later discovered, is attached to each of the sugar rings in the chain, sticking out at the side. The ribose pentagon gives the overall molecule its name, ribonucleic acid, or RNA; an almost identical type of molecule (only identified in the late 1920s) in which the sugar units each have one less oxygen atom (H where the ribose has OH) is called deoxyribonucleic acid (DNA). The one other difference between RNA and DNA is that although each of them contains only four of the bases, RNA contains G, A, C and U, while DNA contains G, A, C and T. It was this discovery that reinforced the idea that nuclein was nothing more than a structural molecule and held back the development of a proper understanding of its role in heredity.

The tetranucleotide hypothesis

The person who was most responsible for this misunderstanding was the Russian-born American Phoebus Levene (1869–1940), who was a founder member of the Rockefeller Institute in New York in 1905, and stayed there for the rest of his career. He played a leading part in identifying the way the building blocks of RNA are linked together, and was actually the person who eventually identified DNA itself, in 1929; but he made an understandable mistake which, thanks to his prestige and influence as a leading biochemist, had an unfortunately wide influence. When he was born (in the same year that Miescher discovered nuclein), in the little town of Sagor, Levene was given the Jewish name Fishel, which was changed to the Russian Feodor when his family moved to St Petersburg when he was two years old. When the family emigrated to the United States in 1891 to escape the latest anti-Jewish pogroms, he changed this to Phoebus in the mistaken belief that that was the English equivalent; by the time he found out that he should have chosen Theodore, there didn't seem much point in changing it again. Levene's understandable mistake resulted from analysis of relatively large amounts of nucleic acid. When this was broken down into its component building blocks for analysis, it turned out to contain almost equal amounts of G, A, C and U (the yeast cells used in this work yielded RNA). This led him to conclude that the nucleic acid was a simple structure made up of four repeating units, linked together in the way we have already described; it even seemed possible that a single molecule of RNA contained just one of each of the four bases. This package of ideas became known as the tetranucleotide hypothesis – but instead of being treated as a hypothesis and tested properly, it was accorded the status of dogma and accepted more or less without question by far too many of Levene's contemporaries and immediate successors. Since proteins were known to be very complicated molecules made up from a large variety of amino acids linked together in different ways, this reinforced the idea that all the important information in the cell was contained in the structure of proteins, and that the nucleic acids simply provided a simple supporting structure that held the proteins in place. There is, after all, very little information in a 'message' that contains only one word, GACU, repeated endlessly. Even by the end of the 1920s, though, evidence was beginning to emerge that would lead to an understanding that there is more to nucleic acid than scaffolding. The first hint emerged in 1928, a year before Levene finally identified DNA itself.

The clue came from the work of Fred Griffith (1881–1941), a British microbiologist working as a medical officer for the Ministry of Health, in London. He was investigating the bacteria that cause pneumonia, and had no intention of seeking out any deep truths about heredity. But just as fruit flies breed faster than pea plants and therefore can, under the right circumstances, show how inheritance works more quickly, so microorganisms such as bacteria reproduce more swiftly than fruit flies, going through many generations in a matter of hours, and can show in a matter of weeks the kind of changes that would only be revealed by many years of work with *Drosophila*. Griffith had discovered that there are two kinds of pneumococci bacteria, one which was virulent and caused a disease that was often fatal, the other producing little or no ill effects. In experiments with mice aimed at finding information which might help the treatment of pneumonia in people, Griffith found that the dangerous form of the pneumococci could be killed by heat, and that these dead bacteria could be injected into mice with no ill effects. But when the dead bacteria were mixed with bacteria from the non-lethal variety of pneumococci, the mixture was almost as virulent to the mice as the pure strain of live virulent pneumococci. Griffith himself did not discover how this had happened, and he died before the true importance of his work became clear (he was killed in an air raid during the Blitz), but this discovery triggered a change of direction by the American microbiologist Oswald Avery (1877–1955), who had been working on pneumonia full time at the Rockefeller Institute in New York since 1913.

During the 1930s, and on into the 1940s, Avery and his team investigated the way in which one form of pneumococci could be transformed into another form in a series of long, cautious and careful experiments. They first repeated Griffith's experiments, then found that simply growing a colony of non-lethal pneumococci in a standard glass dish (a Petri dish) which also contained dead, heat-treated cells from the virulent strain was sufficient to transform the growing colony into the virulent form. Something was passing from the dead cells into the living pneumococci, being incorporated into their genetic structure and transforming them. But what? The next step was to break cells apart by alternately freezing and heating them, then use a centrifuge to separate out the solid and liquid debris that resulted. It turned out that the transforming agent, whatever it was, was in the liquid fraction, not the insoluble solids, narrowing down the focus of the search. All of this work kept various people in Avery's lab busy until the mid-1930s. It was at this point that Avery, who had previously overseen the work in his lab but not been directly involved in these experiments, decided on an all-out attack to identify the transforming agent, which he carried out with the aid of two younger researchers, the Canadian born Colin MacLeod (1909–1972) and, from 1940, Maclyn McCarty (1911–), from South Bend, Indiana.

Partly because of Avery's insistence on painstaking attention to detail, partly because of the disruption caused by the Second World War and partly because what they found was so surprising that it seemed hard to believe,[1] it took until 1944 for Avery, MacLeod and

1. After all, it flew in the face of the tetranucleotide hypothesis, and Levene was a towering, influential figure at the Rockefeller until his death in 1940.

McCarty to produce their definitive paper identifying the chemical substance responsible for the transformation that had first been observed by Griffith in 1928. They proved that the transforming substance was DNA – not, as had been widely assumed, a protein. But even in that 1944 paper, they did not go so far as to identify DNA with the genetic material, although Avery, now 67 (a remarkable age for someone involved in such a fundamental piece of scientific research) did speculate along those lines to his brother Roy.[1]

The Chargaff rules

The implications, though, were clear for those who had eyes to see, and in another passing on of the torch, the publication of the Avery, MacLeod and McCarty paper in 1944 stimulated the next key step, which was taken by Erwin Chargaff (1905–). Chargaff was born in Vienna, where he gained his PhD in 1928, the year of Griffith's discovery, spent two years at Yale, then returned to Europe, where he worked in Berlin and Paris, before settling permanently in America in 1935 and spending the rest of his career at Columbia University. Accepting the evidence that DNA could convey genetic information, Chargaff realized that DNA molecules must come in a great variety of types, with a more complicated internal structure than had previously been appreciated. Using the new techniques of paper chromatography (familiar in its simplest incarnation from schoolday experiments where inks are spread out into their component colours as they travel at different speeds through blotting paper) and ultraviolet spectroscopy, Chargaff and his colleagues were able to show that although the composition of DNA is the same within each species they studied, it is different in detail from one species to the next (although still, of course, DNA). He suggested that there must be as many different kinds of DNA as there are species. But as well as this variety on the large scale, he also found that there is a degree of uniformity underlying this complexity of DNA molecules. The four different bases found in molecules of DNA come in two varieties. Guanine and adenine are each members of a chemical family known as purines, while cytosine and thymine are both pyrimidines. What became known as the Chargaff rules were published by him in 1950. They said that, first, the total amount of purine present in a sample

1. See Judson.

of DNA (G + A) is always equal to the total amount of pyrimidine present (C + T); second, the amount of A is the same as the amount of T, while the amount of G is the same as the amount of C. These rules are a key to understanding the famous double-helix structure of DNA. But in order to appreciate just how this structure is held together, we need to take stock of the developments in chemistry that followed the quantum revolution.

The chemistry of life

Starting out with the work of Niels Bohr, and culminating in the 1920s, quantum physics was able to explain the patterns found in the periodic table of the elements and give insight into why some atoms like to link up with other atoms to make molecules, while some do not. The details of the models depend on calculations of the way energy is distributed among the electrons in an atom, which is always in such a way as to minimize the overall energy of the atom, unless the atom has been excited by some energetic influence from outside. We do not need to go into the details here, but can jump straight to the conclusions, which were clear even on Bohr's model of the atom, although they became more securely founded with the developments of the 1920s. The most important difference is that where Bohr originally thought of electrons as like tiny, hard particles, the full quantum theory sees them as spread-out entities, so that even a single electron can surround an atomic nucleus, like a wave.

The quantum properties of electrons only allow certain numbers of electrons to occupy each energy level in an atom, and although it is not strictly accurate, you can think of these as corresponding to different orbits around the nucleus. These energy states are sometimes known to chemists as 'shells', and although several electrons may occupy a single shell, you should envisage each individual electron as being spread out over the entire volume of the shell. It turns out that full shells, in the sense that they have the maximum number of electrons allowed, are energetically favoured over partly full shells. Whatever element we are dealing with, the lowest energy state for the individual atoms (the shell 'nearest the nucleus') has room for just two electrons in it. The next shell has room for eight electrons, and so has the third shell, although we then run into complications which are beyond the scope of the present book. A hydrogen atom has a single proton in its nucleus and, therefore, a single electron in its only occupied shell.

Energetically, this is not so desirable a state as having two electrons in the shell, and hydrogen can achieve at least a kind of halfway house to this desirable state by linking up with other atoms in such a way that it gets at least a share of a second electron. In, for example, molecules of hydrogen (H_2) one electron from each atom contributes to a pair shared by, and surrounding, both nuclei, giving an illusion of a full shell. But helium, with two electrons in its only occupied shell, is in a very favourable energetic state, atomic nirvana, and does not react with anything.

Moving up the ladder of complexity, lithium, the next element, has three protons in its nucleus (plus, usually, four neutrons) and therefore three electrons in its cloud. Two of these slot into the first shell, leaving one to occupy the next shell on its own. The most obvious feature of an atom to another atom, determining its chemical properties, is the outermost occupied shell – in this case, the single electron in the outermost occupied shell – which is why lithium, eager to give away a share in this lone electron in a way we describe shortly, is highly reactive and has similar chemical properties to hydrogen. The number of protons in a nucleus is the atomic number of that particular element. Adding protons to the nucleus and electrons to the second shell (and ignoring the neutrons, which play essentially no part in chemistry at this level) takes us up to neon, which has ten protons and ten electrons in all, two in the innermost shell and eight in the second shell. Like helium, neon is an inert gas – and by now you can see where the repeating pattern of chemical properties for elements eight units apart in the periodic table comes from. Just one more example will suffice. Adding yet another proton and electron takes us up from neon to sodium, which has two closed inner shells and a single electron out on its own; and sodium, with atomic number 11, has similar chemical properties to lithium, with atomic number 3.

Covalent bond model and carbon chemistry

The idea of bonds forming between atoms as they shared pairs of electrons to complete effectively closed shells was developed, initially on a qualitative basis, by the American Gilbert Lewis (1875–1946), in 1916. It is known as the covalent bond model, and is particularly important in describing the carbon chemistry that lies at the heart of life, as the simplest example shows. Carbon has six protons in its nucleus (and six neutrons, as it happens), plus six electrons in its cloud. Two of these electrons, as usual, sit in the innermost shell, leaving four to occupy the second shell – exactly half the number required to make a full shell. Each of these four electrons can pair up with the electron offered by an atom of hydrogen, so that a molecule of methane (CH_4) is formed in which the carbon atom in the middle has the illusion of a full shell of eight electrons and each of the four hydrogen atoms on the outside has the illusion of a full shell of two electrons. If there were five electrons in the outer shell, the central atom would only need to make three bonds to complete its set; if it only had three electrons, it could only make three bonds, however much it might 'want' to make five. Four bonds is the maximum any atom can make,[1] and bonds are stronger for shells closer to the central nucleus, which is why carbon is the compound maker *par excellence*. Replace one or more of the hydrogen atoms with something more exotic – including, perhaps, other carbon atoms or phosphate groups – and you begin to see why carbon chemistry has so much potential to produce a wide variety of complex molecules.

The ionic bond

There is, though, another way in which atoms can form bonds, which brings us back to lithium and sodium. They can both form bonds in this way, but we shall use sodium as an example, because this kind of bond is found in a very common everyday substance – common salt, NaCl. The bond is known as an ionic bond, and the idea was developed by several people as the nineteenth century turned into the twentieth century, although most of the credit for the basis of the idea probably belongs to the Swede Svante Arrhenius (1859–1927), who received the Nobel prize for his work on ions in solution in 1903. Sodium, as we have seen, has two full inner shells and a single electron out on its own. If it could get rid of that lone electron, it would be left with an arrangement of electrons similar to that of neon (not quite identical to neon, because the extra proton in the sodium nucleus means that it holds on to the electrons a tiny bit more tightly), which is favoured energetically. Chlorine, on the other hand, has no fewer than 17 electrons in its cloud (and, of course, 17 protons in its nucleus), arranged in two full shells and a third shell of

1. Under normal circumstances; there are always exceptions, but this is not the place to discuss them.

seven electrons, with that one 'hole' where another electron could fit. If a sodium atom gives up an electron completely to a chlorine atom, both of them achieve nirvana, but at the cost of being left with an overall electric charge – positive for the sodium, negative for the chlorine. The resulting ions of sodium and chlorine are held together by electric forces in a crystalline array, which is rather like a single huge molecule – molecules of NaCl do not exist as independent units in the way molecules of H_2 or CH_4 do.

In quantum physics, though, things are seldom as clear cut and straightforward as we would like them to be, and chemical bonds are best thought of as a mixture of these two processes at work, with some more covalent but with a mixture of ionic, some more ionic with a mixture of covalent and some more or less 50:50 (even in molecules of hydrogen, you can envisage one hydrogen atom giving up its electron completely to the other). But all of these images are no more (and no less) than crutches for our imagination. What matters is that the energies involved can be calculated, with great accuracy. Indeed, within a year of Schrödinger publishing his quantum mechanical wave equation, and just a year before Griffith's key work with pneumococci, in 1927 two German physicists, Walter Heitler (1904–1981) and Fritz London (1900–1954), had used this mathematical approach to calculate the change in overall energy when two hydrogen atoms, each with its own single electron, combine to form one molecule of hydrogen with a pair of shared electrons. The change in energy that they calculated very closely matched the amount of energy which chemists already knew, from experiment, was required to break the bonding between the atoms in a hydrogen molecule. Later calculations, made as the quantum theory was improved, gave even better agreement with experiment. The calculations showed that there was no arbitrariness in the arrangement of electrons in atoms and atoms in molecules, but that the arrangements which are most stable in the atoms and molecules are always the arrangements with the least energy. This was crucially important in making chemistry a quantitative science right down at the molecular level; but the success of this approach was also one of the first, and most powerful, pieces of evidence that quantum physics applies in general, and in a very precise way, to the atomic world, not just to isolated special cases like the diffraction of electrons by crystals.

The person who put all of the pieces together and made chemistry a branch of physics was the American Linus Pauling (1901–1994). He was another of those scientists who was the right man, in the right place, at the right time. He gained his first degree, in chemical engineering, from Oregon State Agricultural College (a forerunner of Oregon State University) in 1922, then studied for a PhD in physical chemistry at Caltech; he was awarded this degree in 1925, the year that Louis de Broglie's ideas about electron waves began to gain attention. For the next two years, exactly during the time that quantum mechanics was being established, Pauling visited Europe on a Guggenheim Fellowship. He worked for a few months in Munich, then in Copenhagen at the Institute headed by Niels Bohr, spent some time with Erwin Schrödinger in Zürich and visited William Bragg's laboratory in London.

Bragg, and in particular his son Lawrence, are also key figures in the story of the discovery of the structure of DNA. The elder Bragg, William Henry, lived from 1862 to 1942, and is always known as William Bragg. He graduated from Cambridge University in 1884, and after a year working with J. J. Thomson moved to the University of Adelaide, in Australia, where his son William Lawrence (always known as Lawrence Bragg) was born. He worked on alpha rays and X-rays, and after returning to England in 1909, working at Leeds University until 1915, then moving to University College London, he developed the first X-ray spectrometer, to measure the wavelength of X-rays. In 1923 he was appointed Director of the Royal Institution, reviving it as a centre of research and establishing the laboratory which Pauling visited a few years later. It was William Bragg who first had the dream of using X-ray diffraction to determine the structure of complex organic molecules, although the technology available to him in the 1920s was not yet up to this task.

Lawrence Bragg (1890–1971) studied mathematics at the University of Adelaide (graduating in 1908) then moved to Cambridge, where he initially continued with mathematics but switched to physics in 1910 at his father's suggestion, graduating in 1912. So Lawrence was just starting as a research student in Cambridge and William was a professor in Leeds when news came from Germany in 1912 that Max von Laue (1879–1960), working at the University of Munich, had observed the diffraction of X-rays by crystals.[1] This is exactly equivalent to the way light is diffracted in the double-slit experiment, but because the wavelengths of X-rays are much shorter than those of light, the spacing between the 'slits' has to be much smaller; it turns out that the spacing between layers of atoms in a crystal is just right to do the job. This work established that X-rays are indeed a form of electromagnetic wave, like light but with shorter wavelengths; the importance of the breakthrough can be gauged by the fact that von Laue received the Nobel prize for the work just two years later, in 1914.

Bragg's law

Chemistry as a branch of Physics

Von Laue's team had found what were certainly complicated diffraction patterns, but were not immediately able to work out details of how the patterns related to the structure of the crystals the X-rays were diffracting from. The Braggs discussed the new discoveries with one another, and each worked on different aspects of the problem. It was Lawrence Bragg who worked out the rules which made it possible to predict exactly where the bright spots in a diffraction pattern would be produced when a beam of X-rays with a particular wavelength struck a crystal lattice with a particular spacing between atoms at a particular angle. Almost as soon as X-ray diffraction had been discovered, it was established that it could be used to probe the structure of crystals, once the wavelengths involved had been measured (which is where William Bragg's spectrometer, built in 1913, would come in). The relationship Lawrence came up with soon became known as Bragg's law, and it made it possible to work in either direction – by measuring the spacing of the bright spots in the pattern you could determine the wavelength of the X-rays if you knew the spacing of atoms in the crystal, and once you knew the wavelength of the X-rays you could use the same technique to measure the spacing between atoms in a crystal, although interpreting the data soon became horribly complicated for complex organic structures. It was this work which showed that in substances

1. To be precise, von Laue devised the experiment, which was actually carried out by Walther Friedrich and Paul Knipping, at the Institute of Theoretical Physics in Munich; this echoes the way Ernest Rutherford devised the experiment carried out by Hans Geiger and Ernest Marsden which revealed the existence of the atomic nucleus.

such as sodium chloride there are no individual molecules (NaCl), but an array of sodium ions and chlorine ions arranged in a geometric pattern. The two Braggs worked together, and published together, over the next couple of years, producing the book *X Rays and Crystal Structure* in 1915 – just twenty years after the discovery of X-rays. The previous year, Lawrence had become a Fellow of Trinity College, but his academic career was interrupted by war service as a technical adviser to the British Army in France; it was while he was there that he learned, in 1915, that he and his father had received the Nobel prize for their work. Lawrence was the youngest person (at 25) to receive the award, and the Braggs are the only father-and-son team to have shared the award for their joint work. In 1919, Lawrence Bragg became professor of physics at Manchester University, and in 1938 he succeeded Rutherford as head of the Cavendish Lab, where he will soon come back into the story of the double helix. When he left Cambridge in 1954, he too became Director of the RI, until he retired in 1966.

Linus Pauling

Pauling had learned about X-ray crystallography as a student, largely from the book written by William and Lawrence Bragg, and had carried out his own first determination of a crystal structure using the technique in 1922 (the crystal was molybdenite). When he returned to the United States, taking up a post at Caltech in 1927 and becoming a full professor there in 1931, he had all the up-to-date ideas about X-ray crystallography at his fingertips, and soon developed a set of rules for interpreting the X-ray diffraction patterns from more complicated crystals. Lawrence Bragg developed essentially the same set of rules at the same time, but Pauling published first, rather to Bragg's chagrin, and to this day the expressions are known as Pauling's rules. This established a rivalry between Pauling and Bragg that was to last into the 1950s, and would play a part in the discovery of the structure of DNA.

At this time, however, Pauling's main interest was in the structure of the chemical bond, which he explained in quantum mechanical terms over the next seven years or so. As early as 1931, following another visit to Europe imbibing the new ideas in quantum physics, he produced a great paper, 'The Nature of the Chemical Bond', which was published in the *Journal of the American Chemical Society*; this laid all the groundwork. It was followed by six more papers elaborating on the theme over the next two years, then by a book rounding everything up. 'By 1935,' Pauling later commented, 'I felt that I had an essentially complete understanding of the nature of the chemical bond.'[1] The obvious thing to do was to move on to using this understanding to elucidate the structure of complex organic molecules, such as proteins (remember that DNA was still not regarded as a very complex molecule in the mid-1930s). These structures yielded to a two-pronged investigation – chemistry and an understanding of the chemical bond told people like Pauling how the subunits of the big molecules were allowed to fit together (in the case of proteins, the subunits are amino acids), while X-ray crystallography told them about the overall shapes of the molecules. Only certain arrangements of the subunits were allowed by chemistry, and only certain arrangements of the subunits could produce the observed diffraction patterns. Combining both pieces of information with model building (sometimes as simple as pieces of paper cut into the shapes of the molecular subunits and pushed around like pieces of a jigsaw puzzle, sometimes more complicated modelling in three dimensions) eliminated many impossible alternatives and eventually, after a lot of hard work, began to reveal the structures of the molecules important to life. An enormous amount of work by researchers such as Pauling himself, Desmond Bernal (1901–1971), Dorothy Hodgkin (1910–1994), William Astbury (1889–1961), John Kendrew (1917–1977), Max Perutz (1914–2002) and Lawrence Bragg enabled biochemists to determine, over the next four decades, the structure of many biomolecules, including haemoglobin, insulin and the muscle protein myoglobin. The importance of this work, both in terms of scientific knowledge and in terms of the implications for improved human healthcare, scarcely needs pointing out; but, like medicine itself, the full story is not one we can go into here. The thread we want to pick up, leading on to the determination of the structure of DNA, is the investigation of the structure of certain proteins by Pauling and by his British rivals; but

1. See Judson. This was no idle boast, but a simple statement of fact; Pauling duly received the Nobel prize for this work in 1954. In 1962, he received the Nobel peace prize, for his work in the campaign for nuclear disarmament.

before we do so there is one more piece of quantum chemistry that needs to be mentioned.

The nature of the hydrogen bond

The existence of so-called 'hydrogen bonds' highlights the importance of quantum physics to chemistry – in particular, the chemistry of life – and brings home the way in which the quantum world differs from our everyday world. Chemists already knew that under certain circumstances it is possible to form links between molecules that involve a hydrogen atom as a kind of bridge. Pauling wrote about this hydrogen bond, which is weaker than a normal covalent or ionic bond, as early as 1928, and returned to the theme in the 1930s, first in the context of ice (where hydrogen bonds form bridges between water molecules) and then, with his colleague Alfred Mirsky, applied the idea to proteins. The explanation of hydrogen bonding requires you to think of the single electron associated with the proton in a hydrogen atom as smeared out in a cloud of electric charge, not as a tiny billiard ball. When the hydrogen atom is involved in forming a conventional bond with an atom such as oxygen, which strongly attracts its electron, the cloud of charge is pulled towards the other atom, leaving only a thin covering of negative charge on the other side of the hydrogen atom. Unlike all other chemically reactive atoms (helium is not chemically reactive), hydrogen has no other electrons in inner shells to help conceal the positive charge on its proton, so some of the positive charge is 'visible' to any other nearby atoms or molecules. This will attract any nearby atom which has a preponderance of negative charge – such as an oxygen atom in a water molecule, which has gained extra negative charge from its two hydrogen atoms. In water molecules, the positive charge on each of the two hydrogen atoms can link in this way with the electron cloud on another water molecule (one for each hydrogen atom), which is what gives ice a very open crystalline structure, with such a low density that it floats on water. The value of Pauling's work on ice is that, once again, he put numbers into all of this, calculating the energies involved[1] and showing that they matched the values revealed in experiments. In his hands, the idea of the hydrogen bond became precise, quantitative science, not a vague,

1. In this case, it was actually the entropy that Pauling was investigating, but the principle is the same.

qualitative idea. In proteins, as Pauling and Mirsky demonstrated in the mid-1930s, when long-chain protein molecules fold up into compact shapes (not unlike the way the toy known as Rubik's snake folds up into compact shapes), they are held in those shapes by hydrogen bonds which operate between different parts of the same protein chain. This was a key insight, since the shape of a protein molecule is vital to its activity in the machinery of the cell. And all this thanks to a phenomenon, the hydrogen bond, which simply cannot be explained properly except in terms of quantum physics. It is no coincidence that our understanding of the molecular basis of life came after our understanding of the rules of quantum mechanics, and once again we see science progressing by evolution, not revolution.

Studies of fibrous proteins

The first great triumph to result from the combination of a theoretical understanding of how the subunits of proteins can fit together and the X-ray diffraction patterns produced by the whole molecules (actually by many whole molecules side by side in a sample), came with the determination of the basic structure of a whole family of proteins, the fibrous kind found in hair, wool and fingernails, at the beginning of the 1950s. What was to be a long road to that triumph began when William Astbury was working in William Bragg's group of crystallographers at the Royal Institution in London in the 1920s. It was here that Astbury started his work on biological macromolecules with X-ray diffraction studies of some of these fibres, providing the first X-ray diffraction pictures of fibrous protein and continuing this line of research after he moved to the University of Leeds in 1928. In the 1930s, he came up with a model for the structure of these proteins that was actually incorrect, but it was Astbury who showed that globular protein molecules (such as haemoglobin and myoglobin) are made up of long-chain proteins (polypeptide chains) that are folded up to make balls.

The alpha-helix structure

Pauling came into the story in the late 1930s, and later recalled how he 'spent the summer of 1937 in an effort to find a way of coiling a polypeptide chain in three dimensions, comparable with the X-ray data reported by Astbury'.[1] But it would take much longer than a single summer to solve the problem. Fibrous

1. See Judson.

proteins looked more promising, but with the Second World War intervening, it was in the late 1940s that both Pauling and his colleagues at Caltech (notably Robert Corey) and Lawrence Bragg (by now head of the Cavendish) and his team in Cambridge closed in on the solution. Bragg's group published first, in 1950 – but it soon turned out that their model was flawed, even though it contained a great deal of the truth. Pauling's team came up with the correct solution in 1951, identifying the basic structure of fibrous protein as being made up of long polypeptide chains wound round one another in a helical fashion, like the strands of string that are wound together to make a rope, with hydrogen bonds playing an important part in holding the coils in shape. This was a spectacular triumph in itself, but the world of biochemistry was almost overwhelmed when the Caltech team published seven separate papers in the May 1951 issue of the *Proceedings of the National Academy of Sciences*, laying out in detail the chemical structure of hair, feathers, muscles, silk, horn and other proteins, as well as the alpha-helix structure, as it became known, of the fibres themselves. The fact that the structure was helical certainly set other people thinking about helices as possible structures for other biological macromolecules, but what was equally important was the overwhelming success of the whole approach used by Pauling, combining the X-ray data, model building and a theoretical understanding of quantum chemistry. As Pauling has emphasized, the alpha-helix structure was determined 'not by direct deduction from experimental observations on proteins but rather by theoretical considerations based on the study of simpler substances'.[1] The example inspired the work of the two people who would very soon determine the structure of DNA itself, snatching the prize from under the noses not just of the Caltech team, but another group working on the problem in London.

It was obvious that Pauling would now turn his attention to DNA, which, as we have seen, had been identified as the genetic material by the 1940s.[2] It is also easy to imagine how Lawrence Bragg, now twice

1. See *Chemistry*.

2. Any lingering doubts were removed at around this time by a brilliant experiment in which the Americans Alfred Hershey and Martha Chase, working at the Cold Spring Harbor Laboratory on Long Island, proved that the genetic material of viruses is composed of DNA.

pipped at the post by Pauling, might have longed for an opportunity for the structure of DNA to be determined at his own laboratory in Cambridge. In fact, this should not have been possible, not for scientific reasons but because of the way the limited funding available for scientific research in Britain, where the economy was still slowly recovering from the effects of the war, restricted the freedom of researchers. There were only two groups capable of tackling the problem of the structure of DNA, one under Max Perutz at the Cavendish, the other under John Randall (1905–1984) at King's College in London, both funded by the same organization, the Medical Research Council (MRC); and there was every reason to avoid a duplication of effort which might result in a waste of limited resources. The result was an understanding (nothing formal, but a well-understood gentlemen's agreement) that the King's team had first crack at DNA. The snag, for anyone who cared about such things, was that the team at King's, headed by Maurice Wilkins (1916–), did not seem to be in any great hurry to complete the work, and was also handicapped by the way in which Rosalind Franklin (1920–1958), a young researcher who produced superb X-ray diffraction photographs from DNA and should have been Wilkins's partner, was largely frozen out by him in a personality clash which seems to have been at least partly based on prejudice against her as a woman.

Francis Crick and James Watson: the model of the DNA double helix

It was the disarray among the team ('team' in name only) at King's that opened a window of opportunity for a brash young American, James Watson (1928–), who turned up in Cambridge in 1951 on a post-doctoral scholarship, fired with a determination to work out the structure of DNA and neither knowing nor caring anything about English gentlemen's agreements. Watson was given space in the same room as a rather older English PhD student, Francis Crick (1916–), who turned out to have a complementary background and approach to Watson, and was soon recruited to the cause. Crick had started out as a physicist, and carried out war work on mines for the Admiralty. But, like many physicists of his generation, he became disillusioned with physics as a result of seeing its application to the war. He was also, again like many of his contemporaries, influenced by a little book called *What is Life?*, written by Erwin Schrödinger and published in 1944, in which the great physicist had looked at the problem of what is now called the genetic code from a physicist's point of view. Although at the time he wrote the book Schrödinger did not know that chromosomes were made of DNA, he spelled out in general terms that 'the most essential part of a living cell – the chromosome fibre – may suitably be called *an aperiodic crystal*', drawing a distinction between an ordinary crystal such as one of common salt, with its endless repetition of a simple basic pattern, and the structure you might see in 'say, a Raphael tapestry, which shows no dull repetition but an elaborate, coherent, meaningful design', even though it is made up of a few colours arranged in different ways. Another way of looking at the storing of information is in terms of the letters of the alphabet, which spell out information in words, or a code such as the Morse code, with its dots and dashes arranged in patterns to represent letters of the alphabet. Among several examples of the way in which information could be stored and passed on in such an aperiodic crystal, Schrödinger noted that in a code similar to the Morse code but with three symbols, not just dot and dash, used in groups of ten, 'you could form 88,572 different "letters"'. It was against this background that the physicist Crick joined the MRC unit at the Cavendish as a research student in 1949, at the late age of 33. For his thesis, he was working on X-ray studies of polypeptides and proteins (and duly received his degree in 1953); but he will always be remembered for the unofficial work he carried out on the side, when he should have been concentrating on his PhD, at the instigation of Watson.

This work was entirely unofficial – indeed, Crick was twice told by Bragg to leave DNA to the King's team, and twice ignored him, only gaining any kind of formal approval from the Cavendish professor in the later stages of the investigation when it seemed as if Pauling was about to crack the puzzle. Although the theoretical insight and practical modelling were important, everything depended on the X-ray diffraction photographs, and the first such images of DNA had been obtained by Astbury only in 1938. These were not improved upon (again, in no small measure because of the hiatus caused by the war) until the 1950s, when Wilkins's group (in particular, Rosalind Franklin, assisted by a research student, Raymond Gosling) took up the subject; indeed, Pauling's work on the structure of DNA was handicapped by only having Astbury's old data to work with. Using data Watson had gleaned from a talk given by Franklin at King's, and which he had not properly understood, the Cavendish pair soon came up with a model for DNA, involving the strands twining around one another with the nucleotide bases (A, C, G and T) sticking out from the sides, which was proudly presented to Wilkins, Franklin and two of their colleagues from London, who were specially invited to Cambridge for the presentation. The model was so embarrassingly bad, and the comments it elicited so acerbic, that even the ebullient Watson retreated into his shell for a time, while Crick went back to his proteins. But in the summer of 1952, in a conversation with the mathematician John Griffith (a nephew of Frederick Griffith, and himself very interested in, and knowledgeable about, biochemistry), Crick tossed out the idea that the nucleotide bases in the DNA molecule might fit together somehow, to hold the molecules together. Mildly interested, Griffith worked out from the shapes of the molecules that adenine and thymine could fit together, linking up through a pair of hydrogen bonds, while guanine and cytosine could also fit together, linking up through a set of three hydrogen bonds, but that the four bases could not pair up in any other way. Crick did not immediately appreciate the importance of this pairing, nor the relevance of the hydrogen bonds, and as a newcomer to biochemistry he was unaware of Chargaff's rules. In a rare piece of serendipity, however, in July 1952 Chargaff himself visited the Cavendish, where he was introduced to Crick and, learning of his interest in DNA, mentioned the way in which samples of DNA always contain equal amounts of A and G, and equal amounts of C and T. This, combined with Griffith's work, clearly suggested that the structure of DNA must involve pairs of long-chain molecules, linked together by AG and CT bridges. It even turns out that the length of a CT bridge formed in this way is the same as the length of an AG bridge formed in this way, so there would be an even spacing between the two molecular chains. But for months the Cavendish team tossed the idea around among themselves, without doing any serious work on it. They were only galvanized into another frantic burst of model building (Watson did most of the model building, Crick provided most of the bright ideas) right at the end of 1952. In December, Peter Pauling, a graduate student at the Cavendish and son of Linus Pauling, received a letter from his father saying that he had worked out the structure of DNA. The news spread gloom in the Watson–Crick camp, but there were no details of the model in the letter. In January 1953, though, Peter Pauling received an advance copy of his father's paper, which he showed to Watson and Crick. The basic structure was a triple helix, with three strands of DNA chains wound round one another. But to their amazement, Crick and Watson (by now a little wiser in the ways of X-ray diffraction patterns) realized that Pauling had made a blunder, and that his model could not possibly match the data being obtained by Franklin.

A few days later, Watson took the copy of Pauling's paper to London to show it to Wilkins, who responded by showing Watson a print of one of Franklin's best photographs, in a serious breach of etiquette, without her knowledge. It was this picture, which could only be interpreted in terms of a helical structure, plus the Chargaff rules and the relationships worked out by John Griffith, that enabled Crick and Watson to produce their famous model of the double helix, with the entwined molecules held together by hydrogen bonds linking the nucleotide bases in the middle, by the end of the first week of March 1953. As it happens, Pauling was not in the race at the time, since he had not yet realized that his triple-helix model was wrong – indeed, he never really thought of there being a race, since he never knew how close his rivals in England were to the goal. But Franklin, at King's, was thinking along very similar lines to Crick and Watson (without the physical model building) and was almost ready to publish her own version of the double helix when the news came from Cambridge. She had actually prepared the first draft of a paper for *Nature* the day before. The burst of activity triggered by Pauling's premature paper had resulted in Crick and Watson snatching the prize from under the nose not of Pauling, but of Franklin. The immediate result was that three papers appeared alongside one another in the issue of *Nature* dated 25 April 1953. The first, from Crick and Watson, gave details of their model, and stressed its relationship to the Chargaff rules, downplaying the X-ray evidence; the second, from Wilkins and his colleagues A. R. Stokes and H. R. Wilson, presented X-ray data which suggested in general terms a helical structure for the DNA molecule; the third, from Franklin and Gosling, gave the compelling X-ray data

39. *Watson, Crick and their model of a molecule of DNA, 1951.*

indicating the kind of double-helix structure for DNA proposed by Crick and Watson, and was (although nobody else knew it at the time) essentially the paper Franklin had been working on when the news came from Cambridge. What nobody also knew at the time, or could have guessed from the presentation of the three papers, is that rather than being just a confirmation of the work by Crick and Watson, the paper by Franklin and Gosling represented a completely independent discovery of the detailed structure of DNA, and that the Crick and Watson discovery was largely based on Franklin's work. It was only much later that it emerged just how the crucial X-ray data got to Cambridge, what a vital role it had played in the model building, and just how badly Franklin had been treated both by her colleague at King's and by Watson and Crick. Franklin herself, happy to be leaving King's in 1953 for the more congenial environment at London's Birkbeck College, never felt hard done by – but then she never knew the whole truth, since she died in 1958, from cancer, at the age of 38. Crick, Watson and Wilkins shared the Nobel prize for Physiology or Medicine just four years later, in 1962.

The genetic code

There are two key features of the double-helix structure of DNA which are important for life, reproduction and evolution. The first is that any combination of bases – any message written in the letters A, C, G and T – can be spelled out along the length of a single strand of DNA. During the 1950s and into the early 1960s, the efforts of many researchers, including Crick (Watson never did anything else to compare with his work with Crick on the double helix) and a team at the Pasteur Institute in Paris, showed that the genetic code is actually written in triplets, with sets of three bases, such as CTA or GGC, representing each of the twenty or so individual amino acids used in the proteins that build and run the body. When proteins are being manufactured by the cell, the relevant part of the DNA helix containing the appropriate gene uncoils, and a string of three-letter 'codons' is copied into a strand of RNA (which raises interesting questions about whether RNA or DNA was the first molecule of life); this 'messenger RNA', whose only essential difference to DNA is that it has uracil everywhere DNA has thymine, is then used as a template to assemble a string of amino acids corresponding to the codons, which are linked together to make the required protein. It keeps doing this until no more of that particular protein is required. The DNA has long since coiled up again, and after enough protein has been manufactured the RNA is disassembled and its components reused. Just how the cell 'knows' when and where to do all this remains to be explained, but the principles of the process were clear by the mid-1960s.

The other important feature of the DNA double helix is that the two strands are, in terms of their bases, mirror images of one another, with every A on either strand opposite T on the other, and every C opposite a G. So if the two strands are unwound, and a new partner is built for each of them from the chemical units available in the cell (as happens prior to cell division[1]), in the two new double helices, one of which from each pair goes into each daughter cell, there will be the same genetic message, with the letters of the code in the same order, and with A opposite T and C opposite G. Although the details of the mechanism are subtle and, again, not yet fully understood, it is immediately obvious that this also provides a mechanism for evolution. During all the copying of DNA that goes on when cells divide, there must occasionally be mistakes. Bits of DNA get copied twice, or bits get left out, or one base (one 'letter' in the genetic code) gets accidentally replaced by another. None of this matters much in the kind of cell division that produces growth, since all that happens is that a bit of DNA in a single cell (probably not even a bit of DNA that that particular cell uses) has been changed. But when reproductive cells are produced by the special process of division that halves the amount of DNA in the daughter cells, not only is there more scope for mistakes to occur (thanks to the extra processes involved in crossing over and recombination), but if the resulting sex cell successfully fuses with a partner and develops into a new individual, all of the DNA, including the mistakes, gets a chance to be expressed. Most of the resulting changes will be harmful, making the new individual less efficient, or at best neutral; but those rare cases when a DNA copying error produces a gene, or gene package, that makes its owner better fitted to its environment are all that Darwinian evolution needs for natural selection to operate.

The genetic age of humankind

From the perspective of our theme of how science has altered humankind's perception of our own place in nature, this is as far as we need to take the story of DNA. A great deal of work has been carried out since the 1960s in determining the composition of genes at the level of DNA codons, and a great deal more has yet to be carried out before we will understand the processes by which some genes control the activity of other genes, and in particular the way genes are 'switched on' as required during the complicated process of

1. The strands do not unwind entirely before copying begins. Instead, as the double helix starts to untwist, new partners begin to build up for each strand, twining around them as the process continues, so that by the time the untwisting of the original helix has finished, the two daughter helices are essentially complete.

development of an adult from a single fertilized egg cell. But to see where we fit in to the tapestry of life, and to see just how accurate Charles Darwin's assessment of man's place in nature was, we can step back from these details and look at the broader picture. From the 1960s onward, as biochemists investigated the genetic material of human beings and other species in more and more detail, it gradually became clear just how closely related we are to the African apes, who Darwin himself regarded as our closest living relatives. By the late 1990s, it had been established that human beings share 98.4 per cent of their genetic material with the chimpanzee and the gorilla, making us, in popular terminology, only 'one per cent human'. From various lines of attack, comparing the genetic material of more or less closely related living species with fossil evidence of when those species split from a common stock, this amount of genetic difference can be used as a kind of molecular clock, and tells us that the human, chimp and gorilla lines split from a common stock just four million years ago.

The fact that such a small genetic difference can produce creatures as different as ourselves and chimps already suggested that the important differences must lie in those control genes that regulate the behaviour of other genes, and this interpretation of the evidence has been supported by the evidence from the human genome project, which completed its mapping of all the DNA in every chromosome of the human genome in 2001. The resulting map, as it is sometimes called, simply lists all the genes in terms of strings of codons, A, T, C and G; it is not yet known what most of the genes actually do in the body. But the immediate key feature of the map is that it shows that human beings have only about 30,000 genes, a much smaller number than anyone had predicted, although the 30,000 genes are capable of making at least 250,000 proteins. This is only twice as many genes as the fruit fly, and just 4000 more than a garden weed called thale cress, so it is clear that the number of genes alone does not determine the nature of the body they build. Human beings do not have many more genes than other species, so the number of genes on its own cannot explain the ways in which we are different from other species. Again, the implication is that a few key genes are different in us, compared with our closest relatives, and that these are affecting the way the other genes operate.

Humankind is nothing special

Underpinning all this, though, is the bedrock fact that none of these comparisons would be possible if all the species being investigated did not use the same genetic code. At the level of DNA and the mechanisms by which the cell operates, involving messenger RNA and the manufacture of proteins, as well as in reproduction itself, there is absolutely no difference between human beings and other forms of life on Earth. All creatures share the same genetic code, and we have all evolved in the same way from primordial forms (perhaps a single primordial form) of life on Earth. There is nothing special about the processes that have produced human beings, compared with the processes that have produced chimpanzees, sea urchins, cabbages or the humble wood louse. And our removal from centre stage is just as profound when we look at the place of the Earth itself in the Universe at large.

15 Outer Space

Measuring the distances of stars – Stellar parallax determinations – Spectroscopy and the stuff of stars – The Hertzsprung–Russell diagram – The colour–magnitude relationship and the distances to stars – The Cepheid distance scale – Cepheid stars and the distances to other galaxies – General theory of relativity outlined – The expanding Universe – The steady state model of the Universe – The nature of the Big Bang – Predicting background radiation – Measuring background radiation – Modern measurements: the COBE satellite – How the stars shine: the nuclear fusion process – The concept of 'resonances' – CHON and humankind's place in the Universe – Into the unknown

Measuring the distances of stars

Our understanding of the Universe at large rests upon two foundations – being able to measure the distances to the stars, and being able to measure the compositions of the stars. As we have seen, the first real understanding of the distances to the stars emerged in the eighteenth century, when Edmond Halley realized that some of the 'fixed' stars had moved since the time they were observed by his predecessors in Ancient Greece. By this time, astronomers had begun to make accurate measurements of distances across the Solar System, using the same process of triangulation that is the basis of surveying. To measure the distance to an object without actually going there, you need to be able to see the object from both ends of a baseline of known length. From the angles made by the lines of sight to the object from each end of the baseline, you can then work out the distance from the geometry of triangles. This technique had already been used to measure the distance to the Moon, our nearest neighbour in space, just 384,400 km away; but for more distant objects you need longer baselines to be able to make accurate measurements. In 1671 the French astronomer Jean Richer (1630–1696) travelled to Cayenne, in French Guiana, where he made observations of the position of Mars against the background of 'fixed' stars at the same time that his colleague in Paris, the Italian-born Giovanni Cassini (1625–1712), made similar observations. This made it possible to work out the distance to Mars, and, by combining this with Kepler's laws of planetary motion, to calculate the distance from the Earth (or any other planet in the Solar System) to the Sun. The figure Cassini came up with for the Sun–Earth distance, 140 million km, was only 7 per cent less than the accepted modern value (149.6 million km), and gave the first accurate indication of the scale of the Solar System. Similar studies of Venus during the transits of 1761 and 1769 (predicted by Halley) led to an improved estimate of the Sun–Earth distance (known as the Astronomical Unit, or AU) of 153 million km, close enough to the modern value for us to leave the later improvements in the measurements as fine tuning, and accept that by the end of the eighteenth century astronomers had a very good idea of the scale of the Solar System.

Stellar parallax determinations

What was highly worrying about this at the time was that it implied almost unimaginable distances to the stars. During any interval of six months, the Earth moves from one side of the Sun to the other, at opposite ends of a baseline 300 million km (or 2 AU) long. Yet the positions of the stars on the night sky do not change when viewed from either end of this enormous baseline. You would expect the nearer stars to seem to move against the background of more distant stars, in the same way that if you hold a finger out at arm's length and close each of your eyes in turn, the position of the finger seems to move against the background of more distant objects (an example of the effect known as parallax). It is easy to calculate how much a star ought to be seen to move when viewed from different places in the Earth's orbit. Astronomers define one parallax second of arc, or parsec, as the distance to a star which would show a displacement of one second of arc on the sky from opposite ends of a baseline 1 AU long.[1] So a star 1 parsec away would show a displacement of 2 seconds of arc from opposite ends of the 300 million km baseline represented by the diameter of the Earth's orbit. From simple geometry, such a star would be 3.26 light years away, 206,265 times as far away from us as the Sun. And yet, no star is close enough to us to show this

1. To give you a feel for these angular sizes, the full Moon covers 31 minutes of arc, just over half a degree; so one second of arc is roughly one sixtieth of one thirtieth, or 1/1800, of the apparent width of the Moon on the sky.

much parallax displacement on the sky as the Earth moves around the Sun.

There were already hints that stars must be at the kinds of distances this simple calculation implied. Christiaan Huygens, for example, tried to estimate the distance to Sirius, the brightest star in the night sky, by comparing its brightness to that of the Sun. To do this, he let sunlight into a darkened room through a pinhole in a screen, adjusting the size of the hole until the pinprick of light looked about the same as the brightness of Sirius – not easy, since obviously he had to look at the Sun in daytime and Sirius at night. Nevertheless, by showing how small a fraction of the Sun's light corresponded to the observed brightness of Sirius, and knowing that the brightness of an object is inversely proportional to the square of its distance, he argued that if Sirius were actually as bright as the Sun it must be 27,664 times further away. The Scot James Gregory (1638–1675) improved on this technique by comparing the brightness of Sirius with the brightness of the planets, which could be seen in the sky at the same time. The calculation was a little more complicated, since it involved working out how the sunlight became attenuated on its way out to the planets, estimating how much of the light was reflected and calculating how the reflected light was attenuated on its way to Earth. But in 1668, Gregory provided an estimate of the distance to Sirius equivalent to 83,190 AU. Isaac Newton updated this calculation, using improved estimates of the distances to the planets, and came up with a distance to Sirius of one million AU, published in his *System of the World* in 1728, the year after he died. The actual distance to Sirius is 550,000 AU, or 2.67 parsecs; but the apparent accuracy of Newton's estimate owes as much to luck as judgement, with several of the inevitable errors resulting from the imperfect data available to him cancelling each other out.

Measuring distances to stars using the triangulation, or parallax, technique required the positions of the stars on the sky (which really means their positions relative to each other) to be measured to very high accuracy. Flamsteed's catalogue, a tremendous achievement in its day, gave positions to an accuracy of only 10 seconds of arc (merely 1/180th of the diameter of the full Moon on the sky). The first distances to stars were only measured in the 1830s, because it was only then that, with improving technology, the measurements became accurate enough to measure the tiny parallax shifts involved – but once the technology was good enough, several astronomers immediately began to make the measurements. The pioneers chose for study stars which they had some reason to think must be relatively close to us – either because they are very bright or because they are seen to move across the sky as the years go by (they have large 'proper motions'), or both. The first person to announce a stellar parallax determination and the associated distance to a star was the German Friedrich Wilhelm Bessel (1784–1846), in 1838. He chose 61 Cygni, a star with a large proper motion, and found its parallax to be 0.3136 seconds of arc, implying a distance of 10.3 light years (modern measurements give a distance of 11.2 light years, or 3.4 parsecs). In fact, the first person to *measure* a stellar parallax was the Scot Thomas Henderson (1798–1874), working in South Africa in 1832; he studied Alpha Centauri, the third brightest star in the night sky, and came up with a parallax of 1 second of arc (later reduced to 0.76 arc seconds, implying a distance of 1.3 parsecs (4.3 light years)). Henderson's results, though, were not published until he returned to England in 1839. Alpha Centauri (which is now known to be a triple system, with three stars in orbit around one another) is the closest star to the Sun, with the largest measured parallax. A year after Henderson's announcement, the German-born astronomer Friedrich von Struve (1793–1864), working at the Pulkova Observatory near St Petersburg, measured the parallax of Vega (also known as Alpha Lyrae); his figure was a little too high, but modern measurements give a parallax of 0.2613 seconds of arc and a distance of 8.3 parsecs (27 light years). The important thing to take away from these measurements is that they are all for stars which are our near neighbours on the cosmic scale. The *nearest* star to the Sun is 7000 times further away than Pluto, generally regarded as the most distant planet in the Solar System. And once you know the true distance to a star, you can work out its true brightness (called the absolute magnitude) by reversing the technique which Huygens, Gregory and Newton applied to Sirius. In this way, we now know that Sirius itself, 2.67 parsecs away from us, is actually much brighter than the Sun, something that Newton and his contemporaries had no way of telling. Even these breakthroughs at the end of the 1830s, however, did no more than indicate the vast scale of the Universe. It wasn't until the end of the nineteenth century that it became possible to measure parallaxes more easily, using photographic plates to record the positions of the stars. Before then, the positions had to be measured by eye, using the crosswires of a telescope, in real time; hardly surprisingly, the rate of new measurements was roughly one a year from 1840 to the end of the century, so that by 1900 only 60 parallaxes were known. By 1950, the distances to some 10,000 stars had been determined (not all of them by parallax[1]), and towards the end of the twentieth century the satellite Hipparcos measured the parallaxes of nearly 120,000 stars, to an accuracy of 0.002 arc seconds.

Spectroscopy and the stuff of stars

In many ways, modern astronomy – astrophysics – only began at the beginning of the twentieth century, precisely because of the application of photographic techniques to preserve images of the stars. As well as giving distances to enough stars for statistical studies of stars to be meaningful, photography also provided a way of recording and preserving images of the spectra of stars, and it was, of course, spectroscopy (developed, as we have seen, only in the 1860s) that enabled astronomers to obtain information about the composition of the stars. One other vital piece of information was needed – the masses of stars. This was supplied by studies of binary systems, in which two stars orbit around one another. For a few nearby binaries, the separation between the stars can be measured in angular terms, and this can be converted into linear distances if the actual distance to the star system is known (as it is for Alpha Centauri). The invaluable Doppler effect[2] in the spectrum of light seen from the stars in the binary system tells astronomers how fast the stars are moving around one another, and together with Kepler's laws (which apply equally as well to stars orbiting one another as they do to planets

1. For example, distances to groups of stars that move together through space in a cluster can be approximately determined geometrically, by measuring the way in which the proper motions of the stars seem to converge at a point on the sky, just as parallel railway lines seem to converge at a point in the distance. There are other statistical techniques that helped indicate the distances to stars, but the details need not concern us here.

2. This squashes light waves from objects moving towards us, shifting features in the spectrum towards the blue end of the spectrum, and stretches waves of light from objects moving away from us, producing a redshift, with the size of the shift in either case indicating the relative speed of the object.

orbiting around stars), this is enough to enable astronomers to work out the masses of the stars. Once again, by the early 1900s there were just enough observations of this kind for the statistics to be meaningful. So it is no surprise that at just that time two astronomers working on opposite sides of the Atlantic Ocean independently put all of the pieces of the puzzle together and came up with the single most important insight into the nature of stars, a diagram which relates the colours of the stars to their brightnesses. It doesn't sound all that impressive, but it is as important to astrophysics as the periodic table of the elements is to chemistry. But, as I hope we have made clear, like most developments in science it was not really a revolutionary development but an evolutionary progression from what had gone before, built on the foundations of improved technology.

The Hertzsprung–Russell diagram

The Dane Ejnar Hertzsprung was born in Frederiksberg on 8 October 1873. He trained as a chemical engineer, graduating from Copenhagen Polytechnic in 1898, and later studied photochemistry, but from 1902 onwards he worked privately (that is, in an unpaid capacity) at the observatory of the University of Copenhagen, learning how to be an observational astronomer and applying his photographic skills to astronomical observations. It was during this time that he discovered the relationship between the brightness of a star and its colour, but he published these results (in 1905 and 1907) in a photographic journal, where they lay unnoticed by professional astronomers around the world. Even so, Hertzsprung's local reputation grew to the point where, in 1909, he was offered a post at the Göttingen Observatory by Karl Schwarzschild (1873–1916), with whom he had been corresponding. When Schwarzschild moved to the Potsdam Observatory later that year, Hertzsprung went with him, staying there until 1919 when he moved to The Netherlands, becoming first a professor at Leiden University and then, in 1935, Director of the Leiden Observatory. Although he officially retired in 1944, Hertzsprung carried on astronomical research back home in Denmark well into his eighties, and died on 21 October 1967, just after his ninety-fourth birthday. He made many contributions to observational astronomy, including studies of proper motions and work on the cosmic distance scale, but nothing to rank with the discovery he made while still technically an amateur.

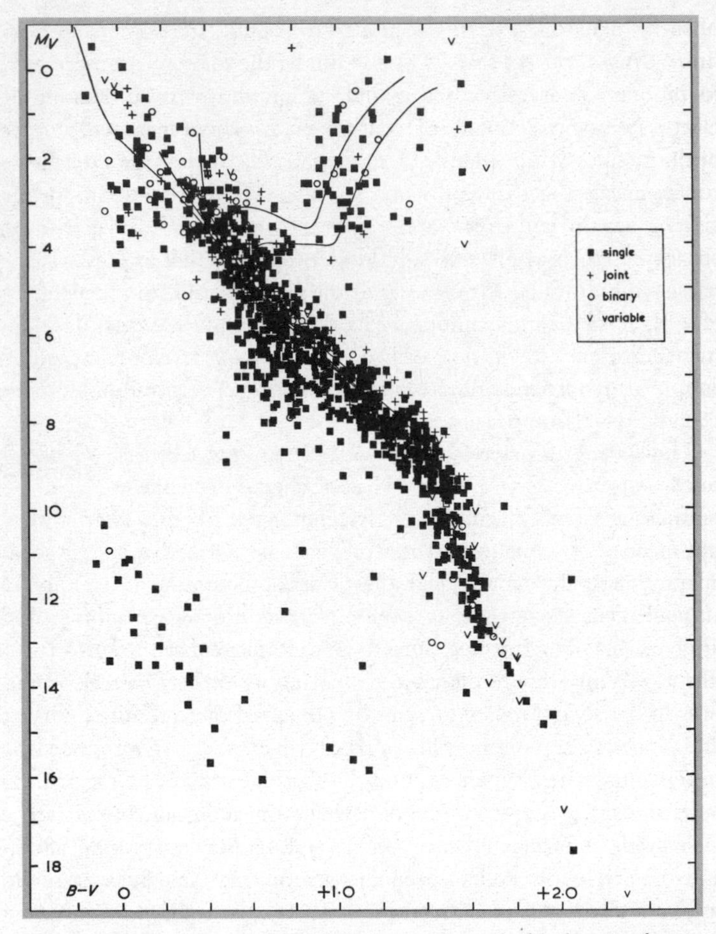

40. *The Hertzprung–Russell diagram relates the brightness of a star (vertical) to its colour (horizontal).*

Henry Norris Russell was born in Oyster Bay, New York, on 25 October 1877. He had a more conventional start to his academic career than Hertzsprung, studying at Princeton and visiting the University of Cambridge before taking up a post as professor of astronomy at Princeton in 1911. It was there that he made essentially the same discovery as Hertzsprung about the relationship between the colours of stars and their brightnesses, but he had the good sense to publish (in 1913) in a journal read by astronomers, and the flash of inspiration to plot the relationship on a kind of graph, now known as the Hertzsprung–Russell (or just HR) diagram, which made the importance of the discovery immediately apparent to his readers.[1] Hertzsprung's contribution to the discovery was quickly recognized, hence the name given to the diagram. Russell was based at Princeton throughout the rest of his working life, although he also made good use of the new telescopes built in California over the next few years. Apart from the HR diagram, he made important contributions to the study of binary stars, and also investigated the composition of the atmosphere of the Sun, using spectroscopy. He retired in 1947 and died in Princeton on 18 February 1957.

The colour–magnitude relationship and the distances to stars

The point about the HR diagram (sometimes called a colour–magnitude diagram, since in astronomy magnitude is another term for brightness) is that the temperature of a star is closely related to its colour. We are not talking just in a qualitative way about the colours of the rainbow here, although it is true that blue and white stars are always intrinsically bright, while some orange and red stars are bright and some are faint[2] (the key observation made by Hertzsprung in the first decade of the twentieth century). Astronomers can do better than this, and put the measurement of colour on a quantitative footing. They define the colour of a star very precisely in terms of the amount of energy it is radiating at different wavelengths, which tells you the temperature of the surface emitting the light. Using the known properties of black-body radiation, the surface temperature of a star can be determined from measurements at just three wavelengths (at a pinch, and slightly less accurately, from just two). But the intrinsic brightness of a star (its absolute magnitude) tells you how much energy the star is radiating overall, regardless of its temperature. It is possible for some red stars to be both cool and bright because they are very big, so that even

1. By then, Hertzsprung had also published his results in a graphical form, in 1911, but yet again in a journal that was rather obscure (to astronomers).

2. *Intrinsically* bright or faint, we emphasize. This is not about the apparent brightness of a star on the sky, but how bright it really is, close up, which we know from its distance.

though each square metre of the surface only glows red, there are an awful lot of square metres letting energy out into the Universe. Small stars can only be as bright if they are blue or white hot, with a lot of energy crossing each square metre of the smaller surface; and small orange stars (like the Sun) are intrinsically less bright than hot stars the same size or large stars with the same temperature. And the bonus comes when masses of stars are included. When the temperatures (or colours) and brightnesses (or magnitudes) of stars are plotted on the HR diagram, most stars lie on a band running diagonally across the diagram, with hot, massive stars about the same size (diameter) as the Sun at one end of the band, and cool, dim stars with less mass than the Sun at the other end. The Sun itself is an average star, roughly in the middle of this so-called main sequence. The large, cool but bright stars (red giants) lie above the main sequence, and there are also some dim, small but hot stars (white dwarfs) below the main sequence. But it was the main sequence itself which gave astrophysicists their first insight into the internal workings of the stars, an insight developed initially largely by the British astronomer Arthur Eddington, often regarded as the first astrophysicist, who was the person who discovered the relationship between the mass of a star and its position on the main sequence.

Eddington was born in Kendal, in the Lake District of England, on 28 December 1882. His father died in 1884 and the family (Arthur had one sister) moved to Somerset, where he was brought up as a Quaker. Eddington studied at Owens College in Manchester (the forerunner of the University of Manchester), then from 1902 to 1905 at the University of Cambridge. He worked at the Royal Greenwich Observatory until 1913, before returning to Cambridge as Plumian Professor of Astronomy and Experimental Philosophy (succeeding George Darwin), and in 1914 also became director of the university observatories. He held these posts until he died, in Cambridge, on 22 November 1944. Skilled as an observer, a brilliant theorist, an able administrator and with a gift for communicating important scientific ideas in clear language to a wide audience (he was the first popularizer of Einstein's theories of relativity in English), Eddington made a profound mark on astronomy in the twentieth century, but is best remembered for two key contributions.

The first came about partly because Eddington was a Quaker and a conscientious objector to war. Einstein's general theory of relativity was first presented by him to the Berlin Academy of Sciences in 1915, and published in Germany the following year, when Britain and Germany were, of course, at war. But a copy of Einstein's paper went to Willem de Sitter (1872–1934) in the neutral Netherlands, and de Sitter passed a copy on to Eddington, who among his other activities was Secretary of the Royal Astronomical Society at that time. In that capacity, he broke the news about Einstein's work to the Society; this was the beginning of Eddington's role as the front man for the general theory in the English-speaking world. Among other things, Einstein's theory predicted that light from distant stars should be bent by a certain amount as it passed close by the Sun, shifting the apparent positions of those stars on the sky. This could be observed during an eclipse. As it happened, a suitable eclipse was due in 1919, but it would not be visible from Europe. In 1917, the Royal Astronomical Society began to make contingency plans to send a pair of expeditions to observe and photograph this eclipse from Brazil and from the island of Principe, off the west coast of Africa, if the war ended in time.

At that time, though, there was little obvious reason why the war should end quickly, and the losses at the front had become so heavy that the British government introduced conscription, with all able-bodied men eligible for the draft. Although 34, Eddington was able-bodied, but clearly much more valuable to Britain as a scientist than as a soldier in the trenches (although the argument that scientists deserve special treatment is not one that we would endorse; *everyone* at the front would have been more use to society back at home). The point was put to the Home Office by a group of eminent scientists, and Eddington was advised that he would be exempt from the draft on the grounds of his value to the scientific community. He replied that if he had not been deferred on these grounds he would have claimed exemption on conscientious grounds anyway, which infuriated the bureaucrats in the Home Office. Their first reaction was that if Eddington wanted to be a conscientious objector he could go and join his Quaker friends in agricultural work, which he was quite prepared to do. But some nifty footwork by the Astronomer Royal, Frank Dyson, saved face all round and persuaded the Home Office to defer Eddington's draft on condition that he would lead an expedition for the government to test Einstein's light-bending prediction. He would have been the ideal choice anyway, having first-hand experience of eclipse studies from Brazil during his time with the Royal Greenwich Observatory; but these machinations make an intriguing background to the fact that Eddington was indeed 'the man who proved Einstein was right'. This time, he went to Principe, but a twin expedition was sent to Brazil, with Eddington in overall charge of processing and analysing the results. The importance of these eclipse observations will become clear shortly; but first, Eddington's other key contribution to science.

As the world returned to normal after the First World War, in the early 1920s Eddington gathered all the data he could find on stellar masses and linked this with data from the HR diagram to show that the brighter stars are the most massive. A main sequence star twenty-five times the mass of the Sun, for example, is 4000 times as bright as the Sun. This made sense. A star holds itself up by the pressure it generates in its interior, counteracting the inward pull of gravity. The more massive it is, the more weight there is pressing inwards and the more pressure it has to generate. It can only do this by burning its fuel – whatever that fuel may be – more quickly, thereby generating more heat, which eventually escapes from the surface of the star as more light for us to see. The physics of what goes on is actually rather simple, for the reasons we mentioned before concerning the fate of complicated structures under conditions of high temperature and pressure, so the temperature at the heart of a star can be calculated from observations of its brightness, mass and size (determined from the brightness if the distance is known, but also, once the relationships were discovered, from its position on the HR diagram). When Eddington put the numbers in, he came up with a profound insight – all main sequence stars have roughly the same central temperature, even though they cover a range in masses from tens of times the mass of the Sun down to a tenth the mass of the Sun. It is as if stars have an inbuilt thermostat; as a ball of gas shrinks under its own weight and gets hotter inside as gravitational energy is converted into heat, nothing happens to halt this process until a critical temperature is reached, when the thermostat switches on an almost inexhaustible (by human standards) supply of energy. And by the 1920s, it was fairly obvious (at least to Eddington) where the energy must be coming from.

In the nineteenth century there had been a sometimes fierce debate between the geologists and evolutionists on one side, and the physicists on the other, about the age of the Earth and the Sun. Quite reasonably, physicists such as William Thomson (Lord Kelvin) pointed out that there was no process known to the science of the time that could keep the Sun shining for the long timescales required to explain the evolution of life on Earth. They were right, but even before the end of the nineteenth century, as we have seen, sources of energy new to science were discovered, in the form of radioactive isotopes. In the early years of the twentieth century this led to speculation that a star like the Sun might be kept hot if it contained radium – just 3.6 grams of pure radium in every cubic metre of the Sun's volume would be enough to do the job, an idea which was discussed by, among others, Eddington's predecessor as Plumian Professor, George Darwin. In fact, the half-life of radium, as was soon appreciated, is much too short for this to work, but it was clear that 'subatomic energy' must hold the key to the longevity of the Sun and stars. With the developments in subatomic physics over the first two decades of the twentieth century, and armed with Einstein's special theory of relativity and his equation $E = mc^2$, as early as 1920 Eddington was able to spell out the implications to the audience at the annual meeting of the British Association for the Advancement of Science:

A star is drawing on some vast reservoir of energy by means unknown to us. This reservoir can scarcely be other than the sub-atomic energy which, it is known, exists abundantly in all matter; we sometimes dream that man will one day learn to release it and use it for his service. The store is well-nigh inexhaustible, if only it could be tapped. There is sufficient in the Sun to maintain its output of heat for 15 billion years.

He went on to justify that assertion:

[Francis] Aston has further shown conclusively that the mass of the helium atom is even less than the masses of the four hydrogen atoms which enter into it – and in this, at any rate, the chemists agree with him. There is a loss of mass in the synthesis amounting to 1 part in 120, the atomic weight of hydrogen being 1.008 and that of helium just 4. I will not dwell on his beautiful proof of this, as you will no doubt be able to hear it from himself. Now mass cannot be annihilated, and the deficit can only represent the mass of the electrical energy set free in the transmutation. We can therefore at once calculate the quantity of energy liberated when helium is made out of hydrogen. If 5 percent of a star's mass consists initially of hydrogen atoms, which are gradually being combined to form more complex elements, the total heat liberated will more than suffice for our demands, and we need look no further for the source of a star's energy.

Eddington was on the right track, but it would take decades for the details of how energy is liberated inside stars to be worked out, partly because of a misunderstanding implicit in that reference to '5 percent' of a star being composed of hydrogen, partly because the full calculations would need quantum mechanics, which was not fully developed until the end of the 1920s. We shall come back to this story in due course; but by the 1920s stellar astronomy had come up with another way to measure distances to at least some of the stars, and a new telescope with which to apply the technique; this combination would soon produce another dramatic change in humankind's view of our place in the Universe. The evidence of the main sequence was that the Sun is just an ordinary star, nothing special in the Milky Way. The evidence that would eventually emerge from the new distance indicators was that the Milky Way itself is nothing special in the Universe.

The colour–magnitude relationship portrayed in the HR diagram itself gives you a guide to the distances to stars. If you measure the colour of a star, then you know where it belongs on the main sequence, and that tells you its absolute magnitude. So all you have to do is measure its apparent magnitude to work out how far away it is. At least in principle. Things aren't quite that easy in practice, chiefly because dust in space, along the line of sight to the star, dims its light (causing 'extinction') and makes it look redder – a process known as reddening, but nothing to do with the redshift. This interferes with our observations of both colour and brightness, although the effects can often be compensated for, at least approximately, by looking at different stars in roughly the same direction in space. But the key step in developing a cosmic distance scale came from a quite different kind of investigation, also going on at about the time Hertzsprung and Russell were developing their ideas about the colour–magnitude relation.

The discovery came as a result of an investigation of the stars of the southern skies, carried out under the direction of Edward Pickering (1846–1919), who became director of the Harvard University Observatory in 1876. Pickering was an inveterate cataloguer, and the inspiration for the next generation of American astronomers, but his single most important contribution to astronomy resulted from a survey of the southern skies carried out for him from Peru by his brother William Pickering (1858–1938). The actual job of cataloguing – recording neatly, using pen and ink in large ledgers, the positions and brightnesses of all the individual stars on the photographic plates sent to Harvard – was carried out by teams of women, who, in those less enlightened days, were generally cheaper to hire than men, and often not regarded as having the intellectual capacity for any more creative work. To his credit, Pickering encouraged some of these women who showed an aptitude for astronomy to move into research proper, giving a few of them an entry into the almost exclusively male academic world of the time. One of these women was Henrietta Swan Leavitt (1868–1921), who joined the Harvard team in 1895 (initially as an unpaid volunteer, such was her enthusiasm for astronomy, although she later became head of the department of photographic photometry). Pickering gave her the task of identifying variable stars in the southern skies, which could only be done by comparing photographic plates of the same region obtained at different times to see if any of the stars had changed their appearance.

Such variations can happen for two reasons. First, it may be because the 'star' is actually a binary system and we are seeing partial eclipses as one star moves in front of the other – and studying binaries, as we have seen, is a key to measuring stellar masses. Second, stars may be intrinsically variable, changing their brightness as a result of some change in their internal structure, and this is interesting in its own right. Some such stars, we now know, swell up and shrink back upon themselves, pulsating in a repeating cycle, with their light output changing regularly as they do so. One such class of pulsating star is known as the Cepheids, after the archetypal example of its kind, a star known as Delta Cephei, which was identified as a variable by the English astronomer John Goodricke in 1784, two years before he died at the age of 21. All Cepheids show a characteristic pattern of repeated brightening and dimming, but some have periods as short as a day or so, while others have periods of more than a hundred days.

The Cepheid distance scale

The photographic plates from Peru that Leavitt was studying in Harvard covered two clouds of stars, known as the Large and Small Magellanic Clouds, which are now known to be small satellite galaxies associated with the Milky Way Galaxy in which we live. During the course of her painstaking work, Leavitt noticed that the Cepheids in the Small Magellanic Cloud (SMC) showed an overall pattern of behaviour in which the brighter Cepheids (averaging their brightness over a whole cycle) went through their cycle more slowly. The initial discovery was reported in 1908, and by 1912 Leavitt had enough data to pin down this period-luminosity relationship in a mathematical formula, established from her study of twenty-five Cepheids in the SMC. She realized that the reason why the relationship showed up is because the SMC is so far away that the stars in it are all effectively at the same distance from us, so that the light from each of them is dimmed by the same amount en route to our telescopes. Of course, there are differences in the distances to individual stars in the SMC, and these may amount to dozens or hundreds of light years, in absolute terms; but at the distance of the SMC, these differences are only a small percentage of the distance from Earth, so they only affect the apparent brightnesses of the stars by a small fraction of the overall dimming caused by their distance from us. Leavitt found a clear mathematical relationship between the apparent brightness of a Cepheid in the SMC and its period, so that, for example, a Cepheid with a period of three days is only one-sixth the brightness of a Cepheid with a period of 30 days. This could only mean that the *absolute* magnitudes of Cepheids are related to one another in the same way, since the distance effect is essentially the same for all of the Cepheids in the SMC. All that was needed now was to find the distance to just one or two Cepheids in our neighbourhood so that their absolute magnitudes could be determined, and then the absolute magnitudes of all other Cepheids (and therefore their distances) could be worked out from the period-luminosity law that Leavitt had discovered.

It was, in fact, Hertzsprung who first measured the distances to nearby Cepheids in 1913, providing the calibration that the Cepheid distance scale needed.[1] As is usually the case in astronomy, however, the observations were beset by difficulties, not least the problems of extinction and reddening. Hertzsprung's calibration implied a distance to the SMC of 30,000 light years (roughly 10,000 parsecs); the modern figure, taking account of reddening and extinction effects that he was unaware of, is 170,000 light years (52,000 parsecs). At such a distance, even if two Cepheids are 1000 light years apart this represents only 0.6 per cent of their distance from us, with a correspondingly small effect on their relative dimming due to their distance. Even Hertzsprung's underestimate was the first indication of just how big space really is. The Cepheid distance scale is, of course, no less important in studies of stars within the Milky Way than it is in studies of the Universe at large. Some clusters of stars, grouped together in space, contain dozens or hundreds of stars with different masses, colours and brightnesses, and if there is just one Cepheid in the cluster then the distance to all those stars is known, with all that that implies for understanding the properties of the stars and, for example, removing the effects of reddening and extinction when plotting them on an HR diagram. But it was in probing beyond the Milky Way that Cepheids altered our appreciation of our place in the Universe.

Cepheid stars and the distances to other galaxies

That probing became possible thanks to the development of a new generation of telescopes, largely as a result of the enthusiasm of one man, George Ellery Hale (1868–1938), an astronomer with a gift for persuading benefactors to part with large sums of money and the administrative skills to see the successful application of that money to the construction of new telescopes and observatories, first at the University of Chicago, then at Mount Wilson in California and finally at Mount Palomar, also in California. The key instrument in this particular phase of the exploration of the Universe was the 100-inch diameter reflector known as the Hooker tele-

1. And it was Russell, with his student Harlow Shapley, of whom more shortly, who first improved these distance estimates by allowing for extinction, in 1914.

scope (after the benefactor who paid for it), completed on Mount Wilson in 1918 and still in use today (or rather, tonight). It was the largest telescope on Earth for 30 years, and transformed our understanding of the Universe, largely in the hands of two men, Edwin Hubble (1889–1953) and Milton Humason (1891–1972).

You mustn't believe everything you read about Hubble, who, to put it politely, exaggerated his own early achievements, making up stories about his prowess as an athlete and pretending that he had once been a successful lawyer. But that doesn't detract from the importance of his work in astronomy.

The first person to use Cepheids to produce a map of the Milky Way Galaxy which resembles the modern one was Russell's former student Harlow Shapley (1865–1972), at the end of the second decade of the twentieth century. He used a 60-inch reflector on Mount Wilson, from 1908 to 1918 the biggest telescope in the world, and was one of the first people to use the 100-inch, but moved on to become director of the Harvard College Observatory in 1921, missing the opportunity to take full advantage of the opportunities opened up by the new telescope. Unknown to Shapley, some of the stars he thought were Cepheids were actually members of a different family, now known as RR Lyrae stars. They behave in a similar fashion to Cepheids (making them important distance indicators in their own right), but with a different period–luminosity relationship. Fortunately, some of the errors introduced into Shapley's calculations by this confusion were cancelled out by the fact that he wasn't making sufficient allowance for extinction. It was already clear by then (and had been increasingly clear since the time of Galileo and Thomas Wright) that the band of light on the night sky known as the Milky Way is a flattened, disc-shaped system containing vast numbers of stars, and that the Sun is just one star among this multitude. It had been widely thought that the Sun lay at the centre of the disc of stars that makes up the Milky Way. But there are also concentrations of stars, roughly spherical systems known as globular clusters, which lie above and below the plane of the Milky Way, occupying a spherical volume of space in which the disc of the Milky way is embedded. By mapping the distances to the globular clusters, Shapley found where this sphere is centred and established that the Sun is not at the centre of the Milky Way. By 1920, his measurements indicated that the Milky Way itself was about 100,000 parsecs across and that the centre of the Milky Way lay 10,000 parsecs (more than 30,000 light years) away from us. His numbers were still plagued by the extinction problem and the confusion of RR Lyrae stars with Cepheids – we now know that he got the distance to the centre of the Milky Way about right (the modern figure is 8000–9000 parsecs), but that he got the overall diameter of the Galaxy too big (we now estimate it is 28,000 parsecs across). The disc of the Milky Way itself is only a couple of hundred parsecs thick – very thin, actually, compared with its diameter. But the numbers are less important than the fact that Shapley had made yet another reduction in the status of our home in space, removing the Sun to an ordinary location in the suburbs of the disc of the Milky Way Galaxy, an unimportant member of a system estimated to contain several hundred billion stars.

At the beginning of the 1920s, though, it was still widely thought that the Milky Way itself dominated the Universe. Although there were other fuzzy patches of light on the night sky (like the Magellanic Clouds), these were thought to be either smaller systems which were satellites of the Milky Way (a bit like super globular clusters), or glowing clouds of gas within the Milky Way. Only a few astronomers, among whom the American Heber Curtis (1872–1942) was the most vociferous, argued that many of these 'spiral nebulae' were actually galaxies in their own right, so far away that individual stars in them could not be resolved even with the best telescopes available,[1] that the Milky Way was much smaller than Shapley estimated and that it was just one 'island universe' among many comparable galaxies scattered across the void.

This is where Hubble comes into the story. In the winter of 1923/4, using the 100-inch Hooker telescope, Hubble was able to resolve individual stars in a large spiral nebula, known as M31, in the direction of the constellation Andromeda (it is sometimes referred to as the

1. Just as the stars of the Milky Way cannot be resolved by the unaided human eye, and were only 'discovered' when Galileo turned his telescope upon them.

Andromeda Nebula or Andromeda Galaxy). Even better, to his surprise he was able to identify several Cepheids in the nebula and calculate its distance, which came out as 300,000 parsecs, almost a million light years; with modern calibration of the Cepheid distance scale and better allowance for problems such as extinction, the Andromeda Galaxy is now known to be even further away, at a distance of 700,000 parsecs. Hot on the heels of this discovery, Hubble found Cepheids in several other similar nebulae, establishing that Curtis was essentially correct. As other techniques were developed to measure distances to galaxies, including observations of exploding stars, supernovae, which all have roughly the same absolute maximum brightness, it eventually became clear that just as there are hundreds of billions of stars in the Milky Way Galaxy, so there are hundreds of billions of galaxies in the visible Universe, which extends for billions of light years in all directions. The Solar System is an insignificant speck within an insignificant speck in all this vastness. But the key step in mapping the Universe is still the magnitude–distance relationship for Cepheids, against which the secondary distance indicators (such as supernovae) are calibrated. And as a result, there was one hangover from the early difficulties posed by problems such as extinction which distorted our view of our place in the Universe right into the 1990s.

As the example of M31 shows, on the distance scale used by Hubble everything seemed to be closer than it really is. For a galaxy of a certain size (say, the same absolute size as the Milky Way itself), the closer it is the bigger the patch of sky it covers. What astronomers actually measure is the angular size of a galaxy on the sky, and if they think it is closer than it really is, they will think that it is smaller than it really is. A child's toy aeroplane in front of your face or a 747 airliner coming in to land can each look the same angular size; but your guess as to how much bigger a 747 is than the toy will depend on how far away you think the aircraft is. The underestimate of their distances meant that the sizes of all galaxies beyond the Milky Way were at first underestimated, and it seemed that the Milky Way Galaxy was the biggest such object in the Universe. Repeated refinements of the distance scale gradually changed this perception over the decades, but it was only in the late 1990s, using Cepheid data obtained by the Hubble Space Telescope to provide accurate distances to a significant number of spiral galaxies similar to the Milky Way, that it was finally established that our galaxy is just average in size.[1]

Following on from his work of 1923–4, with the aid of Milton Humason in the late 1920s and early 1930s, Hubble extended his measurements of distances to galaxies out into the Universe as far as possible using the 100-inch. Although direct Cepheid distances could only be measured for a handful of relatively nearby galaxies, with the distances to those galaxies known, he could calibrate other features of galaxies in general, such as supernovae or the brightness of particular features in spirals, and use them as secondary indicators to give distances to more remote galaxies in which Cepheids could not be resolved even with the 100-inch. It was while carrying out this survey that Hubble made the discovery with which his name will always be associated – that there is a relationship between the distance to a galaxy and the redshift in the spectrum of light from it.

The preponderance of redshifts in the light from 'nebulae' had actually been discovered by Vesto Slipher (1875–1969) in the second decade of the twentieth century, working at the Lowell Observatory (in Flagstaff, Arizona) with a 24-inch refracting telescope. His work in obtaining spectra of such faint objects photographically using this telescope was at the cutting edge of technology at the time, and Slipher was convinced that these diffuse nebulae must be composed of many individual stars because of the similarity between their spectra and the spectra of stars in general. But his equipment was not up to the task of resolving individual stars in these nebulae, so he could not take the step that would be taken by Hubble in the 1920s, and could not measure distances to the nebulae he studied. By 1925, Slipher had measured 39 redshifts in nebulae, but found just two blueshifts. Only four redshifts, and no blueshifts, had been measured by other astronomers in systems that Slipher had not studied first, although many of his results had been confirmed by other observers. The natural interpretation of these data was that they were a result of the Doppler effect, with most nebulae moving rapidly away from us, and just

1. I was a member of the team that finally established that the Milky Way is just an ordinary galaxy; my colleagues were Simon Goodwin, now at the University of Cardiff, and Martin Hendry, now at the University of Glasgow.

two moving towards us. Hubble and Humason began by measuring distances to nebulae that had first been observed spectroscopically by Slipher, as well as taking their own spectroscopic data (it was Humason who actually did this) to test their own apparatus and confirm Slipher's results. Then they extended this kind of investigation to other galaxies. Apart from the very few objects already known, no blueshifts were found.[1] They discovered that the distance of a galaxy is proportional to its redshift, a phenomenon reported in 1929 and now known as Hubble's law. To Hubble, the value of the discovery was as a distance indicator – now, he (or Humason) only had to measure the redshift of a galaxy and they could infer the distance to it. But the significance of the discovery went far deeper than that, as a few other astronomers were quick to realize.

General theory of relativity outlined

The explanation for the discovery made by Hubble and Humason came from Einstein's general theory of relativity, which, as we have seen, had been published in 1916. The feature which makes this theory 'general' (as opposed to the restricted nature of the 'special' theory of relativity) is that it deals with accelerations, not just with objects moving in straight lines at constant speed. But Einstein's great insight was to appreciate that there is no distinction between acceleration and gravity. He said that this insight came to him while sitting at his desk in the Patent Office in Bern one day, when he realized that a person falling from a roof would be weightless and would not feel the pull of gravity – the acceleration of their downward motion cancels out the feeling of weight, because the two are *exactly* equal. We've all experienced the equivalence of acceleration and gravity in a lift – as the lift starts to move upward, we are pressed to the floor and feel heavier; when the lift stops, we feel lighter while it is decelerating and, in the case of express lifts, may rise up on our toes as a result. Einstein's genius was to find a set of equations to describe both acceleration and gravity in one package – as well as all of the special theory of relativity and, indeed, all of Newtonian mechanics, as special cases of the general theory. It is by no means true, in spite

1. The couple of blueshifted galaxies, one of which is Andromeda, are very close to us on a cosmic scale and moving our way under the influence of gravity; this overwhelms the universal expansion on such relatively local scales.

of what the newspaper headlines screamed in the wake of Eddington's eclipse expedition, that Einstein's theory 'overturned' Newton's work; Newtonian gravity (in particular, the inverse square law) is still a good description of the way the Universe works except under extreme conditions, and any better theory has to reproduce all of the successes of Newtonian theory, plus more besides, just as if a better theory than Einstein's is ever developed it will have to explain everything that the general theory explains, and more besides.

It took Einstein ten years to develop the general theory from the foundation of the special theory, although he did plenty of other things in those years from 1905 to 1915, leaving the Patent Office in 1909 to become a full-time academic at the University of Zürich, and devoting a lot of his efforts to quantum physics until about 1911, when he worked briefly in Prague before taking up a post at the ETH in Zürich (where he had been such a lazy student) and then settling in Berlin in 1914. The key to the mathematics which underpinned the general theory of relativity was given to him when he was in Zürich in 1912 by an old friend, Marcel Grossmann (1878–1936), who had been a fellow student at the ETH, where he lent Einstein his lecture notes to copy up when Einstein couldn't be bothered to attend classes. By 1912, Einstein had accepted Hermann Minkowski's neat representation of the special theory of relativity in terms of the geometry of flat, four-dimensional spacetime. Now, he needed a more general form of geometry to go with his more general form of physics, and it was Grossmann who pointed him towards the work of the nineteenth-century mathematician Bernhard Riemann (1826–1866), who had studied the geometry of curved surfaces and had developed the mathematical tools to describe this kind of geometry (called non-Euclidean geometry, since Euclid dealt with flat surfaces) in as many dimensions as you cared to choose.

This kind of mathematical investigation of non-Euclidean geometry had a long pedigree. Early in the nineteenth century Karl Friedrich Gauss (1777–1855) worked on the properties of geometries in which, for example, parallel lines can cross one another (the surface of the Earth is an example, as shown by lines of longitude, which are parallel at the equator and cross at the poles). Gauss didn't publish all of his work, much of which only became known after his death, although he did coin the term which translates as 'non-Euclidean geometry'. Some of his achievements in this area were rediscovered by the Hungarian Janos Bolyai (1802–1860) and the Russian Nikolai Lobachevsky (1793–1856), working independently of one another, in the 1820s and 1830s; but, like the then-unknown work of Gauss, these models dealt only with specific cases of non-Euclidean geometry, such as the geometry of the surface of a sphere. Riemann's outstanding contribution was to find, and present in a lecture given at the University of Göttingen in 1854, a general mathematical treatment which was the footing for the whole of geometry, allowing a range of different mathematical descriptions of a range of different geometries, which are all equally valid and with the familiar Euclidean geometry of everyday life as just one example. These ideas were introduced to the English-speaking world by the British mathematician William Clifford (1845–1879), who translated Riemann's work (which had only been published in 1867, a year after Riemann's early death, from tuberculosis) and used it as the basis for a speculation that the best way to describe the Universe at large is in terms of curved space. In 1870, he read a paper to the Cambridge Philosophical Society in which he talked of 'variation in the curvature of space' and made the analogy that 'small portions of space *are* in fact of nature analogous to little hills on a surface which is on the average flat; namely, that the ordinary laws of geometry are not valid in them'. Today, following Einstein, the analogy is made the other way around – concentrations of matter, such as the Sun, are seen as making little dimples in the spacetime of an otherwise flat Universe.[1] But it is a salutary reminder of the way science progresses, piece by piece, not through the work of isolated individuals, that Clifford was making his version of this analogy nine years before Einstein was born. Clifford himself died (also of TB) in the year Einstein was born, 1879, and never developed his ideas fully. But when Einstein came on the scene the time was clearly ripe for the general theory, and his contribution, although inspired, is not the isolated act of genius it is often portrayed.

1. Einstein's theory predicts the exact size of the dimples, and therefore to what degree light is bent as it follows a line of least resistance passing near an object like the Sun, which is why Eddington's eclipse expedition of 1919 was so important.

The general theory of relativity describes the relationship between spacetime and matter, with gravity as the interaction that links the two. The presence of matter bends spacetime, and the way material objects (or even light) follow the bends in spacetime is what shows up to us as gravity. The snappiest summary of this is the aphorism 'matter tells spacetime how to bend; spacetime tells matter how to move'. Naturally, Einstein wanted to apply his equations to the biggest collection of matter, space and time there is – the Universe. He did so as soon as he had completed the general theory, and published the results in 1917. The equations he found had one bizarre and unexpected feature. In their original form, they did not allow for the possibility of a static universe. The equations insisted that space itself must either be stretching as time passed, or shrinking, but could not stand still. At that time, remember, the Milky Way was thought to be essentially the entire Universe, and it showed no signs of either expanding or contracting. The first few redshifts of nebulae had been measured, but nobody knew what that meant, and in any case Einstein was unaware of Slipher's work. So he added another term to his equations to hold the universe they described still. Usually represented by the Greek letter lambda (λ), this is often referred to as the cosmological constant, and in Einstein's own words 'that term is necessary only for the purpose of making possible a quasi-static distribution of matter, as required by the fact of the small velocities of the stars'. In fact, it is wrong to refer to 'the' cosmological constant. The equations set up by Einstein allowed you to choose different values of the lambda term, some of which would make the model universe expand faster, at least one of which would hold it still and some of which would make it shrink. But Einstein thought that he had found a unique mathematical description of matter and spacetime which matched the known Universe of 1917.

The expanding Universe

As soon as the equations of the general theory were made public, however, other mathematicians used them to describe different model universes. Also in 1917, Willem de Sitter, in Holland, found a solution to Einstein's equations which describes a universe expanding exponentially fast, so that if the distance between two particles doubles after a certain time, it quadruples in the next equal time interval, increases eight times as much in the next time interval, sixteen-fold in the next interval, and so on. In Russia, Aleksandr Friedmann (1888–1925) found a whole family of solutions to the equations, some describing expanding universes and some describing contracting universes, and published the results in 1922 (somewhat to Einstein's irritation, since he had hoped his equations would provide a unique description of the Universe). And the Belgian astronomer Georges Lemaître (1894–1966), who was also an ordained priest, independently published similar solutions to Einstein's equations in 1927. There were some contacts between Hubble and Lemaître, who visited the United States in the mid-1920s, and was present at the meeting in 1925 where the discovery of Cepheids in the Andromeda Nebula was announced (on behalf of Hubble, who was not present) by Henry Norris Russell. Lemaître also corresponded with Einstein. One way and another, by the beginning of the 1930s, when Hubble and Humason published redshifts and distances for nearly a hundred galaxies, showing that redshift is proportional to distance, it was not only clear that the Universe is expanding, but there was already a mathematical description – actually, a choice of such cosmological models – to describe the expansion.

It is important to spell out that the cosmological redshift is not caused by galaxies moving through space, and is not, therefore, a Doppler effect. It is caused by the space between the galaxies stretching as time passes, exactly in the way that Einstein's equations described, but Einstein refused to believe, in 1917. If space stretches while light is en route to us from another galaxy, then the light itself will be stretched to longer wavelengths, which, for visible light, means moving it towards the red end of the spectrum.[1] The existence of the observed redshift–distance relation (Hubble's law) implies that the Universe was smaller in the past, not in the sense that galaxies were crammed together in a lump in a sea of empty space, but because there was no space either between the galaxies or 'outside' them – there was no outside. This in turn implies a beginning to the Universe – a concept repugnant to many astronomers in the 1930s, including Eddington,

1. You can picture this by drawing a wavy line on a fat elastic band and then stretching the elastic band.

but one which the Roman Catholic priest Lemaître embraced wholeheartedly. Lemaître developed the idea of what he called the Primeval Atom (or sometimes, the Cosmic Egg), in which all of the matter in the Universe was initially in one lump, like a superatomic nucleus, which then exploded and fragmented, like a colossal fission bomb. The idea gained popular attention in the 1930s, but most astronomers went along with Eddington in thinking that there could not really have been a beginning to the Universe, and what is now known as the Big Bang model[1] only became part of mainstream astronomy (and then only a small part) in the 1940s, following the work of the ebullient Russian émigré George Gamow (1904–1968) and his colleagues at George Washington University and Johns Hopkins University, in Washington DC.

Apart from the difficulty many astronomers had at first in accepting the idea that the Universe had a beginning, in the 1930s and 1940s there was another problem with this straightforward interpretation of the observations made by Hubble and Humason (and soon followed up by other astronomers, although the Mount Wilson team retained the technological advantage of the 100-inch telescope). Still plagued by the observational problems we have mentioned, and the confusion between Cepheids and other kinds of variable stars, the distance scale worked out by Hubble at the beginning of the 1930s was, we now know, in error by roughly a factor of ten. This meant that he thought the Universe was expanding ten times faster than we now think. Using the cosmological equations derived from the general theory of relativity (in their simplest form, these solutions correspond to a model of the Universe developed by Einstein and de Sitter, working together in the early 1930s, and known as the Einstein–de Sitter model), it is straightforward to calculate how long it has been since the Big Bang from the redshift–distance relation. Because Hubble's data implied that the Universe was expanding ten times too fast, such calculations based on those data gave an age of the Universe only one tenth as large as the modern value, as low as 1.2 billion years – and that is scarcely

1. A term actually coined by the astronomer Fred Hoyle in the 1940s as a term of derision for a model he abhorred.

a third of the well-determined age of the Earth. Clearly something was wrong, and until the age question was resolved it was hard for many people to take the idea of the Primeval Atom seriously.

The steady state model of the Universe

Indeed, this age problem was one of the reasons why Fred Hoyle (1915–2001), Herman Bondi (1919–) and Thomas Gold (1920–), in the 1940s, came up with an alternative to the Big Bang, known as the steady state model. In this picture, the Universe was envisaged as eternal, always expanding, but always looking much the same as it does today because new matter, in the form of atoms of hydrogen, is continuously being created in the gaps left behind as galaxies move apart, at just the right rate to make new galaxies to fill the gaps. This was a sensible and viable alternative to the Big Bang model right through the 1950s and into the 1960s – it is, after all, no more surprising that matter should be created steadily, one atom at a time, than it is to suggest that all the atoms in the Universe were created in one event, the Big Bang. But improving observations, including the new techniques of radio astronomy developed in the second half of the twentieth century, showed that galaxies far away across the Universe, which we see by light (or radio waves) which left them long ago, are different from nearby galaxies, proving that the Universe is changing as time passes and galaxies age. And the age question itself was gradually resolved as better telescopes became available (notably the 200-inch reflector on Mount Palomar, completed in 1947 and named in honour of Hale) and the confusion between Cepheids and other kinds of variable stars was resolved. It took a long time to narrow down the uncertainty in these still-difficult measurements of the expansion rate of the Universe to an uncertainty of 10 per cent – indeed, this was only achieved in the late 1990s, with the aid of the Hubble Space Telescope.[1] But by the end of the twentieth century the age of the Universe had been determined reasonably accurately, as somewhere between 13 billion and 16 billion years. This is comfortably older than anything whose age we can measure, including

1. The very latest data suggest that the Universe may now have begun to expand more rapidly, presumably because there is, after all, a cosmological constant. This does not affect these calculations of the age of the Universe significantly, and discussion of this work in progress lies outside the scope of the present book.

the Earth itself and the oldest stars.[1] But all that lay far in the future when Gamow and his colleagues began the scientific investigation of what went on in the Big Bang itself.

The nature of the Big Bang

Gamow had actually been one of Friedmann's students in the 1920s, and also visited the University of Göttingen, the Cavendish Laboratory and Niels Bohr's Institute in Copenhagen, where he made significant contributions to the development of quantum physics. In particular, he showed how quantum uncertainty could enable alpha particles to escape from radioactive atomic nuclei during alpha decay, by a process known as tunnelling. The alpha particles are held in place by the strong nuclear force, and in these nuclei they have nearly enough energy to escape, but not quite, according to classical theory. Quantum theory, however, says that an individual alpha particle can 'borrow' enough energy to do the job from quantum uncertainty, since the world is never quite sure how much energy it has. The particle escapes, as if it had tunnelled its way out of the nucleus, and then repays the borrowed energy before the world has time to notice it had ever been borrowed. In one of the many oversights of the Nobel committee, Gamow never received the ultimate prize for this profound contribution to our understanding of nuclear physics.

Gamow's background in nuclear and quantum physics coloured the way he and his student Ralph Alpher (1921–) and Alpher's colleague Robert Herman (1922–1997) investigated the nature of the Big Bang. Alongside his post at George Washington University, in the 1940s and early 1950s Gamow was a consultant at the Applied Physics Laboratory of Johns Hopkins University, where Alpher worked from 1944 onwards full time while studying for his bachelor's degree, master's and finally PhD (awarded in 1948) in the evenings and at weekends at George Washington University. Herman had a more

1. This is actually a very profound discovery. The age of the Universe is essentially calculated from the general theory of relativity, and deals with the laws of physics on the very large scale; the ages of stars are, as we shall see below, essentially calculated from the laws of quantum mechanics, physics on the very small scale. Yet the age of the Universe comes out to be just enough older than the ages of the oldest stars to allow the time required for the first stars to form after the Big Bang. This agreement between physics on the largest and smallest scales is an important indication that the whole of science is built on solid foundations.

conventional academic background, with a PhD from Princeton, and joined the Johns Hopkins Laboratory in 1943, initially, like Alpher, involved in war work. Also like Alpher, he did his work on the early Universe in his own time, technically as a hobby. Under Gamow's supervision, Alpher investigated for his doctorate the way in which more complicated elements could be built up from simple elements under the conditions that they assumed must have existed in the Big Bang, when the entire observable Universe was packed into a volume no bigger across than our Solar System is today. The chemical elements we, and the rest of the visible Universe, are made of have to come from somewhere, and Gamow guessed that the raw material for their manufacture was a hot fireball of neutrons. This was at a time when the first nuclear bombs had recently been exploded, and when the first nuclear reactors were being constructed. Although a great deal of the information on how nuclei interact with one another was classified, there was an expanding data bank of unclassified information about what happens to different kinds of material when irradiated with neutrons from such reactors, with nuclei absorbing neutrons one by one to become the nuclei of heavier elements, and getting rid of excess energy in the form of gamma radiation. Sometimes, unstable nuclei would be created in this way, and would adjust their internal composition by emitting beta radiation (electrons). Although the raw material of the Universe was assumed to be neutrons, neutrons themselves decay in this way to produce electrons and protons, which together make the first element, hydrogen. Adding a neutron to a hydrogen nucleus gives a nucleus of deuterium (heavy hydrogen), adding a further proton makes helium-3, and adding another neutron as well makes helium-4, which can also be made by the fusion of two helium-3 nuclei and the ejection of two protons, and so on. Nearly all the deuterium and helium-3 is converted into helium-4, one way or another. Alpher and Gamow looked at all of the available neutron-capture data for different elements and found that the nuclei formed most easily in this way turned out to be those of the most common elements, while nuclei that did not form readily in this way corresponded to rare elements. In particular, they found that this process would produce an enormous amount of helium compared with other elements, which matched up with observations of the composition of the Sun and stars that were becoming available around that time.

Predicting background radiation

As well as providing Alpher with the material for his PhD dissertation, this work formed the basis of a scientific paper published in the journal *Physical Review*. When the time came to submit the paper, Gamow, an inveterate joker, decided (overriding Alpher's objections) to add the name of his old friend Hans Bethe (1906–) as co-author, for the sole reason that he liked the sound of the names Alpher, Bethe, Gamow (alpha, beta, gamma). To his delight, by a coincidence the paper appeared in the issue of *Physical Review* dated 1 April 1948. That publication marks the beginning of Big Bang cosmology as a quantitative science.

Soon after the alpha-beta-gamma paper (as it is usually referred to) was published, Alpher and Herman came up with a profound insight into the nature of the Big Bang. They realized that the hot radiation which filled the Universe at the time of the Big Bang must still fill the Universe today, but that it would have cooled by a quantifiable amount as it expanded along with the universal expansion of space – you can think of this as an extreme redshift, stretching the wavelengths of the original gamma rays and X-rays far into the radio part of the electromagnetic spectrum. Later in 1948, Alpher and Herman published a paper reporting their calculation of this effect, assuming that this background radiation is black-body radiation at the appropriate temperature. They found that the temperature of the background radiation today should be about 5 K – that is, roughly −268 °C. At the time, Gamow did not accept the validity of this work, but after about 1950 he became an enthusiastic supporter of the idea and referred to it in several of his popular publications,[1] often getting the details of the calculation wrong (he was never good at sums) and without giving proper credit to Alpher and Herman. The result is that he is often incorrectly given credit for the prediction of the existence of this background radiation, when that credit belongs entirely to Alpher and Herman.

Nobody took much notice of the prediction at the time. People who knew about it mistakenly thought that the available technology of

1. Such as his book *The Creation of the Universe*.

Measuring background radiation

radio astronomy was not good enough to measure such a weak hiss of radio noise coming from all directions in space; people with access to the technology seem to have been unaware of the prediction. In the early 1960s, however, two radio astronomers working with a horn antenna at the Bell Laboratories research station near Holmdel, New Jersey, found that they were plagued by faint radio noise coming from all directions in space, corresponding to black-body radiation with a temperature of about 3 K. Arno Penzias (1933–) and Robert Wilson (1936–) had no idea what they had discovered, but just down the road at Princeton University a team working under Jim Peebles (1935–) was actually building a radio telescope specifically intended to search for this echo of the Big Bang – not because of the pioneering work of Alpher and Herman, but because Peebles had independently carried out similar calculations. When news came to Princeton of what the Bell researchers had found, Peebles was quickly able to explain what was going on. The discovery, published in 1965, marked the moment when most astronomers started to take the Big Bang model seriously as a plausible description of the Universe in which we live, rather than as some kind of abstract theoretical game. In 1978, Penzias and Wilson shared the Nobel prize for the discovery, an honour they perhaps deserved rather less than Alpher and Herman, who did not receive the prize.[1]

Modern measurements: the COBE satellite

Since then, the cosmic microwave background radiation has been observed in exquisite detail by many different instruments, including the famous COBE satellite, and has been confirmed to be perfect black-body radiation (the most perfect black-body radiation ever seen) with a temperature of just 2.725 K. This is the most powerful single piece of evidence that there really was a Big Bang – or, to put it in more scientific language, that the visible Universe experienced an extremely hot, dense phase about 13 billion years ago. Cosmologists in the twenty-first century are tackling the puzzle of how this superhot fireball of energy came into

1. I've always suspected that the Nobel committee, like many other people, thought that the prediction had been made by Gamow, who was dead by 1978, and Nobel prizes are never awarded posthumously. There is no other obvious reason why Alpher and Herman were ignored.

existence in the first place, but we shall not describe these still-speculative ideas here, and will end our discussion of the history of cosmology at the point where there is overwhelming evidence that the Universe as we know it did emerge from a Big Bang – if you want to put a date on this, the announcement of the COBE results in the spring of 1992 is as good as any. Indeed, having made predictions that have been proved correct by observation, the Big Bang model is now entitled to be given the name Big Bang theory.

But what exactly was it that emerged from the Big Bang? As Alpher and Herman refined their calculations further, they soon discovered that there was a major problem with their whole scheme of manufacturing elements (nucleosynthesis) by the repeated addition of neutrons, one at a time, to nuclei. It soon turned out that there are no stable nuclei with masses of 5 units or 8 units on the atomic scale. Starting out with a sea of protons and neutrons (now thought to have been manufactured out of pure energy in the Big Bang fireball (in line with $E = mc^2$), it is easy to make hydrogen and helium, and modern versions of the calculations pioneered by Gamow's team tell us that a mixture of roughly 75 per cent hydrogen and 25 per cent helium could be made in this way in the Big Bang. But if you add a neutron to helium-4, you get an isotope so unstable that it spits out the extra neutron before it has time to interact further and make a stable nucleus. A very little lithium-7 can be made by rare interactions in which a helium-3 nucleus and a helium-4 nucleus stick together, but under the conditions existing in the Big Bang fireball the next step is to produce a nucleus of beryllium-8, which immediately breaks into two helium-4 nuclei. If you could only make hydrogen and helium (and tiny traces of lithium-7 and deuterium) in the Big Bang, then all the other elements must have been manufactured somewhere else. That 'somewhere' – the only possible alternative place – is the insides of stars. But an understanding of just how this happens emerged only gradually, starting with the realization, in the late 1920s and 1930s, that the Sun and stars are not made of the same mixture of elements as the Earth.

The idea that the Sun is basically made of the same kind of stuff as the Earth, but hotter, had a long pedigree, and represented the fruits of the first known attempt to describe heavenly bodies in what we would now call scientific terms, rather than treating them as gods. It goes back to the Greek philosopher Anaxagoras of Athens, who lived in the fifth century BC. Anaxagoras got his ideas about the composition of the Sun when a meteorite fell near Aegospotami. The meteorite was red hot when it reached the ground, and it had come from the sky, so Anaxagoras reasoned that it came from the Sun. It was made largely of iron, so he concluded that the Sun was made of iron. Since he knew nothing about the age of the Earth, or how long it would take a large ball of red-hot iron to cool down, or whether there might be a form of energy keeping the Sun shining, the idea that the Sun was a ball of red-hot iron was a good working hypothesis in those days (not that many people took Anaxagoras seriously at the time). When people started thinking about nuclear energy as the source of the Sun's heat at the beginning of the twentieth century, the realization that the radioactive decay of a relatively small amount of radium could keep the Sun shining (if only for a relatively short time) encouraged the idea that most of the Sun's mass might consist of heavy elements. As a result, when a few astronomers and physicists started to investigate how nuclear fusion might provide the energy that keeps the Sun and stars hot, they started out by investigating processes in which protons (nuclei of hydrogen) fuse with nuclei of heavy elements, on the assumption that heavy elements were common and protons were rare inside the stars. Even Eddington, with his prescient comments in 1920 about converting hydrogen into helium, was still only suggesting that 5 per cent of a star's mass might start out in the form of hydrogen.

The process by which protons penetrate heavy nuclei is the opposite of the process of alpha decay, in which an alpha particle (helium nucleus) escapes from a heavy nucleus, and it is governed by the same rules of quantum tunnelling discovered by Gamow. Gamow's calculations of the tunnel effect were published in 1928, and just a year later the Welsh astrophysicist Robert Atkinson (1889–1982) and his German colleague Fritz Houtermans (1903–1966), who had previously worked with Gamow, published a paper describing the kind of nuclear reactions that might occur inside stars as protons fused with heavy nuclei. Their paper began with the words 'Recently Gamow demonstrated that positively charged particles can penetrate the atomic nucleus even if traditional belief holds their energy to be inadequate.'

This is the key point. Eddington, in particular, had used the laws of physics to calculate the temperature at the heart of the Sun from its mass, radius and the rate at which it is releasing energy into space. Without the tunnel effect, this temperature – about 15 million K – is too low to allow nuclei to come together with sufficient force that they overcome their mutual electrical repulsion and stick together. In the early 1920s, when physicists first calculated the conditions of temperature and pressure required for protons to fuse together to make helium, this seemed to many an insuperable problem. In his book *The Internal Constitution of the Stars*, published in 1926 just as the quantum revolution was taking place, Eddington replied that 'we do not argue with the critic who urges that the stars are not hot enough for this process; we tell him to go and find a *hotter place*'. This is usually interpreted as Eddington telling his critics to go to Hell. It was the quantum revolution, and tunnelling in particular, which soon showed that Eddington was right to stick to his guns, and nothing indicates more clearly the interdependence of different scientific disciplines. Progress in understanding the internal workings of the stars could only be made once the quantum properties of entities such as protons were beginning to be understood.

But even Atkinson and Houtermans, as we have seen, were still assuming in 1928 that the Sun was rich in heavy elements. Just around the time they were carrying out their calculations, however, spectroscopy became sophisticated enough to cast doubt on this assumption. In 1928, the British-born astronomer Cecilia Payne (later Cecilia Payne Gaposchkin; 1900–1979) was working for her PhD at Radcliffe College, under the supervision of Henry Norris Russell. Using spectroscopy, she discovered that the composition of stellar atmospheres is dominated by hydrogen, a result so surprising that when she published her results, Russell insisted that she include a caveat to the effect that the observed spectroscopic features could not really be taken as implying that stars are made of hydrogen, but must be due to some peculiar behaviour of hydrogen under stellar conditions, enhancing its appearance in the spectra. But at about the same time, the German Albrecht Unsöld (1905–1995) and the young Irish astronomer William McCrea (1904–1999) independently established that the prominence of hydrogen lines in stellar spectra indicates that there are a million times more hydrogen atoms present in the atmospheres of stars than there are atoms of everything else put together.

How the stars shine: the nuclear fusion process

All of these pieces of work, coming together at the end of the 1920s, marked the beginning of the development of an understanding of what keeps the stars shining. It still took some years for astrophysicists to pin down the most likely nuclear interactions to explain the process, and slightly longer for them to appreciate fully just how much hydrogen dominates the composition of the visible Universe. This was partly because of an unfortunate coincidence. As astrophysicists developed mathematical models to describe the internal structure of the stars in more detail in the 1930s, they found that these models worked – in the sense that they predicted the existence of balls of hot gas with the same sort of size, temperature and mass as the stars – either if the composition of the hot objects is roughly two-thirds heavy elements and one-third hydrogen (or a mixture of hydrogen and helium), or if their composition is at least 95 per cent hydrogen and helium, with just a trace of heavy elements. With either mixture, but no other, the properties of the hot balls of gas predicted by the equations would match up with those of real stars. Having only just realized that there is more than a trace of hydrogen inside the stars, at first the astrophysicists naturally plumped for the option with two-thirds heavy elements, and this meant that for a almost a decade they concentrated on investigating interactions in which protons tunnel into heavy nuclei. It was only after they discovered the detailed processes which can turn hydrogen into helium that they realized that heavy elements are rare in stars, and that hydrogen and helium together make up 99 per cent of star stuff.

As is so often the case with scientific ideas whose time has come, the key interactions involved in the nuclear fusion processes which keep the stars shining were identified independently by different researchers at about the same time. The principal contributions came from the German-born Hans Bethe, then working at Cornell University, and Carl von Weizsäcker (1912–), working in Berlin, in the final years of the 1930s. They identified two processes which could operate at the temperatures known to exist inside stars, making allowance for quantum processes such as tunnelling, to convert hydrogen into helium, with the appropriate release of energy. One of these, known as the proton–proton chain, turns out to be the dominant interaction in stars like the Sun. It involves two protons coming together, with a positron being ejected, to make a nucleus of deuterium (heavy hydrogen).[1] When another proton fuses with this nucleus, it forms helium-3 (two protons plus one neutron), and when two helium-3 nuclei come together and eject two protons, the result is a nucleus of helium-4 (two protons plus two neutrons). The second process operates more effectively at the slightly higher temperatures found in the hearts of stars at least one and a half times as massive as the Sun, and in many stars both processes are at work. This second process, the carbon cycle, operates in a loop, and it requires the presence of a few nuclei of carbon, involving protons tunnelling into these nuclei in the way Atkinson and Houtermans suggested. Because the process operates in a loop, these heavy nuclei emerge at the end of the cycle unchanged, effectively acting as catalysts. Starting with a nucleus of carbon-12, the addition of a proton makes unstable nitrogen-13, which spits out a positron to become carbon-13.[2] Adding a second proton makes nitrogen-14, while adding a third proton to the nitrogen-14 nucleus makes unstable oxygen-15, which ejects a positron to become nitrogen-15. Now comes the finale – with the addition of a fourth proton the nucleus ejects a whole alpha particle and reverts to being carbon-12, the starting ingredient. But an alpha particle is simply a nucleus of helium-4. Once again, the net effect is that four protons have been converted into a single nucleus of helium, with a couple of positrons and a lot of energy ejected along the way.

These processes were identified shortly before the beginning of the Second World War, and further progress in understanding the internal workings of the stars had to await the return of normal conditions in the late 1940s. But these studies then benefited enormously from the wartime effort to understand nuclear interactions in connection with research into nuclear weapons and the development of the first nuclear reactors. As the appropriate information was declassified, it helped

1. Many of these interactions also involve the ejection of neutrinos, but for simplicity we shall not go into such detail.
2. Ejecting a positron converts one of the protons in the nucleus into a neutron.

astrophysicists to work out the rates at which interactions like the ones we have just described could go on inside the stars. And, as the work by Alpher, Herman and Gamow highlighted, the problem of the 'mass gaps' for the manufacture of heavier elements step by step from hydrogen and helium, in the 1950s several astronomers looked at the problem of how the heavy elements (which, after all, had to come from somewhere) might be manufactured inside stars. One idea that was aired was the possibility that three helium-4 nuclei (three alpha particles) could come together essentially simultaneously, forming a stable nucleus of carbon-12 without having to manufacture the highly unstable beryllium-8 as an intermediate step. The key insight came from the British astronomer Fred Hoyle in 1953. Rather in the way that 'classical' physics said that two protons could not fuse under the conditions inside a star like the Sun, the simplest understanding of nuclear physics said that such 'triple-alpha' interactions could occur, but would be far too rare to make sufficient amounts of carbon during the lifetime of a star. In most cases, such triple collisions ought to smash the particles apart, not combine them in a single nucleus.

The concept of 'resonances'

The proton fusion puzzle was solved by quantum tunnelling; Hoyle suggested, on the basis of no other evidence than the fact that carbon exists, a comparably profound solution to the triple alpha puzzle – that the nucleus of carbon-12 must possess a property known as a resonance, which greatly increased the probability of three alpha particles fusing. Such resonances are states of higher-than-usual energy. If the base energy of the nucleus is likened to the fundamental note played on a guitar string, resonances can be likened to higher notes played on the same string, with only certain notes (certain harmonics) being possible. There was nothing mysterious about the idea of resonances when Hoyle made his suggestion – but there was no way to calculate in advance what resonances carbon-12 ought to have, and in order for the trick to work, carbon-12 had to have a resonance with a certain very precise energy, corresponding to a very pure note. Hoyle persuaded Willy Fowler (1911–1995), an experimental physicist working at Caltech, to carry out experiments to test for the existence of such a resonance in the carbon-12 nucleus. It turned up exactly where Hoyle had predicted. The existence of this resonance allows three alpha particles to merge smoothly together, instead of colliding in an impact which smashes them apart. This creates an energetic nucleus of carbon-12 which then radiates away its excess energy and settles into the basic energy level (known as the ground state). This was the key discovery which explained how elements heavier than helium can be manufactured inside stars.[1] Once you have carbon nuclei to work with, you can make heavier elements still by adding more alpha particles (going from carbon-12 to oxygen-16 to neon-20, and so on) or by the kind of drip-feed addition of protons discussed by people such as Atkinson and Houtermans and, in a different context, by Alpher and Herman (this kind of process is also at work in the carbon cycle). Hoyle, Fowler and their British-born colleagues Geoffrey Burbidge (1925–) and Margaret Burbidge (1922–) produced the definitive account of how the elements are built up in this way inside stars in a paper published in 1957.[2] Following this work, astrophysicists were able to model in detail the internal workings of the stars, and by comparing these models with observations of real stars, to determine the life cycles of stars and work out, among other things, the ages of the oldest stars in our Galaxy.

This understanding of nuclear fusion processes operating inside stars explained how all the elements up to iron can be manufactured from the hydrogen and helium produced in the Big Bang. Even better, the proportions of the different elements predicted to be produced in this way match the proportions seen in the Universe at large – the amount of carbon relative to oxygen, or neon relative to calcium, or whatever. But it cannot explain the existence of elements heavier than iron, because iron nuclei represent the most stable form of everyday matter, with the least energy. To make nuclei of even heavier elements – such as gold, or uranium, or lead – energy has to be put in to force the nuclei to fuse together. This happens when stars rather more massive than the Sun reach the end of their lives and run out of nuclear fuel which can generate heat (by the kind of interactions we have just described) to hold them up. When their fuel runs out, such stars

1. And it's worth noting that it was made less than half a century after Rutherford had identified alpha radiation as helium nuclei.

2. With the name of the authors listed alphabetically as Burbidge, Burbidge, Fowler and Hoyle, this paper is known to all astronomers as 'B^2FH'.

collapse dramatically in upon themselves, and as they do so, enormous amounts of gravitational energy are released and converted into heat. One effect of this is to make the single star shine, for a few weeks, as brightly as a whole galaxy of ordinary stars, as it becomes a supernova; another is to provide the energy which fuses nuclei together to make the heaviest elements. And a third effect is to power a huge explosion in which most of the material of the star, including those heavy elements, is scattered through interstellar space, to form part of the raw material of new stars, planets and possibly people. The theoretical models describing all this were developed by many people in the 1960s and 1970s, drawing on observations of supernovae (which are rather rare events) in other galaxies. Then, in 1987, a supernova was seen to explode in our near neighbour, the Large Magellanic Cloud – the closest supernova to us to have been seen since the invention of the astronomical telescope. With a battery of modern telescopes turned upon the event for months, analysing it in every detail with observations at all possible wavelengths, the processes unfolding in this supernova were seen to match closely the predictions of those models, effectively slotting into place the last piece in our understanding of the basics of how stars work. They were, to astronomers who had seen this understanding develop during the span of a single human lifetime, the most important and exciting discoveries concerned with the origin of the elements, confirming that the theoretical model is broadly correct.

CHON and humankind's place in the Universe

This leads us to what is, in my view, the most profound discovery of the whole scientific endeavour. Astronomers are able to calculate with great accuracy how much material of different kinds is manufactured inside stars and scattered into space by supernovae and lesser stellar outbursts. They can confirm these calculations by measuring the amount of different kinds of material in clouds of gas and dust in space, the raw material from which new stars and planetary systems form, using spectroscopy. What they find is that apart from helium, which is an inert gas that does not take part in chemical reactions, the four most common elements in the Universe are hydrogen, carbon, oxygen and nitrogen, collectively known by the acronym CHON. This is an ultimate truth revealed by a process of enquiry that began when Galileo first turned his telescope

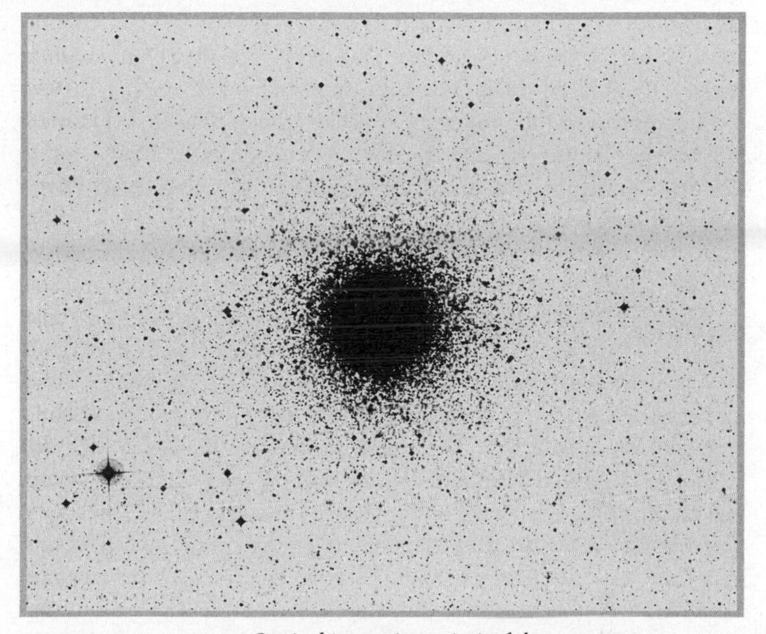

41. *Optical image (negative) of the globular star cluster NGC 362.*

towards the sky, and ended with those observations of the supernova of 1987. Another line of investigation, which for centuries seemed to have nothing to do with the scientific study of the stars, began a little earlier, when Vesalius started to put the study of the human body on a scientific footing. The ultimate truth revealed by this line of enquiry, culminating with the investigation of DNA in the 1950s, is that there is no evidence of a special life force, but that all of life on Earth, including ourselves, is based on chemical processes. And the four most common elements involved in the chemistry of life are hydrogen, carbon, oxygen and nitrogen. We are made out of exactly the raw materials which are most easily available in the Universe. The implication is that the Earth is not a special place, and that life forms based on CHON are likely to be found across the Universe, not just in our Galaxy but in others. It is the ultimate removal of humankind from any special place in the cosmos, the completion of the process that began with Copernicus and *De Revolutionibus*. The Earth is an ordinary planet orbiting an ordinary star in the suburbs of an average galaxy. Our Galaxy contains hundreds of billions of stars, and there are hundreds of billions of galaxies in the visible Universe, all of them full of stars like the Sun and laced with clouds of gas and dust rich in CHON. Nothing could be further removed from the pre-Renaissance idea that the Earth was at the centre of the Universe, with the Sun and stars orbiting around it, and humankind as the unique pinnacle of creation, qualitatively different from the 'lesser' forms of life.

Into the unknown

But do these discoveries mean, as some have suggested, that science is about to come to an end? Now that we know how life and the Universe work, is there anything left except to fill in the details? I believe that there is. Even filling in the details will be a long job, but science itself is now undergoing a qualitative change. The analogy I have used before, but which I cannot improve upon, is with the game of chess. A small child can learn the rules of the game – even the complicated rules like the knight's move. But that does not make the child a grandmaster, and even the greatest grandmaster who ever lived would not claim to know everything there is to know about the game of chess. Four and a half centuries after the publication of *De Revolutionibus*, we are in the situation of that small child who has just learned the rules of the game. We are just beginning to make our first attempts to play the game, with developments such as genetic engineering and artificial intelligence. Who knows what the next five centuries, let alone the next five millennia, might bring.

Coda: The Pleasure of Finding Things Out

Science is a personal activity. With very few exceptions, scientists throughout history have plied their craft not through a lust for glory or material reward, but in order to satisfy their own curiosity about the way the world works. Some, as we have seen, have taken this to such extremes that they have kept their discoveries to themselves, happy in the knowledge that they have found the solution to some particular puzzle, but feeling no need to boast about the achievement. Although each scientist – and each generation of scientists – exists and works in the context of their time, building on what has gone before with the aid of the technology available to them, it is as individual people that they make their own contribution. It has therefore seemed natural to me to use an essentially biographical approach to the history of science (at least, for my first attempt at such a history), in the hope of teasing out something of what makes a scientist tick, as well as revealing how one scientific advance led to another. I am aware that this is not an approach that is much favoured by historians today, and that any professional historians who have read this far may accuse me of being old-fashioned, or even reactionary. But if I am old-fashioned, it is because I choose to be so, not because I am unaware that I am out of step. I am also aware that there are almost as many approaches to the study of history as there are historians, and each approach can shed light on the subject. Few, if any, historians would claim that one person's view (or interpretation) of history reveals 'the' truth about history, any more than a single snapshot of a human being reveals everything about that person. But perhaps there is something about my own approach to the history of science which may provide food for thought even for the professionals.

Although the process of doing science is a personal activity, science itself is essentially impersonal. It involves absolute, objective truths. A confusion between the process of doing science and science itself has led to the popular myth of the scientist as a cold-blooded, logical machine. But scientists can be hot-blooded, illogical and even mad while still pursuing the search for ultimate truth. By some criteria, Isaac Newton was insane, both in his single-minded obsession with a succession of interests (science, alchemy, religion) and in the intensity of his personal vendettas, while Henry Cavendish was decidedly odd. So it is important to make the distinction between what is subjective, and therefore open to debate, in the present volume and what is objective and unarguably true.

I do not claim that this is the last word on the history of science – no single volume could be. It is subjective, like all histories; but it is written from the perspective of someone who has been involved professionally in scientific research, not as a professional historian, which has both advantages and disadvantages. The most important insight this provides, as I hope the book makes clear, is that I reject the Kuhnean idea of 'revolutions' in science, and see the development of the subject in essentially incremental, step-by-step terms. The two keys to scientific progress, it seems to me, are the personal touch and building gradually on what has gone before. Science is made by people, not people by science, and my aim has been to tell you about the people who made science, and how they made it. Closely related to this view of science is the idea that science is to some extent divorced from the economic and social upheavals going on in the world at large, and also that it really is about a search for objective truth.

Historians or sociologists who have no training in or experience of scientific research sometimes suggest that scientific truth is no more valid than artistic truth, and that (to put it crudely) Albert Einstein's general theory of relativity might go out of fashion just as much of the painting done by Victorian artists later went out of fashion. This is absolutely not so. Any description of the Universe that supercedes Einstein's theory must both go beyond the limitations of that theory and include all the successes of the general theory within itself, just as the general theory includes Newton's theory of gravity within itself. There will never be a successful description of the Universe which says that Einstein's theory is wrong in any of the areas where it has already been tested. It is a factual, objective truth that, for example, light gets 'bent' by a certain amount when it passes near a star like the Sun, and the general theory will always be able to tell you how much it gets bent. At a simpler level, like many other scientific facts the inverse square law of gravity is an ultimate truth, in a way that no historical account of how that law was discovered can ever be 'the' truth. Nobody will ever know the extent to which Newton's thinking on gravity was influenced by observing the fall of an apple; by the time he told the story, Newton himself might not have remembered the details correctly. But we can all know what law of gravity he discovered. So my account is personal and subjective in its interpretation of the evidence about how scientific truths were discovered; but it is impersonal and objective in describing what those scientific truths are. You may or may not agree with my opinion that Robert Hooke was maligned by Newton; but either way you would still have to accept the truth of Hooke's law of elasticity.

If a specific example is needed of the inverse argument, the way scientific truth cannot be distorted to fit the way we might want the world to be, it is only necessary to point to the distortion of the study of genetics under the Stalinist regime in the USSR half a century ago. Trofim Denisovich Lysenko (1898–1976) gained great favour and influence under that regime because his ideas about genetics and heredity offered a politically correct view of the biological world, while the Mendelian principles of genetics were regarded as incompatible with the principles of dialectical materialism. They may well be; but the fact remains that Mendelian genetics provides a good description of how heredity works, while Lysenkoism does not – and this had disastrous repercussions at a very practical level through Lysenko's influence on Soviet agricultural practice.

One of the strangest arguments that I have seen put forward – apparently seriously – is that using a word such as 'gravity' to describe the cause of the fall of an apple from a tree is no less mystical than invoking 'God's will' to explain why the apple falls, since the word 'gravity' is just a label. Certainly it is – in the same way that the words 'Beethoven's Fifth' are not a piece of music, but only a label which indicates a piece of music, and an alternative label, such as the Morse code symbols for the letter V, could just as easily be used to indicate the same piece of music. Scientists are well aware that words are merely labels which we use for convenience, and that a rose by any other name would smell as sweet. That is why they deliberately choose to use a nonsense word, quark, as the label for a fundamental entity in particle theory, and why they use the names of colours (red, blue and green) to identify different kinds of quark. They do not suggest that quarks really are coloured in this way. The difference between the scientific description of how apples fall and the mystical description of how apples fall is that, whatever the name you ascribe to the phenomenon, in scientific terms it can be described by a precise law (in this case, the inverse square law) and that the same law can be applied to the fall of an apple from a tree, the way the Moon is held in orbit around the Earth and so on out into the Universe. To a mystic, there is no reason why we should expect the way an apple falls from a tree to bear any relation to, say, the way that a comet moves past the Sun. But the word 'gravity' is simply a shorthand expression for the whole suite of ideas incorporated in Newton's *Principia* and Einstein's general theory of relativity. To a scientist, the word 'gravity' conjures up a rich tapestry of ideas and laws, in the same way that to the conductor of a symphony orchestra the words 'Beethoven's Fifth' conjure up a rich musical experience. It is not the label that matters, but the underlying universal law, giving a predictive power to science. We can say with confidence that planets (and comets) orbiting other stars are also under the influence of the inverse square law, whether you ascribe that law to 'gravity' or to 'God's will'; and we can be sure that any intelligent beings inhabiting those planets will measure the same inverse square law, although undoubtedly they will call it by a different name from the one we use.

There is no need for me to labour the point. It is because there are ultimate truths out there that science hangs together so well. And what motivates the great scientists is not the thirst for fame or fortune (although that can be a seductive lure for the less-than-great scientists), but what Richard Feynman called 'the pleasure of finding things out', a pleasure so satisfying that many of those great scientists, from Newton to Cavendish and from Charles Darwin to Feynman himself, have not even bothered to publish their findings unless pressed by their friends to do so, but a pleasure that would hardly exist if there were no truths to discover.

Bibliography

J. A. Adhémar, *Révolutions de la mer* (published privately by the author, Paris, 1842).

Elizabeth Cary Agassiz, *Louis Agassiz, his life and correspondence* (Houghton Mifflin & Co, Boston, 1886; published in two volumes).

Ralph Alpher and Robert Herman, *Genesis of the Big Bang* (OUP, Oxford, 2001).

Angus Armitage, *Edmond Halley* (Nelson, London, 1966).

Isaac Asimov, *Asimov's New Guide to Science* (Penguin, London, 1987).

John Aubrey, *Brief Lives* (ed. by Andrew Clark), vols I and II (Clarendon Press, Oxford, 1898).

Ralph Baierlein, *Newton to Einstein* (CUP, Cambridge, 1992).

Nora Barlow (ed.), *The Autobiography of Charles Darwin, 1809–1882, with original omissions restored* (William Collins, London, 1958).

A. J. Berger, J. Imbrie, J. Hays, G. Kukla and B. Saltzman (eds.), *Milankovitch and Climate* (Reidel, Dordrecht, 1984).

W. Berkson, *Fields of Force* (Routledge, London, 1974).

David Berlinski, *Newton's Gift* (The Free Press, New York, 2000).

A. J. Berry, *Henry Cavendish* (Hutchinson, London, 1960).

Mario Biagioli, *Galileo, Courtier* (University of Chicago Press, Chicago, 1993).

P. M. S. Blackett, E. Bullard and S. K. Runcorn, *A Symposium on Continental Drift* (Royal Society, London, 1965).

W. Bragg and G. Porter (eds.), *The Royal Institution Library of Science*, volume 5 (Elsevier, Amsterdam, 1970).

S. C. Brown, *Benjamin Thompson, Count Rumford* (MIT Press, Cambridge, MA, 1979).

Janet Browne, *Charles Darwin: voyaging* (Jonathan Cape, London, 1995).

Leonard C. Bruno, *The Landmarks of Science* (Facts on File, New York, 1989).

John Campbell, *Rutherford* (AAS Publications, Christchurch, New Zealand, 1999).

G. M. Caroe, *William Henry Bragg* (CUP, Cambridge, 1978).

Carlo Cercignani, *Ludwig Boltzmann* (OUP, Oxford, 1998).
S. Chandrasekhar, *Eddington* (CUP, Cambridge, 1983).
John Robert Christianson, *On Tycho's Island* (CUP, London, 2000).
Frank Close, *Lucifer's Legacy* (OUP, Oxford, 2000).
Lawrence I. Conrad, Michael Neve, Vivian Nutton, Roy Porter and Andrew Wear, *The Western Medical Tradition: 800 BC to AD 1800* (CUP, Cambridge, 1995).
Alan Cook, *Edmond Halley* (OUP, Oxford, 1998).
James Croll, *Climate and Time in their Geological Relations* (Daldy, Isbister, & Co., London, 1875).
J. G. Crowther, *British Scientists of the Nineteenth Century* (Kegan Paul, London, 1935).
J. G. Crowther, *Founders of British Science* (Cresset Press, London, 1960).
J. G. Crowther, *Scientists of the Industrial Revolution* (Cresset Press, London, 1962).
William Dampier, *A History of Science*, 3rd edn (CUP, Cambridge, 1942).
Charles Darwin, *The Origin of Species by Means of Natural Selection*, reprint of the first edition of 1859 plus additional material (Pelican, London, 1968; reprinted in Penguin Classics, 1985).
Charles Darwin and Alfred Wallace, *Evolution by Natural Selection* (CUP, Cambridge, 1958).
Erasmus Darwin, *Zoonomia*, Part 1 (J. Johnson, London, 1794).
Francis Darwin (ed.), *The Life and Letters of Charles Darwin* (John Murray, London, 1887). An abbreviated version is still available as *The Autobiography of Charles Darwin and Selected Letters* (Dover, New York, 1958).
Francis Darwin (ed.), *The Foundations of the Origin of Species: two essays written in 1842 and 1844 by Charles Darwin* (CUP, Cambridge, 1909).
Richard Dawkins, *The Blind Watchmaker* (Longman, Harlow, 1986).
René Descartes, *Discourse on Method and the Meditations*, translated by F. E. Sutcliffe (Penguin, London, 1968).
Adrian Desmond and James Moore, *Darwin* (Michael Joseph, London, 1991).
Ellen Drake, *Restless Genius: Robert Hooke and his earthly thoughts* (OUP, New York, 1996).
Adrian Desmond, *Huxley* (Addison Wesley, Reading, MA, 1997).
Stillman Drake, *Galileo at Work* (Dover, New York, 1978).
Stillman Drake, *Galileo* (OUP, Oxford, 1980).
J. L. E. Dryer, *Tycho Brahe* (Adam & Charles Black, Edinburgh, 1899).
A. S. Eddington, *The Internal Constitution of the Stars* (CUP, Cambridge, 1926).
A. S. Eddington, *The Nature of the Physical World* (CUP, Cambridge, 1928).
Margaret 'Espinasse, *Robert Hooke* (Heinemann, London, 1956).

John Evelyn, *Diary* (ed. E. S. de Beer) (OUP, London, 1959).
C. W. F. Everitt, *James Clerk Maxwell* (Scribner's, New York, 1975).
J. J. Fahie, *Galileo: his life and work* (John Murray, London, 1903).
Otis Fellows and Stephen Milliken, *Buffon* (Twayne, New York, 1972)
Georgina Ferry, *Dorothy Hodgkin* (Granta, London, 1998).
Richard Feynman, *QED: the strange theory of light and matter* (Princeton University Press, Princeton, NJ, 1985).
Richard Fifield (ed.), *The Making of the Earth* (Blackwell, Oxford, 1985).
Antony Flew, *Malthus* (Pelican, London, 1970).
Tore Frängsmyr (ed.), *Linnaeus: the man and his work* (University of California Press, Berkeley, 1983).
Galileo Galilei, *Galileo on the World Systems* (abridged and translated from the *Dialogue* by Maurice A. Finocchiaro) (University of California Press, 1997).
George Gamow, *The Creation of the Universe* (Viking, New York, 1952).
G. Gass, Peter J. Smith and R. C. L. Wilson (eds.), *Understanding the Earth*, 2nd edn (MIT Press, Cambridge, MA, 1972).
J. Geikie, *The Great Ice Age*, 3rd edn (Stanford, London, 1894; 1st edn published by Isbister, London, 1874).
Wilma George, *Biologist Philosopher: a study of the life and writings of Alfred Russel Wallace* (Abelard-Schuman, New York, 1964).
William Gilbert, *Loadstone and Magnetic Bodies, and on The Great Magnet of the Earth*, translated from the 1600 edition of *De Magnete* by P. Fleury Mottelay (Bernard Quaritch, London, 1893).
C. C. Gillispie, *Pierre-Simon Laplace* (Princeton University Press, Princeton NJ, 1997).
H. E. Le Grand, *Drifting Continents and Shifting Theories* (CUP, Cambridge, 1988).
Frank Greenaway, *John Dalton and the Atom* (Heinemann, London, 1966).
John Gribbin, *In Search of Schrödinger's Cat* (Bantam, London, 1984).
John Gribbin, *In Search of the Double Helix* (Penguin, London, 1995).
John Gribbin, *In Search of the Big Bang* (Penguin, London, 1998).
John Gribbin, *The Birth of Time* (Weidenfeld & Nicolson, London, 1999).
John Gribbin, *Stardust* (Viking, London, 2000).
John and Mary Gribbin, *Richard Feynman: a life in science* (Viking, London, 1994).
John Gribbin and Jeremy Cherfas, *The First Chimpanzee* (Penguin, London, 2001).
John Gribbin and Jeremy Cherfas, *The Mating Game* (Penguin, London, 2001).
Howard Gruber, *Darwin on Man* (Wildwood House, London, 1974).
Thomas Hager, *Force of Nature: the life of Linus Pauling* (Simon & Schuster, New York, 1995).

Marie Boas Hall, *Robert Boyle and Seventeenth-Century Chemistry* (CUP, Cambridge, 1958).

Marie Boas Hall, *Robert Boyle on Natural Philosophy* (Indiana University Press, Bloomington, 1965).

Rupert Hall, *Isaac Newton* (Blackwell, Oxford, 1992).

Harold Hartley, *Humphry Davy* (Nelson, London, 1966).

Arthur Holmes, *Principles of Physical Geology* (Nelson, London, 1944).

Robert Hooke, *Micrographia* (Royal Society, London, 1665).

Robert Hooke, *The Posthumous Works of Robert Hooke* (ed. Richard Waller) (Royal Society, London, 1705).

Robert Hooke, *The Diary of Robert Hooke* (eds. Henry Robinson and Walter Adams) (Taylor & Francis, London, 1935).

Ken Houston (ed.), *Creators of Mathematics: the Irish connection* (University College Dublin Press, 2000).

Jonathan Howard, *Darwin* (OUP, Oxford, 1982).

Michael Hunter (ed.), *Robert Boyle Reconsidered* (CUP, Cambridge, 1994).

Hugo Iltis, *Life of Mendel* (Allen & Unwin, London, 1932).

John Imbrie and Katherine Palmer Imbrie, *Ice Ages* (Macmillan, London, 1979).

James Irons, *Autobiographical Sketch of James Croll, with memoir of his life and work* (Stanford, London, 1896).

Bence Jones, *Life & Letters of Faraday* (Longman, London, 1870).

L. J. Jordanova, *Lamarck* (OUP, Oxford, 1984).

Horace Freeland Judson, *The Eighth Day of Creation* (Jonathan Cape, London, 1979).

C. Jungnickel and R. McCormmach, *Cavendish: the experimental life* (Bucknell University Press, New Jersey, 1996).

F. B. Kedrov, *Kapitza: life and discoveries* (Mir, Moscow, 1984).

Hermann Kesten, *Copernicus and his World* (Martin Secker & Warburg, London, 1945).

Geoffrey Keynes, *A Bibliography of Dr Robert Hooke* (Clarendon Press, Oxford, 1960).

Desmond King-Hele, *Erasmus Darwin* (De La Mare, London, 1999).

David C. Knight, *Johannes Kepler and Planetary Motion* (Franklin Watts, New York, 1962).

W. Köppen and A. Wegener, *Die Klimate der Geologischen Vorzeit* (Borntraeger, Berlin, 1924).

Helge Kragh, *Quantum Generations* (Princeton University Press, Princeton, NJ, 1999).

Ulf Lagerkvist, *DNA Pioneers and Their Legacy* (Yale University Press, New Haven, 1998).

H. H. Lamb, *Climate: present, past and future* (Methuen, London, volume 1 1972, volume 1 1977).
E. Larsen, *An American in Europe* (Rider, New York, 1953).
A.-L. Lavoisier, *Elements of Chemistry*, translated by Robert Kerr (Dover, New York, 1965; facsimile of 1790 edition).
Cherry Lewis. *The Dating Game* (CUP, Cambridge, 2000).
James Lovelock, *Gaia* (OUP, Oxford, 1979).
James Lovelock, *The Ages of Gaia* (OUP, Oxford, 1988).
E. Lurie, *Louis Agassiz* (University of Chicago Press, 1960).
Charles Lyell, *Principles of Geology* (Penguin, London, 1997; originally published in three volumes by John Murray, London, 1830–33).
Charles Lyell, *Elements of Geology* (John Murray, London, 1838).
Katherine Lyell (ed.), *Life, Letters and Journals of Sir Charles Lyell, Bart.* (published in two volumes, John Murray, London, 1881).
Maclyn McCarty, *The Transforming Principle* (Norton, New York, 1985).
Douglas McKie, *Antoine Lavoisier* (Constable, London, 1952).
H. L. McKinney, *Wallace and Natural Selection* (Yale University Press, New Haven, 1972).
Frank Manuel, *Portrait of Isaac Newton* (Harvard University Press, Cambridge, MA, 1968).
Ursula Marvin, *Continental Drift* (Smithsonian Institution, Washington DC, 1973).
James Clerk Maxwell, *The Scientific Papers of J. Clerk Maxwell* (ed. W. D. Niven) (CUP, Cambridge, 1890).
Jagdish Mehra, *Einstein, Physics and Reality* (World Scientific, Singapore, 1999).
Milutin Milankovitch, *Durch ferne Welten und Zeiten* (Köhler & Amalang, Leipzig, 1936).
Ruth Moore, *Niels Bohr* (MIT Press, Cambridge, MA 1985).
Yuval Ne'eman and Yoram Kirsh, *The Particle Hunters*, 2nd edn (CUP, Cambridge, 1996).
J. D. North, *The Measure of the Universe* (OUP, Oxford, 1965).
Robert Olby, *The Path to the Double Helix* (Macmillan, London, 1974).
C. D. O'Malley, *Andreas Vesalius of Brussels 1514–1564* (University of California Press, Berkeley, 1964).
Henry Osborn, *From the Greeks to Darwin* (Macmillan, New York, 1894).
Dorinda Outram, *Georges Couvier* (Manchester University Press, Manchester, 1984).
Dennis Overbye, *Einstein in Love* (Viking, New York, 2000).
H. G. Owen, *Atlas of Continental Displacement: 200 million years to the present* (CUP, Cambridge, 1983).

Abraham Pais, *Subtle is the Lord . . .* (OUP, Oxford, 1982).
Abraham Pais, *Inward Bound: of matter and forces in the physical world* (OUP, Oxford, 1986).
Linus Pauling and Peter Pauling, *Chemistry* (Freeman, San Francisco, 1975).
Samuel Pepys, *The Shorter Pepys* (selected and edited by Robert Latham; Penguin, London, 1987).
Roger Pilkington, *Robert Boyle: father of chemistry* (John Murray, London, 1959).
John Playfair, *Illustrations of the Huttonian Theory of the Earth* (facsimile reprint of the 1802 edition, with an introduction by George White) (Dover, New York, 1956).
Franklin Portugal and Jack Cohen, *A Century of DNA* (MIT Press, Cambridge, MA, 1977).
Lawrence Principe, *The Aspiring Adept* (Princeton University Press, Princeton, NJ, 1998).
Bernard Pullman, *The Atom in the History of Human Thought* (OUP, Oxford, 1998).
Lewis Pyenson and Susan Sheets-Pyenson, *Servants of Nature* (HarperCollins, London, 1999).
Susan Quinn, *Marie Curie* (Heinemann, London, 1995).
Peter Raby, *Alfred Russel Wallace* (Chatto & Windus, London, 2001).
Charles E. Raven, *John Ray* (CUP, Cambridge, 1950).
James Reston, *Galileo* (Cassell, London, 1994).
Colin A. Ronan, *The Cambridge Illustrated History of the World's Science* (CUP, Cambridge, 1983).
S. Rozental (ed.), *Niels Bohr* (North-Holland, Amsterdam, 1967).
Jósef Rudnicki, *Nicholas Copernicus* (Copernicus Quatercentenary Celebration Committee, London, 1943).
Anne Sayre, *Rosalind Franklin & DNA* (Norton, New York, 1978).
Stephen Schneider and Randi Londer, *The Coevolution of Climate & Life* (Sierra Club, San Francisco, 1984).
Erwin Schrödinger, *What is Life?* and *Mind and Matter* (CUP, Cambridge, 1967) (collected edition of two books originally published separately in, respectively, 1944 and 1958).
J. F. Scott, *The Scientific Work of René Descartes* (Taylor & Francis, London, 1952).
Steven Shapin, *The Scientific Revolution* (University of Chicago Press, London, 1966).
John Stachel (ed.), *Einstein's Miraculous Year* (Princeton University Press, Princeton, NJ, 1998).

Frans A. Stafleu, *Linnaeus and the Linneans* (A. Oosthoek's Uitgeversmaatschappij NV, Utrecht, 1971).

Tom Standage, *The Neptune File* (Allen Lane, London, 2000).

G. P. Thomson, *J. J. Thomson* (Nelson, London, 1964).

J. J. Thomson, *Recollections and Reflections* (Bell & Sons, London, 1936).

Norman Thrower (ed.), *The Three Voyages of Edmond Halley* (Hakluyt Society, London, 1980).

Conrad von Uffenbach, *London in 1710* (trans. and ed. W. H. Quarrell and Margaret Mare) (Faber & Faber, London, 1934).

Alfred Russel Wallace, *My Life* (Chapman & Hall, London; originally published in two volumes, 1905; revised single-volume edition 1908).

James Watson, 'The Double Helix', in Gunther Stent (ed.), *The Double Helix* 'critical edition' (Weidenfeld & Nicolson, London, 1981).

Alfred Wegener, *The Origin of Continents and Oceans* (Methuen, London, 1967) (translation of the fourth German edition, published in 1929).

Richard Westfall, *Never at Rest: a biography of Isaac Newton* (CUP, Cambridge, 1980).

Richard Westfall, *The Life of Isaac Newton* (CUP, Cambridge, 1993) (this is a shortened and more readable version of *Never at Rest*).

Michael White: *Isaac Newton: the last sorcerer* (Fourth Estate, London, 1997).

Michael White and John Gribbin, *Einstein: a life in science* (Simon & Schuster, London, 1993).

Michael White and John Gribbin, *Darwin: a life in science* (Simon & Schuster, London, 1995).

A. N. Whitehead, *Science and the Modern World* (CUP, Cambridge, 1927).

Peter Whitfield, *Landmarks in Western Science* (British Library, London, 1999).

L. P. Williams, *Michael Faraday* (Chapman, London, 1965).

David Wilson, *Rutherford* (Hodder & Stoughton, London, 1983).

Edmund Wilson, *The Cell in Development and Inheritance* (Macmillan, New York, 1896).

Leonard Wilson, *Charles Lyell* (Yale University Press, New Haven, 1972).

Thomas Wright, *An Original Theory of the Universe* (Chapelle, London, 1750) (facsimile edition, edited by Michael Hoskin, Macdonald, London, 1971).

W. B. Yeats, 'Among School Children' in, for example, *Selected Poetry* (ed. Timothy Webb) (Penguin, London, 1991).

David Young, *The Discovery of Evolution* (CUP, Cambridge, 1992).

Arthur Zajonc, *Catching the Light* (Bantam, London, 1993).

Index